Phototherapeutic Approaches to Neurodegeneration

Hagera Dilnashin • Surya Pratap Singh
Editors

Phototherapeutic Approaches to Neurodegeneration

Current Trends and Prospects

Springer

Editors
Hagera Dilnashin
Department of Biosciences
& Bioengineering
Indian Institute of Technology-Roorkee
Roorkee, Uttarakhand, India

Surya Pratap Singh
Department of Biochemistry
Institute of Science
Banaras Hindu University
Varanasi, Uttar Pradesh, India

ISBN 978-981-95-4750-0 ISBN 978-981-95-4751-7 (eBook)
https://doi.org/10.1007/978-981-95-4751-7

This Springer imprint is published by the registered company Springer Nature Singapore Pte Ltd.
The registered company address is: 152 Beach Road, #21-01/04 Gateway East, Singapore 189721, Singapore

Foreword

In recent decades, the global prevalence of neurodegenerative disorders has emerged as a critical health crisis, affecting millions and placing immense strain on healthcare systems and caregivers alike. Despite growing advancements in neuroscience and pharmacotherapy, definitive treatments for conditions such as Alzheimer's disease, Parkinson's disease, and other related disorders remain largely elusive. It is in this context that *Phototherapeutic Approaches to Neurodegeneration: Current Trends and Prospects* presents itself as a timely and transformative contribution to the field.

This volume presents a compelling synthesis of traditional medicinal knowledge and contemporary scientific insights. It is thoughtfully divided into three interlinked sections, each designed to guide readers through a multifaceted understanding of plant-based therapies in the context of neurodegeneration.

The first section, *Pathology and Phototherapeutic Intervention*, lays the foundation by exploring the biological underpinnings of neurodegenerative diseases and the ways in which plant-derived compounds may modulate these complex pathways. The second section, *Bioactive Photomedicinal Compounds*, delves into the chemical nature, mechanisms of action, and therapeutic efficacy of specific phytochemicals, offering readers a rich insight into their translational potential. The final section, *Photomedicine: Traditional Knowledge to Modern Scientific Assessment*, bridges historical wisdom with empirical rigor, emphasizing the relevance of ethnopharmacology in informing future therapeutic development.

What sets this volume apart is its commitment to evidence-based innovation rooted in nature. It transcends disciplinary boundaries, inviting neuroscientists, pharmacologists, ethnobotanists, and clinicians to collaborate in reimagining neurotherapeutics through a botanical lens.

By emphasizing both mechanistic detail and holistic understanding, this work empowers researchers to explore plant-based therapies not as supplements but as serious contenders in the fight against neurodegeneration. For students, it offers a gateway into interdisciplinary exploration; for seasoned investigators, it provides new frontiers for discovery.

In illuminating the convergence of nature and neuroscience, this book reaffirms that some of our most promising solutions may emerge not from synthetic novelty but from re-examining the profound intelligence of the natural world.

Central University of Gujarat
Vadodara, Gujarat, India

Rama Shanker Dubey,

Preface

Neurodegenerative disorders, including Alzheimer's disease and Parkinson's disease, affect millions worldwide, posing an immense challenge to global healthcare systems. Despite the rapid advancements in medical science, existing pharmacological treatments often fall short of providing sustained relief to a significant portion of the affected population. As the demand for more effective therapies grows, researchers are increasingly turning their attention to alternatives particularly phytomedicine, which has shown promising potential for neuroprotection.

Phytochemicals have a long-standing history in traditional medicine and are now emerging as potent candidates for mitigating the complex pathologies associated with neurodegeneration. This book, *Phototherapeutic Approaches to Neurodegeneration: Current Trends and Prospects*, offers a comprehensive exploration of the therapeutic potential of plant-based compounds in combating neurological and neurodegenerative disorders. It not only delves into the molecular mechanisms underlying their neuroprotective properties, such as anti-oxidant, anti-excitotoxic, and anti-inflammatory effects, but also emphasizes the value of ethnopharmacological knowledge in identifying and developing these natural compounds.

Divided into three sections, this volume covers a wide range of topics, from the pathological basis of neurodegeneration and phytotherapeutic interventions, to bioactive phytomedicinal compounds, and finally, the journey of traditional knowledge into modern scientific validation. Through this structure, the book presents a holistic view of how phytomedicines are reshaping therapeutic strategies for neurodegenerative diseases.

We hope that this book will serve as a valuable resource for researchers, clinicians, and students alike, inspiring future studies and applications of phytotherapeutics in the ongoing fight against neurodegenerative disorders.

Roorkee, Uttarakhand, India — Hagera Dilnashin
Varanasi, Uttar Pradesh, India — Surya Pratap Singh

Contents

Editors and Contributors

About the Editors

Hagera Dilnashin is a post-doctoral fellow at IIT Roorkee and an accomplished researcher in neurodegenerative disorders, with a focus on Parkinson's disease. Her work explores herbal medicines and bioactive compounds as alternative therapeutic strategies, bridging traditional knowledge with modern neuroscience.

She has authored over 25 research articles and book chapters in leading journals such as *Neurotoxicity Research*, *ACS Chemical Neuroscience*, and *Journal of Neurochemistry*. As an editor for Bentham Science Publishers and a peer reviewer for journals including Springer Nature, Taylor & Francis, and *Oxidative Medicine and Cellular Longevity*, she upholds the highest standards of scientific rigor.

A member of the Indian Academy of Neuroscience and the International Parkinson and Movement Disorder Society, she actively fosters collaboration within the global scientific community. In 2025, she was awarded the prestigious IBRO/Dana Brain Awareness Week Grant for her outreach efforts in neuroscience education.

Beyond academia, she is a passionate thematic philatelist. Her exhibit on "Crane" earned a gold medal at KASHIPEX 2019, organized by India Post. Her interdisciplinary contributions and commitment to public engagement make her a dynamic force in both research and science advocacy.

Surya Pratap Singh is currently working as a Professor at Banaras Hindu University. He is a highly learned and distinguished scientist in the field of biochemistry and neuroscience. His academic background includes a Ph.D. in Biochemistry from Banaras Hindu University, India, and post-doctoral fellowships from prestigious institutions like All India Institute of Medical Science (India), National Institute of Neuroscience (Japan), University of Illinois at Chicago (USA), and Johns Hopkins Medical Institute (USA), which highlights his commitment to continuous learning and research. His focus on neurobiology, particularly in areas like neurodegenerative diseases such as Parkinson's disease, neuroinflammation, and neuroprotective interventions, demonstrates his dedication to studying critical aspects of the human nervous system and finding potential solutions for neurological disorders.

His active involvement in various national and international scientific societies and organizations, such as The Endocrine Society (USA), Society for Reproductive Biology and Comparative Endocrinology (SRBCE), International Society of Neurochemistry (ISN), Indian Academy of Neuroscience (IAN), and Association of Clinical Biochemists of India (ACBI), further reflects his engagement in the scientific community and his efforts to contribute to the advancement of these fields.

Contributors

Kaushalya Anuradha Molecular Diagnostics and Vaccines, Sri Lanka Institute of Biotechnology (SLIBTEC), Pitipana, Homagama, Sri Lanka

Disha Arora Faculty of Pharmaceutical Sciences, The ICFAI University, Baddi, Himachal Pradesh, India

Ghulam Md Ashraf Department of Biomedical Sciences, Gulf Medical University, Suzhou, Ajman, United Arab Emirates

Anuj Babbar Department of Molecular Biology and Genetic Engineering, School of Bioengineering and Biosciences, Lovely Professional University, Phagwara, Punjab, India

Malakpogu Ravindra Babu School of Medical and Allied Sciences, Galgotias University, Greater Noida, Uttar Pradesh, India

Tharangika Bowange Laboratory Management, Sri Lanka Institute of Biotechnology (SLIBTEC), Homagama, Sri Lanka

Kokkonda Jackson Sugunakara Chary Department of Molecular Biology and Genetic Engineering, School of Bioengineering and Biosciences, Lovely Professional University, Phagwara, Punjab, India

Rupal Dhariwal Department of Life Sciences, Parul Institute of Applied Sciences, Parul University, Vadodara, Gujarat, India

Research and Development Cell, Parul University, Vadodara, Gujarat, India

Hagera Dilnashin Department of Biosciences & Bioengineering, Indian Institute of Technology-Roorkee, Roorkee, Uttarakhand, India

Sumit Durgapal Department of Pharmaceutics, Uttaranchal Institute of Pharmaceutical Sciences, Uttaranchal University, Dehradun, Uttarakhand, India

Chathurika Elvitigala Agrobiotech, Sri Lanka Institute of Biotechnology (SLIBTEC), Homagama, Sri Lanka

Sivamani Ganesan PG & Research Department of Zoology and Biotechnology, A.V.V.M. Sri Pushpam College, Poondi, Thanjavur, Tamil Nadu, India

Lakmal Gonawala Molecular Diagnostic and Vaccine, Sri Lanka Institute of Biotechnology (SLIBTEC), Homagama, Sri Lanka

Peshala Gunasekara Molecular Diagnostics and Vaccines, Sri Lanka Institute of Biotechnology (SLIBTEC), Pitipana, Homagama, Sri Lanka

Anshu Gupta Department of Pharmacology, Maharishi Markandeshwar Institute of Medical Sciences and Research, Mullana, Ambala, India

Anushka Jain Department of R&D, Quality Assurance, Pure & Cure Healthcare Pvt. Ltd. (A Subsidiary of Akums Drugs & Pharmaceuticals Ltd.), SIDCUL, Haridwar, Uttarakhand, India

Mukul Jain Department of Life Sciences, Parul Institute of Applied Sciences, Parul University, Vadodara, Gujarat, India

Research and Development Cell, Parul University, Vadodara, Gujarat, India

Bhuwan Chandra Joshi Faculty of Pharmacy, Maharaja Agrasen Himalayan Garhwal University, Pokhra, Pauri Garhwal, Uttarakhand, India

Rupa Joshi Department of Pharmacology, Maharishi Markandeshwar Institute of Medical Sciences and Research, Mullana, Ambala, India

Divya Sri Kathiresan Department of Human Genetics and Molecular Biology, Bharathiar University, Coimbatore, Tamil Nadu, India

Adnan Khan School of Medical and Allied Sciences, Galgotias University, Greater Noida, Uttar Pradesh, India

Archana Kumari School of Pharmacy and Emerging Sciences, Baddi University of Emerging Sciences and Technology, Baddi, Solan, Himachal Pradesh, India

Neeraj Kumar Parul College of Pharmacy and Research, Parul University, Ahmedabad, India

Sachin Kumar Department of Pharmacology, Delhi Pharmaceutical Sciences and Research University (DPSRU), New Delhi, Delhi, India

Héctor Eduardo López-Valdés Departamento de Fisiología, Facultad de Medicina, Universidad Nacional Autónoma de México (UNAM), México City, Mexico

Aanchal Loshali Department of Pharmacognosy and Phytochemistry, School of Pharmaceutical Education and Research, Jamia Hamdard, New Delhi, India

Yashi Mahendra University Institute of Engineering and Technology, Maharshi Dayanand University, Rohtak, Haryana, India

Hilda Martínez-Coria Departamento de Fisiología, Facultad de Medicina, Universidad Nacional Autónoma de México (UNAM), México City, Mexico

Nuwan S. Medawaththa Bioinformatics, Sri Lanka Institute of Biotechnology (SLIBTEC), Homagama, Sri Lanka

Ravinesh Mishra School of Pharmacy and Emerging Sciences, Baddi University of Emerging Sciences and Technology, Baddi, Solan, Himachal Pradesh, India

N. Arun Nagendran Department of Zoology, Thiagarajar College, Madurai, Tamil Nadu, India

Juan Angel Ortiz-Lara Departamento de Fisiología, Facultad de Medicina, Universidad Nacional Autónoma de México (UNAM), México City, Mexico

Nil Patil Department of Life Sciences, Parul Institute of Applied Sciences, Parul University, Vadodara, Gujarat, India

Research and Development Cell, Parul University, Vadodara, Gujarat, India

Dananjaya Perera Biotherapeutics, Sri Lanka Institute of Biotechnology (SLIBTEC), Pitipana, Homagama, Sri Lanka

Vinayaga Moorthi Puthamohan Department of Human Genetics and Molecular Biology, Bharathiar University, Coimbatore, Tamil Nadu, India

Sachini Raveesha Quality Assurance, Sri Lanka Institute of Biotechnology (SLIBTEC), Homagama, Sri Lanka

Rakesh Sahu School of Pharmacy, Sharda University, Greater Noida, Uttar Pradesh, India

Anuradha Sharma Department of Molecular Biology and Genetic Engineering, School of Bioengineering and Biosciences, Lovely Professional University, Phagwara, Punjab, India

Bhartendu Sharma School of Pharmacy and Emerging Sciences, Baddi University of Emerging Sciences and Technology, Baddi, Solan, Himachal Pradesh, India

Priya Sharma School of Pharmacy and Emerging Sciences, Baddi University of Emerging Sciences and Technology, Baddi, Solan, Himachal Pradesh, India

Jhanvi Soni Parul Institute of Pharmacy, Parul University, Vadodara, Gujarat, India

Shriyansh Srivastava School of Medical and Allied Sciences, Galgotias University, Greater Noida, Uttar Pradesh, India

Mohd. Tariq Department of Academics, Sumandeep Vidyapeeth (Deemed to be University), Vadodara, Gujarat, India

Department of Biotechnology, Graphic Era (Deemed to be University), Dehradun, Uttarakhand, India

Sushmita Uniyal Department of Pharmacology, School of Pharmaceutical Sciences, Shri Guru Ram Rai University, Dehradun, Uttarakhand, India

Nalaka Wijekoon Molecular Diagnostic and Vaccine, Sri Lanka Institute of Biotechnology (SLIBTEC), Homagama, Sri Lanka

Abbreviations

Aβ	Amyloid beta
AChEIs	Acetylcholinesterase inhibitors
AD	Alzheimer's disease
AGuIX	Activation and guidance of irradiation by X-ray
AhR	Aryl hydrocarbon receptor
ALS	Amyotrophic lateral sclerosis
AKT	Protein kinase B
AMPA	α-Amino-3-hydroxy-5-methyl-4-isoxazolepropionic acid
AP-1	Activator protein-1
APP	Amyloid precursor protein
AR	Androgen
ASC	Apoptosis-associated speck-like protein with a C-terminal caspase recruitment domain
ATD	Amino-terminal domain
ATF	Activating transcription factor 3
Atg 13	Autophagy related gene 13
ATN	Amyloid, tau, and neurodegeneration
ATP	Adenosine triphosphate
BA	Bioavailability
BACE1	β-Site amyloid precursor protein cleaving enzyme 1
BBB	Blood-brain barrier
BBR	Berberine
BPSD	Dementia's behavioral, psychological, and cognitive symptoms
CALs	Chronic active lesions
CAT	Catalase
cAMP	Cyclic adenosine monophosphate
CCO	Cytochrome c oxidase
CCL2	C-C motif chemokine ligand 2
Ce6	Chlorin e6
c-FLIP	Cellular FLICE inhibitory protein
ChEI	Cholinesterase-inhibitor
CHOP	C/EBP homologous protein
CLU	Clusterin
CNS	Central nervous system

COX-2	Cyclooxygenase-2
CR1	Complement receptor 1
CSF-1R	Colony-stimulating factor-1 receptor
CXCL10	C-X-C motif chemokine ligand
DAM	Disease-associated microglia
DAMPs	Damage-associated molecular patterns
DDIT4	DNA-damage-inducible transcript 4
DHA	Docosahexaenoic acid
4E-BP1	4E-binding protein 1
EBNA1	Epstein-Barr virus nuclear antigen 1
Egr-1	Early growth response-1
EGF	Epidermal growth factor
EGFR	Epidermal growth factor receptor
ELAM-1	Endothelial leukocyte adhesion molecule-1
EpRE	Electrophile response element
ER	Endoplasmic reticulum
ER-α	Estrogen receptors
ERAD	ER-associated degradation
ERK	Extracellular signal-regulated kinase
ETC	Electron transport chain
EVs	Extracellular vesicles
FAK	Focal adhesion kinase
FDA	Food and Drug Administration
FGF	Fibroblast growth factor
FTD	Frontotemporal dementia
Gα	Alpha subunit G protein
Gβγ	Beta gamma subunit protein G
GCL	Glutamyl cysteine ligase
GFAP	Glial fibrillary acidic protein
GM-CSF	Granulocyte–macrophage colony-stimulating factor
GPCRs	G protein-coupled receptors
GPX	Glutathione peroxidase
GSH-Px	Glutathione peroxidase
GSK3β	Glycogen synthase kinase 3 beta
GSS	Glutathione synthetase
HD	Huntington's disease
HDL	High-density lipoproteins
HER2	Human epidermal growth factor receptor 2
HGF	Hepatocyte growth factor
HIF-1	Hypoxia-inducible factor-1
HMGB1	High-mobility group box 1
HNE	4-Hydroxy-2-nonenal
HO-1	Heme oxygenase-1
HR	Homologous recombination
IAP	Inhibitor of apoptosis proteins

ICAM-1	Intracellular adhesion molecule-1
IKK	I-kappaB kinase
IL-1β	Interleukin-1β
iNOS	Inducible nitric oxide synthase
IP3K	Phosphoinositol 3-kinase
iPSC	Induced pluripotent stem cells
JAK	Janus kinase
5-LOX	5-Lipoxygenase
L-DOPA	Levodopa
LDL	Low-density lipoproteins
LEDs	Light-emitting diodes
LLLT	Low-level laser therapy
LOAEL	Lowest observed adverse effect level
LPO	Lipid peroxidation
LPS	Lipopolysaccharide
LTP	Long-term potentiation
MAOs	Monoamine oxidases
MAPKs	Mitogen-activated protein kinases
MDA	Malondialdehyde
mHTT	Mutant huntingtin
MMPs	Matrix metalloproteinases
MMSE	Mini-mental state examination
MoCA	Montreal cognitive assessment
MOF	Metal-organic framework
MRP	Multi-drug resistance protein
MS	Multiple sclerosis
MSN	Medium spiny neuron
MSR	Methionine sulfoxide reductase
mTOR	Mechanistic target of rapamycin
mTORC	Mechanistic target of rapamycin complex
NDDs	Neurodegenerative diseases
NF-κB	Nuclear factor κB
NO	Nitric oxide
NOAEL	No observed adverse effect level
NOX	NADPH oxidases
Nrf2	Nuclear factor erythroid 2-related factor 2
NTF	Neurofibrillary tangles
NIR	Near-infrared
NLRP3	NOD-like receptor protein 3
NMDA	N-methyl-D-aspartate
NMDARs	NMDA-receptors
NP	Neuritic plaques
NQO1	NAD(P)H: quinone oxidoreductase 1
OS	Oxidative stress
PBM	Photo-biomodulation

PD	Parkinson's disease
p38MAPK	p38 mitogen-activated protein kinase
p-JNK1	Phospho-c-Jun N-terminal kinase 1
PLGA	Poly (lactic-co-glycolic acid)
PM	Plasma membranes
p-NF-κB	Phosphorylated-nuclear factor kappa B
PDK1	Phosphoinositide-dependent protein kinase 1
PINK1	PTEN-induced kinase 1
PIP2	Ptdln-3-4,5-diphosphate
PIP3	Ptdln-3-4,5-trisphosphate
PKC	Protein kinase C
PP2A	Phosphatase 2A
PPAR-γ	Peroxisome proliferator-activated receptor-gamma
PSEN1	Presenilin 1
PSEN2	Presenilin 2
PTEN	Phosphatase and tensin homolog
PUFAs	Polyunsaturated fatty acids
RAGE	Receptor for advanced glycation end products
RNS	Reactive nitrogen species
ROS	Reactive oxygen species
RTKs	Receptor tyrosine kinases
S6K1	p70 S6 kinase
SCAs	Spinocerebellar ataxias
SESN2	Sestrin 2
SNpc	Substantia nigra pars compacta
SOD	Superoxide dismutase
STATs	Signal transducers and activators of transcription
TFEB	Transcription factor EB
sTREM2	Soluble triggering receptor expressed on myeloid cells 2
TBI	Traumatic brain injury
TGF-β	Transforming growth factor-β
ThT	Thioflavin T
TLRs	Toll-like receptors
TNF-α	Tumor necrosis factor-α
TPPS	Meso-tetra(4-sulfonatophenyl)porphyrin
TREM2	Triggering receptor expressed on myeloid cells 2
TRP	Transient receptor potential
TRX	Thioredoxin
TSC1	Tuberous sclerosis complexes 1
TSPO	Translocator protein
uPA	Urokinase-type plasminogen activator
uPAR	Urokinase-type plasminogen activator receptor
UPR	Unfolded protein response
VCAM-1	Vascular cell adhesion molecule-1

VEGF	Vascular endothelial growth factor
VLDL	Very low-density lipoproteins
ZnPc	Zinc phthalocyanine

Part I

Pathology and Phototherapeutic Intervention

Photopharmacological Agents in the Management of Dementia of Alzheimer's Disease and Other Aetiologies

1

Mukul Jain, Mohd. Tariq, Nil Patil, Rupal Dhariwal, and Ghulam Md Ashraf

Abstract

Dementia types such as Alzheimer's disease (AD), frontotemporal dementia, vascular dementia, and Lewy body dementia are multifactorial progressive diseases with loss of neurons and progressive cognitive deterioration. Conventional pharmacotherapies, despite decades of research, typically lack the ability to impact disease-defining outcomes in part because of suboptimal blood-brain barrier penetration, off-target effects, and the inability to exert spatiotemporal control. A new field in medicinal chemistry, called photopharmacology, aims to bypass these problems using photoswitchable molecules to switch their biological activity to a predetermined level with specific wavelengths of light. They are typically azobenzene, spiropyran, or diarylethene-based reversible activation or inhibition of target proteins, including ion channels, G-protein-coupled receptors, and enzymes related to neurodegeneration. Photopharmacological interventions have demonstrated possibilities to alter cholinergic transmission, effectively inhibit beta-secretase, cripple amyloid-beta assembly, and downregulate tau hyperphosphorylation in the context of Alzheimer's disease, therefore targeting

M. Jain · N. Patil · R. Dhariwal
Department of Life Sciences, Parul Institute of Applied Sciences, Parul University, Vadodara, Gujarat, India

Research and Development Cell, Parul University, Vadodara, Gujarat, India

M. Tariq
Department of Academics, Sumandeep Vidyapeeth (Deemed to be University), Vadodara, Gujarat, India

Department of Biotechnology, Graphic Era (Deemed to be University), Dehradun, Uttarakhand, India

G. M. Ashraf (✉)
Department of Biomedical Sciences, Gulf Medical University, Ajman, United Arab Emirates

H. Dilnashin, S. P. Singh (eds.), *Phototherapeutic Approaches to Neurodegeneration*,
https://doi.org/10.1007/978-981-95-4751-7_1

the fundamental pathological process. Similarly, with other patterns of dementia, there may be opportunities to modulate neurotransmitter systems under light control (e.g., dopaminergic and glutamatergic), which may enable recovery of a synaptic homeostasis without global inhibition of neurons. The possibility of deep-brain phototherapeutic interventions has also increased in recent years after the integration of nanotechnology and fiber-optic or near-infrared light delivery systems. In addition, temporal precision is possible with photopharmacology, thus providing on-demand therapy whereby systemic exposure is reduced, which minimizes the risk of adverse effects. Although promising, clinical translation of the technology is hampered by limitations in the penetration of light by human brain tissue, phototoxicity, in vivo stability of photo switches, and the creation of implantable light-transmission devices. Ongoing research into red-shifted and two-photon activatable switches, biodegradable light guides, and non-invasive wearable devices is paving the way for broader application in neurodegenerative disease management. This abstract underscores photopharmacology as a transformative strategy in dementia therapy, capable of reshaping the therapeutic landscape by offering reversible, localized, and temporally controlled interventions. With continued interdisciplinary collaboration across chemistry, neuroscience, optics, and bioengineering, photopharmacological therapeutics may soon transition from conceptual promise to clinical reality in managing AD and related dementias.

Keywords

Photo pharmacology · Alzheimer's disease · Dementia management · Light-activated therapeutics · Neurodegeneration · Beta-amyloid · Tau pathology · Blood–brain barrier

1.1 Introduction

Dementia is one of the most daunting twenty-first century public health problems, and its prevalence is increasing alarmingly across the world as the age of the population continues to rise. The current globalized figure in dementia is above 55 million people living with dementia, according to the latest Global burden of disease report with the number expected to almost triple in 2050 (WHO 2021). Alzheimer's disease (AD) constitutes the vast majority of diagnoses (almost 60–70%) of all dementia subtypes, representing not only the prevalence but also the disproportionate effect on the socioeconomic and caregiver burden. The decades-long pathological course of AD prior to the development of clinical symptoms highlights the necessity to implement therapeutic strategies at an early stage (Javaid et al. 2021). However, even after 30 years of research, the pharmacological space is limited, with only a few FDA-approved medications ((predominantly) symptomatic cholinesterase inhibitors (donepezil, rivastigmine, and galantamine) and memantine, an antagonist of the NMDA receptor) demonstrating mild, transient cognitive improvement (Cacabelos et al. 2024). More recent disease-modifying antibodies such as aducanumab and lecanemab have brought hope but are still subject to debate because of cost, risk-benefit profile, and non-affordability to many patients.

Part of this therapeutic stagnation can be explained by the complexity of the AD pathobiology, the difficulty in delivering any drugs across the blood–brain barrier (BBB), and the heterogeneity in clinical presentation of the disease. Another relatively recent area of interest is photopharmacology, a field that combines the specificity of light-based control with pharmacological regulation of biological targets. Photopharmacological agents, unlike conventional systemic drugs, are developed to be activated or deactivated in situ with specific wavelengths of light and therefore allow spatiotemporal control of dosing and target engagement (Ganzoni et al. 2025). In theory, such control would enable off-target effects to be minimized, systemic toxicity to be reduced, and highly localized modulation of neural networks suspected of involvement in memory and cognition. In the context of dementia, in which subsets of specific regions of the brain, including the hippocampus, entorhinal cortex, and association cortices, are selectively affected on a neuronal level, photopharmacology provides a method of selectively targeting the disease midway, possibly even prior to irreversible brain loss, as evidenced in some of the earliest functional brain defects experienced by those affected by the disease (Mondal and Mandal 2022).

The new direction ranges between small-molecule photoswitches, photocaged prodrugs, light-activated peptides, and light-controlled delivery systems. As demonstrated in the discovery of Parkin-guided photopharmacology, with increased improvements in optical equipment, minimally invasive light delivery vehicles, and enhanced knowledge on CNS pharmacokinetics, photopharmacology can be used to connect the translational gap between preclinical findings and patient-centered treatment (Sortino et al. 2024). This review will explore the molecular basis of dementia pathophysiology, outline the core principles of photopharmacology, describe current photopharmacological agents under investigation, and discuss technological integrations such as nanomedicine. Finally, we will critically assess the preclinical and clinical evidence, identify key challenges, and map the trajectory for future research and clinical translation.

1.2 Pathophysiology of Dementia

Dementia is a syndrome rather than a single disease with a wide spectrum of etiologies, which have a common path, resulting in progressive cognitive impairment and functional dependence (Elahi and Miller 2017). At the molecular level, all subtypes have unique neuropathology signatures but common pathology, including protein misfolding, synaptic malfunction, mitochondrial dysfunction, oxidative stress, and chronic neuroinflammation. (Fig. 1.1). AD is a disease that presents with two typical proteinopathies which include the amyloid-beta (Aβ) plaques and the neurofibrillary tangles (NFTs) tau. Sequential processing of amyloid precursor protein (APP) by β-site amyloid precursor protein cleaving enzyme 1 and γ-secretase leads to production of aggregation-prone Aβ42 species and activation of the

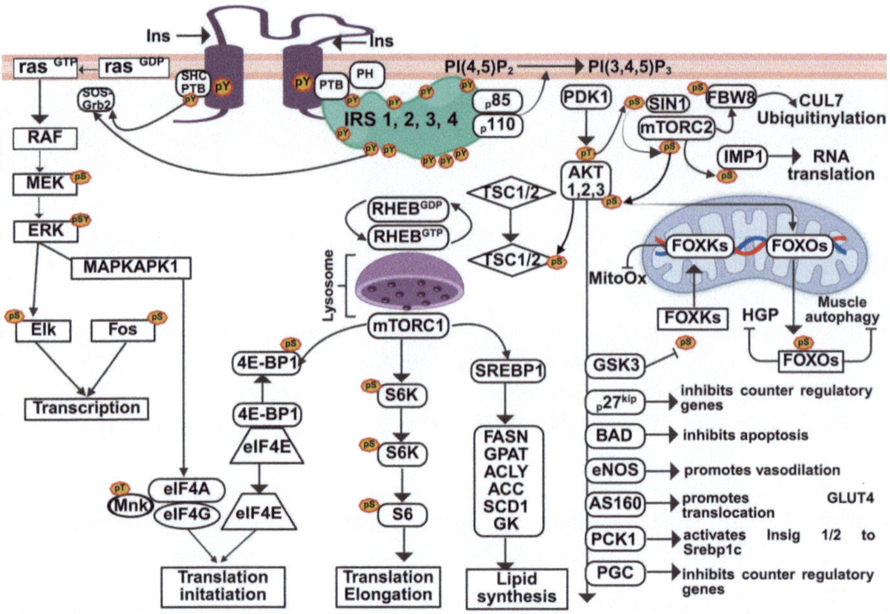

Fig. 1.1 Molecular interplay between type 2 diabetes and Alzheimer's disease. (http://creativecommons.org/licenses/by-nc/4.0)

amyloidogenic pathway (DeTure and Dickson 2019). These oligomers, however, are now considered to be the main neurotoxic species, rather than mature plaques, that alter synaptic integrity by NMDA receptor hyperactivation, due to calcium dyshomeostasis, and internalization of AMPA receptors. The dysregulated kinases, like GSK-3b and CDK5, contribute to Tau pathology, which accelerates microtubule destabilization, axonal transport alterations, and pro-inflammatory microglia activation. The formation of Aβ-tau hypothetically explains this mechanism by suggesting that Aβ accumulation activates a positive feedback loop of tau spreading through prion-like mechanisms, exacerbating neuronal damage.

As neuroinflammation closely relates to AD, it has increasingly become a central contributor and not a subsequent effect of AD pathology. Stimulation of the microglial microenvironment through TREM2 and NLRP3 inflammasome signaling encourages cytokines (IL-1, TNF-alpha, and IL-6), oxidative stress, and synaptic phagocytic clearance, which is a dysfunctional response that contributes to cognitive degradation (Heneka et al. 2015). The astrocytic reactivity also contributes to derailing glutamate homeostasis and blood brain barrier leading to a vicious cycle of toxicity.

In addition to AD, vascular dementia is caused by long-term cerebral hypoperfusion disorders, small-vessel disease, or the aftermath of stroke, resulting in the degeneration of white matter and cholinergic impairments (Inoue et al. 2023). Lewy body dementia is characterized as a 1-synuclein emergence in the cortical neurons that leads to parkinsonism and hallucinations as well as cognition that fluctuates

(Combi et al. 2021). Frontotemporal dementia (FTD) encompasses tauopathies and TDP-43 proteinopathies, often with strong genetic contributions (MAPT, GRN, and C9orf72 mutations). Mixed dementia, particularly AD plus vascular pathology, highlights the interplay between neurodegenerative and vascular mechanisms, complicating diagnosis and treatment (De Boer et al. 2021). Pathophysiology, at the systems level, encompasses early loss of hippocampal-cortical connectivity, impaired long-term potentiation (LTP), affected oscillatory dynamics including the theta and gamma bands, and the basis of episodic memory encoding (Table 1.1). The challenge for therapeutics, photopharmacology included, is to intervene in these multifactorial processes with precision and adaptability.

1.3 Principles of Photopharmacology

Photopharmacology is based on the idea that light is a source of an external stimulus, which can either intensify or reduce the action of drugs in a reversible and controllable way. Photopharmacological agents possess molecular-level designs containing structural or chemical modifications that occur when photons are absorbed to change the affinity, efficacy, or selectivity of the agent toward a target. Photoswitches such as azobenzenes, diarylethenes, and stilbenes are molecules that can reversibly interconvert between two states of different conformation (e.g., cis and trans) in response to light of particular wavelengths (Ryazantsev et al. 2021). Each isomer factor has various pharmacological effects that allow switching between the active and inactive states with a high time resolution. The cis–trans isomerization alters steric hindrance and polarity and determines where the molecule can fit in a target binding pocket. The alternative strategy is through photocaging of compounds, whereby a biologically active drug, which has a photolabile protecting group (e.g., nitrobenzyl, coumarin, BODIPY derivatives), is inactivated. On radiating, the caging group is cleaved, and consequently the active drug is released. This approach is particularly valuable for delivering cytotoxic or neuroactive agents to precise regions, thereby minimizing systemic exposure (Vorobev and Moskalensky 2020). Photorelease systems extend the photocaging concept into nanocarriers or polymeric scaffolds that release drug payloads upon light activation. The primary challenge, in the framework of CNS therapy, is the choice of wavelength: light in the near-infrared (NIR, 650–900 nm) regime better penetrates biological tissues since it scatters and is absorbed by hemoglobin and water less efficiently, but most photoswitches are optimized to work with UV-visible light. Solutions to address this issue involve two-photon excitation and upconversion nanoparticles to transform NIR locally to shorter wavelengths (Park et al. 2015).

The spatiotemporal control offered by photopharmacology has profound implications for CNS therapeutics. Drug activation is localizable to areas of pathology while leaving intact functionality in unaffected circuits, and the activation can be timed to coincide with behavioral or cognitive tasks to increase neuroplasticity. Importantly, the ability to reversibly switch drugs on and off reduces the risk of cumulative toxicity, a significant advantage in chronic diseases like dementia.

Table 1.1 Pathophysiology of major dementia subtypes

Dementia type	Primary pathology	Key molecular features	Clinical hallmarks	Common overlap with AD	Ref.
Alzheimer's disease	NFTs & amyloid plaques	Amyloid-β42, hyperphosphorylated tau, neuroinflammation	Memory loss, executive dysfunction, language deficits	Yes	Liu et al. (2024)
Vascular dementia	Cerebrovascular lesions	Small vessel disease, infarcts, microbleeds	Stepwise decline, focal neurological signs	Often coexists with amyloid pathology	Vinters et al. (2018)
Lewy body dementia	α-Synuclein inclusions	Lewy bodies, dopaminergic loss	Visual hallucinations, REM sleep behavior disorder, parkinsonism	Sometimes with AD pathology	Hansen et al. (2019)
FTD	Neuronal loss in frontal and temporal lobes	TDP-43, tau, FUS proteins	Behavioral changes, language deficits	Minimal amyloid pathology	Fieldhouse et al. (2023)
Mixed dementia	Combination of two or more above	Amyloid + vascular lesions common	Mixed symptom profile	Yes	Jellinger and Attems (2007)

1.4 Photopharmacological Agents for Dementia

The development of photopharmacological agents for dementia represents one of the most conceptually transformative approaches in neurotherapeutics. In contrast to traditional pharmacology, in which temporal and spatial exposure to drugs is defined by systemic pharmacokinetics and biodistribution, photopharmacology opens the prospect of remote, reversible, spatially localized drug activation or deactivation in the CNS (Lerch et al. 2016). Theoretically, this mitigates two of the most persistent challenges in dementia therapeutics: off-target effects and dynamic regulation of neuronal network states (Table 1.2).

1.4.1 Light-Controlled Cholinesterase Inhibitors

A key characteristic of AD and certain other dementias is cholinergic deficits, most notably the loss of basal forebrain cholinergic neurons and a drop in cortical acetylcholine concentrations. Classical acetylcholinesterase inhibitors (AChEIs), donepezil, rivastigmine, and galantamine, have limited symptomatic efficacy and have systemic cholinergic side effects (e.g., nausea and bradycardia). The creation of photoswitchable AChEIs in which an azobenzene or diarylethene moiety is incorporated into the scaffold of the AChEIs represents a standard shift in photopharmacology (Hansen et al. 2008). Under a specific wavelength (typically 365–420 nm for azobenzenes), the trans–cis isomerization changes the inhibitor's steric conformation, modulating its binding affinity to the AChE active site gorge. In a preclinical prototype, trans-azobenzene analogues of donepezil showed an inhibition >80% in the cis state and only approximately 20% in the trans state, creating off-on controllability of cholinergic potentiation in hippocampal slices. On the molecular level, the photoregulation probably changes the interaction of the π-π stacking with Trp84 and Tyr337 moieties in AChE, which interferes with the formation of a catalytic triad (Beharry 2012). Such an approach would enable clinicians to turn on cholinergic potentiation only when a patient is carrying out a cognitive task or during rehabilitation exercise sessions, which may limit 24/7 systemic exposure.

1.4.2 NMDA Receptor Modulators

The glutamatergic system and especially NMDA receptors (NMDARs) contribute to both synaptic plasticity and excitotoxicity. The low-potency NMDA antagonist memantine is the only glutamate-targeting AD drug approved, but uncontrollable in time and able to depress physiological LTP when administered chronically (Murrough et al. 2017). Photopharmacological NMDA modulators combine photoswitchable ligands with affinity to either a subunit of NMDAR extracellular glutamate-binding GluN2 or an intracellular channel vestibule PCP (Goudet et al. 2018). This has been demonstrated, for example, by engineering azobenzene-tethered ifenprodil analogues to specifically inhibit native cis-configured NMDARs

Table 1.2 Examples of photoactive compounds for dementia targets

Target	Compound name	Photopharmacology type	Light wavelength (nm)	Preclinical model	Outcome	Ref.
Acetylcholinesterase	Azo-donepezil	Photoswitch	365	In vitro cholinesterase assays	Reversible inhibition	Chen et al. (2014)
NMDA receptor	Azobenzene-based memantine	Photoswitch	380–500	Rat hippocampal neurons	Light-controlled excitotoxicity prevention	Berlin et al. (2016)
Amyloid-β aggregation	Photo-caged curcumin	Photocage	405	SH-SY5Y cells	Reduced aβ aggregation upon light exposure	Dong et al. (2015)
Tau protein	Azobenzene-tau inhibitor	Photoswitch	390	Tau-overexpressing neurons	Reduced tau aggregation	Inaba et al. (2023)
Microglial activation	Photo-responsive indomethacin	Photorelease	450	LPS-induced microglia	Light-triggered anti-inflammatory effect	Jiang et al. (2013)

containing the GluN2B subunit. MD studies show that the cis form induces an allosteric, closed conformation in the amino-terminal domain, lowering the probability of channel opening, and that the trans form recovers normal activity by adopting a geometry capable of moving out of the hydrophobic cleft (Ye et al. 2020). This control may be used to temporally inhibit excitotoxic bursts in the hippocampal network after microglia activation, but not limit basal synaptic transmission when photoactivation is absent.

1.4.3 Amyloid-β Aggregation Inhibitors

The formation of amyloid-beta (Aβ) occurs over multiple stages: nucleation, oligomerization generating oligomeric Aβ species, fibril extension, and plaque formation, wherein the oligomers are recognized as the most synaptotoxic entities. The photoresponsive Aβ aggregation inhibitors are intended to target early oligomerization processes only when they are required and avoid undesirable interference with physiological APP processing. Other designs involve the photocaged analogues of curcumin, in which an o-nitrobenzyl group is cleaved under UV or visible light, releasing the active inhibitor (Zhang et al. 2021). Once activated, these molecules insert aromatic rings into β-sheet interfaces, disrupting hydrogen bonding between residues such as Asp23-Lys28 and Phe19-Phe20, which stabilize Aβ fibril cores. This approach has been tested in cell-free Thioflavin T assays, showing a >60% reduction in fibril growth when light activation was applied during the nucleation window but a negligible effect if activation occurred after fibril maturation—emphasizing the spatiotemporal dependency of efficacy.

1.4.4 Tau-Targeted and Anti-inflammatory Photoagents

Tau pathology, involving hyperphosphorylation and aggregation into paired helical filaments, correlates more strongly with cognitive decline than Aβ plaque load. Photopharmacology here explores two angles: light-activated kinase inhibitors targeting GSK-3β or CDK5, major tau kinases, with photocaged ATP-competitive inhibitors that can be triggered during early NFT formation (Tiwari and Tolwinski 2023). Photo agents that stabilize microtubules by derivatizing paclitaxel with photoswitchable moieties to favor microtubule integrity in instances of dysfunctional tau (Soliman 2023). Simultaneously, photo-activatable anti-inflammatory reagents are in development that could be used to regulate immune microglial states. As an example, a photoswitchable COX-2 inhibitor can be pulsed to suppress the release of IL-1B and TNF-alpha only during inflammatory flares, precluding long-term immunosuppression (Bulic et al. 2011).

1.4.5 Nanotechnology Integration

The blood–brain barrier (BBB) remains the most formidable obstacle to CNS pharmacotherapy (Table 1.3). Nanotechnology offers a dual advantage to photopharmacology: (i) improved delivery of photoagents across the BBB and (ii) enabling light-controlled release through photothermal or photo-cleavable linkers.

1.4.6 Photoresponsive Nanocarriers

Photoresponsive polymers or ligands can be conjugated to nanocarriers to deliver cargo via exposure to light. For example, gold nanorods coated with azobenzene-containing polymers can undergo conformational collapse under NIR light, expelling their drug payload. The NIR window (650–900 nm) is particularly important for CNS applications, as it offers deeper tissue penetration with minimal phototoxicity (Di Pietro et al. 2017). At the nanoscale, precise engineering of hydrodynamic diameter (~50–80 nm) optimizes both BBB permeability and circulation half-life. Ligand functionalization with transferrin, lactoferrin, or anti-TfR antibodies enhances receptor-mediated transcytosis into brain parenchyma.

1.4.7 BBB Crossing Strategies

The photoagents could be incorporated into liposomes, polymer nanoparticles, or dendrimers that can be surface-modified to take advantage of endogenous transport (e.g., GLUT1 targeting with glucose moieties). More sophisticated methods include cell based delivery, including monocyte- or exosome-based carriers that intrinsically localize to inflammatory foci in the brain (Gaillard et al. 2012). An exciting direction is to integrate photopharmacology with focused ultrasound and

Table 1.3 Nanotechnology-assisted photo-pharmacology for CNS delivery

Nanocarrier type	Photo response mechanism	BBB crossing strategy	Example application	Ref.
Gold nanoparticles	Plasmonic heating	Receptor-mediated transcytosis	NIR-triggered release of cholinesterase inhibitor	EL Alami et al. (2020)
Liposomes	Photo-cleavable lipids	PEGylation + ligand targeting	Delivery of NMDA antagonist with light release	Maleeva et al. (2025)
Polymeric micelles	Photosensitive linkers	pH-sensitive + light	Curcumin delivery for amyloid inhibition	Mirzaie et al. (2019)
Upconversion nanoparticles	NIR-to-UV conversion	Transferrin conjugation	Activating UV-responsive photoswitches in deep tissue	Zhan et al. (2013)

microbubble technology, which transiently opens the BBB in a region-specific fashion prior to subsequent light-based activation of the drug.

1.4.8 Combined Photopharmacology–Photodynamic Approaches

In hybrid systems, a single nanocarrier can deliver both a photopharmacological agent and a photosensitizer. With exposure to light, the photosensitizer forms reactive oxygen species, which may either localize oxidative signals (useful at low concentrations in neurogenesis) or drive neurotoxic aggregates into the clearance mechanism, whereas the photopharmacological molecule modulates receptors of the neurons. This synergistic light-based therapy could offer multimodal control over pathological cascades (Table 1.4; Fig. 1.2).

1.5 Preclinical and Clinical Insights

1.5.1 In Vitro Neuronal Models

Primary cultures of neurons, or those produced by reprogramming adult human-induced pluripotent stem cells, can be useful model systems to train photopharmacological agents before evaluating efficacy in organisms or cell lines in which the underlying cell type is poorly characterized or functionally distinct, due to their controlled environments in terms of light exposure and molecular readout (Gudenschwager et al. 2021). Patch-clamp electrophysiology has the potential to measure the light-induced modulation of receptor currents and super-resolution microscopy to observe receptor redistribution after photoregulation. Neuroinflammation-related photomodulation can be examined in co-cultures of microglia and astrocytes, particularly with agents against inflammatory mediator targets.

1.5.2 In Vivo Dementia Models

Rodent models that either express human APP/PS1 mutations, tau P301L mutations, or combinations of both serve as a means to evaluate pharmacokinetics and behavioral assessments. The chronic photostimulation of selected brain areas (e.g., hippocampus and prefrontal cortex) is possible with fiber-optic implants or miniaturized wireless LEDs (Rodríguez-Fernández et al. 2024). Task-specific cognitive enhancement can be provided by behavioral performance measures like the Morris water maze and novel object recognition, which can be temporally correlated to the photoactivation of a drug of interest.

Table 1.4 Nanotechnology–photopharmacology integration

Nanocarrier type	Photoactivation strategy	BBB penetration mechanism	Example	Advantages	Challenges	Ref.
Gold nanoparticles (AuNPs)	Plasmonic heating with NIR	Receptor-mediated transcytosis	AuNP–photo-curcumin	High light-to-heat conversion	Potential toxicity	Sharma et al. (2021)
Liposomes	Photocleavable lipids	PEGylation + transferrin receptor targeting	Liposome–photo-donepezil	High drug loading	Stability in circulation	Rajput and Butani (2022)
Polymeric nanoparticles	Photo-crosslinked release	GLUT1-mediated transport	PEG–PLGA–photo-ibuprofen	Controlled release	Complex synthesis	Eroglu et al. (2019)
Carbon dots	Photoluminescence-based activation	Adsorptive-mediated transport	CD–photo-GABA	Deep brain penetration	Scale-up issues	Veo et al. (2024)

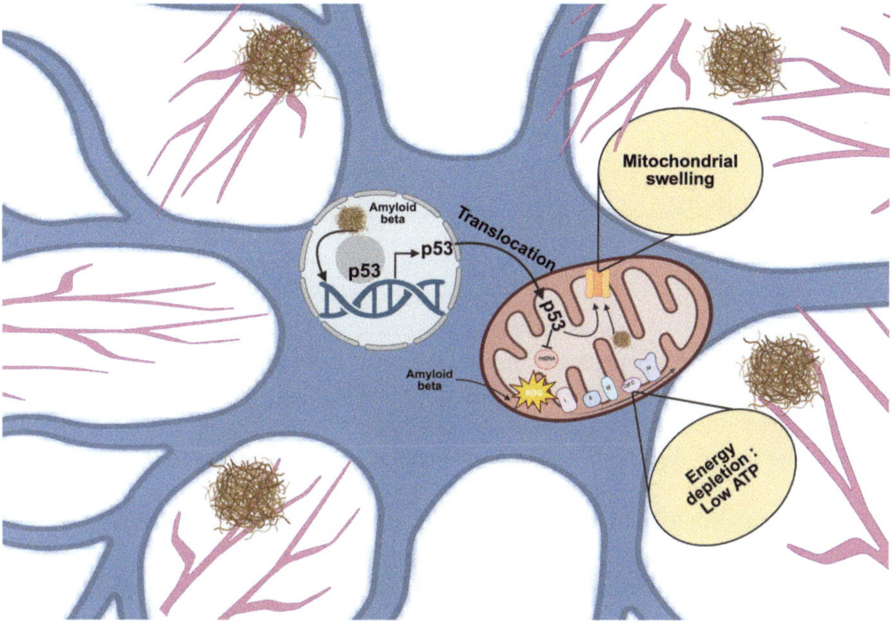

Fig. 1.2 Oxidative Stress, Mitochondrial Dysfunction, and Neuroinflammation Cascade. (http://creativecommons.org/licenses/by-sa/4.0/)

1.5.3 Early Clinical Trials and Safety

Although photopharmacological dementia treatment has not advanced past Phase II in humans, human studies of intracranial light application in other central nervous system conditions (e.g., acute and chronic pain, epilepsy) are illustrating feasibility and safety. Safety considerations include phototoxicity, thermal damage, and immune responses to nanocarriers. Regulatory agencies will require extensive toxicology on both the drug and the light delivery system.

1.6 Challenges and Future Directions

The shortest-term challenge is the delivery of light into deep brain tissue. Optogenetic-style fiber optics, wireless microLED implants, and transcranial NIR delivery alongside upconversion nanoparticles are in active development. Each must balance invasiveness with focality. Drug design and selectivity pose their own molecular challenges. It is necessary that photoswitches should be stable in physiological conditions, reversible in large cycles, and have no photodegradation that may generate toxic byproducts. The distribution of photostationary states of the drug must ensure the inactive form predominates in darkness in order to reduce baseline effects. Integration with optogenetics and neuromodulation could allow multimodal control where light both activates a drug and stimulates neuronal circuits. This could be particularly powerful in network-level dysfunction seen in dementia. Finally, ethical and translational considerations include equitable access to such advanced therapies, patient consent for device implantation, and long-term monitoring in a population that may have progressive cognitive decline.

1.7 Conclusion

Conclusively, dementia is a thorny clinical challenge because it is multifactorial and progressive neuronal loss, and there is minimal response to existing pharmacotherapies. Development of photopharmacology is a paradigm shift in therapeutic approach since it holds promise of surmounting most of the major obstacles in dementia therapy. Leveraging photoswitchable molecules, this method can result in disease-relevant targets being modulated in a reversible, localized, and time-specific manner, including the neurotransmitter systems and key pathological processes, including amyloid-beta aggregation and tau hyperphosphorylation. Association with nanotechnology, fiber-optic tool development, and NIR-coupled light delivery offers tremendous opportunity to extend the application of deep-brain interventions, with reduced systemic exposure and potential systemic side effects. Notwithstanding the still-existing obstacles, especially in the penetration of light into the brain and phototoxicity and biocompatibility of devices, there exists a trend of constant innovation in red-shifted switches, two-photon activation, and biodegradable light guides, which underscores a continuing effective translational path. Notably, this discipline excels when the intersection between chemistry, neuroscience, optical engineering, and bio-nanotechnology takes place, and this makes interdisciplinary actions paramount. With such current momentum, photopharmacology has the development potential to take a concept in search to viable clinical modality type and ultimately change the paradigm of therapeutic interventions in AD and other dementias by providing a previously unheard-of control and specificity in the capacity to modulate the disease.

References

Beharry A (2012) Characterization of Azobenzene-based photo-switches and their evaluation for in vivo applications. Doctoral dissertation

Berlin S, Szobota S, Reiner A, Carroll EC, Kienzler MA, Guyon A et al (2016) A family of photoswitchable NMDA receptors. Elife 5:e12040

Bulic B, Ness J, Hahn S, Rennhack A, Jumpertz T, Weggen S (2011) Chemical biology, molecular mechanism and clinical perspective of γ-secretase modulators in Alzheimer's disease. Curr Neuropharmacol 9(4):598–622

Cacabelos R, Martínez-Iglesias O, Cacabelos N, Carrera I, Corzo L, Naidoo V (2024) Therapeutic options in Alzheimer's disease: from classic acetylcholinesterase inhibitors to multi-target drugs with pleiotropic activity. Life 14(12):1555

Chen X, Wehle S, Kuzmanovic N, Merget B, Holzgrabe U, König B et al (2014) Acetylcholinesterase inhibitors with photoswitchable inhibition of β-amyloid aggregation. ACS Chem Neurosci 5(5):377–389

Combi R, Salsone M, Villa C, Ferini-Strambi L (2021) Genetic architecture and molecular, imaging and prodromic markers in dementia with Lewy bodies: state of the art, opportunities and challenges. Int J Mol Sci 22(8):3960

De Boer EMJ, Orie VK, Williams T, Baker MR, De Oliveira HM, Polvikoski T et al (2021) TDP-43 proteinopathies: a new wave of neurodegenerative diseases. J Neurol Neurosurg Psychiatry 92(1):86–95

DeTure MA, Dickson DW (2019) The neuropathological diagnosis of Alzheimer's disease. Mol Neurodegener 14(1):32

Di Pietro P, Caporarello N, Anfuso CD, Lupo G, Magrì A, La Mendola D, Satriano C (2017) Immobilization of neurotrophin peptides on gold nanoparticles by direct and lipid-mediated interaction: a new multipotential therapeutic nanoplatform for CNS disorders. ACS Omega 2(8):4071–4079

Dong X, Zhang Z, Zhao D, Liu Y, Meng Y, Zhang Y et al (2015) Ultraviolet light triggers the conversion of Cu2+-bound Aβ42 aggregates into cytotoxic species in a copper chelation-independent manner. Sci Rep 5(1):13897

El Alami A, Lagarde F, Huo Q, Zheng T, Baitoul M, Daniel P (2020) Acetylcholine and acetylcholinesterase inhibitors detection using gold nanoparticles coupled with dynamic light scattering. Sensors Int 1:100007

Elahi FM, Miller BL (2017) A clinicopathological approach to the diagnosis of dementia. Nat Rev Neurol 13(8):457–476

Eroglu I, Gultekinoglu M, Bayram C, Erikci A, Ciftci SY, Ayse Aksoy E, Ulubayram K (2019) Gel network comprising UV crosslinked PLGA-b-PEG-MA nanoparticles for ibuprofen topical delivery. Pharm Dev Technol 24(9):1144–1154

Fieldhouse JL, Singleton EH, van Engelen MPE, Van't Hooft JJ, de Boer SC, Froeling VE et al (2023) Decreased emotion recognition and reduced focus on facial hallmarks in behavioral variant frontotemporal dementia compared to primary psychiatric disorders and controls. Eur J Neurol 30(8):2222–2229

Gaillard PJ, Appeldoorn CC, Rip J, Dorland R, van der Pol SM, Kooij G et al (2012) Enhanced brain delivery of liposomal methylprednisolone improved therapeutic efficacy in a model of neuroinflammation. J Control Release 164(3):364–369

Ganzoni RL, Bournons SS, Carreira EM, De Bundel D, Smolders I (2025) A bright future for photopharmaceuticals addressing central nervous system disorders: state of the art and challenges toward clinical translation. Med Res Rev 45:1323

Goudet C, Rovira X, Llebaria A (2018) Shedding light on metabotropic glutamate receptors using optogenetics and photopharmacology. Curr Opin Pharmacol 38:8–15

Gudenschwager C, Chavez I, Cardenas C, Gonzalez-Billault C (2021) Directly reprogrammed human neurons to understand age-related energy metabolism impairment and mitochondrial dysfunction in healthy aging and neurodegeneration. Oxidative Med Cell Longev 2021(1):5586052

Hansen RA, Gartlehner G, Webb AP, Morgan LC, Moore CG, Jonas DE (2008) Efficacy and safety of donepezil, galantamine, and rivastigmine for the treatment of Alzheimer's disease: a systematic review and meta-analysis. Clin Interv Aging 3(2):211–225

Hansen D, Ling H, Lashley T, Holton JL, Warner TT (2019) Clinical, neuropathological and genetic features of Lewy body dementias. Neuropathol Appl Neurobiol 45(7):635–654

Heneka MT, Carson MJ, El Khoury J, Landreth GE, Brosseron F, Feinstein DL et al (2015) Neuroinflammation in Alzheimer's disease. Lancet Neurol 14(4):388–405

Inaba H, Sakaguchi M, Watari S, Ogawa S, Kabir AMR, Kakugo A et al (2023) Reversible photocontrol of microtubule stability by spiropyran-conjugated tau-derived peptides. Chembiochem 24(8):e202200782

Inoue Y, Shue F, Bu G, Kanekiyo T (2023) Pathophysiology and probable etiology of cerebral small vessel disease in vascular dementia and Alzheimer's disease. Mol Neurodegener 18(1):46

Javaid SF, Giebel C, Khan MA, Hashim MJ (2021) Epidemiology of Alzheimer's disease and other dementias: rising global burden and forecasted trends. F1000Res 10:425

Jellinger KA, Attems J (2007) Neuropathological evaluation of mixed dementia. J Neurol Sci 257(1–2):80–87
Jiang X, Ni Y, Liu T, Zhang M, Ren H, Xu G (2013) Inhibition of LPS-induced retinal microglia activation by naloxone does not prevent photoreceptor death. Inflammation 36(1):42–52
Lerch MM, Hansen MJ, van Dam GM, Szymanski W, Feringa BL (2016) Emerging targets in photopharmacology. Angew Chem Int Ed 55(37):10978–10999
Liu Y, Tan Y, Zhang Z, Yi M, Zhu L, Peng W (2024) The interaction between ageing and Alzheimer's disease: insights from the hallmarks of ageing. Transl Neurodegener 13(1):7
Maleeva G, Matera C, Roda S, Colleoni A, De Amici M, Gorostiza P (2025) Molecular tools to study and control dopaminergic neurotransmission with light. Med Res Rev 45:1407
Mirzaie Z, Ansari M, Kordestani SS, Rezaei MH, Mozafari M (2019) Preparation and characterization of curcumin-loaded polymeric nanomicelles to interference with amyloidogenesis through glycation method. Biotechnol Appl Biochem 66(4):537–544
Mondal T, Mandal B (2022) Targeted protein degradation as a therapeutic avenue for Alzheimer's disease. Royal Society of Chemistry
Murrough JW, Abdallah CG, Mathew SJ (2017) Targeting glutamate signalling in depression: progress and prospects. Nat Rev Drug Discov 16(7):472–486
Park YI, Lee KT, Suh YD, Hyeon T (2015) Upconverting nanoparticles: a versatile platform for wide-field two-photon microscopy and multi-modal in vivo imaging. Chem Soc Rev 44(6):1302–1317
Rajput A, Butani S (2022) Donepezil HCl liposomes: development, characterization, cytotoxicity, and pharmacokinetic study. AAPS PharmSciTech 23(2):74
Rodríguez-Fernández L, Zorzo C, Arias JL (2024) Photobiomodulation in the aging brain: a systematic review from animal models to humans. Geroscience 46(6):6583–6623
Ryazantsev MN, Strashkov DM, Nikolaev DM, Shtyrov AA, Panov MS (2021) Photopharmacological compounds based on azobenzenes and azoheteroarenes: principles of molecular design, molecular modelling, and synthesis. Russ Chem Rev 90(7):868
Sharma S, Zvyagin AV, Roy I (2021) Theranostic applications of nanoparticle-mediated photoactivated therapies. J Nanotheranostics 2(3):131–156
Soliman A (2023) Impact of small molecules and peptides on tau microtubule interaction and tau aggregation in primary peripheral neurons and models of Alzheimer's and Parkinson's disease. Universität Osnabrück
Sortino R, Gerwe H, Pérez Pérez NF, Matera C, Riefolo F, Decker M, Gorostiza P (2024) Light-activated drugs to restore neuronal activity in neurological disorders
Tiwari P, Tolwinski NS (2023) Using optogenetics to model cellular effects of Alzheimer's disease. Int J Mol Sci 24(5):4300
Veo B, Wang D, DeSisto J, Pierce A, Vibhakar R (2024) Single-cell multi-omics analysis identifies metabolism-linked epigenetic reprogramming as a driver of therapy-resistant medulloblastoma. Research Square, rs-3
Vinters HV, Zarow C, Borys E, Whitman JD, Tung S, Ellis WG et al (2018) Vascular dementia: clinicopathologic and genetic considerations. Neuropathol Appl Neurobiol 44(3):247–266
Vorobev AY, Moskalensky AE (2020) Long-wavelength photoremovable protecting groups: on the way to in vivo application. Comput Struct Biotechnol J 18:27–34
World Health Organization (2021) Global status report on the public health response to dementia. World Health Organization, Geneva
Ye N, Qin W, Tian S, Xu Q, Wold EA, Zhou J, Zhen XC (2020) Small molecules selectively targeting sigma-1 receptor for the treatment of neurological diseases. J Med Chem 63(24):15187–15217
Zhan Q, He S, Qian J, Cheng H, Cai F (2013) Optimization of optical excitation of upconversion nanoparticles for rapid microscopy and deeper tissue imaging with higher quantum yield. Theranostics 3(5):306
Zhang J, Gao F, Ma Y, Xue T, Shen Y (2021) Identification of early-onset photoreceptor degeneration in transgenic mouse models of Alzheimer's disease. Iscience 24(11):103327

Photodrugs: Role of Redox Regulation in Neurodegenerative Diseases

2

Yashi Mahendra and Hagera Dilnashin

Abstract

Neurodegenerative disorders constitute a heterogeneous group of chronic, progressive conditions characterized by the selective and irreversible degeneration of neurons in the central and peripheral nervous systems. Prominent among these are Alzheimer's disease (AD), Parkinson's disease (PD), and Huntington's disease (HD), each marked by distinct pathological hallmarks yet sharing common underlying mechanisms, including protein misfolding, mitochondrial dysfunction, and, most critically, oxidative stress.

Oxidative stress arises from a persistent imbalance between the generation of reactive oxygen species (ROS) and the cellular antioxidant defense mechanisms, leading to a disruption in redox homeostasis. The resultant accumulation of ROS aggravates neuronal damage through lipid peroxidation, protein oxidation, and DNA fragmentation, further perpetuating neuroinflammation and apoptotic signaling cascades. The failure of endogenous redox-regulatory systems underscores the urgent need for therapeutic interventions capable of restoring oxidative balance while minimizing off-target effects.

In this context, photo-drugs, a novel class of photo-responsive compounds, emerge as a compelling therapeutic strategy, lifting the principles of photobiomodulation (PBM) to modulate cellular redox states with high spatiotemporal precision. PBM operates through the targeted delivery of low-energy light at specific wavelengths (typically within the red to near-infrared spectrum,

Y. Mahendra
University Institute of Engineering and Technology, Maharshi Dayanand University, Rohtak, Haryana, India

H. Dilnashin (✉)
Department of Biosciences & Bioengineering, Indian Institute of Technology-Roorkee, Roorkee, Uttarakhand, India
e-mail: hagera.pd@bt.iitr.ac.in

H. Dilnashin, S. P. Singh (eds.), *Phototherapeutic Approaches to Neurodegeneration*,
https://doi.org/10.1007/978-981-95-4751-7_2

600–1000 nm), which interacts with mitochondrial chromophores, particularly cytochrome c oxidase, to enhance electron transport chain efficiency, attenuate ROS production, and promote ATP synthesis. Unlike conventional pharmacological agents, which often exhibit systemic toxicity and limited blood-brain barrier permeability, photo-drugs offer a non-invasive, multi-targeted approach capable of precise neuromodulation with minimal adverse effects.

This chapter sheds light on the mechanistic basis by which photo-drugs can rectify redox dysregulation in neurodegenerative pathologies, highlighting their potential as a next-generation therapeutic paradigm. By integrating advances in photochemistry, molecular biology, and neuropharmacology, we start exploring the translational prospects of photo-drugs in mitigating neuronal degeneration, thereby paving the way for safer, more efficacious treatments in clinical neurology.

Through a critical examination of preclinical and emerging clinical evidence, we aim to delineate how photo-biomodulation-based interventions could redefine therapeutic strategies, ultimately contributing to the advancement of human health in the face of neurodegenerative decline.

Keywords

Neurodegenerative diseases · Photo-drugs · Photo-biomodulation · Redox regulation · Neuropharmacology

2.1 Overview of Neurodegenerative Diseases

The word "neurodegenerative" is derived from two components "neuro-nerve/neural cells" and "degeneration process of losing structure or function over time." Neurodegenerative diseases, caused as a result of progressive loss of neurons, encompass a group of disorders posing a grave threat to human health, which are degrading the conditions essential for mobility, coordination, strength, sensation, and cognition. Over one in three people globally today face different neurological conditions. For example, about 55 million people globally were affected by dementia in 2019, with projections reaching 139 million by 2050. Others include Alzheimer's disease (AD), Parkinson's disease (PD), Ataxia, Huntington's disease (HD), Multiple System Atrophy, and many other memory and cognitive loss disorders (Cleveland Clinic 2025).

This complex heterogeneous group of chronic disorders is caused by a combination of genetic, environmental, and lifestyle factors. Huntington's disease, which is caused by inherited mutations, and risk genes that increase susceptibility without guaranteeing disease onset, as seen in Alzheimer's and Parkinson's, are key genetic influences. Environmental factors like exposure to toxins and air pollution also play a role, while lifestyle choices such as smoking and high-glycemic diets elevate risk. Chronic conditions like diabetes and obesity exacerbate these diseases through inflammation. Protein misfolding and neuroinflammation are hallmark features, often triggered by environmental or genetic factors, highlighting the need for integrated prevention and treatment strategies. And, nevertheless, stress, high anxiety issues, and overthinking are the most important factors causing neurodegenerative diseases. How? Let's see (Flores-Dorantes et al. 2020).

Chronic stress elevates ROS levels through increased metabolic activity, mitochondrial dysfunction, and heightened inflammatory cytokine release. The brain, with its high oxygen consumption and lipid-rich environment, is particularly vulnerable to oxidative damage caused by this imbalance. Elevated ROS levels impair neuronal function, disrupt neurotrophic factor production (e.g., BDNF, which stands for brain-derived neurotrophic factor), and contribute to neuroinflammation, all of which play an essential key role in aggravating anxiety-related behaviors and cognitive dysfunction (Dash et al. 2024).

2.2 Importance of Redox Regulation in Disease Progression

Reactive oxygen species (ROS) are versatile molecules that are a two-way street in biological systems. While at low concentrations, ROS act as key signaling molecules shaping physiological cell functions; excessive accumulation or dysfunction of antioxidant systems leads to supraphysiological concentrations, causing potentially harmful outcomes (Chen et al. 2017).

ROS arise naturally as byproducts of cellular metabolism, predominantly in mitochondria, where the electron transport chain (ETC) can leak electrons, leading to the formation of superoxide. Other sources include monoamine oxidases (MAOs) and NADPH oxidases (NOX). MAOs, anchored to the outer mitochondrial membrane, catalyze the oxidative deamination of monoamines, leading to H_2O_2 production. The NOX family of enzymes produces superoxide anions through the oxidation of NADPH. The mammalian brain is particularly vulnerable to ROS-mediated toxicity. It has relatively poor antioxidant defenses to cope with the redox burden imposed by the elevated oxygen consumption rate and metabolic activity, as the brain represents only 2% of total body weight but accounts for 20% of total oxygen consumption. Many features of neurodegenerative diseases can be traced back to oxidative stress, where ROS damage cellular components, including proteins, lipids, and DNA. This oxidative stress influences both the onset and progression of brain demise (Song et al. 2024).

In Alzheimer's disease, ROS contribute to amyloid-beta production and tau phosphorylation, while in Parkinson's disease, ROS alter dopaminergic neurons and promote alpha-synuclein aggregation. Consequently, maintaining redox balance is critical for proper neuronal function and development, and its dysregulation can lead to pathological consequences. Understanding the sources of ROS and their impact on neuronal cells is crucial for developing therapeutic strategies targeting neurodegenerative diseases. Even though several therapeutic strategies to target the diseases are present worldwide, new technological advancements are paving the way to a better future, of which photo-biomodulation is a key strategy (Calabresi et al. 2023).

2.3 Emerging Role of Photo-Drugs and Photo-Biomodulation (PBM)

Photo-drugs, also known as photoactivatable drugs, are compounds designed to be activated or modified by light. This approach offers precise spatial and temporal control over drug activity, potentially reducing systemic side effects and improving therapeutic efficacy. Upon exposure to specific wavelengths of light, these drugs undergo a structural change that activates their pharmacological properties. Photobiomodulation (PBM), also known as Low-Level Laser Therapy (LLLT), employs light from lasers or light-emitting diodes (LEDs) to treat various disorders and has garnered increasing interest recently (Yang et al. 2023).

Existing therapeutic drugs for neurodegenerative disease (NDD) primarily offer symptomatic relief, often accompanied by side effects that limit their long-term use. PBM and photo-drugs have emerged as a promising alternative, sparking exponential interest in their application across various neurodegenerative conditions in animal studies. PBM's foundation lies in the principle that incident light excites electrons in chromophores, such as mitochondrial cytochrome c oxidase (CCO), causing them to transition to higher-energy orbits. This process enhances electron transport, mitochondrial membrane potential, and ATP generation. Furthermore, PBM can influence gated channels, like transient receptor potential (TRP) channels, modulating pain and other sensory stimuli. PBM can also mitigate the elevated levels of reactive oxygen species (ROS) associated with neurodegenerative illnesses by enhancing mitochondrial membrane potential, ATP, cyclic adenosine monophosphate (cAMP), and nitric oxide (NO), thereby activating cellular networks and reducing oxidative stress. Its complex mechanisms involve activating multiple pathways (ERK, Akt/GSK3β/β-catenin, Src/Syk/PI3K/Akt, and cAMP/PKA/SIRT1), promoting cell survival, inhibiting inflammatory processes, and reducing apoptosis. This multifaceted approach, involving mitochondrial stimulation, modulation of ion channels, reduction of oxidative stress, and activation of cellular signaling pathways, positions PBM as a potential therapy for managing neurodegenerative disorders, especially in drug-refractory cases (Lamptey et al. 2022).

2.4 Neurodegenerative Diseases and Redox Imbalance

Neurodegenerative diseases (NDDs), including Alzheimer's and Parkinson's disease, are characterized by progressive neuronal loss and cognitive or motor decline, with oxidative stress (OS) playing a central role in their pathophysiology. OS arises from an imbalance between reactive oxygen species (ROS) production and antioxidant defenses, leading to cellular damage, disrupted signaling pathways, and impaired metabolic homeostasis. In NDDs, elevated ROS levels contribute to neuronal dysfunction, synaptic damage, and neurodegeneration by inducing mitochondrial dysfunction, lipid peroxidation, protein misfolding, and DNA damage. This redox imbalance not only accelerates neuronal death but also impairs cellular repair mechanisms, exacerbating disease progression.

Mitochondrial dysfunction is a key feature in NDDs, as affected neurons exhibit increased ROS production and reduced oxidative phosphorylation. In addition to this, redox-sensitive pathways such as Nrf2-Keap1 and NF-κB are disrupted in NDDs, influencing inflammatory responses and antioxidant defenses. Environmental factors like neurotoxic metals and pesticides further amplify OS in these conditions. Understanding the pathophysiology is important to understand the therapeutic intervention of the disease progression (Calabresi et al. 2023; Yang et al. 2023; Lamptey et al. 2022).

2.4.1 Alzheimer's Disease

Alzheimer's disease (AD), a prevalent form of dementia, is characterized by progressive neuroinflammation and cognitive decline, affecting millions globally. The pathogenesis of AD is complex, with two primary hallmarks: amyloid plaques and neurofibrillary tangles. Amyloid plaques are extracellular deposits composed of beta-amyloid (Aβ) peptides, which are derived from the amyloid precursor protein (APP). APP, a transmembrane protein, is cleaved by α-, β-, and γ-secretases. Normally, APP can be cleaved by α- or β-secretase without producing harmful byproducts. However, sequential cleavage by β- and γ-secretase results in the production of Aβ peptides, particularly Aβ42. Aβ42 is prone to aggregation, leading to the formation of amyloid plaques that disrupt neuronal function and trigger neuroinflammation. Genetic factors also influence Aβ production, as mutations in genes encoding APP and presenilins (components of γ-secretase) are associated with early-onset familial AD (Kamatham et al. 2024).

Neurofibrillary tangles, the second hallmark of AD, are intracellular aggregates of hyperphosphorylated tau protein. Tau protein normally stabilizes microtubules, which are essential for axonal transport. In AD, however, tau becomes hyperphosphorylated, causing it to detach from microtubules and aggregate into tangles. These tangles disrupt axonal transport, leading to synaptic dysfunction and neuronal death. The formation and accumulation of neurofibrillary tangles follow a predictable regional pattern, beginning in the hippocampus and spreading to other cortical areas as the disease progresses. This staging of neurofibrillary tangles is an important criterion for the neuropathological diagnosis of AD (Kamatham et al. 2024).

Neuroinflammation also contributes significantly to AD pathogenesis. The presence of amyloid plaques and neurofibrillary tangles activates microglia and astrocytes, triggering the release of pro-inflammatory cytokines, chemokines, and reactive oxygen species (ROS). This chronic neuroinflammatory response exacerbates neuronal damage, promotes synaptic dysfunction, and contributes to the overall progression of AD. In addition to these key pathological features, other factors such as vascular dysfunction and impaired glucose metabolism may also play a role in AD pathogenesis. The neurovascular hypothesis suggests that dysfunction of the neurovascular unit, including impaired clearance of Aβ across the blood-brain barrier (BBB), can contribute to AD. Similarly, impaired glucose metabolism and insulin resistance in the brain have been implicated in AD, potentially through mechanisms involving oxidative stress and inflammation (Dash et al. 2024; Lamptey et al. 2022; Kamatham et al. 2024).

Early signs include memory loss, particularly difficulty recalling recent events or conversations. As the disease advances, individuals may experience confusion and disorientation, often struggling with time and place recognition. Difficulty in completing familiar tasks, such as cooking or managing finances, becomes evident. Language problems also emerge, with individuals finding it hard to follow or join conversations due to word-finding difficulties (Dash et al. 2024; Lamptey et al. 2022).

Behavioral changes are common, including mood swings, increased anxiety, depression, and irritability. Social withdrawal often occurs as individuals become aware of their declining abilities. In later stages, severe cognitive impairment leads to an inability to recognize loved ones or perform basic daily activities like eating or dressing. Physical symptoms such as impaired coordination and balance may develop, increasing the risk of falls. Ultimately, AD can lead to complete dependency on caregivers due to the loss of functional autonomy (Kamatham et al. 2024).

2.4.2 Parkinson's Disease

Parkinson's disease (PD) is a progressive neurodegenerative condition characterized by the loss of dopaminergic neurons in the substantia nigra pars compacta (SNpc). Most cases of PD are idiopathic, meaning there is no known cause, and disease development is influenced by a combination of environmental factors, genetic predisposition, mediators of inflammation, abrogation of the autosomal-lysosomal autophagy system, and endoplasmic reticulum stress. PD is the second most prevalent neurodegenerative ailment in industrialized nations, affecting approximately 1% of adults over 60 and about 3% of individuals over 80 (Kouli et al. 2018).

The exact pathogenetic processes driving the gradual loss of dopaminergic cells in the substantia nigra of people with PD are not yet fully understood. Genetic mutations have been linked to familial forms of PD, but they only account for a small percentage of cases. These genes include α-synuclein (SNCA), leucine-rich repeat kinase 2 (LRRK2), parkin (PRKN), PTEN-induced kinase 1 (PINK1), and DJ-1. The protein α-synuclein is a major component of Lewy bodies, which are intraneuronal cytoplasmic inclusions that are a hallmark of PD. Accumulation and aggregation of misfolded α-synuclein are thought to contribute to neuronal dysfunction and cell death (Lamptey et al. 2022; Kouli et al. 2018).

Mitochondrial dysfunction, oxidative stress, and neuroinflammation are also implicated in the pathogenesis of PD. Impaired mitochondrial function can lead to increased production of reactive oxygen species (ROS), which can damage cellular components and contribute to neuronal death. Neuroinflammation, mediated by activated microglia and astrocytes, can further exacerbate neuronal damage in PD. Environmental factors, such as exposure to pesticides and heavy metals, have also been linked to an increased risk of developing PD, potentially through mechanisms involving oxidative stress and mitochondrial dysfunction (Kouli et al. 2018).

The loss of dopaminergic neurons in the SNpc leads to a deficiency of dopamine in the striatum, which is a brain region involved in motor control. This dopamine deficiency results in the characteristic motor symptoms of PD, including resting tremor, rigidity, bradykinesia (slow movement), and postural instability. Non-motor symptoms, such as depression, anxiety, sleep disturbances, and cognitive impairment, are also common in PD and can significantly impact quality of life (Dash et al. 2024; Kouli et al. 2018).

2.5 Role of Reactive Oxygen Species (ROS) and Oxidative Stress

Reactive oxygen species (ROS) play a complex and multifaceted role in neurodegenerative diseases, acting as both signaling molecules and mediators of cellular damage. The imbalance between ROS production and antioxidant defenses leads to a state of oxidative stress, which is implicated in the pathogenesis of Alzheimer's disease (AD) and Parkinson's disease (PD) (Yang et al. 2023).

In AD, ROS are generated through several mechanisms, including amyloid-beta (Aβ) aggregation, mitochondrial dysfunction, and neuroinflammation. The accumulation of Aβ plaques triggers neuroinflammation, activating microglia and astrocytes, which release ROS as part of their immune response. Some research also suggests that Aβ fragments may directly generate free radicals, assaulting neurons, which is called Amyloid-beta (Aβ) aggregation. Impaired mitochondrial function, common in AD, leads to increased ROS production due to inefficient electron transport and oxidative phosphorylation, which is the core of mitochondrial dysfunction. The inflammatory response itself contributes to ROS production by immune cells, further exacerbating oxidative stress in the brain (Calabresi et al. 2023; Yang et al. 2023).

In PD, ROS generation is linked to dopamine metabolism, which is deficient in PD and can generate ROS as a byproduct. Similar to AD, mitochondrial dysfunction in dopaminergic neurons increases ROS production and can be caused by genetic mutations in genes like α-synuclein, which can lead to protein misfolding and aggregation, triggering oxidative stress and ROS production (Yang et al. 2023; Kouli et al. 2018).

2.6 Mitochondrial Dysfunction and Its Impact on Neuronal Health

Mitochondrial dysfunction represents a critical pathological feature across a spectrum of neurodegenerative diseases (NDDs), including Alzheimer's disease (AD) and Parkinson's disease (PD). Neurons, with their high energy demands and intricate signaling pathways, are particularly vulnerable to disruptions in mitochondrial function. The consequences of these disruptions reverberate throughout the cell, impacting everything from energy production and redox balance to calcium homeostasis and apoptosis (Klemmensen et al. 2023).

In NDDs, impaired mitochondrial function manifests as decreased ATP production, increased generation of reactive oxygen species (ROS), and impaired calcium buffering capacity. The electron transport chain (ETC), located within the inner mitochondrial membrane, is a major site of ROS generation. As discussed earlier, damaged or inefficient ETC complexes leak electrons, leading to the formation of superoxide radicals, which can then be converted to other ROS. This chronic oxidative stress damages cellular components, worsening mitochondrial dysfunction and triggering a vicious cycle of neuronal demise. In both Alzheimer's and Parkinson's disease, neurons struggle to clear out cytotoxic alpha-synucleins, contributing to greater damage (Yang et al. 2023; Klemmensen et al. 2023).

Mitochondrial dysfunction directly impairs neuronal health through multiple mechanisms. Reduced ATP production compromises neuronal function by disrupting ion gradients, neurotransmitter synthesis, and axonal transport. Increased ROS generation damages lipids, proteins, and DNA, leading to oxidative stress and inflammation. The compromised ability to buffer calcium can dysregulate synaptic transmission and trigger excitotoxicity. Ultimately, severe mitochondrial dysfunction triggers apoptotic signaling cascades, leading to neuronal death and contributing to the progressive neurodegeneration observed in NDDs (Klemmensen et al. 2023).

Figure 2.1 explains the common mechanisms and causes of neurodegenerative diseases, which include environmental factors, metabolic stress, genetic contributions, neurovascular coupling, and neuroinflammation (Cleveland Clinic 2025; Flores-Dorantes et al. 2020).

The entangled relationship between mitochondrial dysfunction, oxidative stress, and neuroinflammation forms a complex and interconnected pathological network in NDDs. Mitochondrial dysfunction contributes to oxidative stress, which in turn activates inflammatory pathways and further impairs mitochondrial function. This cyclical process amplifies neuronal damage and accelerates disease progression, highlighting the critical role of mitochondrial health in maintaining neuronal survival and function. Therapeutic strategies aimed at restoring mitochondrial function, reducing oxidative stress, and suppressing neuroinflammation hold significant promise for slowing or preventing neurodegeneration in NDDs (Klemmensen et al. 2023).

2.7 Photo-Drugs

Photo-pharmacology is an innovative field centered on using light to control the activity of drugs, offering significant advantages in spatial and temporal precision over traditional pharmacology. Conventional drugs often lack specificity, affecting both diseased and healthy tissues, which can lead to unwanted side effects. Photo-pharmacology aims to overcome these challenges by designing drugs that can be activated or deactivated by specific wavelengths of light, limiting their action to

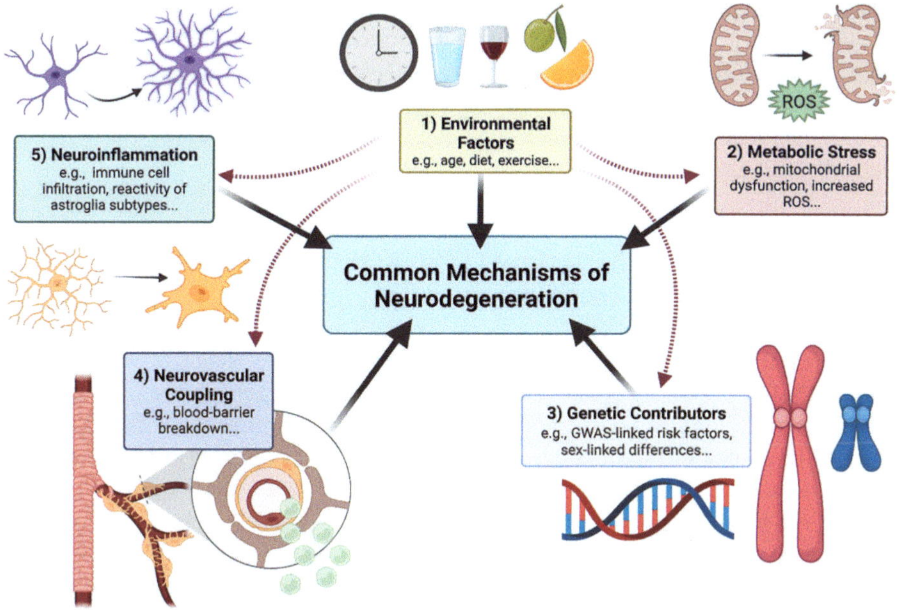

Fig. 2.1 Mechanisms of neurodegeneration

precisely illuminated areas. This targeted approach has the potential to minimize off-target effects and enhance therapeutic efficacy (Menéndez-Velázquez and García-Delgado 2023).

The core principle of photo-pharmacology involves creating photo-responsive drugs that undergo structural changes when exposed to light. These drugs fall into two primary categories: irreversible and reversible. Irreversible photo-drugs often utilize molecular cages, known as photocages, that encapsulate the active compound. When exposed to light, the cage is broken through a process called photolysis, releasing the drug to exert its pharmacological effect. Once activated, however, these drugs cannot be switched back to an inactive state (Menéndez-Velázquez and García-Delgado 2023).

Reversible photo-drugs, on the other hand, are engineered to undergo a molecular change under light exposure, enabling them to switch between active and inactive states and back again by employing different wavelengths of light or through thermal relaxation. One common type of reversible photo-drug is based on photo-switchable molecules, compounds that can exist in two or more distinct states. By attaching a photo-switchable group to a drug molecule, the drug's activity can be toggled on or off by exposing it to light of different wavelengths (Calabresi et al. 2023; Menéndez-Velázquez and García-Delgado 2023).

To achieve light-mediated drug activation deep within tissues, spectral conversion techniques can be employed. Near-infrared (NIR) light, which can penetrate biological tissues more effectively than UV or visible light, can be converted into the UV-visible range to activate specific photo-drugs. One promising strategy uses rare-earth materials for up-conversion, converting NIR light into higher-energy UV-visible light. Organic materials can also be used in down-shifting to fine-tune the light spectrum for optimal drug activation. These strategies allow for deeper tissue penetration and more precise control over drug activity, expanding the potential clinical applications of photo-pharmacology (Menéndez-Velázquez and García-Delgado 2023).

2.8 Mechanisms of Action: Light-Activated Modulation of Biological Processes

Photo-pharmacology 's mechanism of action relies on the selective activation of drugs by light, providing spatial and temporal control not achievable with conventional pharmacology. When a photo-drug is administered, it remains inactive until exposed to a specific wavelength of light. This allows the drug to be distributed throughout the body without causing unwanted side effects, as it will only be activated in the illuminated area (Wijtmans et al. 2022).

The activation process depends on the type of photo-drug used. Irreversible photo-drugs often employ molecular cages, also known as photocages, that encapsulate the active drug. Upon light exposure, the photocage undergoes photolysis, breaking apart and releasing the drug. This mechanism provides spatial selectivity but lacks temporal control, as the drug remains active until metabolized or excreted (Wijtmans et al. 2022).

Reversible photo-drugs offer both spatial and temporal control. These drugs contain photo-switchable molecules that undergo reversible structural changes when exposed to different wavelengths of light or by thermal relaxation. Only one of the molecular states exhibits pharmacological activity. By alternating the light source, the drug can be switched between active and inactive states, allowing for precise control over its therapeutic effect (Wijtmans et al. 2022).

The effectiveness of photo-pharmacology depends on the light's ability to penetrate biological tissues. Since UV and visible light have limited penetration, spectral conversion techniques are often used. Near-infrared (NIR) light, which penetrates tissues more effectively, can be converted to UV-visible wavelengths using luminescent materials. Up-conversion materials, such as rare-earth elements, absorb NIR light and re-emit it as higher energy.

Figure 2.2 explains the functioning of photo-drugs on various neurodegenerative disorders and the results (Cleveland Clinic 2025; Dash et al. 2024; Chen et al. 2017). UV-visible light. Additionally, organic luminescent materials can be used for down-shifting, adjusting the emitted light to match the specific activation wavelengths of the photo-drug. This dual-spectral conversion technique allows for deep-tissue activation of photo-drugs, expanding their therapeutic potential (Wijtmans et al. 2022).

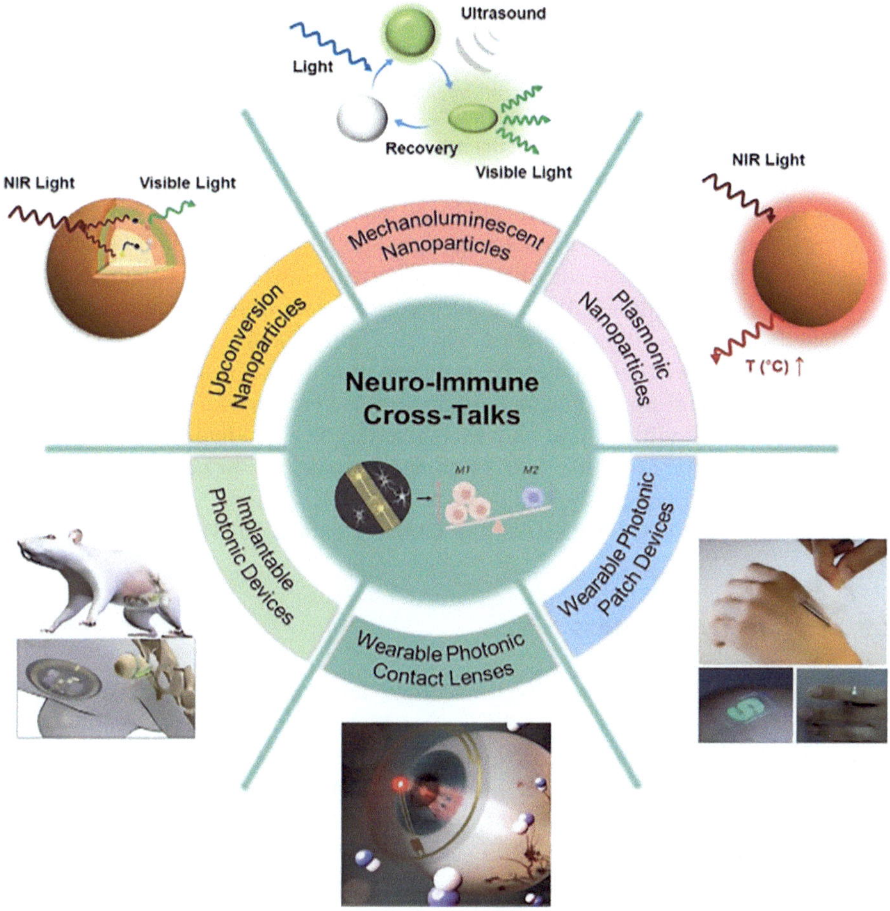

Fig. 2.2 Photo-drugs and NDDs

2.9 Advantages Over Traditional Therapies

Photo-pharmacology offers transformative advantages over conventional therapeutic approaches, primarily through its ability to achieve spatiotemporal control and non-invasiveness. These features address critical limitations of traditional pharmacology, such as systemic toxicity, off-target effects, and drug resistance, while enabling unprecedented precision in treatment (Menéndez-Velázquez and García-Delgado 2023).

2.9.1 Spatiotemporal Control

Traditional drugs lack spatial and temporal selectivity, often acting systemically and persistently, which can lead to collateral damage to healthy tissues. Photo-pharmacology overcomes this by restricting drug activity to specific locations and time frames. Spatial control is achieved by activating drugs only in illuminated regions, ensuring therapeutic effects are localized to diseased tissues. For example, photo-switchable inhibitors or photocaged chemotherapeutics remain inert until exposed to light at the target site, minimizing harm to surrounding healthy cells.

Temporal control allows clinicians to precisely time drug activation, aligning treatment windows with biological rhythms or disease-specific conditions. Reversible photo-switches enable dynamic toggling between active and inactive states, permitting real-time adjustment of drug efficacy. This dual control is particularly valuable in conditions like cancer, where targeted irradiation of tumors can activate cytotoxic agents while sparing adjacent tissues, or in neurological disorders, where timed light pulses could modulate neural activity without prolonged drug exposure (Menéndez-Velázquez and García-Delgado 2023).

2.9.2 Non-invasiveness

Light-based activation in photo-pharmacology leverages the biological window (visible to near-infrared light, 650–900 nm), which penetrates tissues deeply without causing significant damage. Advanced delivery systems, such as fiber optics or wireless subdermal devices, enable light to reach internal organs without invasive surgery. For instance, endoscopic light sources can activate photo-drugs in gastrointestinal tumors, while implantable LED arrays allow repeated, non-invasive modulation of neural circuits. Unlike systemic drugs that require metabolization or invasive injections, photo-pharmacology relies on external or minimally invasive light delivery, reducing procedural risks, infection rates, and recovery times (Fig. 2.3) (Wijtmans et al. 2022).

2.9.3 Reduced Systemic Toxicity and Drug Resistance

By localizing drug activity, photo-pharmacology minimizes systemic exposure, thereby lowering the risk of adverse effects. For example, traditional chemotherapies indiscriminately target rapidly dividing cells, causing hair loss, gastrointestinal distress, and immunosuppression. In contrast, light-activated chemotherapeutics remain inactive until precisely triggered at tumor sites, sparing healthy cells. Similarly, antibiotics activated only at infection sites reduce collateral damage to beneficial microbiota, curbing the development of antimicrobial resistance (Menéndez-Velázquez and García-Delgado 2023; Wijtmans et al. 2022).

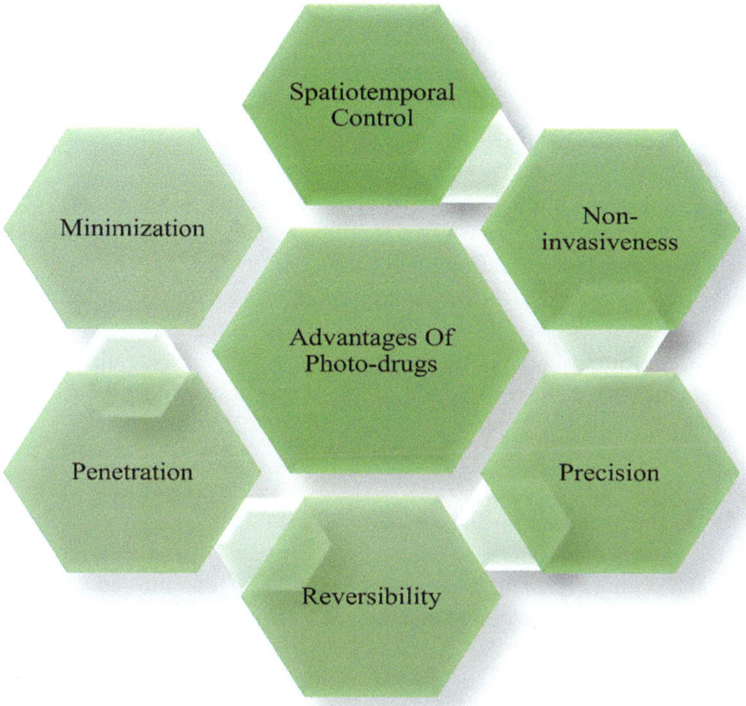

Fig. 2.3 Advantages of photo-drugs

2.9.4 Enhanced Therapeutic Precision

The integration of spectral conversion techniques further refines precision. Rare-earth up-conversion nanoparticles can transform deeply penetrating near-infrared light into UV-visible wavelengths, enabling activation of photo-drugs in tissues several centimeters thick. This approach is being explored for treating deep-seated tumors or neurodegenerative pathologies, where traditional light delivery is impractical (Menéndez-Velázquez and García-Delgado 2023; Wijtmans et al. 2022).

2.10 Photo-Biomodulation (PBM) Therapy

2.10.1 Overview of PBM and Its Therapeutic Applications

Photo-biomodulation (PBM) is a non-invasive therapeutic approach that uses specific wavelengths of light, primarily in the red and near-infrared spectrum, to modulate biological processes and improve cellular function. PBM has evolved from basic light therapy techniques into a sophisticated field within photomedicine, leveraging advancements in light sources and photochemical reactions to achieve precise therapeutic outcomes (Woo 2024).

PBM is applied across various medical fields, including neurology, dermatology, and sports medicine. In neurodegenerative diseases like Alzheimer's and Parkinson's, PBM has demonstrated potential in reducing oxidative stress and neuroinflammation while enhancing neuronal survival. In dermatology, it is used to treat skin conditions such as psoriasis and acne by promoting collagen synthesis and reducing inflammation. Athletes benefit from PBM for muscle recovery and pain relief due to its ability to accelerate tissue healing (Yang et al. 2023; Woo 2024).

One of the key advantages of PBM is its non-invasive nature. Unlike surgical or pharmacological interventions, PBM does not require incisions or systemic drug administration, minimizing risks such as infections or adverse drug reactions. Furthermore, its ability to target specific tissues using focused light ensures localized effects without harming surrounding areas (Woo 2024).

2.10.2 Mechanisms of PBM in Redox Regulation

Photo-biomodulation (PBM) exerts its therapeutic effects on redox regulation through two interconnected mechanisms: activation of cytochrome c oxidase (CCO) and modulation of reactive oxygen species (ROS). These processes synergistically restore mitochondrial function and mitigate oxidative stress, which are critical in neurodegenerative diseases (NDDs) and other conditions characterized by redox imbalance (Wijtmans et al. 2022; Woo 2024).

CCO, a key enzyme in the mitochondrial electron transport chain (ETC), serves as the primary chromophore for red to near-infrared light (650–1200 nm). Absorption of light by CCO enhances its catalytic activity, facilitating the transfer of electrons to molecular oxygen and boosting ATP synthesis. This activation improves electron transport chain efficiency, reducing electron leakage from the ETC and minimizing the formation of superoxide radicals (O_2^-), a primary ROS. Enhanced CCO activity stabilizes the mitochondrial membrane potential, preventing proton leakage and subsequent ROS overproduction. Light-induced dissociation of nitric oxide (NO) from CCO promotes vasodilation, improving oxygen and nutrient delivery to tissues. NO also acts as a signaling molecule, indirectly regulating ROS levels through redox-sensitive pathways (Menéndez-Velázquez and García-Delgado 2023; Woo 2024).

PBM exhibits a biphasic effect on ROS, balancing their dual roles as damaging oxidants and signaling molecules. By enhancing mitochondrial efficiency, PBM lowers baseline ROS production, which is particularly beneficial in NDDs, where chronic oxidative stress damages lipids, proteins, and DNA. Subtherapeutic light doses induce a controlled, transient ROS spike, activating antioxidant defense systems, such as the Nrf2-Keap1 pathway, which upregulates enzymes like superoxide dismutase (SOD) and glutathione peroxidase (GP_X), fostering long-term redox resilience. PBM also suppresses pro-inflammatory cytokine release, like TNF-α and

IL-6, and inhibits NF-$_{\kappa}$B activation, reducing secondary ROS generation from immune cells (Chavda et al. 2021; Woo 2024).

The interplay between CCO activation and ROS modulation creates a feedback loop that sustains mitochondrial health. NO-mediated vasodilation improves tissue oxygenation, further supporting ETC function, while antioxidant upregulation counters residual oxidative damage. In neurodegenerative contexts, this dual action mitigates synaptic dysfunction, neuronal apoptosis, and neuroinflammation, hallmark features of diseases like Alzheimer's and Parkinson's (Chen et al. 2017; Woo 2024).

2.11 Impact of Photo-Drugs on Neurodegenerative Diseases

2.11.1 Effects on Alzheimer's Disease: Cognitive Improvement, Reduced Neuroinflammation

The exploration of light-based therapies for Alzheimer's disease has opened new pathways for targeting its complex pathology. A promising development involves a compound called CRANAD-147, derived from curcumin and modified with a light-sensitive component. This approach focuses on disrupting amyloid-beta (Aβ) aggregates, which are central to Alzheimer's progression. By leveraging light activation, the therapy aims to combine precision with minimal harm to healthy tissues.

2.11.1.1 Cognitive Improvement

Studies in transgenic mice engineered to mimic Alzheimer's pathology reveal that combining CRANAD-147 with targeted light exposure—whether from LEDs or a specialized molecular light source—slows the buildup of Aβ plaques in the brain. While direct measurements of cognitive improvement in these experiments aren't detailed, reducing Aβ accumulation is a well-established strategy for addressing memory loss and cognitive decline in Alzheimer's. The logic is straightforward: fewer plaques mean less disruption to neural networks, which could translate to preserved mental clarity and slower disease progression.

Alzheimer's isn't just about plaques and tangles—it's also marked by chronic inflammation in the brain. Activated immune cells, like microglia, release inflammatory molecules that exacerbate damage to neurons. Light-based therapies are known to dampen inflammatory responses. By targeting Aβ clusters with precision, CRANAD-147 could indirectly calm this inflammatory storm. Imagine a therapy that not only clears toxic aggregates but also soothes the brain's overactive immune response, creating a dual-action defense against neurodegeneration.

2.11.1.2 The Light as a Molecular Switch

Traditional drugs often act broadly, affecting both diseased and healthy cells. Photopharmacology flips this script. CRANAD-147 remains inert until a specific wavelength of light activates it, ensuring the drug only works where and when needed. This is like having a spotlight that illuminates only the problematic areas of the brain. The "molecular light" used in this approach—generated by a companion probe—penetrates deeper into tissues than conventional light sources, making it possible to reach Aβ deposits that would otherwise stay hidden (Kuang et al. 2023).

This precision matters because Alzheimer's doesn't strike uniformly. Aβ plaques cluster in regions critical for memory and cognition, such as the hippocampus. By focusing therapy on these hotspots, the treatment avoids unnecessary exposure to healthy brain areas, reducing side effects. Over time, this targeted strategy could help preserve neuronal function and delay the disease's advance (Kuang et al. 2023).

2.11.1.3 Mechanism of Action and Implications

A photo-pharmacological approach is known to be studied where CRANAD-147, a photolabile compound, undergoes structural changes upon irradiation. This process alters the properties and neurotoxicity of Aβ species, effectively slowing their accumulation. The use of "molecular light" emitted by ADLumin-4—a chemiluminescence probe—is particularly innovative. Unlike external light sources, which struggle to penetrate deep tissues, molecular light is generated directly within the brain, ensuring that the therapeutic effects reach even the most secluded Aβ deposits (Kuang et al. 2023).

This method's dual advantage lies in its precision and adaptability. By activating the drug only in diseased regions, it minimizes collateral damage to healthy neurons. Additionally, the ability to toggle drug activity with light allows for dynamic adjustments, aligning treatment with the brain's fluctuating needs. For example, during early-stage Alzheimer's, intermittent light exposure could prevent plaque formation, while in advanced stages, sustained activation might break down existing aggregates (Kuang et al. 2023).

2.11.1.4 Redefining Alzheimer's Therapy

The implications of this approach extend beyond plaque reduction. By integrating light's spatial and temporal control, CRANAD-147 represents a shift from blunt, systemic treatments to tailored interventions (Kuang et al. 2023).

2.11.2 Effects on Parkinson's Disease: Dopamine Regulation, Mitochondrial Protection

Parkinson's disease (PD) is a debilitating neurodegenerative disorder characterized by the progressive loss of dopaminergic neurons in the substantia nigra, a brain region crucial for motor control. This neuronal loss leads to a deficiency in dopamine, a neurotransmitter essential for coordinating movement, resulting in the hallmark motor symptoms of PD, including tremors, rigidity, bradykinesia (slowed movement), and postural instability. Traditional treatments have limitations, prompting the exploration of innovative approaches like photo-pharmacology, offering the potential for more targeted and effective therapies (Zhou et al. 2023).

2.11.2.1 Current Treatment Limitations

Conventional Parkinson's drugs, primarily levodopa (L-DOPA), aim to replenish dopamine levels in the brain. While L-DOPA can effectively alleviate motor symptoms in the early stages of the disease, its long-term use is associated with significant drawbacks. Over time, patients often develop motor complications, such as dyskinesias (involuntary, erratic movements) and motor fluctuations (periods of "on" and "off" states), which can severely impair their quality of life. The action of traditional drugs is limited by spatial specificity and inaccurate distribution, which reduces therapeutic efficiency. Furthermore, as PD progresses, the effectiveness of L-DOPA diminishes, requiring increased dosages or the addition of other medications, further increasing the risk of side effects (Table 2.1) (Menéndez-Velázquez and García-Delgado 2023; Zhou et al. 2023).

2.11.2.2 Opto-Pharmacology and MRS7145

Opto-pharmacology, also known as photo-pharmacology, is the emerging field that utilizes light to control drug activity with high precision. This approach involves designing drugs that are inactive until exposed to a specific wavelength of light, allowing for targeted activation in specific tissues or brain regions. MRS7145 is a photoactive drug that has shown effectiveness in animal models. According to research, MRS7145 is a photo-sensitive derivative from SCH442416, a selective antagonist of the adenosine A2A receptor. Some A2A antagonist receptors came up as potential drugs to fight Parkinson, since they take part in the involved mechanisms in controlling the movement. This precise control offers the potential to minimize off-target effects and improve therapeutic outcomes compared to traditional systemic drug administration (Zhou et al. 2023).

Table 2.1 Differences between conventional and photo-drugs

Aspect	Photo-drugs	Conventional drugs
Activation mechanism	Require specific light wavelengths (e.g., UV/visible) for activation	Activate via biochemical interactions (no external energy source required)
Delivery method	Often require invasive light delivery systems (e.g., optical fibers)	Administered orally, intravenously, or topically
Regulatory pathway	Subject to FDA drug approval but face unique challenges (e.g., light safety)	Rigorous FDA testing for safety/ efficacy before market approval
Safety profile	Risk of phototoxicity (skin/eye damage) and tissue-specific effects	Side effects linked to systemic exposure (e.g., organ toxicity)
Primary applications	Targeted therapies (e.g., cancer, neuroinflammation, skin conditions) [user data]	Broad-spectrum use (infections, chronic diseases, symptom management)
Drug design	Optimized for light sensitivity, tissue penetration, and photostability	Focus on pharmacokinetics (absorption, metabolism) and receptor binding
Side effects	Light-dependent reactions (e.g., oxidative stress, photosensitivity)	Dose-dependent toxicity or allergic reactions unrelated to light

2.11.2.3 Mechanism of Action

MRS7145 is activated by a specific wavelength of light (405 nm), considered harmless for the body. When the striated bodies of laboratory animals are irradiated with violet light, the active drug is released and blocks the adenosine A2A receptor. The blocking of adenosine receptors has a stimulant effect on the activity of dopamine (pro-dopaminergic action). Adenosine A2A receptors are G protein-coupled receptors that modulate dopamine signaling in the striatum, a key brain region involved in motor control. By selectively blocking A2A receptors in the striatum, MRS7145 enhances dopamine neurotransmission, thereby alleviating motor symptoms associated with Parkinson's disease (Zhou et al. 2023).

2.11.2.4 Spatial and Temporal Precision

One of the key advantages of MRS7145 is its ability to be activated with light at any wished moment, that is, with a high spatiotemporal resolution. Traditional drugs are often distributed throughout the body, affecting both diseased and healthy tissues, and their effects can persist for extended periods. In contrast, MRS7145 can be activated specifically with light at any wished moment, that is, with a high spatiotemporal resolution. The intensity and length of radiation could be regulated with an electronic device. This level of control is crucial in PD, where the degeneration of dopaminergic neurons is localized to specific brain regions. The use of light to activate MRS7145 allows for fine-tuned manipulation of neural circuits to set the functioning of those with therapeutic and neuroprotective purposes, leading to more effective and neuroprotective treatments (Zhou et al. 2023).

2.12 Advantages and Challenges

Photo-pharmacology presents a promising, yet still nascent, approach to treating neurodegenerative diseases (NDDs). It aims to address the limitations of conventional therapies by utilizing light to control drug activity with greater precision. This section outlines the potential advantages and existing challenges of photo-drugs, drawing upon current research in Parkinson's disease (PD) and broader insights into NDD treatments.

2.13 Advantages of Photo-Drugs in NDDs

2.13.1 Enhanced Spatial Specificity

Traditional drugs often lack the ability to target specific areas within the brain, leading to widespread effects and potential side effects. Photo-drugs, however, can be designed to activate only when exposed to light of a specific wavelength. This allows for precise targeting of affected brain regions, minimizing the impact on healthy tissues. For example, in PD, a photo-drug could be activated selectively in the substantia nigra or striatum, the primary sites of dopaminergic neuron loss and motor dysfunction, respectively. And as already stated, "PBM is a non-invasive technique for treating brain disorders." (Zhou et al. 2023; Niazi 2023).

2.13.2 Temporal Control

Traditional drugs also have a limited capacity for temporal control, with their effects persisting for extended periods regardless of the patient's immediate needs. Photo-drugs, on the other hand, offer the potential for dynamic control over drug activity. By controlling the timing and duration of light exposure, clinicians can fine-tune drug effects to match the patient's specific needs at a given moment. This is particularly valuable in managing the fluctuating motor symptoms of PD, where drug levels need to be carefully adjusted throughout the day (Niazi 2023).

2.13.3 Potential for Reduced Side Effects

The precise spatial and temporal control afforded by photo-drugs could translate into fewer side effects compared to traditional systemic medications. By limiting drug activity to the targeted brain regions and controlling the duration of exposure, it may be possible to minimize off-target effects on other organ systems. "PBM is a non-invasive technique for treating brain disorders, and no side effects have been documented during animal research or human treatment till date." (Zhou et al. 2023; Niazi 2023).

2.13.4 Potential for Neuroprotection

While the primary goal of many current PD treatments is to manage symptoms, some photo-drugs may also offer the potential for neuroprotection, slowing down the progression of the disease (Niazi 2023).

2.14 Challenges of Photo-Drugs in NDDs

2.14.1 Light Delivery to the Brain

One of the major challenges in applying photo-drugs to NDDs is delivering light to the targeted brain regions. The skull and brain tissue can scatter and absorb light, limiting its penetration depth. Due to severe human crises regarding the treatment of Parkinson's disease (PD), many cell sources were considered for this treatment, but human embryonic cells were ultimately implemented. Even though ethical and logistical issues provided a significant barrier to this work. Current methods for light delivery include implantable optical fibers or light-emitting devices, which require invasive surgical procedures. Developing non-invasive light delivery techniques, such as transcranial photo-biomodulation or focused ultrasound-mediated light delivery, is a critical area of research (Niazi 2023).

2.14.2 Photo-Drug Design and Development

Designing and synthesizing effective photo-drugs is another significant challenge. The ideal photo-drug should be highly selective for its target, exhibit strong light sensitivity, and have favorable pharmacokinetic properties (e.g., good blood-brain barrier permeability, minimal toxicity). In addition to this, the wavelength of light used to activate the drug should be compatible with tissue penetration and minimize the risk of phototoxicity (Table 2.2) (Niazi 2023).

2.14.3 Treating Neuroinflammation in PD

One of the main causes of PD is neuroinflammation. This is actually being tested to see if photo-drugs are actually improving the patients' current condition or not. While photo-drugs provide neuroprotection, their role in PDs is yet to be studied on animal models.

2.14.4 Long-Term Efficacy and Safety

The long-term efficacy and safety of photo-drugs for NDDs remain largely unknown. Clinical trials are needed to assess the effects of prolonged light exposure on brain tissue, the potential for phototoxicity, and the development of drug resistance.

Table 2.2 Challenges in NDDs and photo-drugs

Challenge	Key considerations
Light delivery to the brain	Skull/brain tissue scatters/absorbs light, limiting penetration depth Current methods (optical fibers/implants) require invasive surgery Non-invasive techniques like transcranial photo-biomodulation are under development
Photo-drug design	Requires high target selectivity, light sensitivity, and blood-brain barrier permeability Activation wavelengths must balance tissue penetration and phototoxicity risks
Treating neuroinflammation	Unclear if photo-drugs improve neuroinflammation in Parkinson's disease Efficacy studies are still limited to animal models
Long-term efficacy/safety	Unknown effects of prolonged light exposure on brain tissue Clinical trials needed to establish optimal dosing protocols
Ethical considerations	Concerns about informed consent and result communication in human trials Requires strict oversight for voluntary participation

Additionally, the optimal treatment protocols, including light dosage, frequency, and duration, need to be established.

2.14.5 Ethical Considerations in Human Testing

While being tested on humans, these drugs may possess ethical concerns in case of neuroinflammation and other conditions. There might be rising concerns in terms of voluntary participations, informed consent, results communications, and other factors which must be kept in mind during successful human trials (Zhou et al. 2023; Niazi 2023).

2.15 Current Gaps and Future Directions

While photo-biomodulation (PBM) shows promise for treating Parkinson's disease (PD), several gaps in research need to be addressed to realize its full potential. These gaps encompass a deeper understanding of PBM's mechanisms of action, optimizing treatment parameters, and addressing challenges related to clinical translation (Niazi 2023).

2.15.1 Incomplete Understanding of PBM'S Mechanisms in PD

The precise mechanisms by which PBM exerts its therapeutic effects in PD remain incompletely understood. The ability to heal injured cells, including the reproduction of nerve cells and the growth of new cells. Numerous scientists have researched this field and determined that no side effects have been documented during animal research or human treatment; however, the specific intracellular signaling pathways and molecular targets are not fully elucidated. A deeper understanding of these mechanisms is crucial for optimizing treatment parameters and identifying potential biomarkers for predicting treatment response. The treatment is designed for Parkinson's disease (PD), with the death of cells in the brain's basal ganglia. Brain cells cannot clear out these aggregates, so the cytotoxic alpha-synucleins continue to damage brain cells (Niazi 2023; Kasowanjete et al. 2023).

Potential Solutions Advanced imaging techniques, such as two-photon microscopy and optogenetics, could visualize and manipulate neuronal activity and intracellular signaling in real-time during PBM. Conducting comprehensive multi-omics analyses (genomics, transcriptomics, proteomics, and metabolomics) could help identify the genes, proteins, and metabolites that are differentially expressed in response to PBM in PD models. Performing detailed in vitro studies using PD-relevant cell models (e.g., dopaminergic neurons derived from induced pluripotent stem cells) could investigate the direct effects of PBM on mitochondrial function, oxidative stress, neuroinflammation, and protein aggregation. Furthermore, implementing animal studies could show how the therapy directly impacts patients, providing insight and improvements to the current research to implement in human cell therapies (Kasowanjete et al. 2023).

2.15.2 Sub-Optimal Treatment Parameters

The optimal treatment parameters for PBM in PD, including wavelength, power density, pulse duration, treatment duration, and frequency, have not been fully established. Fair treatment improves motor function and reduces inflammation in the neural pathways. However, the optimal parameters may vary depending on the stage of the disease, the specific brain region targeted, and individual patient characteristics (Kasowanjete et al. 2023).

Potential Solutions Conducting systematic dose-response studies in preclinical PD models could determine the optimal range of PBM parameters for maximizing therapeutic efficacy and minimizing potential side effects. Developing personalized treatment protocols based on individual patient characteristics, such as disease severity, genetic background, and baseline brain activity, as assessed by neuroimaging techniques, would be beneficial. Implementing adaptive PBM strategies that dynamically adjust treatment parameters based on real-time monitoring of patient response, such as changes in motor function, neuroinflammation, or brain activity, could also improve outcomes (Kasowanjete et al. 2023).

2.15.3 Clinical Translation Challenges

Translating the promising preclinical findings of PBM in PD to successful clinical outcomes faces several challenges, including the invasiveness of getting to PBM, which might include ethical issues that cause a severe barrier in scientific research. Autologous stem cells (obtained from the patients) are an alternative option; however, the procedures used to obtain these cells are somewhat invasive and associated with secondary complications. In the case of induced pluripotent stem cells, it should be noted that using viral vectors to transmit Yamanaka factors is problematic (Skidmore and Barker 2023).

Developing and validating non-invasive light delivery techniques, such as transcranial photo-biomodulation (tPBM), that can effectively deliver light to deep brain structures without requiring surgery, is critical. Combining PBM with targeted drug delivery strategies to enhance the penetration and efficacy of therapeutic agents in the brain could also improve outcomes. Conducting well-designed, randomized, placebo-controlled clinical trials with clearly defined endpoints and rigorous data analysis would help assess the efficacy and safety of PBM in PD patients. Furthermore, focusing on neuroinflammation, one of the possible reasons for PD, and researching its effects could improve treatment. It has been observed that many people with Parkinson have wide-ranging deposits of immunoglobulin G on dopaminergic neurons and that these individuals immunolabel Lewy bodies (Kasowanjete et al. 2023).

Addressing these research gaps and implementing the potential solutions outlined above will be crucial for advancing the field of PBM in PD and translating its potential benefits to patients in need.

2.16 Future Directions

Photo-drugs, as of today, hold significant potential for revolutionizing the treatment of neurodegenerative diseases (NDDs). Future directions will likely focus on overcoming current limitations and expanding the therapeutic applications of these light-activated compounds.

2.16.1 Enhanced Light Delivery Systems

A primary focus will be on developing more effective and less invasive methods for delivering light to targeted brain regions. The current reliance on implanted optical fibers presents surgical risks and limits the widespread adoption of photo-drug therapies. Promising avenues include: (Kurul et al. 2025).

Transcranial Photo-Biomodulation (tPBM) Research will continue to optimize tPBM techniques to enhance light penetration through the skull and brain tissue. This may involve using specific wavelengths, pulsed light patterns, and advanced optical devices to maximize light delivery to deep brain structures (Kurul et al. 2025).

Nanoparticle-Mediated Light Delivery Nanoparticles can be engineered to absorb light and convert it into other forms of energy, such as heat or fluorescence, directly at the target site. These nanoparticles can be designed to cross the blood-brain barrier and accumulate in specific brain regions, allowing for localized light activation of photo-drugs (Lin et al. 2024).

Focused Ultrasound Focused ultrasound can be used to temporarily disrupt the blood-brain barrier in a targeted manner, allowing for enhanced delivery of both photo-drugs and light to specific brain regions (Kurul et al. 2025).

2.16.2 Development of Next-Generation Photo-Drugs

Future research will focus on designing and synthesizing more potent and selective photo-drugs with improved pharmacokinetic properties. This includes:

Optimized Photo-Switches Developing photo-switches that exhibit faster switching speeds, higher light sensitivity, and greater stability in biological environments.

These improved photo-switches will allow for more precise and dynamic control over drug activity (Lin et al. 2024).

Targeted Photo-Drugs Engineering photo-drugs that selectively bind to specific proteins or receptors implicated in NDD pathology. This can be achieved by conjugating photo-drugs to antibodies, peptides, or small molecules that have a high affinity for the desired target (Kurul et al. 2025).

Multimodal Photo-Drugs Designing photo-drugs that combine multiple therapeutic mechanisms into a single molecule. For example, a photo-drug could be designed to both inhibit a specific enzyme and promote the release of neurotrophic factors upon light activation (Kong et al. 2023).

2.16.3 Addressing Neuroinflammation

Given the significant role of neuroinflammation in the pathogenesis of many NDDs, future photo-drug research should prioritize developing compounds that can modulate inflammatory responses in the brain. This could involve (Kong et al. 2023):

Photo-Drugs Targeting Microglia Microglia, the resident immune cells of the brain, play a critical role in neuroinflammation. Photo-drugs could be designed to selectively activate or inhibit microglia, depending on the specific disease context.

Photo-Drugs Modulating Cytokine Release Cytokines, such as TNF-α and IL-1β, are key mediators of neuroinflammation. Photo-drugs could be developed to block the production or release of these pro-inflammatory cytokines (Kong et al. 2023).

Photo-Drugs Promoting Resolution of Inflammation In addition to suppressing inflammation, it is also important to promote the resolution of inflammation and restore tissue homeostasis. Photo-drugs could be designed to stimulate the production of anti-inflammatory cytokines, such as IL-10, or to enhance the clearance of inflammatory debris (Kurul et al. 2025).

2.16.4 Personalized Photo-Drug Therapies

As our understanding of the genetic and molecular underpinnings of NDDs improves, future photo-drug therapies will become more personalized. This will involve genetic screening, biomarker development, and adaptive treatment strategies (Kurul et al. 2025).

2.16.5 Clinical Trials and Regulatory Approval

The ultimate success of photo-drugs for NDDs will depend on rigorous clinical trials and regulatory approval. Future clinical trials should focus on well-defined patient populations, utilize objective outcome measures, and assess long-term safety and efficacy.

By addressing all these future directions, photo-drugs have the potential to transform the treatment landscape for neurodegenerative diseases, offering more targeted, effective, and personalized therapies for millions of patients worldwide (Kurul et al. 2025).

2.17 Conclusion

Photo-drugs represent a paradigm shift in the treatment of neurodegenerative diseases (NDDs), moving beyond symptomatic relief toward targeted, light-activated modulation of cellular processes. The limitations of current NDD therapies, often accompanied by systemic side effects and limited efficacy, highlight the need for innovative approaches that can restore cellular function with greater precision. The future of photo-drugs lies in their ability to selectively modulate redox states with high spatiotemporal control, offering a multi-targeted strategy capable of precise neuromodulation with minimal adverse effects.

Further research should prioritize optimizing light delivery methods to enhance penetration and minimize invasiveness. This includes exploring transcranial photobiomodulation (tPBM) techniques and developing targeted drug delivery systems that can effectively transport both the photo-drug and activating light to affected brain regions. Furthermore, the design of novel photo-drugs with improved selectivity, light sensitivity, and pharmacokinetic properties is crucial. These next-generation compounds should be tailored to target specific pathological hallmarks of NDDs, such as protein misfolding, mitochondrial dysfunction, and neuroinflammation.

Ultimately, the success of photo-drugs in NDDs hinges on rigorous preclinical and clinical evaluation. Studies are needed to assess the long-term efficacy and safety of these therapies, as well as to identify biomarkers that can predict treatment response and monitor disease progression. By addressing these challenges and building upon the promising foundation of current research, photo-drugs hold the potential to revolutionize the treatment of NDDs, offering new hope for patients facing these debilitating conditions.

References

Calabresi P, Mechelli A, Natale G, Volpicelli-Daley L, Di Lazzaro G, Ghiglieri V (2023) Alpha-synuclein in Parkinson's disease and other synucleinopathies: from overt neurodegeneration back to early synaptic dysfunction. Cell Death Dis 14(3). https://doi.org/10.1038/s41419-023-05672-9

Chavda V, Chaurasia B, Garg K, Deora H, Umana GE, Palmisciano P, Scalia G, Lu B (2021) Molecular mechanisms of oxidative stress in stroke and cancer. Brain Disord 5:100029. https://doi.org/10.1016/j.dscb.2021.100029

Chen Q, Wang Q, Zhu J, Xiao Q, Zhang L (2017) Reactive oxygen species: key regulators in vascular health and diseases. Br J Pharmacol 175(8):1279–1292. https://doi.org/10.1111/bph.13828

Cleveland Clinic (2025) Neurodegenerative diseases. Retrieved April 10, 2025, from https://my.clevelandclinic.org/health/diseases/24976-neurodegenerative-diseases

Dash UC, Bhol NK, Swain SK, Samal RR, Nayak PK, Raina V, Panda SK, Kerry RG, Duttaroy AK, Jena AB (2024) Oxidative stress and inflammation in the pathogenesis of neurological disorders: mechanisms and implications. Acta Pharm Sin B 15(1):15–34. https://doi.org/10.1016/j.apsb.2024.10.004

Flores-Dorantes MT, Díaz-López YE, Gutiérrez-Aguilar R (2020) Environment and gene association with obesity and their impact on neurodegenerative and neurodevelopmental diseases. Front Neurosci 14. https://doi.org/10.3389/fnins.2020.00863

Kamatham PT, Shukla R, Khatri DK, Vora LK (2024) Pathogenesis, diagnostics, and therapeutics for Alzheimer's disease: breaking the memory barrier. Ageing Res Rev 101:102481. https://doi.org/10.1016/j.arr.2024.102481

Kasowanjete P, Kumar SSD, Houreld NN (2023) A review of photobiomodulation on PI3K/AKT/mTOR in wound healing. J Photochem Photobiol 19:100215. https://doi.org/10.1016/j.jpap.2023.100215

Klemmensen MM, Borrowman SH, Pearce C, Pyles B, Chandra B (2023) Mitochondrial dysfunction in neurodegenerative disorders. Neurotherapeutics 21(1):e00292. https://doi.org/10.1016/j.neurot.2023.10.002

Kong AH, Wu AJ, Ho OK, Leung MM, Huang AS, Yu Y, Zhang G, Lyu A, Li M, Cheung K (2023) Exploring the potential of APTAmers in targeting neuroinflammation and neurodegenerative disorders: opportunities and challenges. Int J Mol Sci 24(14):11780. https://doi.org/10.3390/ijms241411780

Kouli A, Torsney KM, Kuan W (2018) Parkinson's disease: etiology, neuropathology, and pathogenesis. Codon Publications eBooks, pp 3–26. https://doi.org/10.15586/codonpublications.parkinsonsdisease.2018.ch1

Kuang S, Zhu B, Zhang J, Yang F, Wu B, Ding W, Yang L, Shen S, Liang SH, Mondal P, Kumar M, Tanzi RE, Zhang C, Chao H, Ran C (2023) A photolabile Curcumin-Diazirine analogue enables phototherapy with physically and molecularly produced light for Alzheimer's disease treatment. Angew Chem Int Ed 62(45). https://doi.org/10.1002/anie.202312519

Kurul F, Turkmen H, Cetin AE, Topkaya SN (2025) Nanomedicine: how nanomaterials are transforming drug delivery, bio-imaging, and diagnosis. Next Nanotechnol 7:100129. https://doi.org/10.1016/j.nxnano.2024.100129

Lamptey RNL, Chaulagain B, Trivedi R, Gothwal A, Layek B, Singh J (2022) A review of the common neurodegenerative disorders: current therapeutic approaches and the potential role of nanotherapeutics. Int J Mol Sci 23(3):1851. https://doi.org/10.3390/ijms23031851

Lin H, Li D, Zhu J, Liu S, Li J, Yu T, Tuchin VV, Semyachkina-Glushkovskaya O, Zhu D (2024) Transcranial photobiomodulation for brain diseases: review of animal and human studies including mechanisms and emerging trends. Neurophotonics 11(01). https://doi.org/10.1117/1.nph.11.1.010601

Menéndez-Velázquez A, García-Delgado AB (2023) A novel photopharmacological tool: dual-step luminescence for biological tissue penetration of light and the selective activation of photodrugs. Int J Mol Sci 24(11):9404. https://doi.org/10.3390/ijms24119404

Niazi SK (2023) Non-invasive drug delivery across the blood–brain barrier: a prospective analysis. Pharmaceutics 15(11):2599. https://doi.org/10.3390/pharmaceutics15112599

Skidmore S, Barker RA (2023) Challenges in the clinical advancement of cell therapies for Parkinson's disease. Nat Biomed Eng 7(4):370–386. https://doi.org/10.1038/s41551-022-00987-y

Song N, Mei S, Wang X, Hu G, Lu M (2024) Focusing on mitochondria in the brain: from biology to therapeutics. Transl Neurodegener 13(1). https://doi.org/10.1186/s40035-024-00409-w

Wijtmans M, Josimovic I, Vischer HF, Leurs R (2022) Optical control of Class A G protein-coupled receptors with photoswitchable ligands. Curr Opin Pharmacol 63:102192. https://doi.org/10.1016/j.coph.2022.102192

Woo K (2024) Photobiomodulation therapy in neurodegenerative diseases: mechanisms, clinical applications, and future directions. Med Lasers 13(2):90–97. https://doi.org/10.25289/ml.24.015

Yang S, Wang L, Loredo A, Wang S, Ada N, Xiao H (2023) Visible light-activated prodrug system with a novel heavy-atom-free photosensitizer. Bioorg Med Chem Lett 91:129365. https://doi.org/10.1016/j.bmcl.2023.129365

Zhou ZD, Yi LX, Wang DQ, Lim TM, Tan EK (2023) Role of dopamine in the pathophysiology of Parkinson's disease. Transl Neurodegener 12(1). https://doi.org/10.1186/s40035-023-00378-6

Neuroinflammatory Modulation and Potential Photocompounds

3

Kaushalya Anuradha, Dananjaya Perera, and Peshala Gunasekara

Abstract

Neuroinflammation is a key factor in the onset and advancement of several neurodegenerative diseases, such as Alzheimer's disease, Parkinson's disease, multiple sclerosis, and amyotrophic lateral sclerosis. The continuous activation of microglia and astrocytes, coupled with the release of proinflammatory cytokines as well as reactive oxygen and nitrogen species, leads to neuronal dysfunction and loss. The constraints of existing therapies for central nervous system inflammation require the investigation of innovative approaches. Photochemical interventions, encompassing natural photoactive phytochemicals and light-based therapies, signify promising alternatives to traditional treatments. Recent breakthroughs in photomedicine have identified numerous photoactive natural chemicals with neuroprotective properties, predominantly sourced from plants.

This chapter investigates the molecular mechanisms through which neuroinflammation influences neurodegenerative pathology and analyzes novel photo compounds that may regulate critical inflammatory signaling pathways, including NF-κB, NLRP3 inflammasome, and MAPK cascades. Recent studies primarily focus on curcumin, resveratrol, hypericin, and other light-activated phytochemicals that exhibit immunomodulatory and antioxidant activities. The focus is on the role of light-assisted delivery systems and photobiomodulation in improving the bioavailability and therapeutic efficacy of these molecules.

K. Anuradha · P. Gunasekara (✉)
Molecular Diagnostics and Vaccines, Sri Lanka Institute of Biotechnology (SLIBTEC), Pitipana, Homagama, Sri Lanka
e-mail: peshala.gunasekara@slibtec.gov.lk

D. Perera
Biotherapeutics, Sri Lanka Institute of Biotechnology (SLIBTEC), Pitipana, Homagama, Sri Lanka

H. Dilnashin, S. P. Singh (eds.), *Phototherapeutic Approaches to Neurodegeneration*,
https://doi.org/10.1007/978-981-95-4751-7_3

Additionally, it examines the potential and challenges associated with the development of phototherapeutic strategies for neuroinflammation and the management of chronic neurodegenerative diseases, incorporating recent findings.

Keywords

Neuroinflammation · Photobiomodulation · Photoactive phytochemicals · Neurodegenerative diseases

3.1 Introduction

3.1.1 Overview of Neuroinflammation in Neurodegenerative Disorders

Chronic neuroinflammation is now recognized as one of the eight defining "hallmarks" of neurodegenerative diseases (NDD), highlighting its central role in pathogenesis (Wilson et al. 2023). Chronic activation of central nervous system (CNS) immune responses is a fundamental aspect of the disease process in disorders like Alzheimer's (AD), Parkinson's (PD), amyotrophic lateral sclerosis (ALS), Huntington's disease (HD), multiple sclerosis (MS), and stroke. Unlike the short-lived, protective acute inflammation that follows infections, injuries, or trauma, chronic neuroinflammation is maladaptive (Jayaraj et al. 2019; Müller and Di Benedetto 2025). It fails to resolve and instead continues tissue damage. In NDDs, persistent stimuli (e.g., misfolded proteins or ongoing degeneration) overwhelm normal resolution mechanisms, resulting in a sustained release of proinflammatory mediators and neuronal injury (Adamu et al. 2024).

Microglia and astrocytes, integral components of the central nervous system, play pivotal roles as the primary immune and homeostatic cells within the brain and spinal cord. Traditionally, their functions have been categorized along a spectrum of pro- or anti-inflammatory responses, often denoted by the M1/M2 classification for microglia and A1/A2 for astrocytes (Kwon and Koh 2020).

Similarly, astrocytes exhibit distinct functional phenotypes. The A1 phenotype, analogous to M1 microglia, is considered a proinflammatory or neurotoxic state. A1 astrocytes are characterized by the upregulation of complement cascade components and can secrete neurotoxic factors that lead to neuronal and oligodendrocyte death. This phenotype is often induced by activated microglia in neurodegenerative diseases like Alzheimer's disease, Parkinson's disease, and multiple sclerosis (Liddelow et al. 2017).

In contrast, A2 astrocytes are viewed as neuroprotective and reparative. They are involved in promoting neuronal survival, synaptogenesis, and blood-brain barrier integrity. A2 astrocytes can produce growth factors, antioxidants, and anti-inflammatory molecules, contributing to tissue repair and functional recovery after CNS injury or stroke. The dynamic interplay and precise regulation of these microglial and astrocytic phenotypes are crucial for maintaining brain health and represent significant therapeutic targets for various neurological conditions (Ding et al. 2021). Understanding the mechanisms that drive these phenotypic shifts is paramount for developing effective strategies to modulate neuroinflammation and promote neuroprotection (Hamby and Sofroniew 2010). Both acute and chronic neuroinflammation differ fundamentally. An acute immune response is usually self-limited and protective. Glia rapidly clear debris and then release anti-inflammatory signals to restore homeostasis. However, when resolution mechanisms fail, the response becomes chronic. At this stage, microglia serve as primary immune sensors, switching between neuroprotective and neurotoxic states and mediating cytokine release and reactive oxygen species (ROS) production (Adamu et al. 2024). Astrocytes modulate synaptic function and glutamate clearance but can become reactive (A1) in response to microglial cytokine signals, thereby promoting neurotoxicity. Peripheral immune cells, including monocytes, T cells, and B cells, infiltrate through a compromised blood-brain barrier (BBB) in diseases like MS (Liddelow et al. 2017). Neurons are targets of inflammation and also emit distress signals [e.g., damage-associated molecular patterns (DAMPs)] that further activate glia. In sum, unresolved neuroinflammation drives synaptic dysfunction, network disruption, and progressive neuron loss (Klegeris 2021).

Recent single-cell RNA sequencing has revealed more diverse and disease-specific glial phenotypes. These phenotypic variations are influenced by factors such as the anatomical location within the CNS, the nature of pathological stimuli, and the type and progression stage of neurodegenerative diseases. Recent single-cell RNA sequencing has revealed more diverse and disease-specific glial phenotypes. These phenotypic variations are influenced by factors such as the anatomical location within the CNS, the nature of pathological stimuli, and the type and progression stage of neurodegenerative diseases. They detect pathological signals via receptors like toll-like receptors (TLR), receptors for advanced glycation end (RAGE), and cGAS, releasing cytokines and other mediators (Kwon and Koh 2020; Olah et al. 2020). Targeting these pathways offers potential strategies for anti-inflammatory therapies in neurodegenerative diseases.

3.1.2 Key Neurodegenerative Diseases Involving Neuroinflammation

3.1.2.1 Alzheimer's Disease (AD)

Extracellular Aβ plaques and intracellular tau tangles are characteristic of AD (Zheng and Wang 2025). Chronic microglial activation upregulates Amyloid Precursor Protein (APP) processing and impairs Aβ degradation, leading to the formation of Aβ monomers (He et al. 2025). These monomers tend to oligomerize and aggregate into Aβ plaques, a process that can be accelerated by genetic mutations in the APP or PSEN1/PSEN2 genes (Lanoiselée et al. 2017). Microglia cluster around plaques and, upon sensing Aβ as a DAMP, secrete TNF-α, IL-1β, IL-6, and ROS (He et al. 2025). This inflammatory environment paradoxically tries to clear plaques while promoting further pathology by increasing Aβ aggregation through elevated expression of APP and IFITM3 and by releasing ions such as Zn^{2+}, thus creating a vicious cycle (Chen et al. 2025; Vavougios et al. 2021).

Chronically activated microglia also release proinflammatory cytokines and toxic products such as reactive oxygen species (ROS), reactive nitrogen species (RNS), and nitric oxide (NO). These amplify the immune response and create a neurotoxic environment that further damages neurons. Damage-Associated Molecular Patterns (DAMPs), released from dying neurons (e.g., ATP, HMGB1, S100B, DNA), further amplify inflammation, forming a second positive feedback loop (He et al. 2025; Zheng and Wang 2025).

Inflammatory cytokines also activate kinases that cause hyperphosphorylation of tau, linking inflammation to the formation of neurofibrillary tangles in the neurons' cytosol. Clinically, AD brains display widespread activated microglia and A1 astrocytes co-localized with pathology. Therefore, in AD, neuroinflammation maintains Aβ/tau accumulation and promotes neuronal damage (Deng et al. 2023; Miao et al. 2019; Wang et al. 2015).

3.1.2.2 Parkinson's Disease (PD)

PD features fibrillar α-synuclein aggregates in neurons called Lewy bodies and a significant loss of dopaminergic neurons in the substantia nigra pars compacta. Neuroinflammation is a notable characteristic of PD, marked by microglia activation and some leukocyte accumulation in the CNS (Roodveldt et al. 2024; Tansey et al. 2022).

Postmortem studies have shown an increased density of HLA-HR+ reactive microglia (McGeer et al. 1988) and reactive astrocytes with proinflammatory and neurotoxic features (Liddelow et al. 2017). Genetic studies also emphasize that immune dysfunctions are associated with PD, with increased expression of immune genes (Triggering Receptor Expressed on Myeloid cells 2, TREM2) and PD-related genes (LRKK2 and Parkin), which are highly expressed in microglia (Guo et al. 2019; Kim et al. 2012; Russo 2019). Meanwhile, PTEN-induced kinase 1 (PINK1) and Parkin are involved in antigen presentation (Matheoud et al. 2016).

Postmortem analysis of PD brains reveals an increased density of microglia and reactive astrocytes, accompanied by multiple leukocyte types, including CD4+ and CD8+ T cells, near Lewy bodies and dopaminergic neurons (Roodveldt et al. 2024; Tansey et al. 2022). Misfolded α-synuclein released from degenerating neurons engages microglial toll-like receptors (TLRs) and NOD-like receptor protein 3 (NLRP3) inflammasome, initiating a sterile inflammatory response (Li et al. 2021). This microglial activation leads to the secretion of proinflammatory cytokines, including TNF-α, IL-1β, and IL-6, which further impair the survival of dopaminergic neurons (Tansey et al. 2022). Generally, astrocytes are capable of taking up extracellular α-synuclein and degrading. Extracellular α-synuclein activates its nuclear factor κB (NF-κB) and p38 mitogen-activated protein kinase (p38MAPK) pathways, resulting in the release of various proinflammatory cytokines ROS and RNS, which leads to the death of dopaminergic neurons (Wang et al. 2023).

Overall, neuroinflammation is an important aspect of PD, showing signs like the presence of leukocytes in the brain, along with a rise in microglia and astrocyte density, and the degeneration of dopaminergic neurons.

3.1.2.3 Multiple Sclerosis (MS)

Multiple sclerosis (MS) is a chronic autoimmune disorder involving immune cell infiltration into the CNS (Thompson et al. 2018). While the initial cause remains unknown, both adaptive (T and B lymphocytes) and innate (macrophages, microglia) immune cells contribute to the disease (Babbe et al. 2000; Prineas and Parratt 2021). A key feature of MS is the dysfunction of regulatory immune mechanisms that normally maintain tolerance, permitting neuroinflammation (Goverman 2011). CNS-reactive T cells activate early, possibly via molecular mimicry, where pathogen peptides resemble CNS antigens (El-Taibany et al. 2025). Studies show mimicry between Epstein-Barr virus nuclear antigen 1 (EBNA1) and CNS protein GlialCAM, suggesting viral infections can trigger autoimmune responses (Bjornevik et al. 2022; Lanz et al. 2022). These T cells are reactivated in the CNS, increasing their pathogenic effects.

CD8+ T cells dominate MS lesions, outnumbering CD4+ cells, with clonal expansion and tissue residency, driving prolonged inflammation and damage (Babbe et al. 2000; Fransen et al. 2020; Machado-Santos et al. 2018). Single-cell analysis shows these tissue-resident CD8+ T cells are activated and cytotoxic with high migratory ability (Fransen et al. 2020). Myeloid cells, including macrophages and microglia, make up over 80% of immune cells in active lesions, correlating with axonal injury and neurological issues (Prineas and Parratt 2021). Recent studies reveal heterogeneity among myeloid subsets with distinct proinflammatory and phagocytic gene signatures (El-Taibany et al. 2025). B cells and plasma cells mostly localize in meningeal and perivascular spaces, acting as antigen-presenting cells and producing inflammatory mediators (Machado-Santos et al. 2018). B cell-rich meninges correlate with cortical pathology and disability, especially in progressive MS (Thompson et al. 2018). These structures support local immune activation and produce MS-related oligoclonal bands (Obermeier et al. 2008). Progressive MS features chronic active lesions (CALs) with a demyelinated core and active microglia/macrophages at the rim, representing up to 60% of lesions, associated with ongoing tissue damage and worsening. These border cells show cytotoxic features and iron deposits, detectable via MRI (Absinta et al. 2019). Microglia are highly diverse, responding to danger signals like myelin debris, amyloid-β, or α-synuclein (Ransohoff 2016b). Initially protective, chronic microglial activation becomes neurotoxic, releasing harmful substances that worsen tissue damage (Salter and Stevens 2017). Genes linked to MS are often expressed in microglia, emphasizing their role in disease (Van Den Bosch et al. 2024). Microglial activation occurs early, even before lymphocyte infiltration, and extends into white matter, measurable with PET imaging, correlating with disability (Airas et al. 2017; Sucksdorff et al. 2017). Astrocytes, influenced by microglial cytokines, transform into neurotoxic A1-type reactive astrocytes, losing their supportive role and releasing inflammatory factors that contribute to neuron and oligodendrocyte death. The interaction between microglia and astrocytes perpetuates CNS inflammation in progressive MS (Liddelow et al. 2017).

3.1.2.4 Neuroinflammation in Traumatic Brain Injury (TBI) and Stroke

Neuroinflammation represents a central pathophysiological mechanism driving secondary brain injury in both TBI and stroke, characterized by complex cascades involving DAMPs, microglial activation, and cytokine networks (Jassam et al. 2017; Kumar and Loane 2012). Following acute CNS injury, massive release of DAMPs, including high-mobility group box 1 protein (HMGB1), adenosine triphosphate (ATP), and heat shock proteins, triggers immediate activation of resident microglia and astrocytes (Wang et al. 1999). While initial inflammatory responses serve essential functions in debris clearance and tissue repair, uncontrolled neuroinflammation becomes a primary driver of secondary injury, contributing to blood-brain barrier disruption, neuronal damage, and progressive neurodegeneration (Loane and Kumar 2016; Simon et al. 2017).

The NLRP3 inflammasome has emerged as a critical mediator of this pathological transition, converting danger signals into potent inflammatory responses through production of IL-1β and IL-18 (de Rivero Vaccari et al. 2016). Clinical evidence demonstrates that cerebrospinal fluid NLRP3 concentrations above 6.63 ng/ml are associated with poorer neurological outcomes at six months post-TBI, highlighting its prognostic utility (Wallisch et al. 2017). Microglial activation, detectable through translocator protein (TSPO) positron emission tomography, can persist for months to years after initial injury, with inflammatory changes correlating with disability progression and cognitive deficits (Ramlackhansingh et al. 2011; Scott et al. 2015).

Key inflammatory biomarkers, including IL-6, IL-10, and glial fibrillary acidic protein (GFAP), demonstrate significant elevations in both TBI and stroke patients, with IL-6 and IL-10 levels correlating with injury severity and imaging findings (Tsitsipanis et al. 2023; Woodcock and Morganti-Kossmann 2013). Serum neurofilament light chain and GFAP levels can remain elevated for weeks to months after injury, predicting ongoing white matter volume loss and functional decline (Okonkwo et al. 2013; Shahim et al. 2016a, 2016b).

The complement system plays a dual role, with chronic complement dysregulation driving sustained neuroinflammation up to two years post-injury, while targeted complement inhibition using CR2-Crry demonstrates neuroprotective effects when administered acutely and shows continued benefit when treatment is extended chronically (Mallah et al. 2021). These findings highlight neuroinflammation as both a vital pathological process and an accessible therapeutic target, with the potential to turn acute CNS injuries from degenerative conditions into manageable disorders through precise interventions.

3.1.3 Current Challenges in Managing Neuroinflammation

Effective management of neuroinflammation remains challenging due to multiple interconnected barriers. The blood-brain barrier restricts >98% of small-molecule drugs and essentially 100% of large-molecule therapeutics, with permeability that varies unpredictably during inflammation, complicating delivery strategies (Banks 2016; Pardridge 2012). Once neuroinflammation is established, BBB disruption intensifies glial activation and leukocyte infiltration, creating self-perpetuating pathological cycles (Sweeney et al. 2018). Neuroinflammatory signaling is inherently double-edged: acute responses are often neuroprotective, while chronic activation accelerates neuronal loss, making indiscriminate suppression potentially counterproductive (Hensley 2010).

These biological complexities contribute to poor therapeutic outcomes—CNS drug success rates are approximately half those of non-CNS drugs (6.2% vs. 13.3%), with Phase II-III failure rates reaching 85% (Cummings et al. 2014; DiMasi et al. 2016). Anti-inflammatory approaches in Alzheimer's disease have consistently missed efficacy endpoints despite biomarker engagement (Cummings et al. 2019). Classic NSAIDs, despite epidemiological promise, produced null or adverse outcomes in prospective trials, largely because enrollment occurred after substantial irreversible neurodegeneration (Imbimbo et al. 2010).

Progress is further hindered by limited stage-specific biomarkers. While candidates like GFAP, sTREM2, and YKL-40 show promise for tracking neuroinflammation, they exhibit stage-dependent variability and cross-reactivity with other conditions, limiting their utility for patient stratification (Benedet et al. 2021). Additionally, patient heterogeneity driven by mixed pathologies, genetics, and comorbidities necessitates precision medicine approaches tailored to specific inflammatory endophenotypes (Hampel et al. 2018). These challenges underscore the need for integrated delivery technologies, validated biomarker panels, and adaptive clinical trial designs.

3.2 Blood-Brain Barrier: The Formidable Gateway

Managing neuroinflammation is challenging due to biological and clinical barriers, especially the blood-brain barrier (BBB), which blocks over 98% of small-molecule drugs and nearly all large-molecule therapeutics from reaching the brain (Banks 2016; Pardridge 2012). While essential for CNS protection, the BBB also prevents therapeutic access (Zhao et al. 2022). Its dynamic nature complicates drug delivery, with regional permeability varying unpredictably across disease stages, making standard strategies difficult (Sweeney et al. 2018; Zhao et al. 2022). Recent brain-targeted delivery methods like receptor-mediated transport, ultrasound, and nanoparticles are emerging but remain largely experimental with translational hurdles (Liu et al. 2025; Timbie et al. 2015).

3.3 The Vicious Cycle of Neuroinflammation

Once neuroinflammation is established, BBB disruption creates a self-perpetuating pathological cascade that intensifies both glial activation and peripheral immune cell infiltration (Zhang et al. 2025). This breakdown triggers the release of matrix metalloproteinases (MMPs) and inflammatory cytokines that further compromise barrier integrity, creating what researchers describe as a "leaky brain" syndrome (Sweeney et al. 2018). The infiltrating peripheral immune cells, including monocytes, neutrophils, and lymphocytes, bring their own inflammatory arsenals, amplifying the local immune response and contributing to chronic neurodegeneration (Prinz and Priller 2017).

3.4 The Double-Edged Nature of Neuroinflammatory Signaling

Perhaps the most conceptually challenging aspect of neuroinflammation management lies in its inherently double-edged nature. Acute inflammatory responses serve critical neuroprotective functions, facilitating debris clearance, promoting tissue repair, and coordinating the resolution of injury-induced damage (Hensley 2010). However, when this acute response transitions to chronic activation, the same cellular and molecular mechanisms that initially provided protection become drivers of neuronal death and tissue destruction (Zhao et al. 2022). This temporal paradox means that indiscriminate anti-inflammatory suppression risks eliminating essential repair mechanisms, potentially causing more harm than benefit (Simon et al. 2017).

3.5 Clinical Translation Failures and Statistical Reality

Biological complexities have led to a decline in CNS drug development, with success rates around 6.2% compared to 13.3% for other drugs and an 85% failure rate in Phase II-III (Cummings et al. 2014; DiMasi et al. 2016). Recent data from 2024 confirm persistently high failure rates, especially in neurological late-stage trials. The Alzheimer's anti-inflammatory approach highlights this issue, with many trials missing efficacy despite biomarker success (Cummings et al. 2019). The failures of NSAIDs show that targeting patients with established disease, where neurodegeneration is irreversible, emphasizes the need for timely intervention.

3.6 Biomarker Limitations and Diagnostic Challenges

Progress toward personalized neuroinflammation therapy is significantly hindered by the lack of robust, stage-specific biomarkers capable of guiding therapeutic decisions. While promising candidates like glial fibrillary acidic protein (GFAP), soluble triggering receptor expressed on myeloid cells 2 (sTREM2), and YKL-40 have emerged as potential indicators of glial activation, they exhibit significant stage-dependent variability and cross-reactivity with other inflammatory conditions (Benedet et al. 2021; Hampel et al. 2018).

Recent advances in the ATI(N) biomarker framework, which broadens the traditional amyloid ("A"), tau ("T"), and neurodegeneration ("N") classification to include inflammatory ("I") markers, mark exciting progress toward more personalized medicine approaches (Benedet et al. 2021). However, these biomarkers still need to achieve greater specificity and a wider dynamic range before they can be effectively used for routine patient care. The development of blood-based inflammatory biomarkers is especially promising, offering the potential for more convenient monitoring, although their full clinical usefulness is still being explored (Hampel et al. 2018).

3.7 Patient Heterogeneity and Precision Medicine Imperatives

Contemporary research has shown unprecedented levels of patient heterogeneity in neuroinflammatory responses, caused by complex interactions between mixed pathologies, genetic polymorphisms, and comorbid conditions (Hampel et al. 2018). Single-cell RNA sequencing studies have identified distinct microglial and astrocytic phenotypes that vary significantly between individuals, even within the same diagnostic category (Jordão et al. 2019). This biological diversity highlights the need for precision medicine approaches that can customize interventions to specific inflammatory endophenotypes rather than employing broad-spectrum anti-inflammatory strategies.

Recent work has started identifying peripheral immune signatures that correlate with disease progression and therapeutic response, offering potential routes for patient stratification (Mark et al. 2025). However, translating these molecular insights into clinically useful biomarker panels remains a major challenge, requiring large-scale validation studies and standardized analytical platforms.

3.8 Rationale for Exploring Photocompounds as Therapeutics

Photochemical compounds are substances sensitive to light that undergo chemical transformations upon light exposure. This extensive category encompasses natural pigments and synthetic materials such as dyes and photoswitches (Klán and Wirz 2009). Their interaction with light can induce diverse reactions, including photodegradation (chemical decomposition), photoisomerization (alteration in atomic configuration), photopolymerization (polymer synthesis), photoexcitation (energy assimilation), and photocatalysis (light-accelerated chemical processes) (Oelgemöller and Shvydkiv 2011; Turro et al. 2010). These compounds are indispensable across various domains, notably biology and medicine (e.g., photosynthesis, oncological therapies), materials science (e.g., photochromic substances), chemistry (e.g., photochemical reactions), and environmental science (e.g., degradation of contaminants) (Braslavsky et al. 1988). The capacity to regulate chemical processes via light through the utilization of these compounds presents considerable potential for innovation.

Photocompounds represent a transformative approach to precision medicine by enabling controlled spatiotemporal drug activation through light-responsive molecular switches (Broichhagen et al. 2015; Szymanski et al. 2013; Velema et al. 2014). Photopharmacology incorporates molecular photoswitches, such as azobenzenes, stilbenes, and spiropyrans, into bioactive compounds, addressing the selectivity challenges of traditional pharmaceuticals that often cause systemic toxicity and narrow therapeutic windows (Kobauri et al. 2023).

Light provides unique advantages as an external stimulus: non-invasive activation with exceptional temporal and spatial resolution, allowing targeted drug activation exclusively at illuminated sites while maintaining inactivity in non-target tissues (Beharry and Woolley 2011). This approach is particularly compelling for neuroinflammatory conditions because conventional drugs have poor selectivity, leading to severe side effects, whereas light-activated compounds remain inactive until photochemically triggered at specific anatomical sites (Alvarez-Lorenzo and Concheiro 2014; Lerch et al. 2016).

The scientific foundation for phototherapeutic development includes reversible drug modulation through photoisomerization reactions, enabling clinicians to precisely regulate drug onset, duration, and termination via wavelength-specific illumination, providing unprecedented therapeutic control (Beharry and Woolley 2011; Kobauri et al. 2023). Photocompounds address key pharmacokinetic challenges by eliminating the requirement for systemic drug circulation, decreasing off-target interactions, and reducing resistance development through localized activation mechanisms (Copeland et al. 2006).

The therapeutic window expands considerably with red and near-infrared light (650–850 nm), which provides adequate tissue penetration (4–6 mm) for clinical applications while avoiding cytotoxic effects of higher-energy radiation (Ash et al. 2017). However, tissue penetration limitations must be considered, particularly for deep CNS applications where modest penetration through cranial barriers remains challenging (Bai et al. 2024).

Recent advances in photosensitizer design include second- and third-generation compounds with enhanced red-shifted absorption spectra and oxygen-independent mechanisms that maintain efficacy in hypoxic environments (Lucky et al. 2015; Teng et al. 2023). These developments have strengthened the rationale for photocompounds as next-generation therapeutics by enabling reliable activation in variable oxygen conditions characteristic of diseased tissues.

This convergence of photochemistry and pharmacology offers a promising approach for achieving spatially and temporally precise therapeutic interventions with minimal systemic toxicity, representing a fundamental shift toward precision medicine applications (Ashley 2016; Kobauri et al. 2023).

3.9 Cellular and Molecular Mechanisms of Neuroinflammation

3.9.1 Cellular Mediators of Neuroinflammation

Neuroinflammation involves a complex interplay of immune cells in the CNS, including microglia, astrocytes, peripheral immune cells, and neurons. These mediators coordinate immune responses to protect or damage neural tissue depending on the context.

3.9.1.1 Microglia

Microglia serve as the CNS's resident immune sentinels. In a healthy brain, they display a ramified "surveying" shape, continuously observing their surroundings (Colonna and Butovsky 2017; Kettenmann et al. 2011). They engulf apoptotic cells and debris, prune surplus synapses during development, and release while serving as the CNS's resident immune sentinels. In a healthy brain, they display a ramified "surveying" shape, continuously observing their surroundings, such as IGF-1, TGF-β, and BDNF, to promote neuronal support (Colonna and Butovsky 2017; Xing et al. 2014). Upon sensing pathogens or injury via receptors such as Toll-like receptors and NOD-like receptor protein 3 (NLRP3), microglia can switch to an activated "amoeboid" state (Heneka et al. 2014). Classically activated microglia (often called M1) upregulate nuclear factor kappa B (NF-κB) and MAPK signaling pathways, producing higher levels of proinflammatory mediators. These include TNF-α, IL-1β, IL-6, several chemokines such as C-C Motif Chemokine Ligand 2 (CCL2) and C-X-C motif chemokine ligand (CXCL10), as well as inducible nitric oxide synthase (iNOS) and reactive oxygen species (ROS). These responses are aimed at fighting off pathogens, but they can also cause damage to the body's own tissues (Cherry et al. 2014; Xing et al. 2014).

These factors contribute to oxidative stress, lead to synaptic loss, and can destabilize the blood-brain barrier (BBB) (Colonna and Butovsky 2017; Heneka et al. 2014). In contrast, alternatively activated M2 microglia are prompted by cytokines like IL-4, IL-10, and TGF-β. They produce anti-inflammatory cytokines and growth factors that facilitate debris clearance and tissue repair (Cherry et al. 2014; Colonna and Butovsky 2017). In reality, microglial activation exists on a spectrum, and the outcomes of the disease really depend on the balance between different phenotypes. When there's a persistent shift toward the harmful M1 state, it can worsen neurodegeneration. On the other hand, M2-skewed signaling tends to be neuroprotective, helping to safeguard the brain (Ransohoff 2016b; Tang and Le 2016). Chronic microglial activation in neurodegenerative diseases has been shown to maintain inflammation and oxidative stress, worsening neuronal injury (Heneka et al. 2014; Shi et al. 2017). For example, chronically activated microglia release MMP-9 and TNF-α, which degrade tight junction proteins and weaken the BBB. This process allows peripheral immune cells to infiltrate and exacerbate inflammation in the CNS (Heneka et al. 2014; Yin et al. 2017).

3.9.1.2 Astrocytes

Astrocytes, the most abundant glial cells, play a vital role in maintaining the health of the CNS. They support the BBB by extending their perivascular end feet and releasing helpful factors like sonic hedgehog, basic FGF, α-dystrobrevin, and TGF-β. These substances work together to reinforce the tight junctions of endothelial cells, promoting a strong and healthy barrier (Abbott et al. 2006). In maintaining balance, astrocytes also uptake excess glutamate, help keep extracellular potassium levels steady, and provide vital metabolic support to neurons. When injury or chronic disease occurs, astrocytes become "reactive," a state where they undergo hypertrophy, increase GFAP expression, and release proinflammatory cytokines like IL-1β, IL-6, and TNF-α, along with chemokines, MMPs (such as MMP-2 and MMP-9), and VEGF. This process can lead to increased permeability of the BBB (Sofroniew 2015) (Liddelow et al. 2017).

A vital part of neuroinflammation is the crosstalk between microglia and astrocytes. When microglia are activated, they release IL-1α, TNF-α, and C1q, which together prompt astrocytes to take on the harmful "A1" phenotype. These A1 astrocytes stop providing their normal support and instead release substances that can directly harm neurons and oligodendrocytes. A1 astrocytes are notably abundant in neurodegenerative conditions like Alzheimer's, Parkinson's, ALS, and MS (Liddelow et al. 2017).

3.9.1.3 Peripheral Immune Cells

Peripheral immune cell infiltration occurs when neuroinflammation damages the BBB. In multiple sclerosis, monocytes from the blood enter the CNS and differentiate into macrophage-like cells, which can either clear debris or, if they become proinflammatory, exacerbating tissue damage (Prinz and Priller 2017). CNS-invading macrophages differ at the molecular level from resident microglia. Peripheral T lymphocytes, especially autoreactive Th1/Th17 CD4+ cells and cytotoxic CD8+ T cells, cause damage to oligodendrocytes and neurons through cytolytic actions and the release of proinflammatory cytokines. B lymphocytes participate by presenting antigens and forming meningeal follicles linked with progressive cortical damage (Sabatino et al. 2019). Regulatory T cells (Tregs), usually protective, exhibit impaired function in chronic neurodegenerative conditions (Lewis et al. 2025).

3.9.1.4 Neurons

While neurons are susceptible to neuroinflammatory damage, they also play an active role in regulating microglia and astrocyte functions (Glass et al. 2010). In a healthy brain, neurons produce inhibitory signals such as fractalkine (CX3CL1), CD200, and neuregulin to prevent glial activation (Biber et al. 2007). Neurodegeneration or injury can cause these essential homeostatic factors to diminish, leading neurons to release DAMPs such as misfolded proteins, mitochondrial DNA, HMGB1, and S100 proteins. These molecules interact with pattern recognition receptors on microglia and astrocytes, continuing the inflammation and creating a feedback loop that speeds up degeneration (Piccinini and Midwood 2010).

3.9.2 Molecular Mediators of Neuroinflammation

Neuroinflammation is driven and regulated by a diverse set of molecular mediators, including cytokines, chemokines, protein aggregates, lipid mediators, and oxidative stress molecules.

3.9.2.1 Proinflammatory Cytokines

Activated microglia and astrocytes produce high levels of proinflammatory cytokines like TNF-α, IL-1β, and IL-6 in response to injury or infection (Heneka et al. 2014). These cytokines boost inflammatory signals through NF-κB/MAPK pathways, elevate iNOS and COX-2 expression, and attract more immune cells. Excessive cytokine production over time causes chronic inflammation, disrupts synaptic activity, triggers oxidative and nitrosative stress, and endangers neuron survival, such as TNF-α and IL-1β, which can impair astrocytic glutamate uptake, leading to excitotoxicity and abnormal tau phosphorylation (Ransohoff 2016a).

3.9.2.2 Anti-inflammatory Cytokines

Anti-inflammatory cytokines like IL-10 and TGF-β help reduce inflammation by inhibiting NF-κB signaling, aiding debris cleanup, and supporting tissue repair. These mediators are mainly produced by M2-like microglia, astrocytes, and Tregs. However, in aging and chronic illnesses, their levels tend to be inadequate, leading to increased harmful inflammation (Cherry et al. 2014).

3.9.2.3 Chemokines

Chemokines like CCL2, CCL5, and CXCL10 play a key role in attracting peripheral immune cells into the CNS during inflammation (D'Mello et al. 2009). For instance, CCL2 is significantly increased in activated glia and promotes the infiltration of CCR2+ monocytes into the brain and spinal cord. The CX3CL1/CX3CR1 pathway, involving neuronal fractalkine and its microglial receptor, helps keep microglia in a resting state; disruption of this pathway is associated with Alzheimer's and Parkinson's diseases (Paolicelli et al. 2011).

3.9.2.4 Protein Aggregates (DAMPs)

Protein aggregates such as extracellular amyloid-β, α-synuclein fibrils, and hyperphosphorylated tau are not only key indicators of neurodegeneration but also serve as danger signals (DAMPs) that trigger microglial activation through TLRs, the receptor for advanced glycation end products (RAGE), and the NLRP3 inflammasome. When microglia fail to clear these aggregates effectively, chronic activation and ongoing inflammation occur. Additionally, neuronal death releases DAMPs like ATP and HMGB1, which further amplify microglial activation and neuroinflammatory responses (Heneka et al. 2014).

3.9.2.5 NLRP3 Inflammasome

The NLRP3 (nucleotide-binding domain, leucine-rich repeat, and pyrin domain-containing protein 3) inflammasome is a critical intracellular multiprotein complex that serves as a key sensor of cellular danger and driver of neuroinflammation (Heneka et al. 2014). This sophisticated molecular platform consists of three primary components: the NLRP3 sensor protein, the ASC adaptor protein, and procaspase-1, which together detect damage-associated molecular patterns (DAMPs) and orchestrate inflammatory responses (Swanson et al. 2019).

Activation and Function

In microglia, NLRP3 inflammasomes respond to diverse neuroinflammatory stimuli, including extracellular ATP, ROS, mitochondrial DNA, and pathological protein aggregates such as amyloid-β and α-synuclein (Heneka 2015). Activation occurs through a two-step process: initial priming via NF-κB-dependent transcriptional upregulation of NLRP3, pro-IL-1β, and pro-IL-18, followed by assembly-triggered conformational changes leading to caspase-1 activation (Bauernfeind et al. 2009; Swanson et al. 2019).

Once assembled, the inflammasome processes inactive cytokine precursors pro-IL-1β and pro-IL-18 into their mature, biologically active forms, which are then released to amplify inflammatory responses throughout the CNS (Dinarello 2009). This process can also trigger pyroptosis, a highly inflammatory form of cell death mediated by gasdermin D cleavage that releases additional DAMPs, further propagating neuroinflammation (Shi et al. 2015).

Role in Astrocyte Transformation

A critical consequence of prolonged microglial NLRP3 activation is the induction of neurotoxic A1 astrocytes. Activated microglia secrete specific cytokines—particularly IL-1α, TNF-α, and complement C1q—that are both necessary and sufficient to transform normally protective astrocytes into neurotoxic phenotypes (Liddelow et al. 2017). Recent studies have confirmed that microglial NLRP3 inflammasome activation directly promotes this A1 transformation through enhanced secretion of these inducing cytokines (Xiao et al. 2022).

These A1 astrocytes lose their supportive functions and instead secrete unidentified neurotoxic factors that directly induce neuronal and oligodendrocyte death, significantly contributing to neurodegeneration in multiple disease contexts (Liddelow et al. 2017). The NLRP3 inflammasome thus represents a central hub linking initial cellular stress to sustained neuroinflammation and glial toxicity, making it an attractive therapeutic target for neurodegenerative diseases.

3.9.2.6 Complement System

The complement system plays a role in neuroinflammation, where microglia produce C1q and astrocytes generate C3. Although C1q may be neuroprotective during acute injury, prolonged complement activation leads to the formation of proinflammatory anaphylatoxins (C3a, C5a), which promote harmful glial activation (Stevens et al. 2007).

3.9.2.7 Lipid Mediators

Lipid mediators regulate both the initiation and termination of inflammation. Proinflammatory eicosanoids such as PGE2 and LTB4 encourage cytokine release and immune cell infiltration (Ricciotti and FitzGerald 2011). In contrast, specialized pro-resolving mediators (SPMs), including resolvins, maresins, and lipoxins, help resolve inflammation and facilitate the clearance of cellular debris (Serhan 2014). Importantly, levels of SPMS are reduced in conditions like Alzheimer's, Parkinson's, and MS, which impairs resolution and prolongs chronic neuroinflammation (Basil and Levy 2016).

3.9.2.8 Extracellular Vesicles

Extracellular vesicles (EVs), including exosomes, facilitate the transfer of inflammatory signals and misfolded protein aggregates between CNS cells, aiding the progression of pathology in neurodegenerative diseases (Asai et al. 2015). While EVs can occasionally play protective roles by removing waste, their proinflammatory effects seem to predominate in chronic disease conditions.

3.9.2.9 Oxidative and Nitrosative Stress Products

Oxidative and nitrosative stress both result from and contribute to neuroinflammation. Glia-produced ROS and reactive nitrogen species (RNS), generated through NADPH oxidase and iNOS, directly harm neurons and act as secondary signals that sustain inflammation. Excess ROS and RNS damage lipids, proteins, and DNA in neurons, impair mitochondrial function, and enhance the activation of redox-sensitive transcription factors like NF-κB and Nrf2, thus reinforcing the cycle of neurodegeneration (Heneka et al. 2014).

3.10 Photoactive Modulation of Neuroinflammatory Pathways and Therapeutic Compounds

Neuroinflammation plays a crucial role in the pathology of many neurodegenerative diseases, involving a complex network of cellular and molecular signaling pathways, notably including the sirtuin family (especially SIRT1 and SIRT3), JAK/STAT signaling, AMPK cascades, NF-κB, and the NLRP3 inflammasome. These interconnected systems regulate the transcription and release of numerous proinflammatory cytokines, such as interleukin-1β (IL-1β), IL-6, and tumor necrosis factor-α (TNF-α), and coordinate microglial activation (Jiao and Gong 2020; Kauppinen et al. 2013; Rothgiesser et al. 2010). Sirtuins are of particular therapeutic interest due to their capacity to modulate neuroinflammation through their deacetylase activity. SIRT1, for example, deacetylates the p65 subunit of NF-κB, thereby suppressing the expression of proinflammatory genes and limiting the propagation of neuroinflammatory cascades (Kauppinen et al. 2013; Rothgiesser et al. 2010).

The JAK/STAT signaling pathway, which involves several JAK isoforms and STAT proteins, operates as a central mediator of cytokine-driven inflammatory responses. Dysregulation of this pathway triggers overactivation of microglia and contributes to the excessive production of neurotoxic inflammatory mediators (Shariq et al. 2018). Moreover, sirtuins modulate JAK/STAT pathways through deacetylation of STAT proteins and indirectly through the upregulation of SOCS proteins, which act as negative regulators of cytokine signaling (Piessevaux et al. 2006). The energy-sensing AMPK pathway also acts as a regulator of inflammation, as its activation not only induces a shift in microglial phenotype from the damaging M1 state to the neuroprotective M2 phenotype but also enhances SIRT1 activity, inhibits NF-κB, and activates the antioxidant response via Nrf2 signaling (Tang and Le 2016; Wiciński et al. 2023; Yaku et al. 2018).

The NLRP3 inflammasome is a multiprotein complex essential to innate immunity, involved in the cleavage and maturation of proinflammatory cytokines. If unchecked, NLRP3 activation can drive chronic inflammatory injury but can be reined in by SIRT1 and SIRT3 through deacetylation mechanisms (Chen et al. 2023). Importantly, interventions such as photobiomodulation (PBM) and photoactive drugs further enrich this regulatory network. PBM using red or near-infrared (NIR) light (630–850 nm) targets mitochondrial cytochrome-c-oxidase, boosting ATP and nitric oxide production, activating survival pathways, suppressing proinflammatory signaling, and promoting the protective M2 phenotype in microglia in animal models and reducing damage after trauma or stroke (de Cardoso et al. 2022; Hamblin 2017; Lu et al. 2017; Wang et al. 2017). Light-activated compounds, or photosensitizers, on the other hand, provide high selectivity and spatiotemporal control over neuroinflammation by generating reactive oxygen species and ablating pathogenic microglia when illuminated, thereby reducing secondary inflammatory surges (Abrahamse and Hamblin 2016).

3.11 Potential Photoactive Compounds as Multiple-Target Neuroprotective Agents

A new wave of neurodegenerative disease therapies leverages the unique bioactivity of natural compounds;curcumin, resveratrol, and various photosensitizersto target cellular stress, inflammation, and protein aggregation pathways in the brain. Innovations in formulation, delivery, and light activation have enabled these molecules to cross the blood-brain barrier, modulate key neuroprotective enzymes such as SIRT1, and orchestrate antioxidant, autophagic, and anti-inflammatory effects. Advanced nanoparticles and light-driven systems further enhance deep-brain targeting and pharmacological efficacy, paving the way for precision neuromodulation and disease clearance.

Photocompounds control neuroinflammation through several targeted mechanisms. By activating SIRT1 or AMPK, compounds like curcumin, resveratrol, quercetin, and berberine reduce NF-κB signaling, STAT3 phosphorylation, and NLRP3 inflammasome assembly, leading to lower expression of proinflammatory cytokines (TNF-α, IL-1β, IL-6) and inhibition of microglial activation. Some photosensitizers enhance ROS generation, stimulating anti-inflammatory responses and modulating microglial phenotypes toward tissue repair. Nanoparticle-based delivery and light-triggered activation precisely regulate glial activity and BBB permeability, controlling the migration of immune cells and limiting neurotoxic cascades, which altogether preserve neural function and mitigate chronic neuroinflammatory damage. These synergistic strategies signal a promising future for treating conditions like Alzheimer's and Parkinson's disease through multi-modal, targeted interventions (Fig. 3.1).

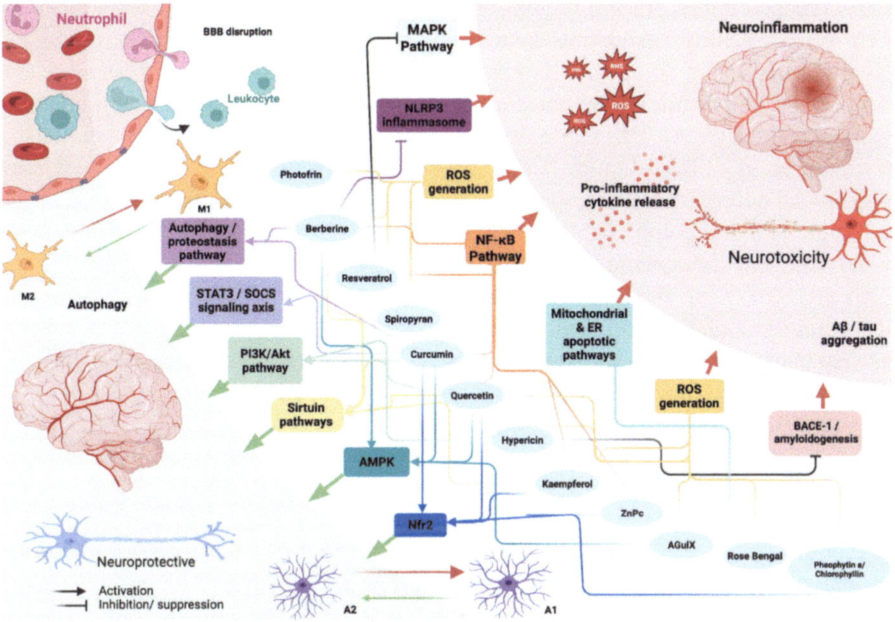

Fig. 3.1 Schematic representation of the modulatory effects of selected photoactive compounds on neuroprotective and neuroinflammatory pathways in the central nervous system. The central column lists all investigated compounds, including resveratrol, curcumin, quercetin, berberine, kaempferol, rose bengal, hypericin, pheophytin a/chlorophyllin, Activation and Guidance of Irradiation by X-ray (AGuIX), zinc phthalocyanine, photofrin, and spiropyran. Arrows from these compounds to the left panel indicate activation of neuroprotective mechanisms—such as SIRT1/SIRT3 and AMPK signaling, SOCS induction, Nrf2 antioxidant response, PI3K/Akt survival pathways, and enhanced autophagy or proteostasis. Arrows pointing from the compounds to the right panel illustrate inhibition of neuroinflammatory processes, including NF-κB, JAK/STAT, NLRP3 inflammasome, MAPK, BACE-1/amyloidogenesis, ROS-mediated damage, and stress-induced apoptosis. Overall, the diagram highlights how these agents can simultaneously promote protective cellular mechanisms and suppress key inflammatory pathways in neurodegenerative disease contexts

3.11.1 Curcumin

Curcumin, the primary bioactive compound from turmeric, demonstrates robust anti-inflammatory and neuroprotective effects via sirtuin modulation, mitochondrial support, and metabolic regulation. Its neurohormetic properties are dose-dependent: physiological doses upregulate sirtuins and activate stress adaptation, while high doses may induce cellular senescence (Grabowska et al. 2017; Kim et al. 2007; Sikora et al. 2010; Yang et al. 2005). Curcumin suppresses STAT3 phosphorylation, increases SOCS protein expression, decreases inflammatory cytokines, and stimulates both SIRT1 and Nrf2 antioxidant responses. These actions are crucial in Alzheimer's and Parkinson's models, conferring improved cognitive and motor outcomes (Aggarwal and Sung 2009; Hassan et al. 2019; Lee et al. 2013; Rahman et al. 2006).

Advanced curcumin formulations such as CurQfen have dramatically improved their bioavailability, achieving better clinical cognitive outcomes in moderate dementia and enhancing pharmacokinetic profiles (Das et al. 2023; T. Krishnareddy et al. 2018). Curcumin also enables improved mitochondrial activity and metabolic resilience through AMPK and SIRT1 activation (Bielak-Zmijewska et al. 2019; Wiciński et al. 2023). Omics-based investigations reveal curcumin orchestrates widespread neuroprotective responses, coordinating antioxidant, neurotrophic, anti-inflammatory, and autophagic effects (Hou et al. 2019).

Of special therapeutic promise, light-activated curcumin analogues (e.g., CRANAD-147) can be triggered with LED or in-situ molecular light, resulting in deep-brain modulation and clearance of amyloid-β in transgenic Alzheimer's models (Kuang et al. 2023). Nanoparticle-based delivery strategies further boost brain exposure and pharmacodynamic outcomes (Qian et al. 2025).

3.11.2 Resveratrol

Resveratrol, a polyphenolic compound with the capability to traverse the blood-brain barrier, stands out for its pleiotropic neuroprotective effects. Its primary therapeutic actions are exerted through SIRT1 activation, which facilitates deacetylation of NF-κB, p53, and PGC-1α, leading to diminished inflammatory gene expression, enhanced mitochondrial biogenesis, and reduced neuronal apoptosis (Bastianetto et al. 2015; Kim et al. 2007; Sarubbo et al. 2018; Wiciński et al. 2023). In models of Alzheimer's disease, resveratrol not only mitigates β-amyloid accumulation and tau phosphorylation but also enhances autophagic clearance pathways, thus reducing toxic protein aggregates (Pasinetti et al. 2015; Vingtdeux et al. 2010).

This compound also exhibits pronounced effects in Parkinson's disease models, protecting against dopaminergic neuron loss, inhibiting the release of proinflammatory cytokines, and fostering anti-inflammatory mediator synthesis (Guo et al. 2019; Jin et al. 2008; Lofrumento et al. 2014). In addition, it reduces oxidative stress markers and elevates antioxidant enzyme activity, further contributing to its neurotherapeutic profile (Khan et al. 2010). Studies have also demonstrated a shift toward the neuroprotective M2 microglial phenotype, a transition critical for resolving chronic neuroinflammation. This is mediated via inhibition of MAPK pathways and suppression of NF-κB nuclear translocation (Li et al. 2021).

Resveratrol's mechanisms extend to AMPK activation, enabling mitochondrial biogenesis, improved energy metabolism, and autophagic clearance of cellular debris, which are pertinent in aging and metabolic dysfunction often present in neurodegenerative diseases (Vingtdeux et al. 2010). While oral resveratrol is generally well-tolerated at doses up to 2000 mg daily, clinical application is complicated by its rapid metabolism and relatively poor bioavailability, mandating advanced delivery systems for optimal efficacy (Pangeni et al. 2014; Turner et al. 2015; Wiciński et al. 2023).

3.11.3 Rose Bengal

Rose Bengal, a xanthene dye, is a potent photosensitizer when activated by green light in the 530–595 nm range. The dye's high quantum yield for singlet oxygen generation underpins its therapeutic value in neurodegeneration, particularly for amyloid and tau pathologies, as it can inhibit Aβ self-assembly, dissolve tau filaments, and rescue deficits in neurobehavioral models (Turbay et al. 2014). Rose Bengal activates multiple cell death and apoptotic pathways, including mitochondrial and ER-stress routes (Panzarini et al. 2011). Advanced strategies such as X-ray–activated nanoparticles have successfully achieved deep-brain photodynamic therapy, overcoming traditional light penetration barriers (Ma et al. 2014). Rose Bengal's clinical versatility is further supported by ongoing trials and tailored drug-delivery systems (Demartis et al. 2021).

3.11.4 Hypericin

Hypericin, a naturally derived anthraquinone, disrupts amyloid aggregates via photodynamic effects that involve both Type I and Type II photochemical pathways and inhibits BACE-1, a key enzyme in amyloidogenesis (Jendželovská et al. 2016; Krupka-Olek et al. 2024; Yuan et al. 2023). Light-activated hypericin not only reduces neuroinflammatory cytokine release but also improves memory outcomes in rodent models (Zhang et al. 2016). It influences multiple signaling axes, including the PI3K/Akt pathway, and demonstrates effective crossing and accumulation within the central nervous system (Huang et al. 2025).

3.11.5 Quercetin

Quercetin, a flavonoid photosensitizer, supports neuroprotection via SIRT1 activation, which regulates numerous beneficial pathways (Cui et al. 2022). Quercetin's multi-pathway actions (SIRT1/AMPK/NF-κB, SIRT1/Nrf2, SIRT1/PI3K/Akt) enhance antioxidant defenses and mitochondrial function. In vivo administration reduces neuroinflammation, cognitive deficits, oxidative stress, and amyloid accumulation (Wiciński et al. 2023). Advanced formulations boost brain entry and light-responsive quercetin release for local disease targeting (Baiomy 2024; Cui et al. 2022).

3.11.6 Berberine

Berberine, an isoquinoline alkaloid from Berberis species, demonstrates multifaceted neuroprotective mechanisms through SIRT1/AMPK pathway activation. Research confirms that berberine activates AMPK, which subsequently enhances SIRT1 activity through NAMPT upregulation and NAD+ regeneration. This dual activation downregulates NF-κB inflammatory signaling and NLRP3 inflammasome pathways while promoting autophagic clearance of damaged proteins. Upon blue light irradiation (420 nm), berberine generates singlet oxygen ($^1O_2$) with a quantum yield of 0.25 in dichloromethane, enabling photodynamic targeting of diseased cells. However, berberine's poor water solubility and rapid metabolism limit bioavailability. Recent berberine-inspired lipid nanoparticles demonstrate remarkable 65.84-fold enhanced cellular uptake and achieve 167% higher brain accumulation compared to standard formulations through dopamine D3 receptor-mediated endocytosis (Bian et al. 2025; DiNicolantonio et al. 2022; Su et al. 2021).

3.11.7 Kaempferol

Kaempferol, a dietary flavonoid, exhibits robust neuroprotective effects through multiple SIRT1-dependent mechanisms. Studies demonstrate kaempferol significantly activates AMPK/SIRT1 signaling pathways, leading to enhanced mitochondrial biogenesis and calcium homeostasis. The compound effectively inhibits TLR4/NF-κB inflammatory cascades, with research showing a 66.3% reduction in TNF-α levels and substantial decreases in proinflammatory cytokine production. Kaempferol promotes microglial polarization from M1 to anti-inflammatory M2 phenotypes through suppression of proinflammatory transcription factors. Additionally, it activates Nrf2 antioxidant pathways, enhancing cellular defense mechanisms. Blood-brain barrier permeability studies indicate kaempferol crosses neural barriers effectively, with PAMPA-BBB assays showing LogPe values of −5.13. Clinical applications demonstrate cognitive protection, with treatments reducing neurological deficit scores and preserving neuronal integrity in ischemic models. Nanoparticle formulations further enhance brain targeting capabilities for therapeutic applications (BinMowyna and AlFaris 2021; Chang et al. 2022; Wang et al. 2020; Yu et al. 2013).

3.11.8 Pheophytin a and Chlorophyllin

Pheophytin a and chlorophyllin represent natural chlorophyll derivatives with exceptional photodynamic therapeutic potential. Pheophytin a demonstrates high singlet oxygen quantum yields, particularly when complexed with DNA structures, showing a 10-fold enhancement in $^1O_2$ generation compared to aqueous solutions. These compounds exhibit favorable tissue penetration characteristics in the near-infrared spectrum, essential for deep-brain applications. Chlorophyllin demonstrates remarkable neuroprotective effects through multiple mechanisms: reduction of cerebral infarct size, enhanced short-term memory, and decreased motor incoordination in ischemic models. The compounds effectively inhibit inflammatory signaling pathways, including NF-κB, while enhancing endogenous antioxidant mechanisms through upregulation of phase 2 cytoprotective genes. Significantly, chlorophyllin successfully crosses the blood-brain barrier in mice, enabling therapeutic brain targeting. Nanoliposomal formulations enhance stability and delivery efficiency. The natural fluorescence properties enable simultaneous imaging and therapeutic applications, supporting theranostic approaches in neurological disorders (Fahey et al. 2005; Martins et al. 2023; Okamoto et al. 2023).

3.11.9 Spiropyran

Spiropyran-based nanovalves represent cutting-edge precision neurotherapeutic engineering for controlled neurotransmitter delivery. These devices utilize photoisomerization between hydrophobic spiropyran and hydrophilic merocyanine forms, triggered by specific wavelengths: UV light (365 nm) opens the valve, while green light (525 nm) closes it. Research demonstrates these nanovalves achieve 96% blockade of glutamate translocation in the closed state and deliver physiologically relevant concentrations (60 $\mu M\ h^{-1}$) upon photostimulation. The systems exhibit exceptional biocompatibility with neuronal cultures and demonstrate functional efficacy in blind retinal explants through patch-clamp studies. Controlled glutamate release mimics synaptic transmission rates, essential for therapeutic neuronal stimulation. The photoswitchable mechanism provides unprecedented temporal and spatial precision in drug delivery. Preclinical studies confirm neuronal recovery and effective neuroinflammation control. These biomimetic neuroprostheses offer revolutionary potential for restoring lost synaptic function in neurodegenerative conditions and enabling light-controlled pharmaceutical interventions (Blanco-Formoso et al. 2025).

3.11.10 Activation and Guidance of Irradiation by X-ray (AGuIX)

AGuIX nanoparticles represent advanced theranostic platforms combining gadolinium-based MRI contrast with therapeutic capabilities. Clinical Phase I trials demonstrate excellent safety profiles with an elimination half-life of 1.3 hours and successful targeting of brain metastases. AGuIX nanoparticles functionalized with porphyrin photosensitizers achieve enhanced tumor selectivity through neuropilin-1 receptor targeting. Remarkably, these nanoparticles activate AMPK-dependent cellular reprogramming, inducing proinflammatory microglial phenotypes that enhance radiotherapy effectiveness. The 10-nm diameter enables efficient blood-brain barrier penetration while maintaining favorable pharmacokinetics. X-ray activation produces therapeutic radiosensitization with minimal systemic toxicity. Recent studies demonstrate AGuIX triggers mitochondrial fragmentation through AMPK activation, leading to beneficial immune responses. Current Phase II trials evaluate efficacy in glioblastoma treatment combined with radiotherapy. The theranostic approach enables real-time imaging guidance while delivering precise therapeutic interventions, representing a significant advancement in personalized brain cancer therapy (Kowolik et al. 2024; Muradova et al. 2024; Verry et al. 2021).

3.11.11 Zinc Phthalocyanine (ZnPc)

ZnPc stands as a premier second-generation photosensitizer with exceptional therapeutic properties for neurodegenerative diseases. ZnPc demonstrates strong near-infrared absorption at 674 nm with high extinction coefficients, enabling deep tissue penetration crucial for brain applications. Research confirms ZnPc effectively targets pathological protein aggregates in Alzheimer's and Parkinson's disease models. Specifically, cationic ZnPc derivatives demonstrate a remarkable ability to modulate amyloid-β aggregation, converting toxic oligomers into non-toxic forms while protecting neurons from cytotoxicity. The compound significantly suppresses NF-κB inflammatory pathways and enhances antioxidant capacity through multiple mechanisms. ZnPc exhibits high singlet oxygen quantum yields (>0.7), ensuring efficient photodynamic activity. Blood-brain barrier penetration is enhanced through cationic modifications, with methylated derivatives showing improved bioavailability. Nanoparticle delivery systems further optimize brain targeting and therapeutic efficacy. Clinical applications demonstrate effective tumor cell elimination at low concentrations (30 nM EC_{50}), indicating high potency with minimal side effects for neurodegenerative disease treatment (Brozek-Pluska et al. 2020; Sheikh et al. 2024).

3.11.12 Photofrin

Photofrin represents the first clinically established photosensitizer with extensive neurological applications. This hematoporphyrin derivative preferentially accumulates in regions with compromised blood-brain barrier integrity, a characteristic advantageous for brain tumor targeting. Red light activation (630–635 nm) generates robust reactive oxygen species production, with clinical studies demonstrating effective tumor cell elimination. Significantly, photofrin induces transient blood-brain barrier disruption at therapeutic doses (2 mg/kg), enhancing drug access to brain parenchyma. Research confirms photofrin's ability to modulate immune responses and inflammatory pathways. The compound demonstrates efficacy in dissolving pathological protein aggregates, including amyloid and tau structures implicated in neurodegenerative diseases. Clinical applications show photofrin treatment results in improved drug penetration across neural barriers. However, prolonged photosensitivity (4–6 weeks) remains a significant limitation. Current research focuses on optimizing dosing regimens and combination therapies to maximize therapeutic benefits while minimizing adverse effects. Photofrin continues as a standard reference for comparing novel photosensitizer efficacy in brain applications (Bartusik-Aebisher et al. 2023; Ji et al. 1994).

In summary, a variety of light-activated compounds from natural plant-based molecules to advanced nanoparticles are making great progress in treating brain diseases. These compounds work by controlling key processes like inflammation, cell cleaning (autophagy), and protecting nerve cells through antioxidant effects. Many of them target important biological pathways that help reduce chronic brain inflammation and support brain health. Using light to precisely activate these drugs at the right place and time, along with improved ways to deliver them to the brain, opens up exciting possibilities for highly targeted treatments. Combining these new compounds with advanced delivery methods is expected to create powerful, multifaceted therapies for complex brain disorders.

References

Abbott NJ, Rönnbäck L, Hansson E (2006) Astrocyte–endothelial interactions at the blood–brain barrier. Nat Rev Neurosci 7(1):41–53

Abrahamse H, Hamblin MR (2016) New photosensitizers for photodynamic therapy. Biochem J 473(4):347–364

Absinta M, Sati P, Masuzzo F, Nair G, Sethi V, Kolb H, Ohayon J, Wu T, Cortese IC, Reich DS (2019) Association of chronic active multiple sclerosis lesions with disability in vivo. JAMA Neurol 76(12):1474–1483

Adamu A, Li S, Gao F, Xue G (2024) The role of neuroinflammation in neurodegenerative diseases: current understanding and future therapeutic targets. Front Aging Neurosci 16:1347987

Aggarwal BB, Sung B (2009) Pharmacological basis for the role of curcumin in chronic diseases: an age-old spice with modern targets. Trends Pharmacol Sci 30(2):85–94

Airas L, Rissanen E, Rinne J (2017) Imaging of microglial activation in MS using PET: research use and potential future clinical application. Mult Scler J 23(4):496–504

Alvarez-Lorenzo C, Concheiro A (2014) Smart drug delivery systems: from fundamentals to the clinic. Chem Commun 50(58):7743–7765

Asai H, Ikezu S, Tsunoda S, Medalla M, Luebke J, Haydar T, Wolozin B, Butovsky O, Kügler S, Ikezu T (2015) Depletion of microglia and inhibition of exosome synthesis halt tau propagation. Nat Neurosci 18(11):1584–1593

Ash C, Dubec M, Donne K, Bashford T (2017) Effect of wavelength and beam width on penetration in light-tissue interaction using computational methods. Lasers Med Sci 32(8):1909–1918

Ashley EA (2016) Towards precision medicine. Nat Rev Genet 17(9):507–522

Babbe H, Roers A, Waisman A, Lassmann H, Goebels N, Hohlfeld R, Friese M, Schröder R, Deckert M, Schmidt S (2000) Clonal expansions of CD8+ T cells dominate the T cell infiltrate in active multiple sclerosis lesions as shown by micromanipulation and single cell polymerase chain reaction. J Exp Med 192(3):393–404

Bai F, Deng Y, Li L, Lv M, Razzokov J, Xu Q, Xu Z, Chen Z, Chen G, Chen Z (2024) Advancements and challenges in brain cancer therapeutics. Exploration 4

Baiomy RF E-s (2024) Quercetin nanoparticles as a therapeutic approach: pharmacological actions and potential applications in therapy. Biotechnologia 105(4):377

Banks WA (2016) From blood–brain barrier to blood–brain interface: new opportunities for CNS drug delivery. Nat Rev Drug Discov 15(4):275–292

Bartusik-Aebisher D, Serafin I, Dynarowicz K, Aebisher D (2023) Photodynamic therapy and associated targeting methods for treatment of brain cancer. Front Pharmacol 14:1250699

Basil MC, Levy BD (2016) Specialized pro-resolving mediators: endogenous regulators of infection and inflammation. Nat Rev Immunol 16(1):51–67

Bastianetto S, Ménard C, Quirion R (2015) Neuroprotective action of resveratrol. Biochim Biophys Acta 1852(6):1195–1201

Bauernfeind FG, Horvath G, Stutz A, Alnemri ES, MacDonald K, Speert D, Fernandes-Alnemri T, Wu J, Monks BG, Fitzgerald KA (2009) Cutting edge: NF-κB activating pattern recognition and cytokine receptors license NLRP3 inflammasome activation by regulating NLRP3 expression. J Immunol 183(2):787–791

Beharry AA, Woolley GA (2011) Azobenzene photoswitches for biomolecules. Chem Soc Rev 40(8):4422–4437

Benedet AL, Milà-Alomà M, Vrillon A, Ashton NJ, Pascoal TA, Lussier F, Karikari TK, Hourregue C, Cognat E, Dumurgier J (2021) Differences between plasma and cerebrospinal fluid glial fibrillary acidic protein levels across the Alzheimer disease continuum. JAMA Neurol 78(12):1471–1483

Bian X, Guo Q, Yau L-F, Yang L, Wang X, Zhao S, Wu S, Qin X, Jiang Z-H, Li C (2025) Berberine-inspired ionizable lipid for self-structure stabilization and brain targeting delivery of nucleic acid therapeutics. Nat Commun 16(1):2368

Biber K, Neumann H, Inoue K, Boddeke HW (2007) Neuronal 'On'and 'Off' signals control microglia. Trends Neurosci 30(11):596–602

Bielak-Zmijewska A, Grabowska W, Ciolko A, Bojko A, Mosieniak G, Bijoch Ł, Sikora E (2019) The role of curcumin in the modulation of ageing. Int J Mol Sci 20(5):1239

BinMowyna MN, AlFaris NA (2021) Kaempferol suppresses acetaminophen-induced liver damage by upregulation/activation of SIRT1. Pharm Biol 59(1):144–154

Bjornevik K, Cortese M, Healy BC, Kuhle J, Mina MJ, Leng Y, Elledge SJ, Niebuhr DW, Scher AI, Munger KL (2022) Longitudinal analysis reveals high prevalence of Epstein-Barr virus associated with multiple sclerosis. Science 375(6578):296–301

Blanco-Formoso M, Galluzzi F, Vacca F, Gianiorio T, Piergentili I, Cook A, Welzen P, van Hest J, Di Marco S, Tantussi F (2025) Spiropyran-based glutamate nanovalve for neuronal stimulation. Mater Horiz

Braslavsky S, Houk K, Verhoeven J (1988) Glossary of terms used in photochemistry. Pure Appl Chem 60:1055–1106

Broichhagen J, Frank JA, Trauner D (2015) A roadmap to success in photopharmacology. Acc Chem Res 48(7):1947–1960

Brozek-Pluska B, Jarota A, Kania R, Abramczyk H (2020) Zinc phthalocyanine photochemistry by Raman imaging, fluorescence spectroscopy and femtosecond spectroscopy in normal and cancerous human colon tissues and single cells. Molecules 25(11):2688

Chang S, Li X, Zheng Y, Shi H, Zhang D, Jing B, Chen Z, Qian G, Zhao G (2022) Kaempferol exerts a neuroprotective effect to reduce neuropathic pain through TLR4/NF-κB signaling pathway. Phytother Res 36(4):1678–1691

Chen H, Deng J, Gao H, Song Y, Zhang Y, Sun J, Zhai J (2023) Involvement of the SIRT1-NLRP3 pathway in the inflammatory response. Cell Commun Signal 21(1):185

Chen L, Shen Q, Liu Y, Zhang Y, Sun L, Ma X, Song N, Xie J (2025) Homeostasis and metabolism of iron and other metal ions in neurodegenerative diseases. Signal Transduct Target Ther 10(1):31

Cherry JD, Olschowka JA, O'Banion MK (2014) Neuroinflammation and M2 microglia: the good, the bad, and the inflamed. J Neuroinflammation 11(1):98

Colonna M, Butovsky O (2017) Microglia function in the central nervous system during health and neurodegeneration. Annu Rev Immunol 35:441–468

Copeland RA, Pompliano DL, Meek TD (2006) Drug–target residence time and its implications for lead optimization. Nat Rev Drug Discov 5(9):730–739

Cui Z, Zhao X, Amevor FK, Du X, Wang Y, Li D, Shu G, Tian Y, Zhao X (2022) Therapeutic application of quercetin in aging-related diseases: SIRT1 as a potential mechanism. Front Immunol 13:943321

Cummings JL, Morstorf T, Zhong K (2014) Alzheimer's disease drug-development pipeline: few candidates, frequent failures. Alzheimer's Res Ther 6(4):37

Cummings J, Lee G, Ritter A, Sabbagh M, Zhong K (2019) Alzheimer's disease drug development pipeline: 2019. Alzheimer's Dementia 5:272–293

Das SS, Gopal PM, Thomas JV, Mohan MC, Thomas SC, Maliakel BP, Krishnakumar I, Pulikkaparambil Sasidharan BC (2023) Influence of CurQfen®-curcumin on cognitive impairment: a randomized, double-blinded, placebo-controlled, 3-arm, 3-sequence comparative study. Front Dementia 2:1222708

de Cardoso FS, Salehpour F, Coimbra NC, Gonzalez-Lima F, Gomes da Silva S (2022) Photobiomodulation for the treatment of neuroinflammation: a systematic review of controlled laboratory animal studies. Front Neurosci 16:1006031

de Rivero Vaccari JP, Dietrich WD, Keane RW (2016) Therapeutics targeting the inflammasome after central nervous system injury. Transl Res 167(1):35–45

Demartis S, Obinu A, Gavini E, Giunchedi P, Rassu G (2021) Nanotechnology-based rose Bengal: a broad-spectrum biomedical tool. Dyes Pigments 188:109236

Deng Q, Wu C, Parker E, Liu T, Duan R, Yang L (2023) Microglia and astrocytes in Alzheimer's disease: significance and summary of recent advances. Aging Dis. 2023. Preprint. https://doi.org/10.14336/AD.2023.0907

DiMasi JA, Grabowski HG, Hansen RW (2016) Innovation in the pharmaceutical industry: new estimates of R&D costs. J Health Econ 47:20–33

Dinarello CA (2009) Immunological and inflammatory functions of the interleukin-1 family. Annu Rev Immunol 27(1):519–550

Ding Z-B, Song L-J, Wang Q, Kumar G, Yan Y-Q, Ma C-G (2021) Astrocytes: a double-edged sword in neurodegenerative diseases. Neural Regen Res 16(9):1702–1710

DiNicolantonio JJ, McCarty MF, Assanga SI, Lujan LL, O'Keefe JH (2022) Ferulic acid and berberine, via Sirt1 and AMPK, may act as cell cleansing promoters of healthy longevity. Open Heart 9(1):e001801

D'Mello C, Le T, Swain MG (2009) Cerebral microglia recruit monocytes into the brain in response to tumor necrosis factorα signaling during peripheral organ inflammation. J Neurosci 29(7):2089–2102

El-Taibany AA, Heydarian P, Porada DA, Seeds MC, Atala A (2025) Multiple sclerosis: etiology in the context of neurovascular unit and immune system involvement and advancements with in vitro blood–brain barrier models. Front Immunol 16:1595276

Fahey JW, Stephenson KK, Dinkova-Kostova AT, Egner PA, Kensler TW, Talalay P (2005) Chlorophyll, chlorophyllin and related tetrapyrroles are significant inducers of mammalian phase 2 cytoprotective genes. Carcinogenesis 26(7):1247–1255

Fransen NL, Hsiao C-C, van der Poel M, Engelenburg HJ, Verdaasdonk K, Vincenten MC, Remmerswaal EB, Kuhlmann T, Mason MR, Hamann J (2020) Tissue-resident memory T cells invade the brain parenchyma in multiple sclerosis white matter lesions. Brain 143(6):1714–1730

Glass CK, Saijo K, Winner B, Marchetto MC, Gage FH (2010) Mechanisms underlying inflammation in neurodegeneration. Cell 140(6):918–934

Goverman JM (2011) Immune tolerance in multiple sclerosis. Immunol Rev 241(1):228–240

Grabowska W, Sikora E, Bielak-Zmijewska A (2017) Sirtuins, a promising target in slowing down the ageing process. Biogerontology 18(4):447–476

Guo Y, Wei X, Yan H, Qin Y, Yan S, Liu J, Zhao Y, Jiang F, Lou H (2019) TREM2 deficiency aggravates α-synuclein–induced neurodegeneration and neuroinflammation in Parkinson's disease models. FASEB J 33(11):12164

Hamblin MR (2017) Mechanisms and applications of the anti-inflammatory effects of photobiomodulation. AIMS Biophys 4(3):337

Hamby ME, Sofroniew MV (2010) Reactive astrocytes as therapeutic targets for CNS disorders. Neurotherapeutics 7(4):494–506

Hampel H, Toschi N, Babiloni C, Baldacci F, Black KL, Bokde AL, Bun RS, Cacciola F, Cavedo E, Chiesa PA (2018) Revolution of Alzheimer precision neurology. Passageway of systems biology and neurophysiology. J Alzheimers Dis 64(s1):S47–S105

Hassan F-u, Rehman M S-u, Khan MS, Ali MA, Javed A, Nawaz A, Yang C (2019) Curcumin as an alternative epigenetic modulator: mechanism of action and potential effects. Front Genet 10:514

He C, Chen B, Yang H, Zhou X (2025) The dual role of microglia in Alzheimer's disease: from immune regulation to pathological progression. Front Aging Neurosci 17:1554398

Heneka M (2015) Neuroinflammation in Alzheimer's disease. Lancet Neurol 14:388–405

Heneka MT, Kummer MP, Latz E (2014) Innate immune activation in neurodegenerative disease. Nat Rev Immunol 14(7):463–477

Hensley K (2010) Neuroinflammation in Alzheimer's disease: mechanisms, pathologic consequences, and potential for therapeutic manipulation. J Alzheimer's Dis 21(1):1–14

Hou Y, Wang J, Feng J (2019) The neuroprotective effects of curcumin are associated with the regulation of the reciprocal function between autophagy and HIF-1α in cerebral ischemia-reperfusion injury. Drug Des Devel Ther 13:1135–1144

Huang S-p, Zhang Z-c, Chen Y, Shang C-j, Zhu M, Li W-h (2025) Review of Chinese medicine intervention in PI3K/AKT pathway to regulate fibrosis. Medicine (Baltimore) 104(28):e42957

Imbimbo BP, Solfrizzi V, Panza F (2010) Are NSAIDs useful to treat Alzheimer's disease or mild cognitive impairment? Front Aging Neurosci 2:1517

Jassam YN, Izzy S, Whalen M, McGavern DB, El Khoury J (2017) Neuroimmunology of traumatic brain injury: time for a paradigm shift. Neuron 95(6):1246–1265

Jayaraj RL, Azimullah S, Beiram R, Jalal FY, Rosenberg GA (2019) Neuroinflammation: friend and foe for ischemic stroke. J Neuroinflammation 16(1):142

Jendželovská Z, Jendželovský R, Kuchárová B, Fedoročko P (2016) Hypericin in the light and in the dark: two sides of the same coin. Front Plant Sci 7:560

Ji Y, Powers SK, Brown JT, Walstad D, Maliner L (1994) Toxicity of photodynamic therapy with photofrin in the normal rat brain. Lasers Surg Med 14(3):219–228

Jin, F., Wu, Q., Lu, Y. F., Gong, Q. H., & Shi, J. S. (2008). Neuroprotective effect of resveratrol on 6-OHDAinduced Parkinson's disease in rats. European journal of pharmacology, 600(1-3), 78–82. https://doi.org/10.1016/j.ejphar.2008.10.005

Jiao F, Gong Z (2020) The beneficial roles of SIRT1 in neuroinflammation-related diseases. Oxidative Med Cell Longev 2020(1):6782872

Jordão MJC, Sankowski R, Brendecke SM, Sagar N, Locatelli G, Tai Y-H, Tay TL, Schramm E, Armbruster S, Hagemeyer N (2019) Single-cell profiling identifies myeloid cell subsets with distinct fates during neuroinflammation. Science 363(6425):eaat7554

Kauppinen A, Suuronen T, Ojala J, Kaarniranta K, Salminen A (2013) Antagonistic crosstalk between NF-κB and SIRT1 in the regulation of inflammation and metabolic disorders. Cell Signal 25(10):1939–1948

Kettenmann H, Hanisch U-K, Noda M, Verkhratsky A (2011) Physiology of microglia. Physiol Rev 91(2):461–553

Khan, M. M., Ahmad, A., Ishrat, T., Khan, M. B., Hoda, M. N., Khuwaja, G., Raza, S. S., Khan, A., Javed, H., Vaibhav, K., & Islam, F. (2010). Resveratrol attenuates 6-hydroxydopamine-induced oxidative damage and dopamine depletion in rat model of Parkinson's disease. Brain research, 1328, 139–151. https://doi.org/10.1016/j.brainres.2010.02.031

Kim D, Nguyen MD, Dobbin MM, Fischer A, Sananbenesi F, Rodgers JT, Delalle I, Baur JA, Sui G, Armour SM (2007) SIRT1 deacetylase protects against neurodegeneration in models for Alzheimer's disease and amyotrophic lateral sclerosis. EMBO J 26(13):3169–3179

Kim B, Yang M-S, Choi D, Kim J-H, Kim H-S, Seol W, Choi S, Jou I, Kim E-Y, Joe E-h (2012) Impaired inflammatory responses in murine Lrrk2-knockdown brain microglia. PLoS One 7(4):e34693
Klán P, Wirz J (2009) Photochemistry of organic compounds: from concepts to practice. John Wiley & Sons
Klegeris A (2021) Regulation of neuroimmune processes by damage-and resolution-associated molecular patterns. Neural Regen Res 16(3):423–429
Kobauri P, Dekker FJ, Szymanski W, Feringa BL (2023) Rational design in photopharmacology with molecular photoswitches. Angew Chem Int Ed 62(30):e202300681
Kowolik E, Szczygieł D, Szczygieł M, Drzał A, Vemuri K, Olsson A-K, Griffioen AW, Nowak-Sliwinska P, Wolnicka-Glubisz A, Elas M (2024) Preclinical photodynamic therapy targeting blood vessels with AGuIX® theranostic nanoparticles. Cancer 16(23):3924
Krupka-Olek M, Bożek A, Czuba ZP, Kłósek M, Cieślar G, Kawczyk-Krupka A (2024) Cytotoxic and immunomodulatory effects of Hypericin as a photosensitizer in photodynamic therapy used on skin cell cultures. Pharmaceutics 16(6):696
Kuang S, Zhu B, Zhang J, Yang F, Wu B, Ding W, Yang L, Shen S, Liang SH, Mondal P (2023) A photolabile curcumin-diazirine analogue enables phototherapy with physically and molecularly produced light for Alzheimer's disease treatment. Angew Chem 135(45):e202312519
Kumar A, Loane DJ (2012) Neuroinflammation after traumatic brain injury: opportunities for therapeutic intervention. Brain Behav Immun 26(8):1191–1201
Kwon HS, Koh S-H (2020) Neuroinflammation in neurodegenerative disorders: the roles of microglia and astrocytes. Transl Neurodegener 9(1):42
Lanoiselée H-M, Nicolas G, Wallon D, Rovelet-Lecrux A, Lacour M, Rousseau S, Richard A-C, Pasquier F, Rollin-Sillaire A, Martinaud O (2017) APP, PSEN1, and PSEN2 mutations in early-onset Alzheimer disease: a genetic screening study of familial and sporadic cases. PLoS Med 14(3):e1002270
Lanz TV, Brewer RC, Ho PP, Moon J-S, Jude KM, Fernandez D, Fernandes RA, Gomez AM, Nadj G-S, Bartley CM (2022) Clonally expanded B cells in multiple sclerosis bind EBV EBNA1 and GlialCAM. Nature 603(7900):321–327
Lee W-H, Loo C-Y, Bebawy M, Luk F, Mason RS, Rohanizadeh R (2013) Curcumin and its derivatives: their application in neuropharmacology and neuroscience in the 21st century. Curr Neuropharmacol 11(4):338–378
Lerch MM, Hansen MJ, van Dam GM, Szymanski W, Feringa BL (2016) Emerging targets in photopharmacology. Angew Chem Int Ed 55(37):10978–10999
Lewis L, Valvi D, Gedaly R, Marti F (2025) Mitochondrial unfolded protein response in regulatory T cell function: a protective mechanism in immune aging. Front Immunol 16:1621759
Li Y, Xia Y, Yin S, Wan F, Hu J, Kou L, Sun Y, Wu J, Zhou Q, Huang J (2021) Targeting microglial α-synuclein/TLRs/NF-kappaB/NLRP3 inflammasome axis in Parkinson's disease. Front Immunol 12:719807
Liddelow SA, Guttenplan KA, Clarke LE, Bennett FC, Bohlen CJ, Schirmer L, Bennett ML, Münch AE, Chung W-S, Peterson TC (2017) Neurotoxic reactive astrocytes are induced by activated microglia. Nature 541(7638):481–487
Liu S, Jin X, Ge Y, Dong J, Liu X, Pei X, Wang P, Wang B, Chang Y, Yu X-a (2025) Advances in brain-targeted delivery strategies and natural product-mediated enhancement of blood–brain barrier permeability. J Nanobiotechnol 23(1):382
Loane DJ, Kumar A (2016) Microglia in the TBI brain: the good, the bad, and the dysregulated. Exp Neurol 275:316–327
Lofrumento, D.D., Nicolardi, G., Cianciulli, A., Nuccio, F.D., Pesa, V.L., Carofiglio, V., Dragone, T., Calvello, R. and Panaro, M.A., 2014. Neuroprotective effects of resveratrol in an MPTP mouse model of Parkinson's-like disease: possible role of SOCS-1 in reducing pro-inflammatory responses. Innate immunity, 20(3), pp.249-260.
Lu Y, Wang R, Dong Y, Tucker D, Zhao N, Ahmed ME, Zhu L, Liu TC-Y, Cohen RM, Zhang Q (2017) Low-level laser therapy for beta amyloid toxicity in rat hippocampus. Neurobiol Aging 49:165–182

Lucky SS, Soo KC, Zhang Y (2015) Nanoparticles in photodynamic therapy. Chem Rev 115(4):1990–2042

Ma L, Zou X, Chen W (2014) A new X-ray activated nanoparticle photosensitizer for cancer treatment. J Biomed Nanotechnol 10(8):1501–1508

Machado-Santos J, Saji E, Tröscher AR, Paunovic M, Liblau R, Gabriely G, Bien CG, Bauer J, Lassmann H (2018) The compartmentalized inflammatory response in the multiple sclerosis brain is composed of tissue-resident CD8+ T lymphocytes and B cells. Brain 141(7):2066–2082

Mallah K, Couch C, Alshareef M, Borucki D, Yang X, Alawieh A, Tomlinson S (2021) Complement mediates neuroinflammation and cognitive decline at extended chronic time points after traumatic brain injury. Acta Neuropathol Commun 9(1):72

Mark JR, Titus AM, Staley HA, Alvarez S, Mahn S, McFarland NR, Wallings RL, Tansey MG (2025) Peripheral immune cell response to stimulation stratifies Parkinson's disease progression from prodromal to clinical stages. Communications Biology 8(1):716

Martins T, Barros AN, Rosa E, Antunes L (2023) Enhancing health benefits through chlorophylls and chlorophyll-rich agro-food: a comprehensive review. Molecules 28(14):5344

Matheoud D, Sugiura A, Bellemare-Pelletier A, Laplante A, Rondeau C, Chemali M, Fazel A, Bergeron JJ, Trudeau L-E, Burelle Y (2016) Parkinson's disease-related proteins PINK1 and Parkin repress mitochondrial antigen presentation. Cell 166(2):314–327

McGeer PL, Itagaki S, Boyes BE, McGeer E (1988) Reactive microglia are positive for HLA-DR in the substantia nigra of Parkinson's and Alzheimer's disease brains. Neurology 38(8):1285–1285

Miao J, Shi R, Li L, Chen F, Zhou Y, Tung YC, Hu W, Gong C-X, Iqbal K, Liu F (2019) Pathological tau from Alzheimer's brain induces site-specific hyperphosphorylation and SDS-and reducing agent-resistant aggregation of tau in vivo. Front Aging Neurosci 11:34

Müller L, Di Benedetto S (2025) Neuroimmune crosstalk in chronic neuroinflammation: microglial interactions and immune modulation. Front Cell Neurosci 19:1575022

Muradova Z, Tannous D, Mostefa-Kara A, Cao-Pham TT, Lamy C, Broutin S, Paci A, Dufort S, Doussineau T, Lux F (2024) Gadolinium-based nanoparticles AGuIX and their combination with ionizing radiation trigger AMPK-dependent proinflammatory reprogramming of tumor-associated macrophages. *bioRxiv* 2024:2001. 2012.575217

Obermeier B, Mentele R, Malotka J, Kellermann J, Kümpfel T, Wekerle H, Lottspeich F, Hohlfeld R, Dornmair K (2008) Matching of oligoclonal immunoglobulin transcriptomes and proteomes of cerebrospinal fluid in multiple sclerosis. Nat Med 14(6):688–693

Oelgemöller M, Shvydkiv O (2011) Recent advances in microflow photochemistry. Molecules 16(9):7522–7550

Okamoto C, Momotake A, Yamamoto Y (2023) Photosensitized generation of singlet oxygen (1O2) from Pyropheophorbide a specifically bound to all parallel-stranded G-Quadruplex DNAs. J Phys Chem B 127(20):4514–4522

Okonkwo DO, Yue JK, Puccio AM, Panczykowski DM, Inoue T, McMahon PJ, Sorani MD, Yuh EL, Lingsma HF, Maas AI (2013) GFAP-BDP as an acute diagnostic marker in traumatic brain injury: results from the prospective transforming research and clinical knowledge in traumatic brain injury study. J Neurotrauma 30(17):1490–1497

Olah M, Menon V, Habib N, Taga MF, Ma Y, Yung CJ, Cimpean M, Khairallah A, Coronas-Samano G, Sankowski R (2020) Single cell RNA sequencing of human microglia uncovers a subset associated with Alzheimer's disease. Nat Commun 11(1):6129

Pangeni R, Sahni JK, Ali J, Sharma S, Baboota S (2014) Resveratrol: review on therapeutic potential and recent advances in drug delivery. Expert Opin Drug Deliv 11(8):1285–1298

Panzarini E, Inguscio V, Dini L (2011) Timing the multiple cell death pathways initiated by rose Bengal acetate photodynamic therapy. Cell Death Dis 2(6):e169–e169

Paolicelli RC, Bolasco G, Pagani F, Maggi L, Scianni M, Panzanelli P, Giustetto M, Ferreira TA, Guiducci E, Dumas L (2011) Synaptic pruning by microglia is necessary for normal brain development. Science 333(6048):1456–1458

Pardridge WM (2012) Drug transport across the blood–brain barrier. J Cereb Blood Flow Metab 32(11):1959–1972

Pasinetti GM, Wang J, Ho L, Zhao W, Dubner L (2015) Roles of resveratrol and other grape-derived polyphenols in Alzheimer's disease prevention and treatment. Biochim Biophys Acta 1852(6):1202–1208

Piccinini A, Midwood K (2010) DAMPening inflammation by modulating TLR signalling. Mediat Inflamm 2010(1):672395

Piessevaux J, Lavens D, Montoye T, Wauman J, Catteeuw D, Vandekerckhove J, Belsham D, Peelman F, Tavernier J (2006) Functional cross-modulation between SOCS proteins can stimulate cytokine signaling. J Biol Chem 281(44):32953–32966

Prineas JW, Parratt JD (2021) Multiple sclerosis: microglia, monocytes, and macrophage-mediated demyelination. J Neuropathol Exp Neurol 80(10):975–996

Prinz M, Priller J (2017) The role of peripheral immune cells in the CNS in steady state and disease. Nat Neurosci 20(2):136–144

Qian H, Lv J, Hu X (2025) Development and evaluation of curcumin nano-niosomes for glioma-targeted therapy. Sci Rep 15(1):10520

Rahman I, Biswas SK, Kirkham PA (2006) Regulation of inflammation and redox signaling by dietary polyphenols. Biochem Pharmacol 72(11):1439–1452

Ramlackhansingh AF, Brooks DJ, Greenwood RJ, Bose SK, Turkheimer FE, Kinnunen KM, Gentleman S, Heckemann RA, Gunanayagam K, Gelosa G (2011) Inflammation after trauma: microglial activation and traumatic brain injury. Ann Neurol 70(3):374–383

Ransohoff RM (2016a) How neuroinflammation contributes to neurodegeneration. Science 353(6301):777–783

Ransohoff RM (2016b) A polarizing question: do M1 and M2 microglia exist? Nat Neurosci 19(8):987–991

Ricciotti E, FitzGerald GA (2011) Prostaglandins and inflammation. Arterioscler Thromb Vasc Biol 31(5):986–1000

Roodveldt C, Bernardino L, Oztop-Cakmak O, Dragic M, Fladmark KE, Ertan S, Aktas B, Pita C, Ciglar L, Garraux G (2024) The immune system in Parkinson's disease: what we know so far. Brain 147(10):3306–3324

Rothgiesser KM, Erener S, Waibel S, Lüscher B, Hottiger MO (2010) SIRT2 regulates NF-κB-dependent gene expression through deacetylation of p65 Lys310. J Cell Sci 123(24):4251–4258

Russo I (2019) The role of LRRK2 on PKA-NFκB pathway in microglia cells: implications for Parkinson's disease. Neural Regen Res 14(10):1713–1714

Sabatino JJ, Zamvil SS, Hauser SL (2019) B-cell therapies in multiple sclerosis. Cold Spring Harb Perspect Med 9(2):a032037

Salter MW, Stevens B (2017) Microglia emerge as central players in brain disease. Nat Med 23(9):1018–1027

Sarubbo F, Esteban S, Miralles A, Moranta D (2018) Effects of resveratrol and other polyphenols on Sirt1: relevance to brain function during aging. Curr Neuropharmacol 16(2):126–136

Scott G, Hellyer PJ, Ramlackhansingh AF, Brooks DJ, Matthews PM, Sharp DJ (2015) Thalamic inflammation after brain trauma is associated with thalamo-cortical white matter damage. J Neuroinflammation 12(1):224

Serhan CN (2014) Pro-resolving lipid mediators are leads for resolution physiology. Nature 510(7503):92–101

Shahim P, Gren M, Liman V, Andreasson U, Norgren N, Tegner Y, Mattsson N, Andreasen N, Öst M, Zetterberg H (2016a) Serum neurofilament light protein predicts clinical outcome in traumatic brain injury. Sci Rep 6:36791

Shahim P, Tegner Y, Gustafsson B, Gren M, Ärlig J, Olsson M, Lehto N, Engström Å, Höglund K, Portelius E, Zetterberg H, Blennow K (2016b) Neurochemical aftermath of repetitive mild traumatic brain injury. JAMA Neurol 73(11):1308–1315

Shariq AS, Brietzke E, Rosenblat JD, Pan Z, Rong C, Ragguett R-M, Park C, McIntyre RS (2018) Therapeutic potential of JAK/STAT pathway modulation in mood disorders. Rev Neurosci 30(1):1–7

Sheikh AM, Tabassum S, Yano S, Abdullah FB, Wang R, Ikeue T, Nagai A (2024) A cationic Zn-Phthalocyanine turns Alzheimer's amyloid β aggregates into non-toxic oligomers and inhibits neurotoxicity in culture. Int J Mol Sci 25(16):8931

Shi J, Zhao Y, Wang K, Shi X, Wang Y, Huang H, Zhuang Y, Cai T, Wang F, Shao F (2015) Cleavage of GSDMD by inflammatory caspases determines pyroptotic cell death. Nature 526(7575):660–665

Shi Y, Yamada K, Liddelow SA, Smith ST, Zhao L, Luo W, Tsai RM, Spina S, Grinberg LT, Rojas JC (2017) ApoE4 markedly exacerbates tau-mediated neurodegeneration in a mouse model of tauopathy. Nature 549(7673):523–527

Sikora E, Scapagnini G, Barbagallo M (2010) Curcumin, inflammation, ageing and age-related diseases. Immun Ageing 7(1):1

Simon DW, McGeachy MJ, Bayır H, Clark RS, Loane DJ, Kochanek PM (2017) The far-reaching scope of neuroinflammation after traumatic brain injury. Nat Rev Neurol 13(3):171–191

Sofroniew MV (2015) Astrogliosis. Cold Spring Harb Perspect Biol 7(2):a020420

Stevens B, Allen NJ, Vazquez LE, Howell GR, Christopherson KS, Nouri N, Micheva KD, Mehalow AK, Huberman AD, Stafford B (2007) The classical complement cascade mediates CNS synapse elimination. Cell 131(6):1164–1178

Su S, Indran KS, Pal S, Le E, Regan A, Jain M, Shah A, Iglesias D, Wang A, Njoo E (2021) Comparative singlet oxygen photosensitizer efficiency of berberine, rose bengal, and methylene blue by time course nuclear magnetic resonance (NMR) monitoring of a photochemical 4+ 2 cycloaddition endoperoxide formation. J Emerg Investig

Sucksdorff M, Rissanen E, Tuisku J, Nuutinen S, Paavilainen T, Rokka J, Rinne J, Airas L (2017) Evaluation of the effect of fingolimod treatment on microglial activation using serial PET imaging in multiple sclerosis. J Nucl Med 58(10):1646–1651

Swanson KV, Deng M, Ting JP-Y (2019) The NLRP3 inflammasome: molecular activation and regulation to therapeutics. Nat Rev Immunol 19(8):477–489

Sweeney MD, Sagare AP, Zlokovic BV (2018) Blood–brain barrier breakdown in Alzheimer disease and other neurodegenerative disorders. Nat Rev Neurol 14(3):133–150

Szymanski W, Beierle JM, Kistemaker HA, Velema WA, Feringa BL (2013) Reversible photocontrol of biological systems by the incorporation of molecular photoswitches. Chem Rev 113(8):6114–6178

T. Krishnareddy N, Thomas JV, Nair SS, N. Mulakal J, Maliakel BP, Krishnakumar I (2018) A novel curcumin-Galactomannoside complex delivery system improves hepatic function markers in chronic alcoholics: a double-blinded, randomized, placebo-controlled study. Biomed Res Int 2018(1):9159281

Tang Y, Le W (2016) Differential roles of M1 and M2 microglia in neurodegenerative diseases. Mol Neurobiol 53(2):1181–1194

Tansey MG, Wallings RL, Houser MC, Herrick MK, Keating CE, Joers V (2022) Inflammation and immune dysfunction in Parkinson disease. Nat Rev Immunol 22(11):657–673

Teng K-X, Niu L-Y, Yang Q-Z (2023) Supramolecular photosensitizer enables oxygen-independent generation of hydroxyl radicals for photodynamic therapy. J Am Chem Soc 145(7):4081–4087

Thompson AJ, Baranzini SE, Geurts J, Hemmer B, Ciccarelli O (2018) Multiple sclerosis. Lancet 391(10130):1622–1636. https://doi.org/10.1016/s0140-6736(18)30481-1

Timbie KF, Mead BP, Price RJ (2015) Drug and gene delivery across the blood–brain barrier with focused ultrasound. J Control Release 219:61–75

Tsitsipanis C, Miliaraki M, Paflioti E, Lazarioti S, Moustakis N, Ntotsikas K, Theofanopoulos A, Ilia S, Vakis A, Simos P (2023) Inflammation biomarkers IL-6 and IL-10 may improve the diagnostic and prognostic accuracy of currently authorized traumatic brain injury tools. Exp Ther Med 26(2):364

Turbay MBE, Rey V, Argañaraz NM, Vieyra FEM, Aspée A, Lissi EA, Borsarelli CD (2014) Effect of dye localization and self-interactions on the photosensitized generation of singlet oxygen by rose bengal bound to bovine serum albumin. J Photochem Photobiol B Biol 141:275–282

Turner RS, Thomas RG, Craft S, Van Dyck CH, Mintzer J, Reynolds BA, Brewer JB, Rissman RA, Raman R, Aisen PS (2015) A randomized, double-blind, placebo-controlled trial of resveratrol for Alzheimer disease. Neurology 85(16):1383–1391

Turro NJ, Ramamurthy V, Scaiano JC (2010) Modern molecular photochemistry of organic molecules, vol 188. University Science Books, Sausalito

Van Den Bosch AM, Van Der Poel M, Fransen NL, Vincenten MC, Bobeldijk AM, Jongejan A, Engelenburg HJ, Moerland PD, Smolders J, Huitinga I (2024) Profiling of microglia nodules in multiple sclerosis reveals propensity for lesion formation. Nat Commun 15(1):1667

Vavougios GD, Nday C, Pelidou S-H, Gourgoulianis KI, Stamoulis G, Doskas T, Zarogiannis SG (2021) Outside-in induction of the IFITM3 trafficking system by infections, including SARS-CoV-2, in the pathobiology of Alzheimer's disease. Brain, behavior, & immunity-health 14:100243

Velema WA, Szymanski W, Feringa BL (2014) Photopharmacology: beyond proof of principle. J Am Chem Soc 136(6):2178–2191

Verry C, Dufort S, Villa J, Gavard M, Iriart C, Grand S, Charles J, Chovelon B, Cracowski J-L, Quesada J-L (2021) Theranostic AGuIX nanoparticles as radiosensitizer: a phase I, dose-escalation study in patients with multiple brain metastases (NANO-RAD trial). Radiother Oncol 160:159 165

Vingtdeux V, Giliberto L, Zhao H, Chandakkar P, Wu Q, Simon JE, Janle EM, Lobo J, Ferruzzi MG, Davies P (2010) AMP-activated protein kinase signaling activation by resveratrol modulates amyloid-β peptide metabolism. J Biol Chem 285(12):9100–9113

Wallisch JS, Simon DW, Bayır H, Bell MJ, Kochanek PM, Clark RS (2017) Cerebrospinal fluid NLRP3 is increased after severe traumatic brain injury in infants and children. Neurocrit Care 27(1):44–50

Wang H, Bloom O, Zhang M, Vishnubhakat JM, Ombrellino M, Che J, Frazier A, Yang H, Ivanova S, Borovikova L (1999) HMG-1 as a late mediator of endotoxin lethality in mice. Science 285(5425):248–251

Wang W-Y, Tan M-S, Yu J-T, Tan L (2015) Role of pro-inflammatory cytokines released from microglia in Alzheimer's disease. Ann Transl Med 3(10):136

Wang X, Tian F, Reddy DD, Nalawade SS, Barrett DW, Gonzalez-Lima F, Liu H (2017) Up-regulation of cerebral cytochrome-c-oxidase and hemodynamics by transcranial infrared laser stimulation: a broadband near-infrared spectroscopy study. J Cereb Blood Flow Metab 37(12):3789–3802

Wang J, Mao J, Wang R, Li S, Wu B, Yuan Y (2020) Kaempferol protects against cerebral ischemia reperfusion injury through intervening oxidative and inflammatory stress induced apoptosis. Front Pharmacol 11:424

Wang Q, Zheng J, Pettersson S, Reynolds R, Tan E-K (2023) The link between neuroinflammation and the neurovascular unit in synucleinopathies. Sci Adv 9(7):eabq1141

Wiciński M, Erdmann J, Nowacka A, Kuźmiński O, Michalak K, Janowski K, Ohla J, Biernaciak A, Szambelan M, Zabrzyński J (2023) Natural phytochemicals as SIRT activators—focus on potential biochemical mechanisms. Nutrients 15(16):3578

Wilson DM, Cookson MR, Van Den Bosch L, Zetterberg H, Holtzman DM, Dewachter I (2023) Hallmarks of neurodegenerative diseases. Cell 186(4):693–714

Woodcock T, Morganti-Kossmann MC (2013) The role of markers of inflammation in traumatic brain injury. Front Neurol 4:18

Xiao T, Ji H, Shangguan X, Qu S, Cui Y, Xu J (2022) NLRP3 inflammasome of microglia promotes A1 astrocyte transformation, neo-neuron decline and cognition impairment in endotoxemia. Biochem Biophys Res Commun 602:1–7

Xing C, Wang X, Cheng C, Montaner J, Mandeville E, Leung W, van Leyen K, Lok J, Wang X, Lo EH (2014) Neuronal production of lipocalin-2 as a help-me signal for glial activation. Stroke 45(7):2085–2092

Yaku K, Okabe K, Hikosaka K, Nakagawa T (2018) NAD metabolism in cancer therapeutics. Front Oncol 8:622

Yang F, Lim GP, Begum AN, Ubeda OJ, Simmons MR, Ambegaokar SS, Chen PP, Kayed R, Glabe CG, Frautschy SA (2005) Curcumin inhibits formation of amyloid β oligomers and fibrils, binds plaques, and reduces amyloid in vivo. J Biol Chem 280(7):5892–5901

Yin J, Valin KL, Dixon ML, Leavenworth JW (2017) The role of microglia and macrophages in CNS homeostasis, autoimmunity, and cancer. J Immunol Res 2017(1):5150678

Yu L, Chen C, Wang L-F, Kuang X, Liu K, Zhang H, Du J-R (2013) Neuroprotective effect of kaempferol glycosides against brain injury and neuroinflammation by inhibiting the activation of NF-κB and STAT3 in transient focal stroke. PLoS One 8(2):e55839

Yuan X, Yan F, Gao LH, Ma QH, Wang J (2023) Hypericin as a potential drug for treating Alzheimer's disease and type 2 diabetes with a view to drug repositioning. CNS Neurosci Ther 29(11):3307–3321

Zhang M, Wang Y, Qian F, Li P, Xu X (2016) Hypericin inhibits oligomeric amyloid β42-induced inflammation response in microglia and ameliorates cognitive deficits in an amyloid β injection mouse model of Alzheimer's disease by suppressing MKL1. Biochem Biophys Res Commun 481(1–2):71–76

Zhang S, Gao Y, Zhao Y, Huang TY, Zheng Q, Wang X (2025) Peripheral and central neuroimmune mechanisms in Alzheimer's disease pathogenesis. Mol Neurodegener 20(1):22

Zhao Y, Gan L, Ren L, Lin Y, Ma C, Lin X (2022) Factors influencing the blood-brain barrier permeability. Brain Res 1788:147937

Zheng Q, Wang X (2025) Alzheimer's disease: insights into pathology, molecular mechanisms, and therapy. Protein Cell 16(2):83–120

Phototherapeutic Compounds: Toward Cognitive Revival in Alzheimer's Disease

4

Divya Sri Kathiresan, Vinayaga Moorthi Puthamohan, and Arun Nagendran N.

Abstract

Among many different neurodegenerative diseases, Alzheimer's Disease (AD) marks the most affected neurodegenerative disease in the world's population, contributing to the utmost suffering of the people and society. Even with the advancements in modern scientific research, the most relatable cause of AD is the amyloid beta (Aβ) plaque formation, and the increasing reactive oxygen species are the ones concentrated, and the therapies are mostly to be implied for their interactive breakdowns. The acetylcholine esterase (AChE), which in involved in the neurotransmitter activity, helps in the cognitive development that is used in many different therapeutic approaches; where some of their combinations aid the best therapeutic approaches for AD, one such is the implication of phototherapeutic compounds. The compounds, like azotacrine, are molecules able to switch shape with exposure to blue and UV light, thereby changing the AChE activity, contributing to the reduction of side effects like hepatotoxicity and ensuring the precise controlling of the enzyme activity. The use of photosensitizers, which comprise the light-activated compounds (such as thioflavin-T derivatives, quinolinium derivatives, curcumin derivatives, boron-dipyrromethene, N- or O-heterocyclic compounds, porphyrins, fullerenes and their derivatives, fused azobenzene–boron complexes, and other transition metal

D. S. Kathiresan
Department of Human Genetics and Molecular Biology, Bharathiar University, Coimbatore, India

V. M. Puthamohan (✉)
Department of Human Genetics and Molecular Biology, Bharathiar University, Coimbatore, India
e-mail: pvmhgmb@buc.edu.in

N. Arun Nagendran
Department of Zoology, Thiagarajar College, Madurai, India

H. Dilnashin, S. P. Singh (eds.), *Phototherapeutic Approaches to Neurodegeneration*,
https://doi.org/10.1007/978-981-95-4751-7_4

compounds), which have the ability to generate singlet oxygen on exposure to light, are now implied in the targeting of Aβ plaques, a promising strategy for AD treatment. Many nanotechnological approaches also contribute to the better phototherapeutic compound discovery, like the polyoxometalates and mesoporous silica nanoparticles coated with a copolymer like poly(N-isopropylacrylamide-co-acrylamide), which are thermally stable, chemically highly porous with biocompatibility and bioavailability, and act as a photothermal agent, making the compound stand out for AD treatment with a phototherapic approach. Some other nanoparticle phototherapy compounds include carbon dots, polymeric nanoparticles, lanthanide-doped upconversion nanoparticles; 2D nanomaterials like graphene oxide, black phosphorus, and carbon nitride aid in AD treatment. Another great improved technique of the phototherapic approach is the photobiomodulation, which uses the low-intensity red light or near-infrared light; it is found to support neuronal health by increasing the ATP production and reducing the toxin production and inflammation rates and is thus used in the treatment of AD, PD, and also some kinds of traumatic injuries. It also suppresses the PERK/eIF2α pathway by triggering the release of exosomes from microglia carrying miR-7670-3p. Overall, this chapter focuses on these innovative phototherapeutic strategies that offer a promising and multifaceted approach for the effective, targeted, and noninvasive treatment of AD, paving the way for next-generation neuroprotective therapies.

Keywords

Neurodegenerative disease · Alzheimer's disease · Phototherapy · Photosensitizers · Targeted treatment

4.1 Introduction

The condition of decrease in intellectual ability is given the term "dementia," where the most prevalent dementia in the developing world is the chronic neurodegenerative disease Alzheimer's disease (AD). The pathophysiological development of AD is through the neuroinflammation, which was first indicated by Oskar Fischer in 1907, which in later periods of research discussed the inflammatory responses nearing the areas of senile plaques by identifying their responsible proteins, astrocytes, microglias, cytokines, and other components contributing to the AD development and prognosis (Heneka et al. 2025; Fischer 1907). AD affecting rate contributes to impacting more than 50 million people worldwide, which is also estimated to increase threefold by 2050 with respect to the higher aging population (Mantzavinos and Alexiou 2017; Zhang et al. 2025a, b). The two forms of AD are the autosomal dominant and the sporadic form, and the sporadic form is found to increase in its disease condition by doubling every five years (Alzheimer's Association 2018). Thirty million people in the world suffer from AD (Ma et al. 2018), with complications of abnormally accumulated amyloid beta plaques (Aβ) around the nerve cells and the high phosphorylated tau protein forming the neurofibrillary tangles (NFT) (Thal et al. 2002; Braak and Braak 1991). The drug treatments available today that are approved are administered through either oral route or transdermal route (Atri 2019; Briggs et al. 2016) belong to either acetylcholine esterase (AChE) inhibitors or anti-glutaminergics (Hsu et al. 2025; Atta et al. 2025), are memantine, galantamine, donepezil, and rivastigmine (Zeng et al. 2025; Kola et al. 2025; Zueva et al. 2025; Piñol-Ripoll and Salas Carrillo 2025). A neurotransmitter responsible for the process of learning and memory, glutamate, when in higher levels, cause the death of neurons; wherein the disease progression, the drugs administrated must encounter the maintaining of the glutamate levels in order to control the functions of cognition and are thus used in the treatment methods of AD (Zhang et al. 2025a, b; El Asmar et al. 2025). Such kinds of drug therapies are not very effective as they cannot address the issues of the diseases when they are in the asymptomatic stages or the early stages (Wu and Fuh 2025; Sperling et al. 2012). They also have another reason for noneffective drug therapy, as they face difficulty in targeting the brain parts accurately as they encounter challenges in crossing the blood–brain barrier (BBB) due to the permeability issues (Donahue and Johanson 2008; Sweeney et al. 2018; Cai et al. 2018).

The alternative treatment offered to AD patients on a nonpharmacological approach is through the maintaining of the cognitive functions and monitoring of their intellectual ability through many therapies and nutritional monitoring and personal care (Rosenberg et al. 2020; Lehtisalo et al. 2019; Raz and Rodrigue 2006), but they are found to provide only shorter periods of disease monitoring rather than a treatment for AD. The human brain changes structurally and functionally around the age of 45 years in terms of the integrity of white matter of the brain and the brain volumes, where the fluidity of brain membranes reduces and the composition of lipids changes in the brain (Colin et al. 2017; Koppelmans et al. 2022; Ju and Tam 2022; Bishop et al. 2010). The phototherapy and photomedicine are emerging fields in the treatment of many diseases, where they use light generated from physical light or light-emitting diode (LED) bulbs (Chen et al. 2023; Wang et al. 2017; Chopra et al. 2025). Nowadays, they are being replaced with molecular lights, which have the self-illuminated molecules following the Cerenkov luminescence (Ciarrocchi and Belcari 2017; Mitchell et al. 2011; Thorek et al. 2014), where they are used in producing the reactive oxygen species (ROS) through photodynamic therapy by ^{18}F-fluorodeoxyglucose-positive emission tomography (^{18}F-FDG PET) (Ottoy et al. 2019; Butt and Bach 2025; Hirakawa et al. 2025) and other phototherapy technique advancements, which will be discussed in this chapter.

4.2 Molecular Pathogenesis of AD

The molecular basis of AD is the accumulation of different types of filamentous Aβ, which marks the pathogenesis of the disease itself, which are the NFT and the neuritic plaques. Another major factor for AD is the deficit in the neurotransmitter, AChE, which has a direct action on the cognitive and memory loss in patients

(Talesa 2001; Lane et al. 2004; Di Lazzaro et al. 2005; Danta et al. 2025). The early onset AD, or the familial AD, is an usual form from the mutations of genes like amyloid precursor protein (*APP*), presenilin 1 (*PSEN1*), and presenilin 2 (*PSEN2),* that causes overproduction of Aβ (Marsh et al. 2025; Najar et al. 2025; Al-kuraishy et al. 2025; Cruchaga et al. 2012). The tau protein, consisting of four primary domains with N-terminal region, proline-rich domain, microtubule-binding region, and the C-terminal region (Kolarova et al. 2012; Goedert 1993; Hong et al. 2025). These are the ones responsible for the production of Aβ sheet formations (Padilla-Godínez et al. 2025; Marien et al. 2025; El Mammeri et al. 2025; Almeida et al. 2025). Their isoform hTau40, with 441 amino acids, produces the tau filaments, which are the major building blocks of the Aβ sheets (Landis and Kunze 2025; Brockhoff et al. 2025). The cholinergic neurons are the base for AD, which reduces, and the reduction in the cortical regions marks the arise of the cholinergic hypothesis of AD, which in turn lead to the development of treatment strategies based on the synaptic, presynaptic, and post synaptic categories (Francis et al. 1999; Craig et al. 2011; Anitha et al. 2025; Scarano et al. 2025). The glial cells and cholinergic neurons are involved in the synthesis of AChE and ACh, which are mostly expressed at the vascular endothelial cells, maintaining the balance of the cholinergic system through the endothelium-mediated process. They change the permeability of the BBB and make provisions for the choline to be transported across and maintain the vascular function (Carpentier et al. 1990; Tomkins et al. 2001; Bashir et al. 2025). The inhibition of them makes alterations in the energy metabolism and leads to the formation of amyloid plaques, neurofibrillary tangles, and also a chance for neurodegeneration in AD patients (Pedder et al. 2025; Alo and Lamers 2025; Karosas et al. 2025).

The disease itself shows many evidences of diversifying immune responses that were evidenced from many postmortem sample investigations of immunohistochemistry and biochemistry procedures in AD brains of patients, which showed abnormal expression of many factors like cytokines, chemokines, integrins, cyclooxygenases, and some of the acute phase proteins (Sowa et al. 2025; Akiyama 1994; Shaheen et al. 2025; Lu et al. 2025; Ghanbarian et al. 2025). There are many studies, including the genome-wide association studies focusing on AD, which put into light the genes of the immune response and inflammation, like clusterin (*CLU*), *CD33* (sialic acid-binding immunoglobulin-like lectin 3), complement receptor 1 (*CR1*), and *TREM2* (triggering receptor expressed on myeloid cells 2) which were found to increase the AD pathogenesis (Lish et al. 2025; Zhao et al. 2025; Vo et al. 2025; Dodd et al. 2025; Yang et al. 2025; Yao et al. 2025). The deacetylating, NAD+-dependent protein family, sirtuins, possesses beneficial qualities for protecting against the neurodegenerative diseases like AD, where the ortholog SIRT1 influences the neuron survival, plasticity of the synaptic junctions, development of the cognitive functions, and neurogenesis process by acetylating the homeostasis of the necessary proteins, which is majorly expressed at the neurons at the hippocampus (Guarente 2006; AlHayani et al. 2025; Zou et al. 2025; Apiraksattayakul et al. 2025). Another major molecular pathology of AD is the oxidative stress, which is the combination of oxidation promoters and the antioxidants that are missing the balance, where in the case of AD the brain cells require more amounts of glucose and oxygen for their normal functioning and often face the oxidative damage (Bellanti et al. 2025; Jones 2006; Floyd and Carney 1992), thereby affecting the CNS (Manton et al. 2004; Goldsteins et al. 2022; Praticò and Delanty 2000). The increased mitochondrial damage is caused by the fluctuations in the antioxidant enzymes by increasing the ROS production (Kathiresan et al. 2025; Reddy et al. 2012; Dumont et al. 2010; Angelova and Abramov 2018), which often occurs during the early AD stages and thereby increases the accumulation process of the Aβ (Kathiresan et al. 2025; Cardoso et al. 2004; Spina et al. 2025).

4.3 Phototherapy and Molecular Level Activation

The photobiomodulation therapy has the ability to activate mitochondria, which influences the mitochondrial complexes I-IV of the electron transport chain and influences the proton gradient and ATP that can defferentially absorb the blue, yellow, and IR light (Zein et al. 2018; Lunova et al. 2020; Hu et al. 2021; Zielke 2014); it also reduces oxidative stress, which is formed by the production of superoxide by the mitochondrial damage process (Ricci et al. 2008; Brand et al. 2004). PT also aids in the process of improving the factors of anti inflammation, as the PT is proved to significantly supress mRNA expression of the pro inflammatory cytokines like interleukins IL-1β, TNF-α, IL-6, and IL-8 (El-Askary et al. 2018; Bogaczewicz et al. 2016; Li et al. 2025); it also improves oxygen saturation at brain; enhances neurogenesis, as PT has shown to improve the neural growth and also aid in upregulating the neurotrophic factors that are derived from brain by the exposure of red light with 670 nm (Valter et al. 2005); activates the NO signaling pathway by helping the busy to use the NO and enhance the permeability of the lymphatic endothelium, which helps in neuronal function through vasodilation (Liu et al. 2008; Samoilova et al. 2008). PT also influences the ion permeability by changing the signal transduction of the Ca ions, which is the major factor for neurotransmitter release (Yadav et al. 2012; Shahriarpanah et al. 2018), as they activate various cells with properties of photoreception and thus influence the release of glutamine, which is responsible for the modulation of GABAergic neurons, cholinergic neurons, and the monomeric systems like dopamine, serotonin, and norepinephrine (Liu et al. 2024; Slominski et al. 2024).

4.4 Phototherapy and AD

The development of the contemporary world uses phototherapy in almost all the diseases, even in the treatment of neurodegenerative disorders, but it started primarily with treating skin disorders. The seeds of phototherapy started with Niels Ryberg Finsen in 1903, gaining him the Nobel Prize in Physiology or Medicine, which was through his experiments on utilizing the radiation of light in managing the lupus (Ding et al. 2024; Allison 2014; Xie et al. 2025), which is developing more diverse ways through the present modern world through photodynamic therapy and combinational therapy wherein the conventional, nonconventional, and novel drugs are used, even employing the nanotechnology in drug discovery, and also through different techniques including ultrasound and precision medicine (Hau Gan et al. 2025; Öztürk et al. 2025; Zhang et al. 2025a, b). Many invasive procedures have adverse effects and induce discomforting feelings in the patients undergoing treatment, whereas phototherapy, the noninvasive technique, uses the exposure of light artificially at specific intensities or wavelengths and is also found to do a promising treatment for various age-related disorders such as AD and PD, also employed in treating acute brain injuries and also psychiatric diseases (Semyachkina-Glushkovskaya et al. 2023; Hamblin and Salehpour 2021; Morries et al. 2015; Bell et al. 2021). The neurological disorders are managed through various phototherapy modes like photothermal therapy, photodynamic therapy, ultraviolet therapy, laser therapy, and several other modes of light therapies (Zeng et al. 2021; Hamblin and Salehpour 2021; Pronin et al. 2018; Cote et al. 2013; Zhou et al. 2024; Gong et al. 2021).

The phototherapy follows different wavelength for different medical applications, where there are a wide range of light spectra, ultra violet light (UV) with 100–400 nm, with subtypes UV-A, UV-B and UV-C, where UV-B and C (Fig. 4.1) (Zhou et al. 2024; Ahirwar et al. 2022; Weichenthal and Schwarz 2005; Slominski et al. 2018) causes changes in the DNA on its exposure (100–315 nm approx); blue light (400–500 nm), which is employed in the treatment of jaundice at the skin exposure (Duteil et al. 2014; Uzunbajakava et al. 2023; Buhl et al. 2023; Kim et al. 2021; de Sousa et al. 2010); green light (500–595 nm), having its application in treating emotional conditions and also studies found them applied in epidermal cell proliferation (Nowak 2008; Wang et al. 2024; Bohnsack and Hirschi 2004); red light (600–740 nm) and infrared light (IR, 740–1100 nm), found to penetrate into deep tissues thus applied in the neurological disorders as it has the ability to cross the cranial region (Umino and Denda 2023; Padalkar and Pleshko 2015; Tsai and Hamblin 2017). Due to the unique nature of the structure of the human brain, the studies from other animal studies cannot be applied directly for human therapeutic approaches; the transcranial surface marks the skull with its varying degrees of translucency, where the temporal, frontal, and the occipital regions of skull have different penetration depths (Hennessy and Hamblin 2016; Montazeri et al. 2021). The near-infrared (NIR) rays are used in phototherapy, as they have the ability to cross through the parietal and frontal regions of the skull, thus exciting the cortex of the brain (Jagdeo et al. 2012; Hong et al. 2014; Guo et al. 2025).

The neuroprotection ability was checked after the phototherapy, which involved the photobiomodulation therapy, and was found to have enhanced the flowing of blood and the oxygen content through the mechanisms of anti-inflammatory pathways and antioxidant pathways, showing the metabolism of the cerebral regions (Abijo et al. 2023; Sunemi et al. 2021). Having a look into the molecular aspects of proteins and their photosensitive nature, they are into many types of visible light and applied in many human disease diagnoses (Fig. 4.2). The inner cellular components of the body comprise many components that are photosensitive; namely, the

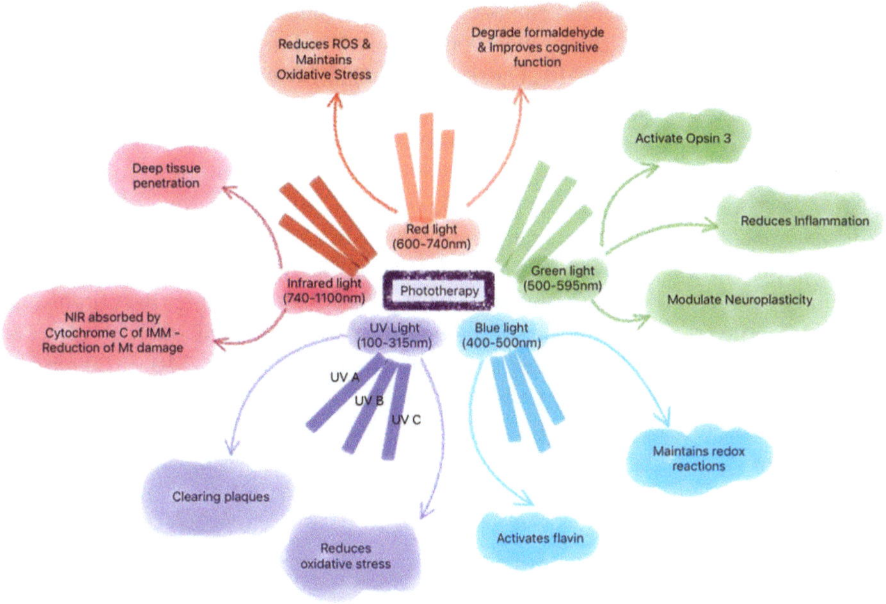

Fig. 4.1 The illustration showing different phototherapy compounds of different light wave intensities, and its therapeutic approach for Alzhiemer disease management. IMM-Inner Mitochondrial Membrane; Mt-Mitochondrial; NIR-Near infra red; ROS-Reactive Oxygen Species; UV-Ultraviolet

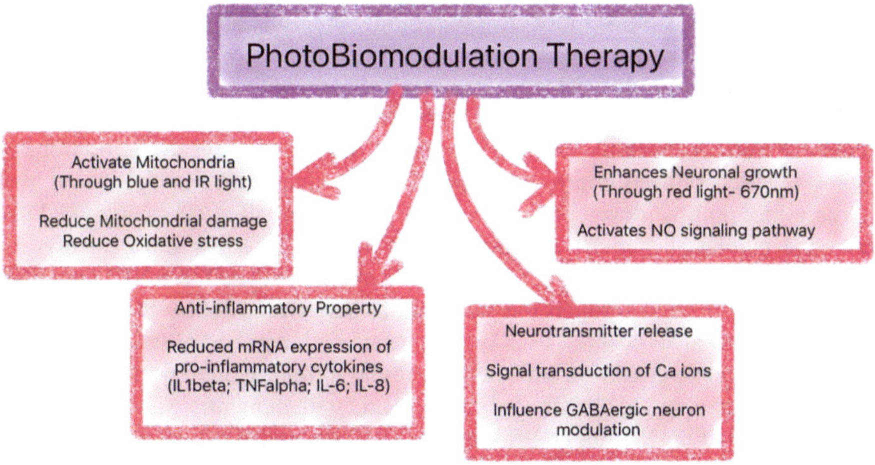

Fig. 4.2 The diagram showing the different properties of photobiomodulation therapy

inner mitochondrial membrane contains the cytochrome C, which can be sensitized through the action of NIR at 760–900 nm and red light with a spectrum of 600–700 nm (Mason et al. 2014; Quirk and Whelan 2016; Soltani et al. 2025); the red light also degrades the formaldehyde by the formaldehyde dehydrogenase enzyme activation and is also found to improve cognitive functions (Gao and Han 2025; Zhou et al. 2025a, b; Zhang et al. 2019a, b). The green light from the visible spectra has the ability to activate opsin 3, which is the protein found in the retina cells that are activated by the connections with opsin and the retinol compounds that will be expressed in the brain (Regazzetti et al. 2018; Musilova and Cortesi 2023; Barreto Ortiz et al. 2018). The blue light with 400–500 nm activates the photosensitive protein, flavin, which takes part in the redox reactions (Losi 2007; Bouly et al. 2007; Lukacs et al. 2022). The therapeutic agents use NIR with 740–1100 nm signals by the cationic channels by activating TRPV1, which is used in the brain disorders (Yoo et al. 2020; Gu et al. 2012; Choi et al. 2025). The visible light with 600–740 nm right light activates the catalase, which in turn clears the ROS and helps in maintaining the oxidative stress (Liebel et al. 2012; Lavi et al. 2010).

4.5 Animal and Human Studies

There are many notable research findings in animal and human AD models and patients using the phototherapy at either transcranial, transocular, or intransal carpal canal with varying therapy times and phototherapy parameters according to the type of phototherapy applied to the models appropriately, which included the different therapeutical effects on AD models. Photobiomodulation therapy carried out using visible light on APP/PS1 mice had showed to reduce the size of senile plaques and also contributed to improved memory with a 5-day therapy duration (Buendía et al. 2022). Using laser as the mode of phototherapy, the animal models of AD and dementia were exposed to 400–630 nm for 14 days to 60 days at the transcranial location, which showed improvement in the microenvironment and also lead to an increase in cognitive development (Wu et al. 2022) in 3xTg-AD mice model systems and increased memory and behavior in TgF344 AD rats and dementia rats (Yang et al. 2022; Sutalangka et al. 2013). In patients with AD, the NIR phototherapy was administered through the transcranial part with 1060–1080 nm NIR for 8 weeks, having an exposure of 1.5 W 6 min/time with a period of twice/day, which showed to improve cognition and was also found to reduce anxiety (Nizamutdinov et al. 2022; Pan et al. 2023); similarly, in patients with dementia, studies were conducted on including NIR and tNIR with 635–810 nm and 1060–1080 nm, showing the improvement of cognition and quality of life (Qi et al. 2021; Salehpour et al. 2019; Nouchi et al. 2021).

4.6 Photosensitizers

The photodynamic therapy, which employs the spatial and temporal selectivity of certain compounds, makes the approach to better precision medicine. One such is the use of photosensitizers in the treatment of AD. The small molecules having the differential properties according to their spatial alignment are differentiated into major categories like derivatives of fullerene, porphyrin, heterocyclic compounds with N or O, and the fluoroboron compounds (Li et al. 2019; Xiong et al. 2024). The increased electron affinity, higher surface volume ratio, and the ability to produce the higher $^1O_2$ production make fullerene and its derivatives to be the best photosensitizers acting as the photodynamic therapy (Hou et al. 2017; Du et al. 2018; Gao et al. 2025; Hou et al. 2022). Fullerenes are found to break down the peptides of Aβ under light exposure (Omar Khudhur et al. 2024; Mi et al. 2024). It is also found to produce ROS, which breaks down the monomers and oligomers of Aβ, thus aiding the effective treatment strategies for AD (Jiang et al. 2023; Anwar et al. 2024). Porphyrins, a smaller molecule of photosensitizers is found to yield high quantum and low energy Q bands, which are also found to inhibit the formation of aggregation of Aβ, which was even more efficiently done with the exposure of photoirradiation (Lee et al. 2015a, b). The compound of porphyrin, meso-tetra(4-sulfonatophenyl) porphyrin (TPPS) is found to regulate the Aβ assembly into fibril formation by the exposure of blue LED, with the formation of ROS, which is also studied in Drosophila with neurodegenerative complications and found to be reduced partially through TPPS (Masuda et al. 2006). Another compound, chlorin e6 (Ce6), with 20 π electrons and 3 pyrroles on the core part, shows high affinity to Aβ40, which reacts to type II photo-oxidation, which reduces the Aβ aggregation and toxicity (Leshem et al. 2019). The investigations on Ce6 found to have the well-formed fibrils during the dark phase and found to have mixed aggregate morphologies with shorter lengths during exposure to photoradiation, thus contributing to anti-amyloidogenic activity (Hirabayashi et al. 2014).

Another set of compounds for photodynamic therapy, which uses small molecules like thioflavin T (ThT), riboflavin, rose bengal, and methylene blue, with reduced side effects and quantum effects to be threefold higher (Abrahamse and Hamblin 2016). It is found that ThT is capable of producing $^1O_2$ through light exposure of 442 nm, whereby the aggregation of Aβ fibers is destroyed (Ozawa et al. 2009). Like the Aβ histidine groups, H6, H13, and H14, and the crosslinking of Aβ was induced through production of $^1O_2$ and was confirmed through a study using HSQC NMR spectroscopy (Taniguchi et al. 2014; Nakamura et al. 2007). On observation of fibrils of beta2 microglobulin through radiation with laser, the immediate formation of K3 fibrils was observed and also was found to stop growing after some time (Ozawa et al. 2009; Ng et al. 2025; Jiang et al. 2023), which proves to be a better treatment option for treating the amyloid plaque formation during the progression of AD, and also it would serve as the preventive method for AD progression. The photosensitizer using rose bengal was excited at a 525 nm targeting Aβ and found to inhibit the changing of Aβ monomers to β sheets, which is the clinical indication of reduction of progression of AD (Lee et al. 2015a, b). Other photosensitizers, like 1,2,4 oxadiazole targeting Aβ40 at 260 nm, interacted with Aβ monomer and formed a nontoxic aggregate, destroying the accumulation of plaques in AD (Mangione et al. 2015); the methyl blue and CRANAD targeting Aβ42 exposed to wavelengths of 630 and 780 nm in Drosophila and mouse models were found methyl blue excited and destroy the Aβ, thereby reducing the cytotoxicity (Lee et al. 2017), and showed selective degradation under light radiation in mouse models (Ni et al. 2018).

4.7 Nanotechnology-Based Phototherapy for AD

The field of nanotechnology makes its contribution in the processing of treatment models for AD by combining with the phototherapeutic compounds and creating a model that would carry the nanoparticle (NP), which will be thermally stable and chemically porous with properties such as biocompatibility, bioavailability, and will also act as a photothermal agent, which would contribute to being the best AD treatment through a phototherapeutic approach (Park et al. 2020; Seaberg et al. 2021). Some of the developing trends of nanotechnological approach is the hybrid materials, where the metal organic frame work (MOF), which is the crystalline organic frame of metal nodes and them bearing the ligands with organic nature, is an active photoactive compound (Furukawa et al. 2013). The porphyrinic Zr MOF NPs were found to inhibit the Aβ aggregations (Yang et al. 2020), which had the action of TCPP porphyrin ligands, which are separated by the Zr cluster that allows fluorescence at 650 nm light irradiation, giving $^1O_2$ yield, contributing to reducing the Aβ aggregation (Wang et al. 2018). Porphyrin's presence of nitrogen atoms binds with the Cu^{2+} ions that are responsible for decreasing the neurotoxicity (Chen and Jiang 2016) and also aids in the Ab of photooxidant by action of Cu^{2+} chelator, which leads to the development of many MOFs with differences in the metal components, like Zr MOF, Al MOF, Ni MOF, and Hf MOF (Wang et al. 2018), among which Hf MOF was found to be the most efficient MOF aiding the $1O_2$ yield (Yu et al. 2019) but a similar Cu^{2+} removal ability. This was combined with LPFFD and checked on the transgenic AD models and PC12 cell lines, showing the photodynamic oxidation process to degrade the Aβ aggregation (Xu et al. 2022). The polyoxometalates also contributed to the clearance of Aβ aggregation, which was investigated through the phosphotungstate, $K_8[P_2CoW_{17}O_{61}]$, structured by Wells-Dawson, that bound with the antibody through the electrostatic interactions, which were crucial for suppressing the amyloid (Gao et al. 2014). The photoactivated system with either UV light penetration or the near-infrared penetration with a range of 650–900 nm would allow deep penetration into the biological tissues effectively (Li et al. 2013).

There are also the metal nanoparticles, which were employed in the AD treatment through phototherapy, through the combination of concepts of photodynamic oxidation and the chiral recognition to modulate the aggregation of Aβ. Some of the NPs are chiral L/DFexCuySe using the penicillamine with D or L type configuration that has the ability to break down Aβ under light (Zhang et al. 2020; Kuk et al. 2017). The D-FexCuySe NPs that are photoactivated reduced the neuronal loss in the live neurons by reducing the Aβ42 adhesion, and it was also found to increase neurotoxicity under the influence of NIR light, and the restoration of memory functions was experimented with AD mice (Zhang et al. 2020; Xu et al. 2022). Another type of NPs is the upconversion NPs, which could convert the NIR light to visible and UV light and have the ability to activate the photoresponsive fluorophores through NIR (Chen et al. 2014). They have the activity of producing the ROS, making the Aβ denaturation, reducing cytotoxicity (Chen and Wang 2020), and biocompatibility even during higher concentrations (Wen et al. 2018). The upconversion NPs in their hexagonal phase were found to dissociate Aβ aggregates in the NIR, which confirms their potential photodynamic property owing to AD therapy (Zhang et al. 2019a, b). There are a wide range of 2D nanomaterials, comprising hybrid materials like black phosphorus, carbon nitride, and graphene oxide (Xie et al. 2021), which have an effective property of generation of ROS by light induction of electrons and hole separation, which also possess the variable energy gap, changeable optical properties, and simple architecture, which allows the preparations and functionalization to be made precise and as per requirements of therapy approaches of AD (Xie et al. 2021; Wang et al. 2019). The self-degradable, widely used 2D-layer nanomaterials have the biophosphorus, which is nontoxic and has unique drug delivery potential and aids in the AD treatment with phototherapy combinations (Yang et al. 2020). The other nanomaterials such as nanosheets of black phosphorous and their quantum dots have the ability to regulate the aggregation of Aβ(1–40) by the absorption of the monomers (Lim et al. 2019). Another such 2D nanosheet is the graphitic carbon nitride nanosheet, which suppresses the beta amyloidosis through exposure to over visible light, with additional advantages like their ultrathin nature, high surface volume ratio, and nontoxic (Li et al. 2016; Wang et al. 2020). They do not have effects when in the dark, but the addition of metal compounds (mostly iron) enhances the ability of ROS production, thereby treating the removal of plaques (Wang et al. 2019), with higher potency during exposure to visible light, owing to AD treatment.

In addition to the phototherapy, the alternate, self-luminescence system is used to replace the light source, which can be used in the action of self-illuminating systems, such as chemiluminescence, bioluminescence, and Cerenkov luminescence, which produces ROS and is used in the deep tissue penetration ailing to the AD treatment (Weickenmeier et al. 2018). Some of the chemiluminescence compounds that have the self-luminescence properties are luminol, methyl acrylate dioxetane, and dicyanomethylchromone dioxetane; the bioluminescence compounds are D-luciferon, cyclic, Akalunine, CTZ, and DTZ (Betthauser et al. 2020; Schöll and Maass 2020). Some of the invasive techniques of BBB permeability are direct intracerebral injection, osmotic opening of BBB, chemical opening of BBB, and disruption of BBB through focused ultrasound; the noninvasive techniques are drug lipidization, functionalization with receptor targeting, nanoparticle-mediated drug delivery, chimeric peptides, monoclonal antibody fusion proteins, inhibition of efflux transporters, and intranasal delivery (Neumann et al. 2006; Xu et al. 2022).

Overall, the progress in designing photosensitive compounds, nanomaterials, and self-luminescent systems highlights how multidisciplinary innovations are reshaping phototherapy for AD. These approaches demonstrate the ability to selectively target amyloid plaques, reduce oxidative stress, and enhance neuronal protection, while advanced nanocarriers and BBB modulation strategies ensure improved delivery and therapeutic precision. By integrating chemistry, nanotechnology, and neuroscience, phototherapeutic platforms are evolving from proof-of-concept studies toward clinically relevant strategies.

4.8 Conclusion

Phototherapeutic strategies represent an innovative and promising frontier in the treatment of Alzheimer's disease. By integrating photoswitchable molecules, light-activated photosensitizers, advanced nanomaterials, and photobiomodulation, these approaches enable precise, noninvasive, and multifunctional targeting of core pathological mechanisms, such as amyloid aggregation, oxidative stress, and neuroinflammation. While still emerging, their ability to combine molecular specificity with controllable light activation holds significant potential to overcome the limitations of current therapies. Continued research into optimizing delivery platforms, light parameters, and long-term safety will be critical to translate these strategies into clinically effective, next-generation neuroprotective treatments for AD.

References

Abijo A, Lee CY, Huang CY, Ho PC, Tsai KJ (2023) The beneficial role of photobiomodulation in neurodegenerative diseases. Biomedicine 11(7):1828

Abrahamse H, Hamblin MR (2016) New photosensitizers for photodynamic therapy. Biochem J 473(4):347–364

Ahirwar D, Khurana D, Chowdhury SR (2022) Measurement of near infrared spectroscopy based biomarkers under in-vitro ischemic stroke condition. In: 2022 44th Annual International Conference of the IEEE Engineering in Medicine & Biology Society (EMBC). IEEE, pp 4366–4369

Akiyama H (1994) Inflammatory response in Alzheimer's disease. Tohoku J Exp Med 174(3):295–303

AlHayani DA, Kubaev A, Uthirapathy S, Mandaliya V, Ballal S, Kalia R et al (2025) Insights into the therapeutic potential of SIRT1-modifying compounds for Alzheimer's disease: A focus on molecular mechanisms. J Mol Neurosci 75(1):29

Al-kuraishy HM, Sulaiman GM, Mohammed HA, Dawood RA, Albuhadily AK, Al-Gareeb AI et al (2025) Alterations in the processing of platelet APP (amyloid beta precursor protein) in Alzheimer disease: the possible nexus. Neuropsychopharmacol Rep 45(1):e12525

Allison RR (2014) Photodynamic therapy: oncologic horizons. Future Oncol 10(1):123–124

Almeida ZL, Vaz DC, Brito RMM (2025) Morphological and molecular profiling of amyloid-β species in Alzheimer's pathogenesis. Mol Neurobiol **62**:4391–4419. https://doi.org/10.1007/s12035-024-04543-4

Alo B, Lamers C (2025) Crossing barriers: advancements in macromolecular therapeutics for neurodegenerative diseases and strategies to overcome the blood–brain barrier. ACS Pharmacol Transl Sci

Alzheimer's Association (2018) Alzheimer's disease facts and figures. Alzheimers Dement 2018(14):367–429

Angelova PR, Abramov AY (2018) Role of mitochondrial ROS in the brain: from physiology to neurodegeneration. FEBS Lett 592(5):692–702

Anitha K, Singh MK, Kohat K, Chenchula S, Padmavathi R, Amerneni LS et al (2025) Recent insights into the neurobiology of Alzheimer's disease and advanced treatment strategies. Mol Neurobiol 62(2):2314–2332

Anwar G, Cao Y, Shi WJ, Niu L, Yan JW (2024) Rational design, synthesis, and evaluation of Rofecoxib-based photosensitizers for the NIR imaging and photo-oxidization of aβ aggregates. ACS Chem Neurosci 15(22):4202–4209

Apiraksattayakul S, Pingaew R, Prachayasittikul V, Ruankham W, Tantimongcolwat T, Prachayasittikul V et al (2025) Neuroprotective potential of aminonaphthoquinone derivatives against amyloid beta-induced neuronal cell death through modulation of SIRT1 and bace1. Neurochem Res 50(1):50

Atri, A. (2019). Current and future treatments in Alzheimer's disease. In Seminars in neurology (Vol. 39, 02, pp. 227–240). Thieme Medical Publishers

Atta AM, Elkalla WS, Omran SA, Elgawish MS (2025) Innovations in acetylcholinesterase inhibitors: emerging strategies for Alzheimer's disease therapy (2019–2024). Egypt Pharma J 24(2):152–170

Barreto Ortiz S, Hori D, Nomura Y, Yun X, Jiang H, Yong H et al (2018) Opsin 3 and 4 mediate light-induced pulmonary vasorelaxation that is potentiated by G protein-coupled receptor kinase 2 inhibition. Am J Phys Lung Cell Mol Phys 314(1):L93–L106

Bashir DJ, Bhat KA, Bashir M (2025) Deciphering Alzheimer's disease: molecular mechanisms, preclinical models and strategies to overcome Blood-Brain-Barrier. Brain Res:149828

Bell KR, Fogelberg D, Barber J, Nakase-Richardson R, Zumsteg JM, Dubiel R et al (2021) The effect of phototherapy on sleep during acute rehabilitation after traumatic brain injury: a randomized controlled trial. Brain Inj 35(2):180–188

Bellanti F, Coda ARD, Trecca MI, Buglio AL, Serviddio G, Vendemiale G (2025) Redox imbalance in inflammation: the interplay of oxidative and reductive stress. Antioxidants 14(6):656

Betthauser TJ, Koscik RL, Jonaitis EM, Allison SL, Cody KA, Erickson CM et al (2020) Amyloid and tau imaging biomarkers explain cognitive decline from late middle-age. Brain 143(1):320–335

Bishop NA, Lu T, Yankner BA (2010) Neural mechanisms of ageing and cognitive decline. Nature 464(7288):529–535

Bogaczewicz J, Malinowska K, Sysa-Jedrzejowska A, Wozniacka A (2016) Medium dose ultraviolet A1 phototherapy and mRNA expression of interleukin 8, interferon γ, and chemokine receptor 4 in acute skin lesions in atopic dermatitis. Adv Dermatol Allergol 33(3):170–175

Bohnsack BL, Hirschi KK (2004) Red light, green light: signals that control endothelial cell proliferation during embryonic vascular development. Cell Cycle 3(12):1506–1511

Bouly JP, Schleicher E, Dionisio-Sese M, Vandenbussche F, Van Der Straeten D, Bakrim N et al (2007) Cryptochrome blue light photoreceptors are activated through interconversion of flavin redox states. J Biol Chem 282(13):9383–9391

Braak H, Braak E (1991) Neuropathological stageing of Alzheimer-related changes. Acta Neuropathol 82(4):239–259

Brand MD, Buckingham JA, Esteves TC, Green K, Lambert AJ, Miwa S et al (2004) Mitochondrial superoxide and aging: uncoupling-protein activity and superoxide production. In: Biochemical Society Symposia, vol 71. Portland Press Limited, pp 203–213

Briggs R, Kennelly SP, O'Neill D (2016) Drug treatments in Alzheimer's disease. Clin Med 16(3):247–253

Brockhoff M, Träuble J, Middya S, Fuchsberger T, Fernandez-Villegas A, Stephens A et al (2025) PseudoSorter: A self-supervised spike sorting approach applied to reveal tau-induced reductions in neuronal activity. Sci Adv 11(11):eadr4155

Buendía D, Guncay T, Oyanedel M, Lemus M, Weinstein A, Ardiles ÁO et al (2022) The transcranial light therapy improves synaptic plasticity in the Alzheimer's disease mouse model. Brain Sci 12(10):1272

Buhl T, Santibanez Santana M, Forkel S, Kromer C, Seidel J, Möbs C et al (2023) Full-body blue light irradiation as treatment for atopic dermatitis: a randomized sham-controlled clinical trial (AD-Blue). J Dtsch Dermatol Ges 21(12):1500–1510

Butt A, Bach H (2025) Advancements in nanotechnology for diagnostics: a literature review, part II: advanced techniques in nuclear and optical imaging. Nanomedicine 20(2):183–206

Cai Z, Qiao PF, Wan CQ, Cai M, Zhou NK, Li Q (2018) Role of blood-brain barrier in Alzheimer's disease. J Alzheimer's Dis 63(4):1223–1234

Cardoso SM, Santana I, Swerdlow RH, Oliveira CR (2004) Mitochondria dysfunction of Alzheimer's disease cybrids enhances Aβ toxicity. J Neurochem 89(6):1417–1426

Carpentier P, Delamanche IS, Le Bert M et al (1990) Seizure-related opening of the blood-brain barrier induced by soman: possible correlation with the acute neuropathology observed in poisoned rats. Neurotoxicology 11:493–508

Chen YZ, Jiang HL (2016) Porphyrinic metal–organic framework catalyzed heck-reaction: fluorescence "turn-on" sensing of Cu (II) ion. Chem Mater 28(18):6698–6704

Chen B, Wang F (2020) Emerging frontiers of upconversion nanoparticles. Trends Chem 2(5):427–439

Chen G, Qiu H, Prasad PN, Chen X (2014) Upconversion nanoparticles: design, nanochemistry, and applications in theranostics. Chem Rev 114(10):5161–5214

Chen YR, Huang WY, Lee TY, Chu H, Chiang KJ, Jen HJ et al (2023) Efficacy of blue LED phototherapy on sleep quality and behavioral and psychological symptoms of dementia: A double-blind randomized controlled trial. Gerontology 69(10):1175–1188

Choi S, Yu E, Park SH, Oh SW, Kwon K, Kim G et al (2025) Protective effect of melatonin against blue light-induced cell damage via the TRPV1–YAP pathway in cultured human epidermal keratinocytes. Biofactors 51(2):e70015

Chopra S, Morrissette K, Fan E, Hendren S, Pavlis M (2025) Photobiomodulation and photodynamic therapy using red LED light in dermatology: A narrative review. Bratislava Medl J:1–13

Ciarrocchi E, Belcari N (2017) Cerenkov luminescence imaging: physics principles and potential applications in biomedical sciences. EJNMMI Phys 4(1):14

Colin J, Thomas MH, Gregory-Pauron L, Pinçon A, Lanhers MC, Corbier C et al (2017) Maintenance of membrane organization in the aging mouse brain as the determining factor for preventing receptor dysfunction and for improving response to anti-Alzheimer treatments. Neurobiol Aging 54:84–93

Cote S, Rodoni L, Miceli E, Concellón A, Civello PM, Vicente AR (2013) Effect of radiation intensity on the outcome of postharvest UV-C treatments. Postharvest Biol Technol 83:83–89

Craig LA, Hong NS, McDonald RJ (2011) Revisiting the cholinergic hypothesis in the development of Alzheimer's disease. Neurosci Biobehav Rev 35(6):1397–1409

Cruchaga C, Chakraverty S, Mayo K, Vallania FL, Mitra RD, Faber K et al (2012) Rare variants in APP, PSEN1 and PSEN2 increase risk for AD in late-onset Alzheimer's disease families. PLoS One 7(2):e31039

Danta CC, Chaudhari P, Nefzi A (2025) Targeting acetylcholinesterase with oxytocin: a new avenue in Alzheimer's disease therapeutics. ACS Chem Neurosci 16(13):2336–2339

de Sousa APC, Santos JN, dos Reis Jr JA, Ramos TA, de Souza J, Cangussú MCT, Pinheiro AL (2010) Effect of LED phototherapy of three distinct wavelengths on fibroblasts on wound healing: a histological study in a rodent model. Photomed Laser Surg 28(4):547–552

Di Lazzaro V, Oliviero A, Pilato F, Saturno E, Dileone M, Marra C et al (2005) Neurophysiological predictors of long term response to AChE inhibitors in AD patients. J Neurol Neurosurg Psychiatry 76(8):1064–1069

Ding L, Gu Z, Chen H, Wang P, Song Y, Zhang X et al (2024) Phototherapy for age-related brain diseases: challenges, successes and future. Ageing Res Rev 94:102183

Dodd RB, Enomoto M, Zhou Y, Sato K, Zhang Y, Chen F et al (2025) CD33 and clusterin interact biophysically and genetically to modulate Alzheimer risk. bioRxiv:2025-07

Donahue JE, Johanson CE (2008) Apolipoprotein E, amyloid-β, and blood-brain barrier permeability in Alzheimer disease. J Neuropathol Exp Neurol 67(4):261–270

Du Z, Gao N, Wang X, Ren J, Qu X (2018) Near-infrared switchable fullerene-based synergy therapy for Alzheimer's disease. Small 14(33):1801852

Dumont M, Lin MT, Beal MF (2010) Mitochondria and antioxidant targeted therapeutic strategies for Alzheimer's disease. J Alzheimer's Dis 20(s2):S633–S643

Duteil L, Cardot-Leccia N, Queille-Roussel C, Maubert Y, Harmelin Y, Boukari F et al (2014) Differences in visible light-induced pigmentation according to wavelengths: a clinical and histological study in comparison with UVB exposure. Pigment Cell Melanoma Res 27(5):822–826

El Asmar K, El Ghoul T, Colle R, Assaf R, Martin S, Trabado S et al (2025) Blood glutamate levels in major depression with antidepressants and benzodiazepines: a 6-month prospective cohort study. Therapies

El Mammeri N, Duan P, Hong M (2025) Pseudo-phosphorylated tau forms paired helical filaments in the presence of high-curvature cholesterol-containing lipid membranes. J Am Chem Soc 147(3):2510–2522

El-Askary AHMAD, Mohamed SA, Mohamed MK, Ahmed AET (2018) Effect of neonatal phototherapy on serum levels of interleukin 1, 6, 8, TNFa and white blood cell count. Med J Cairo Univ 86(December):3767–3772

Fischer P (1907) Miliary necrosis with glandular growths of neurofibrils, a regular change in the cerebral cortex in senile dementia. Month J Psychiatry Neurol 22(4):361–372

Floyd RA, Carney JM (1992) Free radical damage to protein and DNA: mechanisms involved and relevant observations on brain undergoing oxidative stress. Ann Neurol 32(S1):S22–S27

Francis PT, Palmer AM, Snape M, Wilcock GK (1999) The cholinergic hypothesis of Alzheimer's disease: a review of progress. J Neurol Neurosurg Psychiatry 66(2):137–147

Furukawa H, Cordova KE, O'Keeffe M, Yaghi OM (2013) The chemistry and applications of metal-organic frameworks. Science 341(6149):1230444

Gao Y, Han M (2025) Efficiency evaluation of commonly used methods to accelerate formaldehyde release and removal in households: A field measurement in bedroom. Build Environ 268:112348

Gao N, Sun H, Dong K, Ren J, Duan T, Xu C, Qu X (2014) Transition-metal-substituted polyoxometalate derivatives as functional anti-amyloid agents for Alzheimer's disease. Nat Commun 5(1):3422

Gao F, Hou Y, Wang Y, Liu L, Yi X, Xia N (2025) Photothermal and photodynamic strategies for diagnosis and therapy of Alzheimer's disease by modulating amyloid-β aggregation. Biosensors 15(8):480

Ghanbarian E, Khorsand B, Petersen KK, Nallapu BT, Sajjadi SA, Lipton RB et al (2025) Cerebrospinal fluid inflammatory cytokines as prognostic indicators for cognitive decline across Alzheimer's disease spectrum. J Alzheimer's Dis:13872877251335915

Goedert M (1993) Tau protein and the neurofibrillary pathology of Alzheimer's disease. Trends Neurosci 16(11):460–465

Goldsteins G, Hakosalo V, Jaronen M, Keuters MH, Lehtonen Š, Koistinaho J (2022) CNS redox homeostasis and dysfunction in neurodegenerative diseases. Antioxidants 11(2):405

Gong L, Zhang X, Ge K, Yin Y, Machuki JOA, Yang Y et al (2021) Carbon nitride-based nanocaptor: an intelligent nanosystem with metal ions chelating effect for enhanced magnetic targeting phototherapy of Alzheimer's disease. Biomaterials 267:120483

Gu Q, Wang L, Huang F, Schwarz W (2012) Stimulation of TRPV1 by green laser light. Evid Based Complement Alternat Med 2012(1):857123

Guarente L (2006) Sirtuins as potential targets for metabolic syndrome. Nature 444:868–874

Guo R, Li D, Li F, Ji L, Liu H, Qiao H et al (2025) Effects of whole-head 810 nm near-infrared therapy on cognitive and neuropsychiatric symptoms in Alzheimer's disease: A pilot study. J Alzheimer's Dis 104(1):52–60

Hamblin MR, Salehpour F (2021) Photobiomodulation of the brain: shining light on Alzheimer's and other neuropathological diseases. J Alzheimer's Dis 83(4):1395–1397

Hau Gan C, Chen HJ, Yu-Syun Fan R, Pan SC, Voon Kong K (2025) A multifunctional Ru-re complex for photodynamic therapy and CO-mediated therapeutics in Alzheimer's disease. Chemistry 31(11):e202403666

Heneka MT, Gauthier S, Chandekar SA et al (2025) Neuroinflammatory fluid biomarkers in patients with Alzheimer's disease: a systematic literature review. Mol Psychiatry 30:2783–2798. https://doi.org/10.1038/s41380-025-02939-9

Hennessy M, Hamblin MR (2016) Photobiomodulation and the brain: a new paradigm. J Opt 19(1):013003

Hirabayashi A, Shindo Y, Oka K, Takahashi D, Toshima K (2014) Photodegradation of amyloid β and reduction of its cytotoxicity to PC12 cells using porphyrin derivatives. Chem Commun 50(67):9543–9546

Hirakawa H, Terao T, Muronaga M, Izumi T, Kohno K (2025) Increasing the uptake of 18F-fluorodeoxyglucose in the right hippocampus after bright light exposure in patients with mood disorders. Asian J Psychiatr 103:104324

Hong G, Diao S, Chang J, Antaris AL, Chen C, Zhang B et al (2014) Through-skull fluorescence imaging of the brain in a new near-infrared window. Nat Photonics 8(9):723–730

Hong X, Huang L, Lei F, Li T, Luo Y, Zeng M, Wang Z (2025) The role and pathogenesis of tau protein in Alzheimer's disease. Biomolecules 15(6):824

Hou L, Yuan Y, Ren J, Zhang Y, Wang Y, Shan X et al (2017) In vitro and in vivo comparative study of the phototherapy anticancer activity of hyaluronic acid-modified single-walled carbon nanotubes, graphene oxide, and fullerene. J Nanopart Res 19(8):286

Hou W, Shi G, Wu S, Mo J, Shen L, Zhang X, Zhu Y (2022) Application of fullerenes as photosensitizers for antimicrobial photodynamic inactivation: a review. Front Microbiol 13:957698

Hsu CW, Hsu TW, Kao YC, Lin YH, Thompson T, Carvalho AF et al (2025) The efficacy and safety of anti-amyloid monoclonal antibody versus acetylcholinesterase inhibitor with an in-depth analysis across genotypes and disease stages: a systematic review and meta-analysis. J Prev Alzheimers Dis:100195

Hu Q, He C, Lu Z, Xu L, Fu Z (2021) Mitochondria and endoplastic reticulum targeting strategy for enhanced phototherapy. ACS Appl Bio Mater 4(4):3015–3026

Jagdeo JR, Adams LE, Brody NI, Siegel DM (2012) Transcranial red and near infrared light transmission in a cadaveric model

Jiang Y, Zeng Z, Yao J, Guan Y, Jia P, Zhao X, Xu L (2023) Treatment of Alzheimer's disease with small-molecule photosensitizers. Chin Chem Lett 34(5):107966

Jones DP (2006) Redefining oxidative stress. Antioxid Redox Signal 8:1865–1879

Ju Y, Tam KY (2022) Pathological mechanisms and therapeutic strategies for Alzheimer's disease. Neural Regen Res 17(3):543–549

Karosas T, Wallace TC, Li M, Pan Y, Agarwal P, Bennett DA et al (2025) Dietary choline intake and risk of Alzheimer's dementia in older adults. J Nutr

Kathiresan DS, Balasubramani R, Marudhachalam K et al (2025) Role of mitochondrial dysfunctions in neurodegenerative disorders: advances in mitochondrial biology. Mol Neurobiol **62**:6827–6855. https://doi.org/10.1007/s12035-024-04469-x

Kim SJ, Lee SH, Suh IB, Jang JW, Jhoo JH, Lee JH (2021) Positive effect of timed blue-enriched white light on sleep and cognition in patients with mild and moderate Alzheimer's disease. Sci Rep 11(1):10174

Kola A, Costanti F, Kahfi J, Emwas AH, Jaremko M, Valensin D (2025) NMR Metabolomic profiling of differentiated SH-SY5Y neuronal cells: amyloid-β toxicity and protective effects of galantamine and lycorine. Cells 14(7):525

Kolarova M, García-Sierra F, Bartos A, Ricny J, Ripova D (2012) Structure and pathology of tau protein in Alzheimer disease. Int J Alzheimers Dis 2012(1):731526

Koppelmans V, Silvester B, Duff K (2022) Neural mechanisms of motor dysfunction in mild cognitive impairment and Alzheimer's disease: A systematic review. J Alzheimer's Dis Reports 6(1):307–344

Kuk S, Lee BI, Lee JS, Park CB (2017) Rattle-structured upconversion nanoparticles for near-IR-induced suppression of Alzheimer's β-amyloid aggregation. Small (Weinheim an der Bergstrasse, Germany) 13(11)

Landis MK, Kunze A (2025) Intra-axonal nanomagnetic forces differentially impact hTau40 transport dynamics in primary cortical and hippocampal neurons. ACS Nano 19(8):7884–7897

Lane RM, Kivipelto M, Greig NH (2004) Acetylcholinesterase and its inhibition in Alzheimer disease. Clin Neuropharmacol 27(3):141–149

Lavi R, Shainberg A, Shneyvays V, Hochauser E, Isaac A, Zinman T et al (2010) Detailed analysis of reactive oxygen species induced by visible light in various cell types. Lasers Surg Med 42(6):473–480

Lee BI, Lee S, Suh YS, Lee JS, Kim AK, Kwon OY et al (2015a) Photoexcited porphyrins as a strong suppressor of β-amyloid aggregation and synaptic toxicity. Angew Chem 127(39):11634–11638

Lee JS, Lee BI, Park CB (2015b) Photo-induced inhibition of Alzheimer's β-amyloid aggregation in vitro by rose bengal. Biomaterials 38:43–49

Lee BI, Suh YS, Chung YJ, Yu K, Park CB (2017) Shedding light on Alzheimer's β-amyloidosis: photosensitized methylene blue inhibits self-assembly of β-amyloid peptides and disintegrates their aggregates. Sci Rep 7(1):7523

Lehtisalo J, Levälahti E, Lindström J, Hänninen T, Paajanen T, Peltonen M et al (2019) Dietary changes and cognition over 2 years within a multidomain intervention trial—The Finnish Geriatric Intervention Study to Prevent Cognitive Impairment and Disability (FINGER). Alzheimers Dement 15(3):410–417

Leshem G, Richman M, Lisniansky E, Antman-Passig M, Habashi M, Gräslund A et al (2019) Photoactive chlorin e6 is a multifunctional modulator of amyloid-β aggregation and toxicity via specific interactions with its histidine residues. Chem Sci 10(1):208–217

Li M, Xu C, Ren J, Wang E, Qu X (2013) Photodegradation of β-sheet amyloid fibrils associated with Alzheimer's disease by using polyoxometalates as photocatalysts. Chem Commun 49(97):11394–11396

Li M, Guan Y, Chen Z, Gao N, Ren J, Dong K, Qu X (2016) Platinum-coordinated graphitic carbon nitride nanosheet used for targeted inhibition of amyloid β-peptide aggregation. Nano Res 9(8):2411–2423

Li L, Chen Y, Chen W, Tan Y, Chen H, Yin J (2019) Photodynamic therapy based on organic small molecular fluorescent dyes. Chin Chem Lett 30(10):1689–1703

Li B, Golovynska I, Stepanov YV, Golovynskyi S, Golovynskyi A, Kolesnik D et al (2025) Transcranial photobiomodulation therapy with 808 nm light changes expression of genes and proteins associated with neuroprotection, neuroinflammation, oxidative stress, and Alzheimer's disease: whole RNA sequencing of mouse cortex and hippocampus. PLoS One 20(7):e0326881

Liebel F, Kaur S, Ruvolo E, Kollias N, Southall MD (2012) Irradiation of skin with visible light induces reactive oxygen species and matrix-degrading enzymes. J Invest Dermatol 132(7):1901–1907

Lim YJ, Zhou WH, Li G, Hu ZW, Hong L, Yu XF, Li YM (2019) Black phosphorus nanomaterials regulate the aggregation of amyloid-β. ChemNanoMat 5(5):606–611

Lish AM, Grogan EF, Benoit CR, Pearse RV, Heuer SE, Luquez T et al (2025) CLU alleviates Alzheimer's disease-relevant processes by modulating astrocyte reactivity and microglia-dependent synaptic density. Neuron 113(12):1925–1946

Liu GS, Wu H, Wu BQ, Huang RZ, Zhao LH, Wen Y (2008) Effect of phototherapy on blood endothelin and nitric oxide levels: clinical significance in preterm infants. World J Pediatr 4(1):31–35

Liu L, Wu Z, Lu Y, Lu W, Su G, Zhou Z (2024) Effects of phototherapy on biopterin, neopterin, tryptophan, and behavioral neuroinflammatory reaction in patients with post-stroke depression. Sci Rep 14(1):18368

Losi A (2007) Flavin-based blue-light photosensors: a photobiophysics update. Photochem Photobiol 83(6):1283–1300

Lu Y, Xu Y, Zhou L, Wang S, Han Y, Wang K, Qin C (2025) Bone marrow mesenchymal stem cells derived cytokines associated with AKT/IAPs signaling ameliorate Alzheimer's disease development. Stem Cell Res Ther 16(1):14

Lukacs A, Tonge PJ, Meech SR (2022) Photophysics of the blue light using flavin domain. Acc Chem Res 55(3):402–414

Lunova M, Smolková B, Uzhytchak M, Janoušková KŽ, Jirsa M, Egorova D et al (2020) Light-induced modulation of the mitochondrial respiratory chain activity: possibilities and limitations. Cell Mol Life Sci 77(14):2815–2838

Ma M, Gao N, Sun Y, Du X, Ren J, Qu X (2018) Redox-activated near-infrared-responsive polyoxometalates used for photothermal treatment of Alzheimer's disease. Adv Healthc Mater 7(20):1800320. https://doi.org/10.1002/adhm.201800320

Mangione MR, Piccionello AP, Marino C, Ortore MG, Picone P, Vilasi S et al (2015) Photo-inhibition of aβ fibrillation mediated by a newly designed fluorinated oxadiazole. RSC Adv 5(21):16540–16548

Manton KG, Volovik S, Kulminski A (2004) ROS effects on neurodegeneration in Alzheimer's disease and related disorders: on environmental stresses of ionizing radiation. Curr Alzheimer Res 1(4):277–293

Mantzavinos V, Alexiou A (2017) Biomarkers for Alzheimer's disease diagnosis. Curr Alzheimer Res 14:1149–1154. https://doi.org/10.2174/1567205014666170203125942

Marien J, Prevost C, Sacquin-Mora S (2025) Across (conformational) space and (relaxation) time: using coarse-grain simulations to probe the intra-and interdomain dynamics of the Tau protein. bioRxiv:2025-07

Marsh JA, Huang G, Bowling K, Renton AE, Ziegemeier E, Ball T et al (2025) Evaluating pathogenicity of variants of unknown significance in APP, PSEN1, and PSEN2. Neurotherapeutics:e00527

Mason MG, Nicholls P, Cooper CE (2014) Re-evaluation of the near infrared spectra of mitochondrial cytochrome c oxidase: implications for non invasive in vivo monitoring of tissues. Biochim Biophys Acta 1837(11):1882–1891

Masuda M, Suzuki N, Taniguchi S, Oikawa T, Nonaka T, Iwatsubo T et al (2006) Small molecule inhibitors of α-synuclein filament assembly. Biochemistry 45(19):6085–6094

Mi J, Liu C, Chen H, Qian Y, Zhu J, Zhang Y et al (2024) Light on Alzheimer's disease: from basic insights to preclinical studies. Front Aging Neurosci 16:1363458

Mitchell GS, Gill RK, Boucher DL, Li C, Cherry SR (2011) In vivo Cerenkov luminescence imaging: a new tool for molecular imaging. Philos Trans R Soc A Math Phys Eng Sci 369(1955):4605–4619

Montazeri K, Farhadi M, Fekrazad R, Akbarnejad Z, Chaibakhsh S, Mahmoudian S (2021) Transcranial photobiomodulation in the management of brain disorders. J Photochem Photobiol B Biol 221:112207

Morries LD, Cassano P, Henderson TA (2015) Treatments for traumatic brain injury with emphasis on transcranial near-infrared laser phototherapy. Neuropsychiatr Dis Treat:2159–2175

Musilova Z, Cortesi F (2023) The evolution of the green-light-sensitive visual opsin genes (RH2) in teleost fishes. Vis Res 206:108204

Najar AH, Sepulveda SE, Gherardelli C, Elahian F, Fontenot-Roberts A, Bajic A et al (2025) Generation of human induced pluripotent stem cell line from a familial Alzheimer's disease patient carrying missense mutations in PSEN1 and MAPT genes. Stem Cell Res 87:103759

Nakamura M, Shishido N, Nunomura A, Smith MA, Perry G, Hayashi Y et al (2007) Three histidine residues of amyloid-β peptide control the redox activity of copper and iron. Biochemistry 46(44):12737–12743

Neumann M, Sampathu DM, Kwong LK, Truax AC, Micsenyi MC, Chou TT et al (2006) Ubiquitinated TDP-43 in frontotemporal lobar degeneration and amyotrophic lateral sclerosis. Science 314(5796):130–133

Ng K, Allam N, Neshatian M, Vaez M, Hirvonen LM, Lam E et al (2025) Effects of ionizing radiation on the biophysical properties of type I collagen fibrils. PLoS One 20(4):e0319777

Ni J, Taniguchi A, Ozawa S, Hori Y, Kuninobu Y, Saito T et al (2018) Near-infrared photoactivatable oxygenation catalysts of amyloid peptide. Chem 4(4):807–820

Nizamutdinov D, Ezeudu C, Wu E, Huang JH, Yi SS (2022) Transcranial near-infrared light in treatment of neurodegenerative diseases. Front Pharmacol 13:965788

Nouchi R, Nouchi H, Dinet J, Kawashima R (2021) Cognitive training with neurofeedback using NIRS improved cognitive functions in young adults: evidence from a randomized controlled trial. Brain Sci 12(1):5

Nowak L (2008) The effect of timed blue-green light on sleep-wake patterns in women with Alzheimer's disease. Wayne State University

Omar Khudhur Z, Maad H, A., A Ghanimi, H., & Abdolmaleki, A. (2024) Fullerene nanoparticle as new therapeutic agent for the nervous system disorders. Nanomed J 11(4):342–359

Ottoy J, Niemantsverdriet E, Verhaeghe J, De Roeck E, Struyfs H, Somers C et al (2019) Association of short-term cognitive decline and MCI-to-AD dementia conversion with CSF, MRI, amyloid-and 18F-FDG-PET imaging. NeuroImage 22:101771

Ozawa D, Yagi H, Ban T, Kameda A, Kawakami T, Naiki H, Goto Y (2009) Destruction of amyloid fibrils of a β2-microglobulin fragment by laser beam irradiation. J Biol Chem 284(2):1009–1017

Öztürk B, Demir H, Günay MS, Akdag Y, Sahin S, Gulsun T (2025) Photodynamic and Photothermal therapies using nanotechnology approach in Alzheimer's disease. Curr Neuropharmacol

Padalkar MV, Pleshko N (2015) Wavelength-dependent penetration depth of near infrared radiation into cartilage. Analyst 140(7):2093–2100

Padilla-Godínez FJ, Vázquez-García ER, Trujillo-Villagrán MI, Soto-Rojas LO, Palomero-Rivero M, Hernández-González O et al (2025) α-synuclein and tau: interactions, cross-seeding, and the redefinition of synucleinopathies as complex proteinopathies. Front Neurosci 19:1570553

Pan WT, Liu PM, Ma D, Yang JJ (2023) Advances in photobiomodulation for cognitive improvement by near-infrared derived multiple strategies. J Transl Med 21(1):135

Park W, Shin H, Choi B, Rhim WK, Na K, Han DK (2020) Advanced hybrid nanomaterials for biomedical applications. Prog Mater Sci 114:100686

Pedder JH, Sonabend AM, Cearns MD, Michael BD, Zakaria R, Heimberger AB et al (2025) Crossing the blood–brain barrier: emerging therapeutic strategies for neurological disease. Lancet Neurol 24(3):246–260

Piñol-Ripoll G, Salas Carrillo M (2025) Clinicians' perspectives of twice-weekly rivastigmine patches for Alzheimer's disease treatment in Spain: the VIITAL 2S study. Patient Prefer Adherence:1105–1118

Praticò D, Delanty N (2000) Oxidative injury in diseases of the central nervous system: focus on Alzheimer's disease. Am J Med 109(7):577–585

Pronin S, Koh CH, Hughes M (2018) Cytotoxicity of ultraviolet-C radiation on a heterogeneous population of human glioblastoma multiforme cells: Meta-analysis. Photodiagn Photodyn Ther 24:158–163

Qi Xiaoming et al. (2021) Gender differences of dementia in response to intensive self-administered transcranial and intraocular near-infrared stimulation. Cureus 13:7 https://doi.org/10.7759/cureus.16188

Quirk BJ, Whelan HT (2016) Effect of red-to-near infrared light on the reaction of isolated cytochrome c oxidase with cytochrome c. Photomed Laser Surg 34(12):631–637

Raz N, Rodrigue KM (2006) Differential aging of the brain: patterns, cognitive correlates and modifiers. Neurosci Biobehav Rev 30(6):730–748

Reddy PH, Tripathi R, Troung Q, Tirumala K, Reddy TP, Anekonda V et al (2012) Abnormal mitochondrial dynamics and synaptic degeneration as early events in Alzheimer's disease: implications to mitochondria-targeted antioxidant therapeutics. Biochim Biophys Acta 1822(5):639–649

Regazzetti C, Sormani L, Debayle D, Bernerd F, Tulic MK, De Donatis GM et al (2018) Melanocytes sense blue light and regulate pigmentation through opsin-3. J Invest Dermatol 138(1):171–178

Ricci C, Pastukh V, Leonard J, Turrens J, Wilson G, Schaffer D, Schaffer SW (2008) Mitochondrial DNA damage triggers mitochondrial-superoxide generation and apoptosis. Am J Phys Cell Phys 294(2):C413–C422

Rosenberg A, Mangialasche F, Ngandu T, Solomon A, Kivipelto M (2020) Multidomain interventions to prevent cognitive impairment, Alzheimer's disease, and dementia: from FINGER to world-wide FINGERS. J Prev Alzheimers Dis 7(1):29–36

Samoilova KA, Zhevago NA, Menshutina MA, Grigorieva NB (2008) Role of nitric oxide in the visible light-induced rapid increase of human skin microcirculation at the local and systemic level: I. diabetic patients. Photomed Laser Surg 26(5):433–442

Salehpour F, Hamblin MR, DiDuro JO (2019) Rapid reversal of cognitive decline, olfactory dysfunction, and quality of life using multi-modality photobiomodulation therapy: case report. Photobiomodulation, photomedicine, and laser surgery, 37(3):159–167. https://doi.org/10.1089/photob.2018.4569

Scarano N, Musumeci F, Casini B, Brullo C, D'Ursi P, Fossa P et al (2025) Alzheimer's disease etiology hypotheses and therapeutic strategies: a perspective. Int J Mol Sci 26(14):6980

Schöll M, Maass A (2020) Does early cognitive decline require the presence of both tau and amyloid-β? Brain 143(1):10–13

Seaberg J, Montazerian H, Hossen MN, Bhattacharya R, Khademhosseini A, Mukherjee P (2021) Hybrid nanosystems for biomedical applications. ACS Nano 15(2):2099–2142

Semyachkina-Glushkovskaya O, Penzel T, Poluektov M, Fedosov I, Tzoy M, Terskov A et al (2023) Phototherapy of Alzheimer's disease: photostimulation of brain lymphatics during sleep: a systematic review. Int J Mol Sci 24(13):10946

Shaheen MA, Elshal LM, Mohamed GM, Reda S (2025) Possible effects of empagliflozin on hippocampal structural changes associated with Alzheimer's disease induced by aluminum chloride in adult male albino rats (Histological and immunohistochemical study). Zagazig Univ Med J 31(1):228–255

Shahriarpanah S, Tehrani FHE, Davati A, Ansari I (2018) Effect of phototherapy on serum level of calcium, magnesium and vitamin D in infants with hyperbilirubinemia. Iran J Pathol 13(3):357

Slominski AT, Zmijewski MA, Plonka PM, Szaflarski JP, Paus R (2018) How UV light touches the brain and endocrine system through skin, and why. Endocrinology 159(5):1992–2007

Slominski RM, Chen JY, Raman C, Slominski AT (2024) Photo-neuro-immuno-endocrinology: how the ultraviolet radiation regulates the body, brain, and immune system. Proc Natl Acad Sci 121(14):e2308374121

Soltani N, Mohindra R, Lin S, Toronov V (2025) Assessing the relationship between cerebral metabolic rate of oxygen and redox cytochrome C oxidase during cardiac arrest and cardiopulmonary resuscitation. Appl Sci 15(3):1542

Sowa A, Malicka M, Biernacka M, Beszłej JA, Leszek J (2025) VEXAS syndrome and Alzheimer's disease—are there connections? Brain Sci 15(6):573

Sperling R, Donohue M, Aisen P (2012) F3-04-01: the A4 trial: anti-amyloid treatment of asymptomatic Alzheimer's disease. Alzheimers Dement 8(4S_Part_12):P425–P426

Spina E, Ferrari RR, Pellegrini E, Colombo M, Poloni TE, Guaita A, Davin A (2025) Mitochondrial alterations, oxidative stress, and therapeutic implications in Alzheimer's disease: a narrative review. Cells 14(3):229

Sunemi SM, Teixeira ILA, Mansano BSDM, de Oliveira HA, Antonio EL, de Souza Oliveira C et al (2021) Post-resistance exercise photobiomodulation therapy has a more effective antioxidant effect than pre-application on muscle oxidative stress. Photochem Photobiol Sci 20(4):585–595

Sutalangka C, Wattanathorn J, Muchimapura S, Thukham-Mee W, Wannanon P, Tong-un T (2013) Laser acupuncture improves memory impairment in an animal model of Alzheimer's disease. Innov Acupunct Med 6(5):247–251

Sweeney MD, Sagare AP, Zlokovic BV (2018) Blood–brain barrier breakdown in Alzheimer disease and other neurodegenerative disorders. Nat Rev Neurol 14(3):133–150

Talesa VN (2001) Acetylcholinesterase in Alzheimer's disease. Mech Ageing Dev 122(16):1961–1969

Taniguchi A, Sasaki D, Shiohara A, Iwatsubo T, Tomita T, Sohma Y, Kanai M (2014) Attenuation of the aggregation and neurotoxicity of amyloid-β peptides by catalytic photooxygenation. Angew Chem Int Ed 53(5):1382–1385

Thal DR, Rüb U, Orantes M, Braak H (2002) Phases of Aβ-deposition in the human brain and its relevance for the development of AD. Neurology 58(12):1791–1800

Thorek DL, Riedl CC, Grimm J (2014) Clinical Cerenkov luminescence imaging of 18F-FDG. J Nucl Med 55(1):95–98

Tomkins O, Kaufer D, Korn A et al (2001) Frequent blood-brain barrier disruption in the human cerebral cortex. Cell Mol Neurobiol 21:675–691

Tsai SR, Hamblin MR (2017) Biological effects and medical applications of infrared radiation. J Photochem Photobiol B Biol 170:197–207
Umino Y, Denda M (2023) Effect of red light on epidermal proliferation and mitochondrial activity. Skin Res Technol 29(9):e13447
Uzunbajakava NE, Tobin DJ, Botchkareva NV, Dierickx C, Bjerring P, Town G (2023) Highlighting nuances of blue light phototherapy: mechanisms and safety considerations. J Biophotonics 16(2):e202200257
Valter K, Bisti S, Gargini C, Di Loreto S, Maccarone R, Cervetto L, Stone J (2005) Time course of neurotrophic factor upregulation and retinal protection against light-induced damage after optic nerve section. Invest Ophthalmol Vis Sci 46(5):1748–1754
Vo N, Rillahan CD, Chatila ZK, Virga DM, Hall JL, Tamucci KA et al (2025) CD33-CD45 interaction reveals a mechanistic link to Alzheimer's disease susceptibility. bioRxiv:2025-07
Wang X, Han M, Wang Q, Zeng Y, Meng Q, Zhang J, Wei X (2017) A light therapy for treating Alzheimer's disease. In: Biophotonics and immune responses XII, vol 10065. SPIE, pp 129–134
Wang J, Fan Y, Tan Y, Zhao X, Zhang Y, Cheng C, Yang M (2018) Porphyrinic metal–organic framework PCN-224 nanoparticles for near-infrared-induced attenuation of aggregation and neurotoxicity of Alzheimer's amyloid-β peptide. ACS Appl Mater Interfaces 10(43):36615–36621
Wang Y, Qiu M, Won M, Jung E, Fan T, Xie N et al (2019) Emerging 2D material-based nanocarrier for cancer therapy beyond graphene. Coord Chem Rev 400:213041
Wang J, Feng Y, Tian X, Li C, Liu L (2020) Disassembling and degradation of amyloid protein aggregates based on gold nanoparticle-modified g-C3N4. Colloids Surf B: Biointerfaces 192:111051
Wang Z, Yang J, Feng Q, Li L, Chen L, Wang J, Wu J (2024) 500nm blue-green light therapy for cognitive function in Alzheimer's continuum and factors affecting efficacy. Alzheimers Dement 20:e090108
Weichenthal M, Schwarz T (2005) Phototherapy: how does UV work? Photodermatol Photoimmunol Photomed 21(5):260–266
Weickenmeier J, Kuhl E, Goriely A (2018) Multiphysics of prionlike diseases: progression and atrophy. Phys Rev Lett 121(15):158101
Wen S, Zhou J, Zheng K, Bednarkiewicz A, Liu X, Jin D (2018) Advances in highly doped upconversion nanoparticles. Nat Commun 9(1):2415
Wu CK, Fuh JL (2025) A 2025 update on treatment strategies for the Alzheimer's disease spectrum. J Chin Med Assoc 88(7):495–502
Wu X, Shen Q, Chang H, Li J, Xing D (2022) Promoted CD4+ T cell-derived IFN-γ/IL-10 by photobiomodulation therapy modulates neurogenesis to ameliorate cognitive deficits in APP/PS1 and 3xTg-AD mice. J Neuroinflammation 19(1):253
Xie J, Wang Y, Choi W, Jangili P, Ge Y, Xu Y et al (2021) Overcoming barriers in photodynamic therapy harnessing nano-formulation strategies. Chem Soc Rev 50(16):9152–9201
Xie M, Liu M, Wen X, Zheng H, Cheng X, Lu BY (2025) Flexibility of blue light LED arrays for neonatal jaundice phototherapy: a case study of indirect measurement. IEEE Instrum Meas Mag 28(5):25–30
Xiong X, Liu J, Wu L, Xiong S, Jiang W, Wang P (2024) Self-assembly strategies of organic small-molecule photosensitizers for photodynamic therapy. Coord Chem Rev 510:215863
Xu Z, Gong Y, Zou Y, Wan J, Tang J, Zhan C, Zhang Q (2022) Dissecting the inhibitory mechanism of the αB-Crystallin domain against Aβ42 aggregation and its effect on Aβ42 protofibrils: a molecular dynamics simulation study. ACS chemical neuroscience, 13(19):2842–2851. https://doi.org/10.1021/acschemneuro.2c00224
Yadav RK, Sethi RS, Sethi AS, Kumar L, Chaurasia OS (2012) The evaluation of the effect of phototherapy on serum calcium level. People's J Sci Res 5(2):1–4
Yang J, Liu W, Sun Y, Dong X (2020) LVFFARK-PEG-stabilized black phosphorus nanosheets potently inhibit amyloid-β fibrillogenesis. Langmuir 36(7):1804–1812
Yang L, Wu C, Parker E, Li Y, Dong Y, Tucker L et al (2022) Non-invasive photobiomodulation treatment in an Alzheimer disease-like transgenic rat model. Theranostics 12(5):2205

Yang Y, Chen S, Gao Y, Huang L, Liu Z, Liu C et al (2025) Complement receptor 1 is a potential Extracerebral factor promoting α-Synuclein pathology. Mol Neurobiol 62(4):4605–4616

Yao R, Ouyang Z, Xu T, Zhu Y, Tang B, Li J et al (2025) Immune system-associated genes in patients with Alzheimer's disease. J Alzheimer's Dis:13872877251364396

Yoo JA, Yu E, Park SH, Oh SW, Kwon K, Park SJ et al (2020) Blue light irradiation induces human keratinocyte cell damage via transient receptor potential vanilloid 1 (TRPV1) regulation. Oxidative Med Cell Longev 2020(1):8871745

Yu D, Guan Y, Bai F, Du Z, Gao N, Ren J, Qu X (2019) Metal–organic frameworks harness cu chelating and photooxidation against amyloid β aggregation in vivo. Chemistry–A. Chem A Eur J 25(14):3489–3495

Zein R, Selting W, Hamblin MR (2018) Review of light parameters and photobiomodulation efficacy: dive into complexity. J Biomed Opt 23(12):120901–120901

Zeng F, Peng K, Han L, Yang J (2021) Photothermal and photodynamic therapies via NIR-activated nanoagents in combating Alzheimer's disease. ACS Biomater Sci Eng 7(8):3573–3585

Zeng Y, Liu B, Zhou L, Zeng W, Tian X, Jiang J, Dai D (2025) Safety issues of Donepezil combined with Memantine in Alzheimer's disease population: real-world pharmacovigilance. Expert Opin Drug Saf:1–11

Zhang J, Yue X, Luo H, Jiang W, Mei Y, Ai L et al (2019a) Illumination with 630 nm red light reduces oxidative stress and restores memory by photo-activating catalase and formaldehyde dehydrogenase in SAMP8 mice. Antioxid Redox Signal 30(11):1432–1449

Zhang Z, Wang J, Song Y, Wang Z, Dong M, Liu L (2019b) Disassembly of Alzheimer's amyloid fibrils by functional upconversion nanoparticles under near-infrared light irradiation. Colloids Surf B: Biointerfaces 181:341–348

Zhang H, Hao C, Qu A, Sun M, Xu L, Xu C, Kuang H (2020) Light-induced chiral iron copper selenide nanoparticles prevent β-amyloidopathy in vivo. Angew Chem Int Ed 59(18):7131–7138

Zhang N, Chai S, Wang J (2025a) Assessing and projecting the global impacts of Alzheimer's disease. Front Public Health 12:1453489. https://doi.org/10.3389/fpubh.2024.1453489

Zhang Y, Qian C, Chen Y, He W, Guo Z (2025b) Phototherapy via modulation of β-amyloid in combating Alzheimer's disease. Aggregate 6(5):e70020

Zhao Y, Guo Q, Tian J, Liu W, Wang X (2025) TREM2 bridges microglia and extracellular microenvironment: mechanistic landscape and therapeutical prospects on Alzheimer's disease. Ageing Res Rev 103:102596

Zhou L, Wang Z, Zhong Q, Song B, Wang Y, Guan T, Liu Q (2024) Ultra-low-dose UV-C photostimulation promotes neural stem cells differentiation via Presenilin 1 mediated notch and β-catenin activation. Mol Neurobiol 61(11):9491–9506

Zhou P, Jia Y, Zhang T, Abudukeremu A, He X, Zhang X et al (2025a) Red light-activated reversible inhibition of protein functions by assembled trap. ACS Synth Biol 14(5):1437–1450

Zhou Y, Yang X, Hong C, Guan T, Xiao P, Yang Y et al (2025b) Periplaneta americana extract protects glutamate-induced nerve cell damage by inhibiting N-methyl-D-aspartate receptor activation. Biology 14(2):193

Zielke A (2014) Photo-excitation of electrons in cytochrome c oxidase as a theory of the mechanism of the increase of ATP production in mitochondria by laser therapy. In: Mechanisms for low-light therapy IX, vol 8932. SPIE, pp 26–31

Zou J, Liu L, Chai F, Shan Q, Wang Y (2025) The role of Sirt1 in regulating inflammatory cytokines and oxidative stress in the transition mechanism from depressive disorder to Alzheimer's disease. Mol Neurobiol:1–17

Zueva I, Belyaev G, Petrov K (2025) Disease-modifying effect of donepezil on APP/PS1 mice at different stages of Alzheimer's disease. Mol Cell Biochem:1–17

Part II

Bioactive Photomedicinal Compounds

Neuroprotective Effects of Lutein: Considering Microglial Activation and Oxidative Stress

5

Anushka Jain, Shriyansh Srivastava, Sachin Kumar, Rakesh Sahu, Mohd. Tariq, and Ghulam Md Ashraf

Abstract

Alzheimer's disease (AD) is a multidimensional neurodegenerative disorder marked by progressive decline in cognition, influenced by various pathogenic pathways like amyloid-β accumulation, tau hyperphosphorylation, prolonged microglial activation, and oxidative stress. Although existing pharmacotherapies offer limited symptomatic relief, recent research highlights the neuroprotective potential of dietary carotenoids like lutein. Lutein is a naturally occurring xanthophyll recognized for its antioxidant and anti-inflammatory abilities, capable of crossing the blood-brain barrier and particularly accumulating in AD-sensitive

A. Jain
Department of R&D, Quality Assurance, Pure & Cure Healthcare Pvt. Ltd. (A Subsidiary of Akums Drugs & Pharmaceuticals Ltd.), SIDCUL, Haridwar, Uttarakhand, India

S. Srivastava (✉)
School of Medical and Allied Sciences, Galgotias University, Greater Noida, Uttar Pradesh, India

S. Kumar
Department of Pharmacology, Delhi Pharmaceutical Sciences and Research University (DPSRU), New Delhi, Delhi, India

R. Sahu
School of Pharmacy, Sharda University, Greater Noida, Uttar Pradesh, India

M. Tariq
Department of Academics, Sumandeep Vidyapeeth (Deemed to be University), Vadodara, Gujarat, India

Department of Biotechnology, Graphic Era (Deemed to be University), Dehradun, Uttarakhand, India

G. M. Ashraf (✉)
Department of Biomedical Sciences, Gulf Medical University, Ajman, United Arab Emirates

H. Dilnashin, S. P. Singh (eds.), *Phototherapeutic Approaches to Neurodegeneration*, https://doi.org/10.1007/978-981-95-4751-7_5

brain regions, including the hippocampal region and occipital cortex. Recent studies indicate that lutein may inhibit microglial activation by downregulating key inflammatory mediators such as TNF-α, IL-1β, and IL-6, as well as obstructing pro-inflammatory pathways like NF-κB and MAPKs. This modulation shifts the microglial type from a damaging M1 state to a neuroprotective M2-like stage. Lutein, simultaneously, activates the Nrf2/HO-1 pathway and neutralizes reactive oxygen species (ROS), thus reducing oxidative stress, a recognized driver of amyloid pathology and neuronal degeneration in AD. Preclinical models have shown that lutein supplementation enhances cognitive performance, lowers amyloid accumulation, and preserves synaptic integrity. Despite the limited availability of human clinical data, observational studies correlate elevated lutein levels with improved cognitive outcomes in aging populations. Innovative delivery technologies, such as nano-formulations, may augment lutein's cerebral bioavailability, hence enhancing its therapeutic efficacy. This chapter thoroughly examines lutein's molecular pathways in reducing microglial activation and oxidative damage, highlighting its potential as a safe, multi-targeted approach in the management of AD.

Keywords

Alzheimer's disease · Lutein · Neurodegenerative · Pathogenic · Oxidative damage · Neuroprotective

5.1 Introduction

AD is perhaps the most prevalent kind of dementia and is a progressive neurological illness that is typically marked by the onset of memory loss and decline in cognitive function, which may eventually affect behavior, speech, visual and spatial orientation, along with motor abilities (DeTure and Dickson 2019). The incidence of dementia is predicted to increase to 131.5 million by 2050 as a result of the world's aging population, particularly in nations with low and middle incomes (Contador et al. 2024). Although histopathologic assurance is necessary for an accurate assessment of dementia and AD, the majority of demographic investigations into AD classify cases using clinical criteria (Lopez and Kuller 2019). Alongside normal aging processes, AD develops gradually into three distinct clinical phases: mild (initial), moderate, and severe (late). Medical history, brain scans, neurological and physical assessments, cognitive evaluations, related family history, and laboratory testing are typically used to make a clinical assessment of AD (Zvěřová 2019). The Montreal Cognitive Assessment (MoCA) and the Mini-Mental State Examination (MMSE) have both been utilized in clinical practice worldwide, especially in primary care, to detect AD in its initial phases (Porsteinsson et al. 2021). The pathogenic process and the signs or characteristics of AD are distinguished in the 2018 NIA-AA research model. For researchers to help identify the various phases of the disease, it classifies AD only in the context of a bio-pathological framework using the amyloid, tau, and neurodegeneration (ATN) parameters for AD biomarkers (Atri 2019a). Dementia caused by AD has no known cure. However, the present strategy for treating AD involves managing symptoms in a variety of ways to maintain quality of life, lessen the financial strain of illness, and prevent chronic clinical loss. The development, implementation, and maintenance of a solid base of psychological training, non-pharmacological and behavioral care plans, and understanding of care objectives and standards are necessary for effective drug therapy with FDA-approved AD drugs, which initially involves monotherapy with a cholinesterase inhibitor (ChEI) and eventually involves additional combination therapy with a ChEI and memantine (Atri 2019b). The neural extracellular deposit of Aβ peptides triggers the build-up of highly phosphorylated tau protein inside cells, thus leading to the formation of neurofibrillary tangles (NFTs). The primary basis of the behavioral and cognitive impairment seen in AD, however, is synapse damage (Abubakar et al. 2022). The pathophysiology of each hypothesis is linked to oxidative stress (OS). The brain's poor antioxidant capacity and high need for oxygen make it particularly susceptible to oxidative damage. Enzymatic antioxidants include methionine sulfoxide reductase (MSR), glutathione peroxidase (GSH-Px), catalase (CAT), and superoxide dismutase (SOD) (Oxidative stress 2025). Since most free radicals are produced as metabolites of the mitochondrial ETC, mitochondria are the primary site of neuronal OS. Redox disparities, neurological damage, DNA instability, pro-inflammatory gene transcription, and the generation of cytokines, such as TNF alpha, IL-1, and IL-6, can all be caused by mitochondrial ROS. A well-known hypothesis of aging that integrates ROS, DNA damage, and mitochondrial ideas has emerged as a result of the mitochondria's pivotal involvement in ROS generation (Ionescu-Tucker and Cotman 2021). Fe2+ and Cu2+ ions enhance the H2O2 produced by Aβ during the aggregation stage. Reactive oxygen species (ROS) can be produced by the electrochemically active Cu2+ ions that are contained within Aβ. These ROS led to lipid peroxidation in neuronal cellular membranes, which impairs the performance of ion channel ATPases and the transporters for glucose. Aβ-induced OS disrupts neuronal ion equilibrium and metabolism, making neurons vulnerable to death (Abeysinghe et al. 2020). Several molecular pathways, including RCAN1, CREB, Nrf2, and PP2A, are implicated in OS-induced AD. In AD patients' brains, ROS causes these pathways to become dysregulated, which eventually results in neurodegeneration. One of these, the Nrf2 pathway, is crucial for regulating biogenesis, mitochondrial functioning, and degradation. In preclinical studies, age-related alterations in Nrf2 function have been observed in conjunction with decreased glutathione production (Dhapola et al. 2024). The role of microglia in the central nervous system (CNS) is another area of interest. Through the integration of various regulatory networks, microglial activity is thought to align with the CNS's advancement, maturation, and death throughout a lifetime. It also plays significant roles in immunological monitoring, neural cell death, developing synaptic pruning, and the regulation of synaptic plasticity (Leng and Edison 2021). Alterations in microglia shape and elevated expression are referred to as microgliosis. One of the most important elements in the development and course of AD may be an aggressive gliosis. The notion that microglia-related pathways are crucial for the risk and pathophysiology of AD has been validated by recent research (Merighi et al. 2022). There are numerous recognized pathways by which activated microglia contribute to the pathophysiology of AD. Activation of microglia starts a chain of events that increases the synthesis of Aβ species. Apoptosis-associated speck-like protein with a C-terminal caspase recruitment domain (ASC), that acts as a fundamental ligand for Aβ aggregation, is released when activated microglia trigger NF-kB, that in turn triggers the production of the NLRP3 inflammasome (Bivona et al. 2023). LPS are one of the neurotoxins that produce microglial activation, neuroinflammation, and neuronal injury. Neurotoxins lead to OS, and the production of ROS triggers the development of TNF-α, IL-1β, and phosphorylated-nuclear factor kappa B (p-NF-kB), which impacts hippocampal neuronal function and results in neuronal apoptosis as well as deficits in memory and cognition. OS and neuroinflammation also affect phospho-c-Jun N-terminal kinase 1 (p-JNK1) expression, which triggers cell death in AD (Kaur et al. 2019). Interruption of aberrant inflammatory mechanisms or activation of microglial cells, which may increase the elimination of Aβ (Influence of microglia and astrocyte activation 2025). The available drugs for AD are designed to reduce symptoms rather than treat the condition completely. The goal of these drugs is to delay the onset of dementia's behavioral, psychological, and cognitive symptoms (BPSD). The four medications that are now approved for sale are galantamine, memantine, donepezil, and rivastigmine. They fall into two distinct classes: anti-glutaminergic and anticholinesterase inhibitors. These drugs are used topically or taken orally. These pharmaceutical treatments might be more beneficial in the early stages before neurodegeneration develops. Other variables also considerably reduce the effectiveness of these treatments, particularly the difficulties in delivering medications to the brain due to the blood-brain barrier's (BBB) restricted permeability, which limits passage from circulation to the CNS (Passeri et al. 2022). Table 5.1 outlines the specific adverse effects associated with each medicine employed for the management of AD.

Food is thought to have a significant part in maintaining optimal physical and mental well-being. This claim is supported by recent research, which is currently concentrating on the restorative and preventive benefits of individual constituents or their combinations in the treatment of different health conditions. Nutraceuticals are foods or products that have properties that promote health, according to the American Nutraceutical Association. These consist of herbal remedies, genetically engineered foods, vitamin supplements, cereals, fruits, drinks, vegetables, and packaged foods (Pandey et al. 2022). Nutraceuticals' neuroprotective function in promoting brain health and preventing neurodegenerative diseases like dementia and AD is well recognized. These different nutraceuticals may include flavonoids, carotenoids, isoflavones, antioxidant vitamins, flavanols, polyphenols, alkaloids, and various others (Gargantiel et al. 2023). About 40 of the more than 700 known compounds in the carotenoid class are found in human blood and bodily tissues. Lycopene, lutein, zeaxanthin, β-cryptoxanthin, α-carotene, and β-carotene are the main carotenoids found in human bodies. The plasma concentrations of the two main carotenoids, lutein and β-carotene, were lower in individuals with mild to advanced AD than in those with mild AD or controls (Mecocci et al. 2014). Dietary

Table 5.1 Brief review of the potential side effects and drawbacks associated with each AD medication

S.No.	Medication	Limitations/adverse events	References
1	Donepezil	High cost Individual reactions of different patients toward the drug may be different Abdominal pain Agitation and aggression Cholecystitis Convulsions Heart block (all types) Haemolytic anemia Hepatitis Neuroleptic malignant syndrome	Limitations of Alzheimer's Disease Medications (2025); The safety and tolerability (2025)
2	Rivastigmine	High cost Tachycardia Hepatitis Seizure Nightmares Stevens-Johnson syndrome	Limitations of Alzheimer's Disease Medications (2025); Mimica and Presečki (2009)
3	Galantamine	High cost Hallucinations Seizures Extrapyramidal disorder	Limitations of Alzheimer's Disease Medications (2025); Mimica and Presečki (2009); Hoffman and Bloemer (2021)
4	Memantine	High cost Agranulocytosis Leukopenia Thrombocytopenia Cardiac failure Pancreatitis Hepatitis Suicidal ideation Acute renal failure	Limitations of Alzheimer's Disease Medications (2025); Hoffman and Bloemer (2021)

carotenoids like lutein have strong anti-inflammatory and antioxidant qualities and show promise as AD preventative agents. In animal models of AD, prior studies have demonstrated the effectiveness of these drugs in lowering the Aβ burden, reducing OS, reducing neuroinflammation, and enhancing cognition. It is still unclear exactly what mechanisms underlie their neurological protection (Kim et al. 2024).

Therefore, this chapter outlines the various mechanistic pathways and roles of lutein in preventing the different pathological variables presented by AD individuals, focusing on microglial activation and oxidative stress. The chapter also includes various nano delivery methods that have been employed to deliver lutein to bypass the BBB and increase its bioavailability. Some of the challenges associated with the use of lutein in AD are also described in this chapter.

5.2 Lutein: A Neuroprotective Carotenoid

Lutein, being a neuroprotective carotenoid, has established its role in various neurodegenerative disorders such as AD. The following sections describe the sources, bioavailability, and accumulation of lutein in different parts of the brain, making it a potential neuroprotective compound.

5.2.1 Natural Sources and Dietary Importance

Lutein is classified as part of the xanthophyll carotenoid class, which is less hydrophobic and contains oxygen compared to the more hydrophobic carotene class, which is completely hydrocarbon. Plant pigments called carotenoids are lipophilic and are carried by lipoproteins in both human and animal plasma (Wang et al. 2007). Humans cannot produce carotenoids; they are mostly found in plants and microorganisms. Therefore, lutein needs to be obtained from the diet (Fig. 5.1). Xanthophylls, including lutein, are most commonly found in green veggies like kale, spinach, broccoli, peas, and lettuce, as well as in the yolk of eggs. It is also present in food items related to Einkorn, Khorasan, durum wheat, and corn (Wang et al. 2007; Johnson 2004; Li et al. 2020). Vegetables have lutein contents that vary widely,

Fig. 5.1 Dietary sources of lutein

ranging from 0.01 to 40 mg/100 g. Spinach (~11.9 mg/100 g) and kale (~39 mg/100 g) have the highest quantity. According to reports, the lutein content of chicken egg yolks is 1622 ± 650 μg/100 g of yolk. Yellow-orange fruits like corn, carrots, papayas, and peaches contain less than 1 mg of lutein and zeaxanthin. Carotenoids give the marigold flower's petals their yellow and orange color. 90% of lutein is made up of these carotenoid pigments. These xanthophylls are more commonly found in vegetables than in fruits. Fruits having the highest concentration include nectarines, blackberries, avocados, gooseberries, raspberries, and kiwi fruits (Mrowicka et al. 2022; Pandey et al. 2025). In the United States, the average adult consumes only 1–2 mg of lutein and zeaxanthin daily, while children consume even less, at 0.3 mg. People in Europe consume approximately 2.2 mg daily, while people in Thailand take about 3 mg daily. In Spain, elderly people in good health consume a little more lutein than younger adults do. The combined amount of these intakes is significantly less than the 10 mg daily that the Chinese and South Korean governments advise, or even the 6 mg daily that researchers advise (Shi et al. 2024). There is increasing evidence that carotenoids in the diet can help prevent several diseases (Johnson 2004). Lutein may help protect against several age-related conditions, such as cataracts, AMD, heart disease, and some types of cancer, according to an ever-increasing body of research. Recent discoveries about lutein's bioavailability and bioaccumulation have prompted scientists to make assumptions about lutein's potential involvement in cognitive function and brain development. The opposite end of the lifetime likewise shows preferential accumulation of lutein in the brain, where it is responsible for making up about 35 percent of all carotenoids in the centenarian's brain, even though it only makes up about 20% of the carotenoids in matched serum (Ribaya-Mercado and Blumberg 2004; Stringham et al. 2019).

5.2.2 Pharmacokinetics and Brain Bioavailability

The small intestine is the location where dietary carotenoids are absorbed. Since they are lipophilic molecules, their absorption proceeds similarly to that of other fat-soluble food substances like vitamin E or dietary lipids. For the best absorption, some fat must be consumed concurrently. The body absorbs lutein and zeaxanthin directly, given that they are in their free forms. Before absorption, the esterified forms need to be hydrolyzed, and this additional process appears to be ineffective and necessitates the concurrent consumption of abnormally high fat (Gazzolo et al. 2021). Since the mechanisms involved in lutein absorption are complicated and involve multiple transport proteins, lipoproteins, and chylomicrons, individual variances in phenotype can greatly affect the bioavailability that results (Bhat and Mamatha 2021). High-density lipoproteins (HDL), low-density lipoproteins (LDL), and, to a lesser extent, very low-density lipoproteins (VLDL) transport carotenoids when lipoprotein lipase releases them from chylomicrons. Hepatocytes and adipose tissue are the primary locations for lutein storage. It is considered that lutein is eliminated by the kidneys and bile (Mrowicka et al. 2022). Because lutein has a poor solubility in water (0.0342 mg/ml), its bioavailability (BA) is very variable and very poor (approximately 10–15%), and it depends entirely on the level of solvent or vehicle (Madaan et al. 2017). The liver, retina, plasma, and adipose tissue all contain lutein and its metabolites (Ahn and Kim 2021). Lutein is capable of getting through the placenta and the blood-retina barrier (Gazzolo et al. 2021). Additionally, it can cross the BBB and reach different parts of the brain (Park et al. 2020). A population-based study of adults investigated the relationship between cognition and lutein levels in their brain tissue. It was discovered that only lutein, out of all the carotenoids, was consistently linked to multiple types of cognitive measures, including executive function, language, learning, and memory, all of which are linked to different parts of the brain. The fact that lutein is linked to multiple cognitive functions is that its concentrations are higher in all brain regions investigated and that these correlations held true even after adjusting for possible confounding variables (Johnson 2014).

5.2.3 Lutein Accumulation in AD-Vulnerable Brain Regions

Different cell types and cellular compartments have different membrane compositions, such as nuclear, myelin, neuronal, and mitochondrial plasma membranes (PM). Each of these membranes serves a different function, many of which are related to cell survival. Thus, lutein accumulation in particular membrane types and brain locations may provide indications about its function (Erdman et al. 2015). In the brain, lutein makes up almost 59% of all carotenoids, with β-carotene (16%), zeaxanthin (13%), β-cryptoxanthin (7%), and lycopene (5%) following in sequence (Ramirez 2025). The selective build-up of lutein inside the brain of a developing infant suggests its potential significance for cognitive growth. From fetal development to old age, lutein is present in the frontal, occipital, and temporal cortices, as well as the hippocampus and cerebellum (Loskutova et al. 2019). In xanthophyll-supplemented monkeys, the maximum lutein levels were detected in the cerebellum, which is crucial for motor coordination and some types of comprehension, as well as the frontal cortex, which is responsible for various aspects associated with enhanced cognition (Vishwanathan et al. 2013). In a different investigation, lutein was found in the rhesus monkeys' cortex, cerebellum, striatum, and hippocampus following oral dosing at 0.25–0.5 μmol/kg. As a result, research into lutein's advantages for brain development and possible defences against neurodegeneration is rapidly gaining light (Park et al. 2020).

5.3 Microglial Activation in AD

Although various mechanistic pathways have been described to understand the development of AD, like the amyloid as well as tau hypotheses, overactivation of microglial cells and their role in neuroinflammation have gained attention. The various mechanisms by which microglial overexpression causes AD, and lutein's role in their management, have been discussed in the sections below.

5.3.1 Microglia and Neuroinflammation in AD Progression

An inflammation process within the central nervous system (CNS) that can be brought on by a number of clinical events, including infection, trauma, ischemia, and toxins, is commonly referred to as neuroinflammation (Leng and Edison 2021). In general, the brain's acute inflammation serves as a defence against infections, stimulation, and harm to the neurological system. Long-term inflammation, on the other hand, gradually turns into chronic inflammation in order to activate the neurological system. A pro-inflammatory response that is consistently stronger than the anti-inflammatory response could be caused by the chronic inflammation's continued release of different inflammatory cytokines. This would cause the inflammation, eventually harming the neurons and resulting in a number of pathological changes in the body (Wang et al. 2023). Astrocytes are neuro-supportive in nature, although microglial cells serve as the brain's first line of defence against immunological reactions (Kaur et al. 2019). The microglia guard the brain's surroundings to maintain neuronal survival and immunological surveillance under normal physiological conditions. Research has demonstrated that Aβ aggregates cause morphological and molecular alterations in microglia, causing a whole spectrum of activation. Disease-associated microglia (DAM) are activated microglia that exhibit overexpression of certain receptors, chemokines, cytokines, and other components (Singh 2022). Microglial activation, which is triggered by both Aβ and tau overexpression, seems to be the link that enables the synergistic impact of Aβ and misfolded tau (Andronie-Cioara et al. 2023). The post-mortem AD brain showed increased release of pro-inflammatory molecules, such as cytokines (IL-1α, IL-1β, IL-6, TNF-α), granulocyte–macrophage colony-stimulating factor (GM-CSF), IFN-α, and the colony-stimulating factor-1 receptor (CSF-1R), caused by the overexpression of microglial cells. These markers demonstrated neurotoxicity and cognitive decline that caused severe late-stage AD (Kaur et al. 2019). The stimulated microglia have also been found to induce the production of the A1 subtype by activated astrocytes, which damages neurons and oligodendrocytes by expressing pro-inflammatory molecules such as TNF, C1q, and IL-1 through pathways that are dependent on nuclear factor κB (NF-κB). Conversely, M2 microglia-generated reactive astrocytes promote the release of factors that inhibit inflammation via the STAT6 pathway (Cai et al. 2022). With increased amounts of glial fibrillary acidic protein (GFAP) and genes linked to inflammatory pathways and responsiveness to toxins, astrocytes are frequently seen to be in a disease-linked state, much like microglia. Interestingly, reactive astrocytes and activated microglia are often found surrounding the senile plaques of AD patients, indicating that they are important in the disease's development. Furthermore, astrocytes must mature into the toxic astrocyte cell A1 to create TNFα, which is produced by neutrophils, macrophages, and microglia (Merighi et al. 2022). Microglia polarize toward the phagocytic phenotype in response to chronic inflammation, absorbing astrocytic fragments and compromising the blood-brain barrier (Andronie-Cioara et al. 2023).

5.3.2 Key Inflammatory Mediators: TNF-α, IL-1β, IL-6

Activating signaling pathways such as COX, Akt/mTOR, NF-κB, caspases, p38 MAPK, and NO cause microglia to produce inflammatory cytokines that are released (Thakur et al. 2023). The measurement of these pro-inflammatory cytokine levels is the final diagnostic stage in Alzheimer's disease. These include measuring the levels of IL-1, IL-6, and TNF-α, which provide the most precise indication of inflammation in the neurons of an AD patient's brain. ELISA and ELISPOT immunological techniques are used to determine these (Twarowski and Herbet 2023). Among the many pro-inflammatory mediators, TNF-α causes the synthesis of IL-1, IL-6, and IL-8, which, if not counterbalanced by anti-inflammatory cytokines like IL-10, promote chronic inflammation. It activates γ-secretase in HEK cells, releases Aβ peptides, and causes mouse astrocytes to express amyloid precursor protein (APP) and BACE1. Chronic inflammation in the brain creates a self-replicating cycle that keeps TNF-α levels high, possibly promoting Aβ formation, leading to neuronal death, and impeding microglia Aβ phagocytosis (The role of TNF-α in Alzheimer's disease 2025). TNFR1 and TNFR2, two receptor subtypes with different signaling pathways, are bound by TNF-α. Once activated, TNFR1 has a variety of effects, including as activation, cell proliferation, antiviral activity, and mainly TNF-α-mediated cytotoxicity and apoptosis. Because of its greater affinity for TNF-α, TNFR2 can initiate signaling cascades that involve pro-inflammatory and pro-survival pathways by activating NF-κB, phosphatidylinositol 3-kinase-dependent signaling pathways, and the cell blocking molecules of apoptosis proteins 1 and 2 (Chang et al. 2017). Long-term NF-κB activation in neurons can cause neurotoxicity, which reduces neuronal health by causing oxidative stress, excitotoxicity, and inflammation. Additionally, NF-κB affects that neuronal survival and death are balanced (Sivamaruthi et al. 2023). Furthermore, the p38 α isoform implicated in the inflammatory response and the MAPK pathway has demonstrated its role in glial cell–mediated neuroinflammation, including microglia and astrocytes (Thakur et al. 2023). The brain's inducible nitric oxide (NO) synthase (NOS) production and activity, as well as the overflow of NO metabolites into the cerebrospinal fluid, are highly stimulated by IL-1β and TNF-α (Azizi et al. 2015). The movement of monocytes and lymphocytes toward the inflammation site is facilitated by secreted inflammatory mediators, which allow them to cross the blood-brain barrier (BBB). This exacerbates CNS inflammation and causes long-term neuronal damage, which eventually results in cognitive decline. Numerous investigations have demonstrated that natural substances may lessen the inflammation in neuronal cells via inhibiting the p38αMAPK and TLR4/NF-κB signaling pathways (Zheng et al. 2023).

5.3.3 Lutein's Role in Modulating Microglial Activity

It has been shown that lutein reduces neuroinflammation by inhibiting NF-κB and triggering other antioxidants (Mirzaei et al. 2024). Furthermore, a study showed that lutein inhibited the expression of TNF-α, IL-1β, cyclooxygenase-2 (COX-2), and inducible NO synthase (iNOS) in lipopolysaccharide-activated microglia, thereby reducing neuroinflammation. When lutein stimulated nuclear factor erythroid 2-related factor 2 (Nrf2), heme oxygenase (HO)-1, and NAD(P)H: quinone oxidoreductase 1 (NQO1), they were elevated (Ahn and Kim 2021). According to a study by Pap et al., lutein has neuroprotective effects on BV-2 microglial cells by increasing IL-10 and reducing TNF-α generation in response to stress stimuli caused by H2O2. Additionally, research demonstrated that lutein might increase IL-10 levels in BV-2 cells without upregulating pro-inflammatory molecules, either individually or in conjunction with H2O2 therapy (Prathyusha et al. 2025). Lipopolysaccharide (LPS)-triggered BV-2 microglial and N2a cells are NF-κB, MAPK, and IRF3 pathway activators that release cytoplasmic NF-κB and activate the generation of the pro-inflammatory gene. Via the inhibition of signaling pathways, such as NF-κB, and the regulation of IL-1β, COX-2, iNOS, and TNF-α production, lutein decreased neuroinflammation (Flieger et al. 2024; Neuroprotective effects of natural compounds 2025). According to the findings by Li et al. on glutamate-mediated apoptosis in HT-22 cells (hippocampal cells), treatment with increasing lutein concentrations significantly decreased MAPK activation, including glutamate-induced phosphorylation of extracellular signal-regulated kinase (ERK), p38, and c-Jun N-terminal kinase (JNK). The results indicated that, by inhibiting the MAPK pathway, lutein may have a protective role toward glutamate-induced cell mortality in HT22 cells (Li et al. 2024).

5.4 Oxidative Stress in AD

Oxidative stress (OS) is another major pathological variable in the progression of AD. Different reactive oxygen species (ROS) generated in the brain can cause cognitive decline. The sections below discuss the effect of OS on AD and how lutein helps in reducing the same.

5.4.1 Sources of ROS in the AD Brain

It has been frequently shown that pathological alterations identified during the clinically silent prodromal phases of AD can be explained by increased OS. A key defining feature of AD is thought to be OS in the cortex and hippocampal regions. ROS play a role in both defense against invasive microbes and cellular signaling. However, excessive ROS generation causes cellular apoptosis and necrosis as well as damage to cellular macromolecules (Bhatt et al. 2021). One of the main producers of ROS in the CNS is mitochondria. AD is known to be linked to elevated levels of 8-hydroxy-2-deoxyguanosine, a hallmark of oxidative DNA damage, as well as more mitochondrial DNA mutations. These point mutations or deletions, which may be brought on by oxidative stress, can lead to apoptotic cell death and mitochondrial breakdown (Dumont and Beal 2011). Redox disarray, toxicity to neurons, faulty genomic structure, pro-inflammatory transcription of genes, and the generation of cytokines like TNF-α, IL-1, and IL-6 can all be caused by mitochondrial ROS. ROS species have the ability to harm and deactivate portions of the electron transport chain in a positive feedback cycle, which increases the electron conversion of oxygen to superoxide (Ionescu-Tucker and Cotman 2021). Another source is microglia, which can produce reactive oxygen species (ROS) in response to Aβ or other substances linked to danger that are released by damaged cells. Oxidative stress (OS) can therefore result from an excess of ROS and a reduction in antioxidant defences. Activated microglia can also create ROS, which can function as a second messenger by triggering distinct signaling pathways. By triggering antioxidant genes such as glutathione synthetase (GSS), thioredoxin (TRX), and others through the Keap1-Nrf2-ARE signaling pathway when ROS levels are high, Nrf2 fights oxidative stress. Although it has frequently been documented that AD causes downregulation of Nrf2-target genes, which results in the production of ROS (Buccellato et al. 2021). Furthermore, several studies have shown that elevated iron and copper levels are primarily associated with the deposition of AD hallmarks, validating the concept that the disruption of metal homeostasis is a crucial component of AD neurodegenerative disorders. ROS are produced by a number of enzymatic processes that mediate the transition metal iron deposition (Fe), which is a component of the ferroptosis (iron dependency) process in brain tissues. Increased oxidative reactions and low-affinity metal binding, which cause excessive Aβ depositions in brain tissues, are closely linked to the concentration of metal ions in the brain that increases with age (Oxidative stress-A 2025).

5.4.2 Lipid Peroxidation and DNA Damage

In AD pathophysiology, aberrant tau phosphorylation is an important marker, and Aβ can interact with transition metals to produce redox-active ions that precipitate in LPO. Lipid peroxidation (LPO) is an essential feature of OS and the generation of free radicals. Specifically, reactive oxygen species (ROS) may harm polyunsaturated fatty acids (PUFAs) of cell membranes, impairing the membrane functioning and/or organization, ultimately producing a group of α, β-unsaturated extremely reactive aldehydes, of which 4-hydroxy-2-noneal (HNE), malondialdehyde (MDA) (Taso et al. 2019). Co-localization of β-amyloid plaques and LPO secondary products was found in human studies using histological examination, and oxidized low-density lipoprotein (LDL) levels are linked to AD (Valgimigli 2023). MDA can interact or crosslink with DNA or proteins in cells and tissues, causing adducts to develop and biomolecular damage. Furthermore, the AD brain has an overabundance of acrolein, which is primarily discovered in NFTs, targeting DNA directly, reacting with the DNA base guanine to generate acrolein–deoxyguanosine (Petrovic et al. 2020). Alterations in DNA have been linked to transcription, its replication, and a variety of signaling channels, ranging from the nucleus to the mitochondrial signaling pathway. DNA carries genetic codes for cell formation, functionality, development, and replication (Wang et al. 2021). Mutations in DNA repair proteins that cause loss of function are especially sensitive to the neurological system.

Increased ROS byproducts, the exclusion of error-free homologous recombination (HR) repair, age-related declines in DNA repair enzyme performance, and an overall increased risk of somatic mutation are outcomes of these characteristics. Glial DNA damage is known to contribute to neurodegeneration (Welch and Tsai 2022).

5.4.3 Lutein's Antioxidant Mechanisms in AD

Numerous experimental investigations have shown the antioxidative properties of lutein in the case of AD. According to a number of studies, lutein may provide protection against AD by lowering oxidative stress, increasing anti-inflammatory responses, preventing neuronal cell apoptosis, enhancing mitochondrial function, and preventing amyloid beta (Aβ) aggregation and acetylcholinesterase activity (Firozjae et al. 2025). Several mechanisms, including reduced oxidative stress and activation of anti-inflammatory pathways, have been suggested to explain the protective roles of lutein and zeaxanthin, but the molecular basis of these effects is yet unknown (Johnson 2012).

Extracellular signal-regulated kinase (ERK) mediated the effect of lutein on Nrf2 activation. Nrf2 is phosphorylated by mitogen-activated protein kinases, such as ERK, which modulate the Nrf2-dependent antioxidant response. Lutein may therefore trigger ERK, which phosphorylates Nrf2 to cause it to go into the nucleus and trigger the production of antioxidant enzymes. These findings showed that lutein protected against inflammation-related neurodegenerative diseases by increasing the expression of Nrf2-target genes (antioxidant enzymes) and lowering the levels of inflammatory mediators (Ahn and Kim 2021). Suppression of Nrf2 activity makes the brain more vulnerable to the damaging effects of OS and inflammatory stimuli. The neuroprotective effects of Nrf2 activation are beneficial for in vitro as well as in vivo neurodegenerative prototypes. As people age, Nrf2 activity declines, which in turn reduces the impact of antioxidant enzyme activity (Lee et al. 2020). According to a study, *Drosophila melanogaster* provided a diet containing lutein had longer survival times and higher levels of antioxidant enzymes like catalase (CAT) and superoxide dismutase (SOD) than the group provided a basal diet. They also had lower levels of the peroxidation product malondialdehyde (MDA) (Ye et al. 2024).

Additionally, lutein protects brain polyunsaturated fatty acids like DHA from lipid peroxidation by scavenging singlet oxygen and peroxyl radicals. Its protective function against cognitive decline is further supported by the significant correlation between lower plasma levels of lutein and greater dementia severity (Wang et al. 2008). Inhibiting DHA oxidation keeps DHA intact so it can be cleaved and transformed into anti-inflammatory chemicals, in addition to maintaining membrane fluidity and structure. Lutein likely acts through other, distinct processes, such as altering the stability and function of membranes and neural connections, even if its potential lipid-protective impact might validate the link between lutein and memory (Nutrition Today 2025).

5.5 Molecular Mechanisms of Lutein in AD

The subsequent sections provide an insight into the different molecular mechanisms of lutein's antioxidant as well as anti-inflammatory roles.

5.5.1 Inhibition of NF-κB and Upregulation of Nrf2/HO-1 Pathway

As previously mentioned, it has been demonstrated that the Aβ peptide's neurodegenerative activity is mediated through NF-κB activation. By controlling the expression of NF-κB and interleukin, lutein inhibits inflammatory reactions (Liu et al. 2017). According to in vivo research, lutein acts through the NF-κB activity pathway to lower lipid peroxidation and the production of pro-inflammatory cytokines. Overall, research conducted both in vitro and in vivo has shown that carotenoids, such as lutein, are highly effective antioxidants and anti-inflammatory substances. They can enhance the functioning of antioxidants, block the NF-κB activity, downregulate pro-inflammatory molecules, and shield several cellular membranes from oxidative damage (Mohammadzadeh Honarvar et al. 2017). Numerous studies have demonstrated that intraperitoneal injections of lipopolysaccharides (LPS) impair memory and learning and increase the formation of Aβ by activating the NFκB signaling pathway. Furthermore, in many neurodegenerative illnesses, including AD, LPS can produce massive levels of ROS, which results in neuronal cell death and memory loss (Wang et al. 2018). Furthermore, p38 MAPK is highly expressed in regions of the CNS that are essential for memory and learning, and it is probably a major player in higher brain processes. Therefore, the pathophysiology of certain neurological illnesses, including AD, may be linked to the malfunction of this route (Lee and Kim 2017). Environmental stressors and inflammatory cytokines are the main sources of activation for p38 MAPKs, one of the three MAPKs found in mammalian cells (Gee et al. 2020). According to a study, lutein inhibited phosphorylated p38 kinase, c-Jun N-terminal kinase (JNK), and Akt kinase, signaling pathways necessary for NF-κB activation and the subsequent expression of the inflammatory genes controlled by NF-κB. This prevented LPS-induced NF-κB activation (Fuad et al. 2020). NF-E2-related factor 2 (Nrf2) is the main transcription factor needed for cellular protection against OS, according to emerging data. As already mentioned, under OS, Nrf2 detaches from Keap1, enters the nucleus, and activates antioxidant protein genes like HO-1. In a study by Wanqiang Wu et al., lutein significantly reduced Keap-1 and activated Nrf2, which led to an increase in HO-1 and NQO-1 levels in BV-2 cells. According to reports, direct stimulation of Nrf2 can be done by focusing on the Nrf2-Keap1 complex or indirectly by Nrf2 being phosphorylated by the MAPK. After stabilizing, Nrf2 moves into the nucleus and triggers its gene targets (Lutein suppresses inflammatory 2015).

5.5.2 Modulation of Mitochondrial Function and Apoptotic Pathways

It has long been known that micronutrients like docosahexaenoic acid (DHA) improve the mobility and functionality of the mitochondrial membrane. Like DHA, lutein is known to promote mitochondrial respiration (Xie et al. 2019). According to one study, taking supplements of both lutein and DHA enhanced cognitive function more than taking supplements of either compound alone (Bivona et al. 2023). Additional cognitive advantages related to spatial working memory and comprehension rate were produced by the combination (Mohn et al. 2017). Due to inadequate production of ATP, impaired mitochondria produce more ROS, which can lead to cell death. Thus, lutein that has built up in the mitochondria could safeguard this organelle from harm (Erdman et al. 2015). Numerous studies have indicated that the amyloid precursor protein (APP) is a substrate for cleavage by caspase-3, which might be a factor in the development of Aβ, synapse loss, and behavioral abnormalities linked to AD (Rana et al. 2018). It was shown that lutein has an inhibitory effect on apoptosis via inhibiting CASP-3 (Fig. 5.2) (Mirzaei et al. 2024).

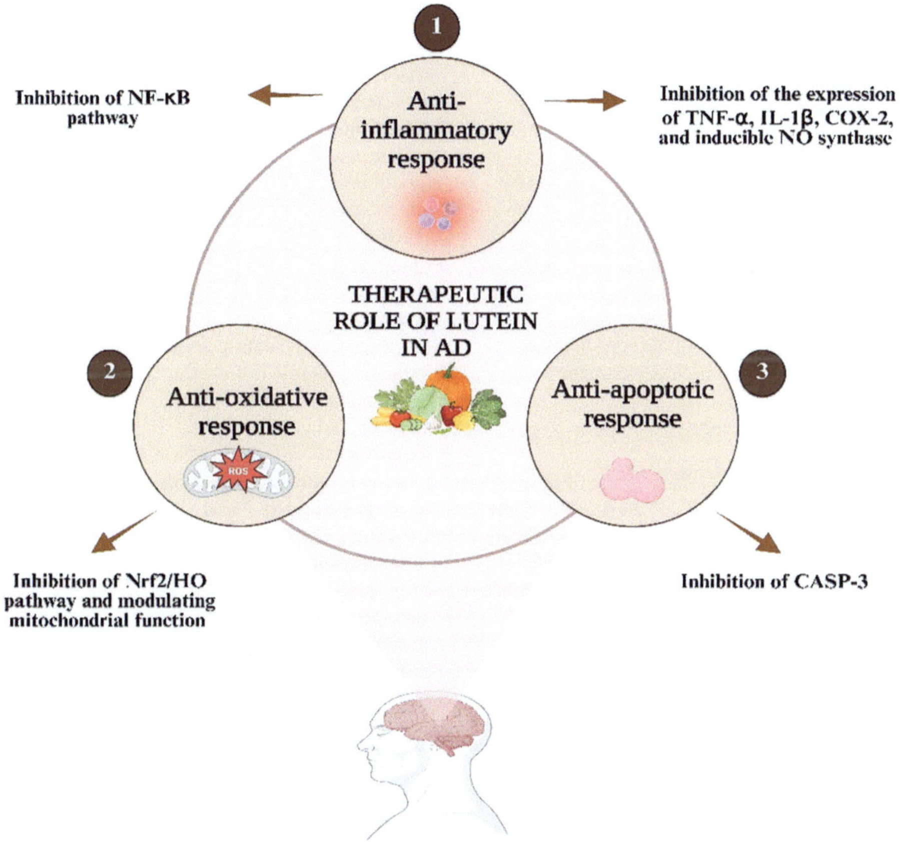

Fig. 5.2 Various neuroprotective mechanisms of action of lutein

5.6 Preclinical and Clinical Evidence of Lutein in AD Models

The established neuroprotective role of lutein has been supported by various preclinical and clinical trials. Although the number of standardized clinical trials is less, the following sections give an account of some of the preclinical and clinical trials that have been reported.

5.6.1 Preclinical Evidence

Different in vivo and in vitro studies have been performed to explain the neuroprotective effects of lutein via different antioxidant and anti-inflammatory mechanisms, as expressed in Table 5.2.

Table 5.2 Representation of various preclinical experimental studies conducted to evaluate the neuroprotective potential of lutein

S.No.	Author/ year	Aim of the study	Key findings	References
1	Da-Sol Kim et.al (2024)	To evaluate the protective effects of zeaxanthin, lutein, and resveratrol against Alzheimer's-like memory deficits (AD-CON) induced by hippocampal infusions of amyloid-β (25-35) in rats	Zeaxanthin and lutein more effectively than resveratrol prevented amyloid-β (25-35)-induced memory impairment by reducing AChE functioning, LPO, and pro-inflammatory cytokines (TNF-α, IL-1β)	Kim et al. (2024)
2	Chirag Patel et al. (2021)	To test the neuroprotective effect of lutein in zebrafish and mice with scopolamine-induced memory damage	Lutein reduced LPO and protein carbonyl levels while increasing GSH, superoxide dismutase, and catalase activity, indicating its role in lowering brain oxidative stress linked to memory and cognitive impairment	Patel et al. (2021)
3	Leila Nazari et al. (2022)	To assess lutein's protective effects on cognition and learning in male AD rats induced by ICV Aβ injection and characterized by impaired passive avoidance, spatial, and object recognition ability	Lutein exerts neuroprotection by modulating Nrf-2 and NF-κB expression in cerebrovascular endothelial cells, reducing oxidative damage markers such as MDA and TOS, and enhancing TAC levels, thereby acting as a potent antioxidant in the brain	Nazari et al. (2022)

(continued)

Table 5.2 (continued)

S.No.	Author/ year	Aim of the study	Key findings	References
4	Ramóna Pap et al. (2021)	To investigate lutein's protective effects against H2O2-induced oxidative stress in BV-2 microglia	Lutein reduced ROS in H_2O_2-treated BV-2 cells, increased SOD2 mRNA and protein expression, and elevated IL-10 levels without inducing pro-inflammatory molecules, highlighting its antioxidant and anti-inflammatory effects	Pap et al. (2021)
4	Tao Liu et al. (2017)	To evaluate the role of lutein in protecting bEND.3 cells from OS and apoptosis induced by the Aβ 25-35 peptide	Lutein pre-treatment enhanced cell viability, reduced ROS and lipid peroxidation, inhibited Aβ25-35-induced NF-κB activation, upregulated Nrf-2, preserved mitochondrial membrane potential, and lowered caspase activity	Liu et al. (2017)
5	Misuzu Matsui et al. (2024)	To investigate the effects of lutein from chicken eggs on reducing amyloid β-induced inflammation in human macrophages (THP-1)	Lutein suppressed IL-1β and TNF expression in Aβ1–42-stimulated THP-1 microglia, reduced IL-1β protein release, and inhibited Aβ-induced inflammatory cytokine secretion	Matsui et al. (2024)
6	Zhenhua Li et al. (2024)	To describe the mechanisms that explain lutein's ability to defend HT22 cells from glutamate-induced cellular death	Lutein promoted Nrf2 nuclear translocation, upregulated HO-1 expression, and prevented glutamate-induced activation of ERK1/2, p38, and JNK MAPKs, with HO-1 inhibition	Li et al. (2024)

5.6.2 Clinical Trials

The potential of supplements to decrease the onset and course of the disease is gaining attention due to their proven ability to modify immune responses, gut microbiota, and inflammatory pathways. Clinical evidence is still lacking, and future studies are highly encouraged to understand the therapies' efficacy and safety as well as their direct human relevance, despite strong preclinical results suggesting they may be involved in regulating neuroinflammation and cognitive decline (Gualtieri et al. 2025) (Table 5.3).

Though some clinical trials, as described above, have been performed on lutein's ability to improve cognition, no relevant explanation for the mechanisms involved in the same has been established, and thus it needs further investigation.

Table 5.3 Summarizes the clinical studies conducted to evaluate the neuroprotective role of lutein

S.No.	Type of the trial	Aim of the trial	Key findings	References
1	Placebo randomized controlled trials ($n = 7$)	To examine the impact of lutein on the 3 cognitive areas of memory, executive function, and complex attention in people who had taken lutein-rich meals or supplements as part of RCT	The general effect size estimated by standardized mean difference (SMD) showed that dietary lutein and its isomers did not mainly enhance cognitive function in complex attention, executive function, or memory, possibly due to the smaller sample size. However, they could preserve intellectual abilities and maintain brain health	Li and Abdel-Aal (2021)
2	Randomized, double-blind, placebo-controlled trial ($n = 90$)	To investigate how supplementing with lutein and zeaxanthin affects cognitive function in individuals who express cognitive impairments	Daily supplementation with 10 + 2 mg lutein and zeaxanthin, respectively, improved visual episodic and learning abilities compared to placebo, though no significant benefits were seen in other cognitive, mood, or physical function measures	Lopresti et al. (2022)
3	Randomized controlled trial and cross-sectional design	To present the neuropharmacological effects of lutein by its possible antioxidant and anti-inflammatory effects	Lutein, in 4/17 investigations, showed enhanced cognitive function. 2/17 studies showed improvements in lipid profiles, and 8/17 studies showed a decrease in the development of AD	Firozjae et al. (2025); Dement (2015)
4	Single-site, double-blind, randomized controlled trial ($n = 44$)	To apply functional magnetic resonance imaging (fMRI) to examine the link between lutein and zeaxanthin and brain functioning	The findings indicated that consuming L and Z improved neuronal response and blood flow to the brain during cognitive function	Lindbergh et al. (2018)

(continued)

Table 5.3 (continued)

S.No.	Type of the trial	Aim of the trial	Key findings	References
5	Randomized controlled trial ($n = 44–91$)	To investigate the effects of carotenoids like lutein on cognition	The results showed that over the duration of a year, consuming 10 mg of lutein can increase inhibitory efficiency in elderly and older individuals as well as visual episodic memory retention in young and old adults	Nouchi et al. (2020)

5.7 Formulation Strategies for Enhanced Delivery

What's a drug alone without being delivered to a patient's body? Lutein can be obtained from dietary sources that have been described above, but the correct amount of lutein for its therapeutic potential in neurodegenerative disorders needs critical points to be covered, like bypassing the BBB. Nanoparticles, in this context, have shown great progress in recent years. This heading highlights the various formulation approaches by which the delivery of lutein can be enhanced.

5.7.1 Nano-Formulations and Lipid-Based Carriers

These substances have significant problems in terms of stability and bioavailability. Fortunately, these molecules can have their biological activity enhanced by structural changes. The use of drug delivery systems is one strategy that enables the carotenoids to overcome these limitations and greatly increase their stability and bioavailability. Thanks to the advancement of nano-drug delivery technology, there are currently a lot of opportunities for the prevention and treatment of AD. At the BBB, this technique interacts well with endothelial microvascular cells and can result in increased drug concentrations in the cerebral parenchyma. Various nano delivery systems have been employed, including lipid-based nano-carriers, inorganic nano-carriers, polymer/biopolymer nano-carriers, and hybrid nano-carriers (Su et al. 2023). A novel liposome system, comprised of egg yolk, phosphatidylcholine, Tween 80, and mixed lipids, was investigated in a series of research on the encapsulation of carotenoids (lutein, β-carotene, lycopene, and canthaxanthin) in liposomes. Carotenoid stability against stresses from the environment during preparation, storage, heating, and exposure to surfactants was enhanced by encapsulation. Depending on the kind and dose of carotenoid, a protective synergistic effect between liposomes and carotenoids was identified. Lutein demonstrated the largest loading capacity, the highest liposome incorporation, and the strongest antioxidant activity (Algan et al. 2022a). However, in terms of bioavailability, a lutein nano-emulsion was made and contrasted with supplements. Nano-emulsions could increase bioavailability even at physiological dosages, as demonstrated by the lutein nano-emulsions, which had a diameter of 150 nm and were noticeably more bioavailable even at 10–40% lower doses than supplements (dos Santos et al. 2018). In a preclinical study, Vishwanathan et al. compared the bioavailability of lutein given by NEs with that of direct supplementation. When compared to the supplements, the NEs caused a 31% and 28% increase in lutein serum concentrations, despite the fact that the real concentration was 10% and 40% lower (Fernandes et al. 2021). Tan et al. (2014), on the other side, evaluated the bio-accessibility of carotenoids, such as lutein, β-carotene, lycopene, and canthaxanthin, by successfully encapsulating them in nano-liposomal structures in addition to nano-emulsions. The maximum bio-accessibility was shown by lutein, which was followed by canthaxanthin, lycopene, and β-carotene. These carotenoids also show increased antioxidant activity as a result of being encapsulated in liposomes, with lutein exhibiting the highest level of this activity (Verardi et al. 2023). Natural biomaterials are a non-invasive treatment option that can reduce the pace of development of AD and include protective qualities. Interestingly, research has shown that cationic compounds that contain chitosan can reduce the oxidative stress linked to AD. Furthermore, it is shown that lutein-functionalized cationic biopolymer nanoparticles can reduce OS without causing any harm (Zivari-Ghader et al. 2024). The nano-system created in Dhas and Mehta's study consisted of a poly (lactic-co-glycolic acid) (PLGA) core that was loaded with lutein and covered in a coating of chitosan. This ionic biopolymer core-shell structure improved lutein's stability and biological absorption while preventing degradation brought on by exposure to light as well as extreme temperatures. Their findings confirmed the system's ability to transfer lutein to the brain through the nose and the nasal mucosa's notable membrane crossing. The antioxidant characteristics of lutein decreased OS, which could improve the efficacy of AD treatment (Daraban et al. 2024).

5.7.2 BBB Targeting Approaches and Safety Considerations

The BBB is the primary reason that solutions for getting bioactive substances into the brain are still poorly developed (Ma et al. 2022). To be precise, because of the blood-brain barrier (BBB), carotenoids like lutein are poorly transported into the CNS and have limited bioavailability (Elkholy et al. 2020). Other nutrients may also be transported by lutein, which exhibits promise as a carrier for brain-targeted delivery systems. Numerous substances have been investigated for brain targeting, such as cationic liposomes that restore cortical monoamine levels and EEG signals and polymer-based nanoparticles that improve cognitive function in mice. Furthermore, trans-nasal delivery has demonstrated efficacy, demonstrating a 4.4-fold increase in lutein delivery to the brain in comparison to its free form (Ma et al. 2022). A well-established non-invasive method, intranasal drug delivery has begun to gain light recently due to its capacity to transport therapeutic agents, as it is, directly to the CSF, circumventing the blood-brain barrier. For instance, chitosan-coated lutein-PLGA NPs effectively transported the drug to the brain and showed a high degree of ability to pass across the nasal mucosal barrier, where lutein has a beneficial impact on nerve cells (Zivari-Ghader et al. 2024). Despite the fully hydrophobic carotenes (α-, β-, and lycopene), lutein and zeaxanthin have partially amphiphilic characteristics due to their hydroxyl groups. Thus, a thorough comprehension of these compounds' in vivo chemistry and biology is necessary for ensuring effective biosorption (Kumar et al. 2024). Comprehensive human-based trials should be carried out to assess the potential of nanotechnology-based techniques for lutein administration. Recent advancements in the realm of nano delivery systems have been made available, with encouraging results. These technologies' success is still largely dependent on their expenses, assurance, and bench-to-bedside migration. The growing effectiveness of these delivery methods may open the door for lutein's clinical translation and therapeutic efficacy to be improved (Algan et al. 2022b). How to select and produce the most cost-effective and efficient NP components is the crucial aspect that needs to be emphasized. Encapsulation, release, and physicochemical qualities, together with method costs, are all significant factors (Maghsoudi et al. 2022). As an example, during the formation of nanoparticles, poly (lactic-co-glycolic acid) (PLGA) is a widely used co-polymer from a variety of biodegradable polymers because of its non-toxic nature, ability to degrade, and biocompatibility, along with its ability to maintain and regulate drug release (Dhas and Mehta 2020). Similarly, Weigel et al. discovered that sodium caseinate formed lutein NEs of about 600 nm, while quillaja saponin, Tween 80, and whey protein extract yielded lutein NEs of 220–250 nm. The stability of the Tween 80-stabilized particles was low (−9 mV zeta potential), while the others were more stable (−42 to −62 mV) (Fernandes et al. 2021).

5.8 Challenges, Limitations, and Future Directions

The translation to human neurodegenerative therapy continues to be challenged by a number of issues, including the absence of a well planned, long-term clinical trial, the main disadvantage of which is a small sample size and an unclear mechanism of action, despite promising preclinical findings on lutein's antioxidant and anti-inflammatory attributes in AD-related models (Li and Abdel-Aal 2021; Davinelli et al. 2021). Uncertainty regarding the best dosage and duration of treatment due to different trial designs and individual differences in bioavailability remains another challenge.

The amount of lutein that needs to be consumed each day has not yet been determined. Since no clinical trials have shown adverse effects linked to high doses of lutein in humans, neither a Lowest Observed Adverse Effect Level (LOAEL) nor a No Observed Adverse Effect Level (NOAEL) has been established (Mrowicka et al. 2022). Multi-center RCTs with biomarker-guided outcomes and uniform dosage schedules must be the main emphasis of future research (Kumar et al. 2025). It appears likely that merely increasing dietary intake of carotenoids may have limited protective benefits on neurons due to the complexity of the pathogenic pathways causing diseases like Alzheimer's. To establish promising treatments for neurodegenerative illnesses, it is believed that future research should be crucial in identifying additional neuroprotective drugs or therapies that work in combination with lutein-like carotenoids (Cho et al. 2018). Lutein and conventional neuroprotective medications may work in combination (Kumar et al. 2025). Even though numerous studies have demonstrated the health advantages of carotenoids like lutein, precise measurements of consumption, metabolism, excretion, and bioavailability are necessary to validate observational linkages to chronic illnesses like AD. In order to guarantee that efficacy trials are supported by reliable analytical data, investigators should be knowledgeable about the compound's chemistry and analysis. Well planned clinical studies that monitor absorption and metabolism and aid in the definition of distinct dose-related responses should be a part of future research (Fig. 5.3) (Zia-Ul-Haq 2021).

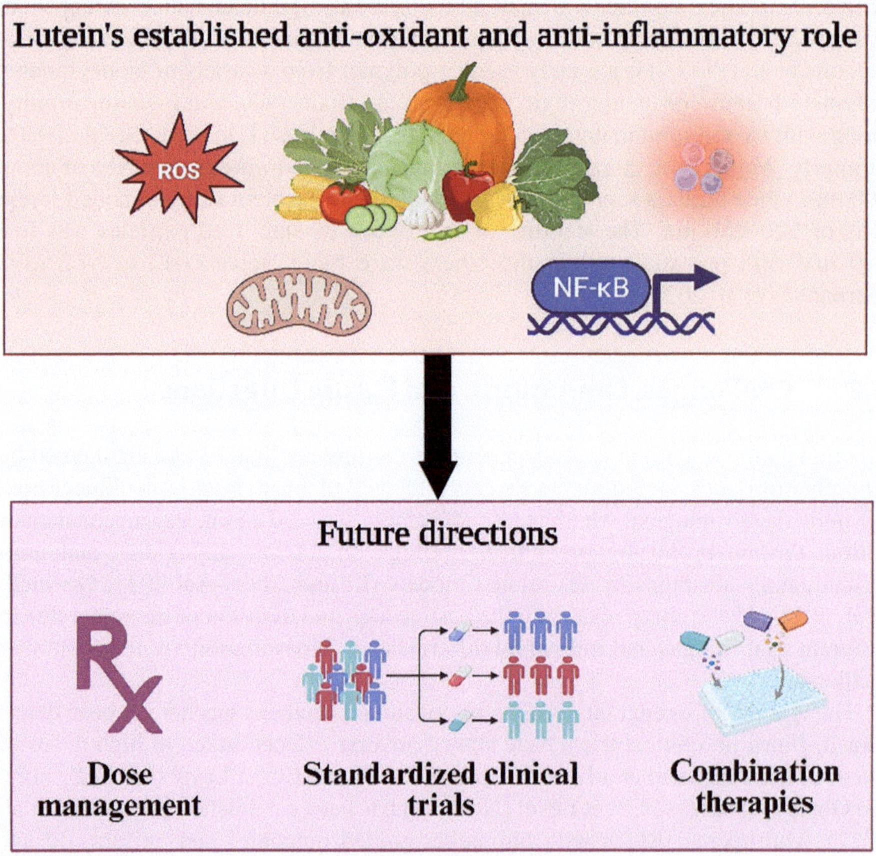

Fig. 5.3 Future directions in the context of lutein's established neuroprotective role

5.9 Conclusion

A dietary xanthophyll carotenoid with strong anti-inflammatory and antioxidant qualities, lutein has drawn increasing interest due to its possible role in neuroprotection. AD is marked by various pathological variables, including microglial overactivation and oxidative stress. Since there is currently no proven treatment for this illness, new therapeutic strategies are required. This chapter has explored the capacity of lutein to inhibit oxidative stress and microglial overactivation and its complex involvement in reducing AD pathogenesis. Lutein reduces inflammatory mediators like interleukins and TNF-α, limits the production of ROS, and preserves neuronal integrity by regulating important signaling pathways, such as NF-κB, MAPK, and the Nrf2/HO antioxidant response. Furthermore, methods for improving the stability, bioavailability, and targeted transport of lutein to the brain were mentioned, including liposomes, solid lipid nanoparticles, polymeric nanoparticles, and nanoemulsions. The chapter concluded by mentioning the gaps in the research of lutein as a neuroprotective compound in AD. Standardizing formulation techniques, guaranteeing long-term safety, and confirming efficacy through efficient clinical studies continue to be difficult tasks despite promising preclinical results. Subsequent investigations ought to concentrate on refining scalable nano-formulation platforms, utilizing sophisticated molecular techniques to clarify exact pathways, and converting these discoveries into efficacious, clinically relevant treatments for AD.

References

Abeysinghe AADT, Deshapriya RDUS, Udawatte C (2020) Alzheimer's disease; a review of the pathophysiological basis and therapeutic interventions. Life Sci 256:117996

Abubakar MB, Sanusi KO, Ugusman A, Mohamed W, Kamal H, Ibrahim NH, et al (2022) Alzheimer's Disease: an update and insights into pathophysiology. Front Aging Neurosci [Internet]. [cited 2025 Aug 10];14. Available from: https://www.frontiersin.org/journals/aging-neuroscience/articles/10.3389/fnagi.2022.742408/full

Ahn YJ, Kim H (2021) Lutein as a modulator of oxidative stress-mediated inflammatory diseases. Antioxidants 10(9):1448

Algan AH, Gungor-Ak A, Karatas A (2022a) Nanoscale delivery systems of Lutein: an updated review from a pharmaceutical perspective. Pharmaceutics 14(9):1852

Algan AH, Gungor-Ak A, Karatas A (2022b) Nanoscale delivery systems of Lutein: an updated review from a pharmaceutical perspective. Pharmaceutics. 14(9):1852

Andronie-Cioara FL, Ardelean AI, Nistor-Cseppento CD, Jurcau A, Jurcau MC, Pascalau N et al (2023) Molecular mechanisms of neuroinflammation in aging and Alzheimer's disease progression. Int J Mol Sci 24(3):1869

Atri A (2019a) The Alzheimer's disease clinical spectrum: diagnosis and management. Med Clin North Am 103(2):263–293

Atri A (2019b) Current and future treatments in Alzheimer's disease. Semin Neurol 39:227–240

Azizi G, Navabi SS, Al-Shukaili A, Seyedzadeh MH, Yazdani R, Mirshafiey A (2015) The role of inflammatory mediators in the pathogenesis of Alzheimer's disease. Sultan Qaboos Univ Med J 15(3):e305–e316

Bhat I, Mamatha BS (2021) Genetic factors involved in modulating lutein bioavailability. Nutr Res 91:36–43

Bhatt S, Puli L, Patil CR (2021) Role of reactive oxygen species in the progression of Alzheimer's disease. Drug Discov Today 26(3):794–803

Bivona G, Iemmolo M, Agnello L, Lo Sasso B, Gambino CM, Giglio RV et al (2023) Microglial activation and priming in Alzheimer's disease: state of the art and future perspectives. Int J Mol Sci 24(1):884

Buccellato FR, D'Anca M, Fenoglio C, Scarpini E, Galimberti D (2021) Role of oxidative damage in Alzheimer's disease and neurodegeneration: from pathogenic mechanisms to biomarker discovery. Antioxidants 10(9):1353

Cai Y, Liu J, Wang B, Sun M, Yang H (2022) Microglia in the neuroinflammatory pathogenesis of Alzheimer's disease and related therapeutic targets. Front Immunol [Internet]. [cited 2025 Jul 22];13. Available from: https://www.frontiersin.org/journals/immunology/articles/10.3389/fimmu.2022.856376/full

Chang R, Yee KL, Sumbria RK (2017) Tumor necrosis factor α inhibition for Alzheimer's disease. J Cent Nerv Syst Dis 9:1179573517709278

Cho KS, Shin M, Kim S, Lee SB (2018) Recent advances in studies on the therapeutic potential of dietary carotenoids in neurodegenerative diseases. Oxidative Med Cell Longev 2018(1):4120458

Contador I, Buch-Vicente B, del Ser T, Llamas-Velasco S, Villarejo-Galende A, Benito-León J et al (2024) Charting Alzheimer's disease and dementia: epidemiological insights, risk factors and prevention pathways. J Clin Med 13(14):4100

Daraban BS, Popa AS, Stan MS (2024) Latest perspectives on Alzheimer's disease treatment: the role of blood-brain barrier and antioxidant-based drug delivery systems. Molecules 29(17):4056

Davinelli S, Ali S, Solfrizzi V, Scapagnini G, Corbi G (2021) Carotenoids and cognitive outcomes: a meta-analysis of randomized intervention trials. Antioxidants 10(2):223

DeTure MA, Dickson DW (2019) The neuropathological diagnosis of Alzheimer's disease. Mol Neurodegener 14(1):32

Dhapola R, Beura SK, Sharma P, Singh SK, HariKrishnaReddy D (2024) Oxidative stress in Alzheimer's disease: current knowledge of signaling pathways and therapeutics. Mol Biol Rep 51(1):48

Dhas N, Mehta T (2020) Cationic biopolymer functionalized nanoparticles encapsulating lutein to attenuate oxidative stress in effective treatment of Alzheimer's disease: a non-invasive approach. Int J Pharm 586:119553

dos Santos PP, Andrade L d A, Flôres SH, Rios A d O (2018) Nanoencapsulation of carotenoids: a focus on different delivery systems and evaluation parameters. J Food Sci Technol 55(10):3851–3860

Dumont M, Beal MF (2011) Neuroprotective strategies involving ROS in Alzheimer disease. Free Radic Biol Med 51(5):1014–1026

Elkholy NS, Shafaa MW, Mohammed HS (2020) Biophysical characterization of lutein or beta carotene-loaded cationic liposomes. RSC Adv 10(54):32409–32422

Erdman JW, Smith JW, Kuchan MJ, Mohn ES, Johnson EJ, Rubakhin SS et al (2015) Lutein and brain function. Foods 4(4):547–564

Fernandes F, Dias-Teixeira M, Delerue-Matos C, Grosso C (2021) Critical review of lipid-based nanoparticles as carriers of neuroprotective drugs and extracts. Nano 11(3):563

Firozjae AA, Shiran MR, Rashidi M (2025) The neuropharmacological and clinical effects of lutein: a systematic review. Horm Mol Biol Clin Investig 46(1):27–38

Flieger J, Forma A, Flieger W, Flieger M, Gawlik PJ, Dzierżyński E et al (2024) Carotenoid supplementation for alleviating the symptoms of Alzheimer's disease. Int J Mol Sci 25(16):8982

Fuad NIN, Sekar M, Gan SH, Lum PT, Vaijanathappa J, Ravi S, et al (2020) Lutein: a comprehensive review on its chemical, biological activities and therapeutic potentials. [cited 2025 Jul 28]. Available from: https://www.phcogj.com/article/1291

Gargantiel M, Faller E, Kumar D, Tiwari P (2023) Nutraceuticals as therapeutic interventions in Alzheimer's disease. One Health Bull 3(1):4

Gazzolo D, Picone S, Gaiero A, Bellettato M, Montrone G, Riccobene F et al (2021) Early pediatric benefit of lutein for maturing eyes and brain—an overview. Nutrients 13:3239.

[Internet]. s Note: MDPI stays neutral with regard to jurisdictional claims in published ...; 2021 [cited 2025 Jul 21]. Available from: https://pdfs.semanticscholar.org/1640/ac0d4103ae-e7d37c672e8807c742dbed456f.pdf

Gee MS, Son SH, Jeon SH, Do J, Kim N, Ju YJ et al (2020) A selective p38α/β MAPK inhibitor alleviates neuropathology and cognitive impairment, and modulates microglia function in 5XFAD mouse. Alzheimer's Res Ther 12(1):45

Gualtieri P, Frank G, Cianci R, Ciancarella L, Romano L, Ortoman M et al (2025) Exploring the efficacy and safety of nutritional supplements in Alzheimer's disease. Nutrients 17(5):922

Hoffman L, Bloemer J (2021) Chapter 5 - Side effects of drugs used in the treatment of Alzheimer's disease. In: Ray SD, editor. Side effects of drugs annual [Internet]. Elsevier. [cited 2025 Aug 17]. pp. 71–77. Available from: https://www.sciencedirect.com/science/article/pii/S0378608021000349

Influence of microglia and astrocyte activation in the neuroinflammatory pathogenesis of Alzheimer's disease: rational insights for the therapeutic approaches - ScienceDirect [Internet]. [cited 2025 Aug 10]. Available from: https://www.sciencedirect.com/science/article/abs/pii/S0967586818309214

Ionescu-Tucker A, Cotman CW (2021) Emerging roles of oxidative stress in brain aging and Alzheimer's disease. Neurobiol Aging 107:86–95

Johnson EJ (2004) A biological role of lutein. Food Rev Int 20(1):1–16

Johnson EJ (2012) A possible role for lutein and zeaxanthin in cognitive function in the elderly12345. Am J Clin Nutr 96(5):1161S–1165S

Johnson EJ (2014) Role of lutein and zeaxanthin in visual and cognitive function throughout the lifespan. Nutr Rev 72(9):605–612

Kaur D, Sharma V, Deshmukh R (2019) Activation of microglia and astrocytes: a roadway to neuroinflammation and Alzheimer's disease. Inflammopharmacology 27(4):663–677

Kim DS, Kang S, Moon NR, Shin BK, Park S (2024) Zeaxanthin and lutein ameliorate Alzheimer's disease-like pathology: modulation of insulin resistance, neuroinflammation, and acetylcholinesterase activity in an amyloid-β rat model. Int J Mol Sci 25(18):9828

Kumar P, Banik SP, Ohia SE, Moriyama H, Chakraborty S, Wang CK et al (2024) Current insights on the photoprotective mechanism of the macular carotenoids, Lutein and Zeaxanthin: safety, efficacy and bio-delivery. J Am Nutr Assoc 43(6):505–518

Kumar R, Pal OP, Kamboj N, Jaiswal A (2025) Phytonutrients in neuroprotection and neurodegenerative disorders. Curr Res Nutr Food Sci J 13(Special Issue Phytonutrients July 2025):72–96

Lee JK, Kim NJ (2017) Recent advances in the inhibition of p38 MAPK as a potential strategy for the treatment of Alzheimer's disease. Mol J Synth Chem Nat Prod Chem 22(8):1287

Lee KH, Cha M, Lee BH (2020) Neuroprotective effect of antioxidants in the brain. Int J Mol Sci 21(19):7152

Leng F, Edison P (2021) Neuroinflammation and microglial activation in Alzheimer disease: where do we go from here? Nat Rev Neurol 17(3):157–172

Li J, Abdel-Aal ESM (2021) Dietary lutein and cognitive function in adults: a meta-analysis of randomized controlled trials. Molecules 26(19):5794

Li LH, Lee JCY, Leung HH, Lam WC, Fu Z, Lo ACY (2020) Lutein supplementation for eye diseases. Nutrients 12(6):1721

Li Z, Cao Z, Chen F, Li B, Jin H (2024) Lutein inhibits glutamate-induced apoptosis in HT22 cells via the Nrf2/HO-1 signaling pathway. Front Neurosci 18:1432969

Limitations of Alzheimer's Disease Medications | NeuroPharmac J [Internet]. [cited 2025 Aug 10]. Available from: https://www.neuropharmac.com/limitations-of-alzheimers-disease-medications/

Lindbergh CA, Renzi-Hammond LM, Hammond BR, Terry DP, Mewborn CM, Puente AN et al (2018) Lutein and Zeaxanthin influence brain function in older adults: a randomized controlled trial. J Int Neuropsychol Soc 24(1):77–90

Liu T, Hong LW, Sheng ZJ, Zheng MF, Wang H (2017) Lutein protects against β-amyloid peptide-induced oxidative stress in cerebrovascular endothelial cells through modulation of Nrf-2 and NF-κb. Cell Biol Toxicol 33(1):57–67

Lopez OL, Kuller LH (2019) Chapter 9 - Epidemiology of aging and associated cognitive disorders: prevalence and incidence of Alzheimer's disease and other dementias. In: Dekosky ST, Asthana S, editors. Handbook of clinical neurology [Internet]. Elsevier. [cited 2025 Aug 10]. pp. 139–48. (Geriatric Neurology; vol. 167). Available from: https://www.sciencedirect.com/science/article/pii/B9780128047668000091

Lopresti AL, Smith SJ, Drummond PD (2022) The effects of Lutein and Zeaxanthin supplementation on cognitive function in adults with self-reported mild cognitive complaints: a randomized, double-blind, placebo-controlled study. Front Nutr [Internet]. [cited 2025 Aug 9];9. Available from: https://www.frontiersin.org/journals/nutrition/articles/10.3389/fnut.2022.843512/full

Loskutova E, Shah K, Flitcroft ID, Setti A, Butler JS, Nolan Y, et al (2019) Lutein and zeaxanthin: the possible contribution, mechanisms of action and implications of modern dietary intake for cognitive development in children. [Internet]. HRB Open Research. [cited 2025 Jul 22]. Available from: https://hrbopenresearch.org/articles/2-8

Lutein suppresses inflammatory responses through Nrf2 activation and NF-κB inactivation in lipopolysaccharide-stimulated BV-2 microglia - Wu - 2015 - Mol Nutr Food Res - Wiley Online Library [Internet]. [cited 2025 Aug 10]. Available from: https://onlinelibrary.wiley.com/doi/abs/10.1002/mnfr.201500109

Ma Y, You T, Wang J, Jiang Y, Niu J (2022) Research progress on construction of lutein-loaded nano delivery system and their improvements on the bioactivity. Coatings 12(10):1449

Madaan T, Choudhary AN, Gyenwalee S, Thomas S, Mishra H, Tariq M et al (2017) Lutein, a versatile phyto-nutraceutical: an insight on pharmacology, therapeutic indications, challenges and recent advances in drug delivery. Pharma Nutr 5(2):64–75

Maghsoudi S, Taghavi Shahraki B, Rabiee N, Fatahi Y, Bagherzadeh M, Dinarvand R et al (2022) The colorful world of carotenoids: a profound insight on therapeutics and recent trends in nano delivery systems. Crit Rev Food Sci Nutr 62(13):3658–3697

Matsui M, Murata T, Kurobe-Takashima Y, Ikeda T, Noguchi-Shinohara M, Ono K et al (2024) Lutein from chicken eggs prevents amyloid β-peptide aggregation in vitro and amyloid β-induced inflammation in human macrophages (THP-1). ACS Omega 9(24):26616–26627

Mecocci P, Tinarelli C, Schulz RJ, Polidori MC (2014) Nutraceuticals in cognitive impairment and Alzheimer's disease. Front Pharmacol [Internet]. [cited 2025 Aug 10];5. Available from: https://www.frontiersin.org/journals/pharmacology/articles/10.3389/fphar.2014.00147/full

Merighi S, Nigro M, Travagli A, Gessi S (2022) Microglia and Alzheimer's disease. Int J Mol Sci 23(21):12990

Mimica N, Presečki P (2009) Side effects of approved antidementives. Psychiatr Danub 21(1):108–113

Mirzaei F, Bhatnagar K, Karingapara AS, Kumar AS, Agbaria L (2024) Carotenoids in Alzheimer's disease and dementia. In: Moradikor N, Chatterjee I, Mohamed W, editors. Nutrition in brain aging and dementia [Internet]. Singapore: Springer Nature; [cited 2025 Jul 22]. p. 193–222. Available from: https://doi.org/10.1007/978-981-97-4117-5_10

Mohammadzadeh Honarvar N, Saedisomeolia A, Abdolahi M, Shayeganrad A, Taheri Sangsari G, Hassanzadeh Rad B et al (2017) Molecular anti-inflammatory mechanisms of retinoids and carotenoids in Alzheimer's disease: a review of current evidence. J Mol Neurosci 61(3):289–304

Mohn ES, Erdman JW Jr, Kuchan MJ, Neuringer M, Johnson EJ (2017) Lutein accumulates in subcellular membranes of brain regions in adult rhesus macaques: relationship to DHA oxidation products. PLoS One 12(10):e0186767

Mrowicka M, Mrowicki J, Kucharska E, Majsterek I (2022) Lutein and Zeaxanthin and their roles in age-related macular degeneration—neurodegenerative disease. Nutrients 14(4):827

Nazari L, Komaki S, Salehi I, Raoufi S, Golipoor Z, Kourosh-Arami M et al (2022) Investigation of the protective effects of lutein on memory and learning using behavioral methods in a male rat model of Alzheimer's disease. J Funct Foods 99:105319

Neuroprotective effects of natural compounds on LPS-induced inflammatory responses in microglia - PMC [Internet]. [cited 2025 Jul 23]. Available from: https://pmc.ncbi.nlm.nih.gov/articles/PMC7344058/

Nouchi R, Suiko T, Kimura E, Takenaka H, Murakoshi M, Uchiyama A et al (2020) Effects of Lutein and Astaxanthin intake on the improvement of cognitive functions among healthy adults: a systematic review of randomized controlled trials. Nutrients 12(3):617

Nutrition Today [Internet]. [cited 2025 Jul 27]. Available from: https://journals.lww.com/nutritiontodayonline/abstract/2017/07000/lutein_and_cognition_across_the_lifespan.7.aspx

Oxidative stress: the core pathogenesis and mechanism of Alzheimer's disease—ScienceDirect [Internet]. [cited 2025 Aug 10]. Available from: https://www.sciencedirect.com/science/article/abs/pii/S1568163722000617

Oxidative stress-A direct bridge to central nervous system homeostatic dysfunction and Alzheimer's disease - Anwar - 2022 - Cell Biochemistry and Function - Wiley Online Library [Internet]. [cited 2025 Jul 25]. Available from: https://analyticalsciencejournals.onlinelibrary.wiley.com/doi/abs/10.1002/cbf.3673

Pandey SN, Singh G, Semwal BC, Gupta G, Alharbi KS, Almalki WH et al (2022) Therapeutic approaches of nutraceuticals in the prevention of Alzheimer's disease. J Food Biochem 46(12):e14426

Pandey S, Gupta A, Mahato DK, Paul V, Tripathi AD, Rasane P et al (2025) Lutein and Zeaxanthin: source, extraction, stability, bioactivity, and functional food applications. Curr Pharm Biotechnol

Pap R, Pandur E, Jánosa G, Sipos K, Agócs A, Deli J (2021) Lutein exerts antioxidant and anti-inflammatory effects and influences iron utilization of BV-2 microglia. Antioxidants 10(3):363

Park HA, Hayden MM, Bannerman S, Jansen J, Crowe-White KM (2020) Anti-apoptotic effects of carotenoids in neurodegeneration. Molecules 25(15):3453

Passeri E, Elkhoury K, Morsink M, Broersen K, Linder M, Tamayol A et al (2022) Alzheimer's disease: treatment strategies and their limitations. Int J Mol Sci 23(22):13954

Patel C, Patel P, Sarkar D, Bulsara J, Soni A, Isapure K et al (2021) Neuroprotective effect of lutein in scopolamine-induced Alzheimer's disease in mice and zebrafish. Rev Bras 31(6):762–771

Petrovic S, Arsic A, Ristic-Medic D, Cvetkovic Z, Vucic V (2020) Lipid peroxidation and antioxidant supplementation in neurodegenerative diseases: a review of human studies. Antioxidants 9(11):1128

Porsteinsson AP, Isaacson RS, Knox S, Sabbagh MN, Rubino I (2021) Diagnosis of early Alzheimer's disease: clinical practice in 2021. J Prev Alzheimers Dis 8(3):371–386

Prathyusha P, Viswanathan G, Tomcy AT, Binitha PP, Bava SV, Sindhu ER (2025) Lutein and inflammation: a comprehensive review of its mechanisms of action. Explor Drug Sci 3:100885

Ramirez M. Lipides et cerveau. [cited 2025 Jul 21]; Available from: https://pdfs.semanticscholar.org/62d8/ecbcd8204df553a2fa3fc6cc10cb2dc4e0b6.pdf

Rana A, Awasthi A, Kumar D, Singh S, Singh S (2018) Alzheimer's disease silent killer of memory: a review on pathological mechanisms. J Alzheimers Neurodegener Dis 4:17

Ribaya-Mercado JD, Blumberg JB (2004) Lutein and Zeaxanthin and their potential roles in disease prevention. J Am Coll Nutr 23(sup6):567S–587S

Shi H, Nolan JM, Flynn R, Prado-Cabrero A (2024) Beyond food colouring: lutein-food fortification to enhance health. Food Biosci 59:104085

Singh D (2022) Astrocytic and microglial cells as the modulators of neuroinflammation in Alzheimer's disease. J Neuroinflammation 19(1):206

Sivamaruthi BS, Raghani N, Chorawala M, Bhattacharya S, Prajapati BG, Elossaily GM et al (2023) NF-κB pathway and its inhibitors: a promising frontier in the management of Alzheimer's disease. Biomedicine 11(9):2587

Stringham JM, Johnson EJ, Hammond BR (2019) Lutein across the lifespan: from childhood cognitive performance to the aging eye and brain. Curr Dev Nutr 3(7):nzz066

Su W, Xu W, Liu E, Su W, Polyakov NE (2023) Improving the treatment effect of carotenoids on Alzheimer's disease through various nano-delivery systems. Int J Mol Sci 24(8):7652

Taso OV, Philippou A, Moustogiannis A, Zevolis E, Koutsilieris M (2019) Lipid peroxidation products and their role in neurodegenerative diseases. Ann Res Hosp [Internet]. [cited 2025 Jul 26];3. Available from: https://arh.amegroups.org/article/view/4663

Thakur S, Dhapola R, Sarma P, Medhi B, Reddy DH (2023) Neuroinflammation in Alzheimer's disease: current progress in molecular signaling and therapeutics. Inflammation 46(1):1–17
The role of TNF-α in Alzheimer's disease: a narrative review [Internet]. [cited 2025 Jul 22]. Available from: https://www.mdpi.com/2073-4409/13/1/54
The safety and tolerability of donepezil in patients with Alzheimer's disease - Jackson - 2004 - Br J Clin Pharmacol - Wiley Online Library [Internet]. [cited 2025 Aug 17]. Available from: https://bpspubs.onlinelibrary.wiley.com/doi/full/10.1111/j.1365-2125.2004.01848.x
Twarowski B, Herbet M (2023) Inflammatory processes in Alzheimer's disease—pathomechanism, diagnosis and treatment: a review. Int J Mol Sci 24(7):6518
Valgimigli L (2023) Lipid peroxidation and antioxidant protection. Biomolecules 13(9):1291
Verardi A, Sangiorgio P, Lopresto CG, Casella P, Errico S (2023) Enhancing carotenoids' efficacy by using chitosan-based delivery systems. Nutraceuticals 3(3):451–480
Vishwanathan R, Neuringer M, Snodderly DM, Schalch W, Johnson EJ (2013) Macular lutein and zeaxanthin are related to brain lutein and zeaxanthin in primates. Nutr Neurosci 16(1). https://doi.org/10.1179/1476830512Y.0000000024
Wang W, Connor SL, Johnson EJ, Klein ML, Hughes S, Connor WE (2007) Effect of dietary lutein and zeaxanthin on plasma carotenoids and their transport in lipoproteins in age-related macular degeneration. Am J Clin Nutr 85(3):762–769
Wang W, Shinto L, Connor WE, Quinn JF (2008) Nutritional biomarkers in Alzheimer's disease: the association between carotenoids, n-3 fatty acids, and dementia severity. J Alzheimer's Dis 13(1):31–38
Wang J, Li L, Wang Z, Cui Y, Tan X, Yuan T et al (2018) Supplementation of lycopene attenuates lipopolysaccharide-induced amyloidogenesis and cognitive impairments via mediating neuroinflammation and oxidative stress. J Nutr Biochem 56:16–25
Wang H, Lautrup S, Caponio D, Zhang J, Fang EF (2021) DNA damage-induced neurodegeneration in accelerated ageing and Alzheimer's disease. Int J Mol Sci 22(13):6748
Wang C, Zong S, Cui X, Wang X, Wu S, Wang L, et al (2023) The effects of microglia-associated neuroinflammation on Alzheimer's disease. Front Immunol [Internet]. [cited 2025 Jul 22];14. Available from: https://www.frontiersin.org/journals/immunology/articles/10.3389/fimmu.2023.1117172/full
Welch G, Tsai L (2022) Mechanisms of DNA damage-mediated neurotoxicity in neurodegenerative disease. EMBO Rep 23(6):e54217
Xie K, Ngo S, Rong J, Sheppard A (2019) Modulation of mitochondrial respiration underpins neuronal differentiation enhanced by lutein. Neural Regen Res 14(1):87
Ye J, Cheng J, Xiong R, Chen H, Huang S, Li H et al (2024) Effects and mechanisms of lutein on aging and age-related diseases. Antioxidants 13(9):1114
Zheng Y, Zhang X, Zhang R, Wang Z, Gan J, Gao Q et al (2023) Inflammatory signaling pathways in the treatment of Alzheimer's disease with inhibitors, natural products and metabolites (Review). Int J Mol Med 52(5):1–42
Zia-Ul-Haq M (2021) Past, present and future of carotenoids research. In: Zia-Ul-Haq M, Dewanjee S, Riaz M, editors. Carotenoids: structure and function in the human body [internet]. Cham: Springer. [cited 2025 Aug 10]. pp. 827–54. Available from: https://doi.org/10.1007/978-3-030-46459-2_28
Zivari-Ghader T, Valioglu F, Eftekhari A, Aliyeva I, Beylerli O, Davran S, et al (2024) Recent progresses in natural based therapeutic materials for Alzheimer's disease. Heliyon [Internet]. [cited 2025 Aug 6];10(4). Available from: https://www.cell.com/heliyon/abstract/S2405-8440(24)02382-X
Zvěřová M (2019) Clinical aspects of Alzheimer's disease. Clin Biochem 72:3–6

Curcumin as a Natural Mammalian Target of Rapamycin and Autophagy Modulator

6

Chathurika Elvitigala, Tharangika Bowange, Nalaka Wijekoon, and Lakmal Gonawala

Abstract

Nearly 40% of all new molecules submitted for Food and Drug Administration approval are either natural products or derivatives. Curcumin derived from the rhizome of *Curcuma longa* exhibits a wide pharmacological spectrum, encompassing antioxidant properties, suppression of neuroinflammatory cytokines, and regulation of cellular stress responses, and regulate autophagy and mammalian target of rapamycin (mTOR) signaling. Disruptions in mTOR signaling pathways are associated with neurodegenerative diseases. In response to oxidative stress, mTOR can modulate processes such as apoptosis and autophagy, often promoting tissue repair under. Additionally, the interaction between mTOR and other signaling cascades is reported to be crucial in human health. Notably, mTOR is linked with key pathways, including NF-κB, Akt, phosphatidylinositol 3-kinase (PI3K), AMPK, TFEB, and ERK1/2, which collectively influence inflammation, programmed cell death, cell survival, and oxidative stress. Curcumin has been reported to suppress the PI3K/Akt/mTOR signaling cascade and activate autophagy via AMPK, ULK1, and Beclin-1 pathways, promoting the degradation of misfolded proteins and damaged organelles. This chapter aims to explore the mechanistic role of curcumin in regulation of mTOR and autophagy pathways promoting neuroprotection.

C. Elvitigala
Agrobiotech, Sri Lanka Institute of Biotechnology (SLIBTEC), Homagama, Sri Lanka

T. Bowange
Laboratory Management, Sri Lanka Institute of Biotechnology (SLIBTEC), Homagama, Sri Lanka

N. Wijekoon · L. Gonawala (✉)
Molecular Diagnostic and Vaccine, Sri Lanka Institute of Biotechnology (SLIBTEC), Homagama, Sri Lanka
e-mail: lakmal.gonawala@slibtec.gov.lk

H. Dilnashin, S. P. Singh (eds.), *Phototherapeutic Approaches to Neurodegeneration*,
https://doi.org/10.1007/978-981-95-4751-7_6

Keywords

Curcumin · Mammalian target of rapamycin · Autophagy

6.1 Introduction

Neurodegenerative diseases, including Alzheimer's disease (AD), Parkinson's disease (PD), and Huntington's disease (HD), represent a group of disorders characterized by the gradual loss of structure and function of neurons. A hallmark feature of these conditions is the abnormal accumulation of misfolded or aggregated proteins such as amyloid β (Aβ), tau, and α synuclein which disrupt neuronal homeostasis and trigger cell death pathways These pathological protein aggregates often arise from impaired proteostasis and defective cellular clearance mechanisms, particularly autophagy dysfunction (Heras-Sandoval et al. 2020).

Autophagy, a highly conserved lysosome-mediated degradation process, is essential for the removal of damaged organelles and toxic protein species. By maintaining cellular quality control, autophagy plays a vital role in neuronal survival, especially in postmitotic neurons of the central nervous system. The mammalian target of rapamycin (mTOR) functions as a central regulator of autophagy, integrating signals from nutrients, growth factors, and cellular stressors to control protein synthesis, metabolism, and degradation pathways. Dysregulation of the mTOR autophagy axis has been strongly implicated in the onset and progression of neurodegenerative diseases, leading to impaired protein clearance, oxidative stress, and heightened inflammatory responses (Heras-Sandoval et al. 2020).

In recent years, naturally occurring bioactive compounds have emerged as potential therapeutic agents for modulating autophagy and mTOR signaling. Curcumin, a polyphenolic compound derived from *Curcuma longa* (turmeric), has shown promising neuroprotective effects in experimental models. It can regulate the PI3K/Akt/mTOR pathway, restore autophagic balance, and reduce neuroinflammation, thereby mitigating neuronal damage. Beyond its autophagy modulating role, curcumin's antioxidant and anti-inflammatory properties enable it to counteract multiple pathogenic processes associated with neurodegeneration (Wang et al. 2025).

This chapter provides an in-depth discussion of the interplay between mTOR signaling and autophagy in neurodegeneration, emphasizing the mechanistic basis and therapeutic promise of curcumin as a modulator of these pathways.

6.2 Autophagy Pathways: From Molecular Mechanisms to Neuronal Clearance

In 1962, Ashford and Porter made the first visual observation of autophagy, a vital cellular function commonly known as "self-eating," in mouse hepatocytes. It is essential for cell maintenance, development, differentiation, and survival. Based on the manner in which the cargo is delivered to the lysosome, autophagy can be divided into three main types: chaperone-mediated autophagy, microautophagy, and macroautophagy (Doherty and Baehrecke 2018; Fang et al. 2017).

Macroautophagy is the most ubiquitous type of autophagy, distinguished by the development of autophagosomes and its ability to eliminate protein aggregates, organelles, and ribosomes. The autophagosome then merges with a lysosome, becoming an autolysosome in which lysosomal acid proteases dismantle the encapsulated material. Multiple upstream signals, encompassing both mTOR-dependent and mTOR-independent pathways, can modulate the macroautophagic process. The serine/threonine kinase mTOR serves as a principal negative regulator of macroautophagy, and its inhibition activates the macroautophagic cascade. In contrast, signaling pathways such the cAMP–Epac–PLC-ε–IP3 cascade and the Ca^{2+} -calpain-G-stimulatory protein α pathway can regulate macroautophagic responses independently of mTOR. (Wu et al. 2015).

6.2.1 CrossTalk between Ubiquitin-Proteasome System and Autophagy Systems

The two principal degradation and recycling mechanisms identified with eukaryotes that have been conserved throughout evolution are ubiquitin-proteasome system (UPS) and autophagy. Recent research indicates that there are connections and crosstalks between these two systems, despite the fact that their operations are not dependent on one another. Cellular materials that accumulate after one degradative system is inhibited must be removed, at least partially, by the other system in order to preserve homeostasis (Kocaturk and Gozuacik 2018).

The UPS and autophagy pathways are both initiated by the endoplasmic reticulum (ER). In endoplasmic reticulum (ER) stress, misfolded and/or unfolded proteins accumulate in the ER lumen due to oxidative stress, disruptions in calcium homeostasis, and circumstances resulting in errors in protein glycosylation or folding. Endoplasmic reticulum (ER) stress has resulted in the evolution of ER-specific stress response mechanisms, including the unfolded protein response (UPR) and the ER-associated degradation pathways (Kocaturk and Gozuacik 2018). The two pathways possess direct or indirect associations with autophagy and the ubiquitin-proteasome system. The UPR signal modulates autophagy via influencing the activities of mTOR complex 1 (mTORC1), AKT, and AMPK. Activated PERK enhances the synthesis of DNA-damage-inducible transcript 4 (DDIT4) and sestrin 2 (SESN2) via facilitating the translation of activating transcription factor 4 (ATF4). SESN2 and DDIT4 directly diminish mTORC1 activity. Concurrently, ATF4 suppresses AKT activity, which subsequently diminishes mTORC1 activity via the DDIT3/TRIB3 pathway. The PERK signaling pathway can concurrently activate AMPK. Activated AMPK not only directly activates ULK1 but also suppresses mTORC1 activity. The UPR can influence the transcription of essential autophagy-related proteins as well as the activity of these proteins (Li et al. 2022; Wu et al. 2015).

6.2.2 UPS and Autophagy in Neurodegenerative Disease

Neural development, the preservation of neuronal homeostasis, and the assurance of neuronal survival are all contingent upon autophagy. Moreover, autophagy is essential in neuronal physiology by regulating synaptogenesis, dendritic spine formation, and axonal homeostasis. Disruption of neuronal autophagy leads to defects in cargo-specific axonal transport, resulting in axonal swelling and dystrophic alterations. Furthermore, due to impaired synaptic pruning mechanisms, there exists an overabundance of dendritic spine density. Autophagy is essential for synapse growth and brain plasticity, with autophagosome biogenesis occurring in proximity to synapses. Autophagic activity influences presynaptic neurotransmission by altering axonal endoplasmic reticulum calcium release in hippocampal neurons, whereas rapamycin-induced macroautophagy rapidly inhibits neurotransmitter release in dopaminergic neurons. Moreover, autophagy is essential for cognitive function, information processing, and memory consolidation. These findings indicate that autophagy is a crucial step for optimal brain function (Ban et al. 2013; Kuijpers et al. 2021; Valencia et al. 2021; Wang et al. 2019). Despite ongoing debate about whether neurodegeneration is a cause or consequence of impaired UPS function, experimental evidence typically associates neurodegeneration with this disease. It is proposed that autophagy contributes to the onset or advancement of certain neurodegenerative disorders (Kocaturk and Gozuacik 2018).

6.2.3 Dysregulation of mTOR-Autophagy Axis in Neurodegenerative Pathologies

6.2.3.1 mTOR-Autophagy

The 289 kDa serine/threonine protein kinase mTOR regulates and impacts protein synthesis, cytoskeletal organization, autophagy, metabolism, cell growth, and cell survival within eukaryotic cells with the aid of two distinct protein complexes, mTOR complex 1 (mTORC1) and mTOR complex 2 (mTORC2). Upstream regulators of the mTOR pathway include growth factors, nutrition, and factors associated with hypoxic stress. mTORC1 and mTORC2 modify microtubule architecture, autophagy, protein synthesis, and cell survival by influencing downstream targets. However, the mechanisms underlying the function of mTORC2 are not as well understood as those of the mTORC1 complex. mTOR is essential in a number of neurodegenerative diseases (Saxton and Sabatini 2017; Wang et al. 2022). Autophagic processes, which eliminate unwanted proteins and maintain homeostasis, are frequently impaired in neurodegenerative diseases. The main signaling pathway that controls autophagy is mTOR. The complex interplay between stimulatory and inhibitory signals tightly controls mTOR activity, which contributes to the various autophagy stages (Maiese 2016; Narayanan et al. 2024; Panwar et al. 2023).

6.2.3.2 Hyperactive mTOR and Inhibition of Autophagy in Alzheimer's Disease

AD pathology is intricately associated with dysregulated mTOR signaling, which intensifies amyloid-beta (Aβ) accumulation and synaptic impairment. Aβ aggregates, due to their hydrophobic properties, activate mTOR through the PI3K/Akt pathway, resulting in the hyperphosphorylation of downstream targets such as p70S6K and p4E-BP1. Hyperactivation of mTOR inhibits autophagy, thereby impairing Aβ clearance and facilitating plaque formation. Decreased PTEN expression in Alzheimer's disease brains exacerbates Akt/mTORC1 signaling by not dephosphorylating PIP3, consequently inhibiting autophagy and aggravating Aβ pathology. Simultaneously, Aβ-induced insulin desensitization impedes Akt activation, establishing a detrimental cycle that perpetuates mTOR hyperactivity and neurodegeneration (Mueed et al. 2019; Perluigi et al. 2021; Schmukler and Pinkas-Kramarski 2020).

6.2.3.3 Impaired Autophagosome-Lysosome Fusion in Parkinson's α-Synucleinopathy

Impaired autophagosome–lysosome fusion is a critical pathogenic mechanism in PD, especially in α-synucleinopathy, where abnormal accumulation of α-synuclein in dopaminergic neurons drives neurodegeneration (Gokuladhas et al. 2025). Mutated or post-translationally modified α-synuclein disrupts multiple steps in the autophagy lysosomal pathway, notably blocking the fusion of autophagosomes with lysosomes, an essential step for autophagosome maturation and substrate degradation. This blockade results in autophagosomes loaded with undegraded α-synuclein, intensifying cellular toxicity and facilitating aggregate propagation. Mechanistically, α-synuclein interferes with SNARE complex formation, decreasing SNAP29 levels and disrupting ykt6–SNAP29 interactions critical for membrane fusion. Additionally, elevated farnesyltransferase activity negatively regulates ykt6, but pharmacological farnesyltransferase inhibition has been shown to restore macroautophagy and fusion efficiency in PD models, highlighting a promising therapeutic avenue (Gokuladhas et al. 2025).

Abnormal mTOR signaling further compounds autophagy impairment in PD. mTOR hyperactivation inhibits autophagy initiation, reduces lysosomal biogenesis, and contributes to defective clearance of α-synuclein aggregates, thereby exacerbating autophagosome lysosome fusion failure. Genetic mutations in PD-related genes such as *LRRK2*, *parkin*, *PINK1*, *GBA*, and *ATP13A2* impair lysosomal acidification, hydrolase trafficking, and overall lysosomal function, creating a feedback loop that promotes α-synuclein accumulation and neurotoxicity. This dysregulated mTOR activity, combined with lysosomal dysfunction, perpetuates impaired autophagic flux and accelerates PD progression, underscoring the importance of targeting both α-synuclein clearance mechanisms and mTOR modulation in therapeutic strategies (Zhang et al. 2025; Zhu et al. 2019).

6.2.3.4 Targeting mTOR-Regulated Autophagy in Huntington's Disease and Spinocerebellar Ataxias

mTOR signaling plays a pivotal role in regulating autophagy in both HD and spinocerebellar ataxias (SCAs), influencing disease progression through its control of protein aggregation and clearance. In HD, caused by mutant huntingtin (mHTT) with expanded polyglutamine repeats, impaired autophagy contributes to toxic protein accumulation. Pharmacological inhibition of mTOR, such as with the small molecule J3, enhances autophagic flux, promoting the selective degradation of soluble and insoluble mHTT, restoring neuronal markers like DARPP-32, and improving motor phenotypes in HD models (Long et al. 2022). Similarly, in Q175 HD knock-in mice, mTOR inhibition facilitated neuronal autophagy, promoted the clearance of mHTT localized in autophagic vacuoles, and increased colocalization with autophagy receptors p62/SQSTM1 and ubiquitin, supporting the role of autophagy in degrading toxic aggregates (Stavrides et al. 2025). In Machado–Joseph disease (SCA3), mutant ataxin-3 aggregates are linked to neurodegeneration, and both mTOR-dependent (rapamycin/CCI-779) and mTOR-independent (lithium chloride) autophagy inducers have shown potential to reduce aggregate burden in experimental systems. However, combined activation of both pathways in SCA3 mice proved neurotoxic, underscoring the complexity of autophagy modulation in vivo (Duarte-Silva et al. 2016). Mechanistically, mTORC1 suppresses autophagy by inhibiting ULK1 activation and TFEB nuclear translocation, and its inhibition can restore clearance of aggregated proteins (Schmeisser and Parker 2019). These findings indicate that while mTOR inhibition-mediated autophagy activation can be beneficial for clearing pathogenic proteins in HD and SCA, therapeutic strategies must carefully balance activation levels to avoid detrimental effects.

6.3 Curcumin's Multimodal Pharmacodynamics

6.3.1 Structural Determinants (Diarylheptanoid and Diketone Moieties)

In nature, diarylheptanoids are widely distributed. The predominant diarylheptanoid, curcumin (1E,6E)-1,7-Bis(4-hydroxy-3-methoxyphenyl)-1,6-heptadiene-3,5-dione, exhibits a broad range of biological activities relevant to human health. It is the main ingredient in *Curcuma longa* (*C. longa*), also known as turmeric, a perennial plant in the Zingiberaceae family that is grown in subtropical and tropical climates all over the world. It was first cultivated in India, Sri Lanka, Indonesia, and Southeast Asia. Throughout South Asia, curries and other dishes frequently contain turmeric, a spice with a golden hue. Its ability to add color, improve flavor, and preserve food through its antioxidant qualities is well known, but its benefits for health are less well known or appreciated. Turmeric's medicinal qualities have long been recognized in Chinese, Siddha, Ayurvedic, and Unani medicine, and it has been used for a variety of purposes in traditional medicine. Additionally, when its rhizomes are dried and crushed, they produce a tasty yellow powder that has been used for millennia in Indian and Chinese medicine for a variety of pharmacological effects, such as the following: hypocholesterolemic, antidiabetic, antithrombotic, antidiarrheal, carminative, antihepatotoxic, diuretic, antirheumatic, hypotensive, larvicidal, antiviral, antityrosinase, insecticidal, antioxidant, antimicrobial, antiprotozoal, anti-inflammatory, antivenom, antiaging, antiangiogenic, antiproliferative, antimalarial, and antitumor activities. It can also be used to treat autoimmune diseases, parasite infections, ulcers, cold and flu symptoms, and skin conditions (Monroy et al. 2013; Vo et al. 2021).

Curcumin, a polyphenolic compound with a molecular weight of 368 g/mol and melting point of 183 °C, features two aromatic rings containing hydroxyl and methoxy groups connected by a seven-carbon β-diketone linker (Zhai et al. 2020). Its tautomeric equilibrium between keto and enol forms is pH dependent: the keto form dominates in neutral/acidic conditions (pH 3–7), while the enol form stabilizes in basic environments (pH ≥8) through resonance-assisted hydrogen bonding (Kah et al. 2023). Solvent polarity further influences stability, with the enol form favored in nonpolar solvents (5–8 kcal/mol more stable than keto) due to extended π-conjugation (Zhai et al. 2020). The molecule exhibits cis-trans isomerism, where the trans-isomer (phenolic-methoxy groups on opposite sides) is marginally more stable than the cis-form. With a log P of ~3.0 and ground-state dipole moment of 10.77 D, curcumin displays poor aqueous solubility but dissolves in polar solvents like ethanol, DMSO, and chloroform, while remaining insoluble in hydrocarbons like hexane (Kah et al. 2023).

The phenolic groups in curcumin's aromatic rings confer antioxidant and anti-inflammatory properties but also contribute to hydrophobicity and instability at physiological pH, where it degrades via autoxidation to bicyclopentadione and alkaline hydrolysis products (e.g., ferulic acid), which exhibit reduced bioactivity (Bertoncini-Silva et al. 2024; Zhai et al. 2020). Commercial curcumin comprises three curcuminoids: diferuloylmethane (80%, primary bioactive component), demethoxycurcumin (17%), and bisdemethoxycurcumin (3%), collectively governing its therapeutic effects (Forouzanfar et al. 2020; Kah et al. 2023). These structural and stability challenges underscore the need for advanced formulation strategies to enhance bioavailability and preserve its pharmacological integrity.

6.3.2 Redox-Modulating Versus Direct Pathway-Targeting Effects of Curcumin

Oxygen reactions support vital functions like NAD^+/NADH recycling and detoxification while preserving electrical equilibrium among macromolecules, making them crucial for cellular homeostasis. Oxidative stress, which is defined by excessive oxidation that harms proteins, lipids, DNA, and RNA, can result from disturbances in this balance, also known as redox modulation (Fraunberger et al. 2016). By donating hydrogen atoms to neutralize reactive oxygen species (ROS) such as superoxide anions ($O_2^{\bullet-}$) and hydroxyl radicals (•OH), curcumin prevents oxidative damage (Zhai et al. 2020; Kah et al. 2023). Its β-diketone moiety promotes antioxidant enzymes like glutathione peroxidase and superoxide dismutase (SOD) and allows hydrogen bonding with proteins and the chelation of harmful metals (e.g., Cu^{2+} and Fe^{3+}) (Genchi et al. 2024; Priyadarsini 2014). These interactions facilitate coordination with DNA, RNA, and carrier proteins and increase curcumin's bioavailability.

The therapeutic potential of curcumin is reinforced by its pleiotropic effects, which are caused by modulation of different molecular targets (Table 6.1) such as transcription factors, inflammatory cytokines, apoptosis-related proteins, adhesion molecules, and growth factors, thereby establishing its role as a multitarget therapeutic agent (Genchi et al. 2024; Zhou et al. 2011).

6.3.3 Mechanistic Interplay: Curcumin as mTOR-Autophagy Interface Modulator

6.3.3.1 AMPK-Mediated mTOR Inhibition

AMPK is a pivotal metabolic sensor present in all eukaryotes that regulates glucose and lipid metabolism in response to changes in dietary availability and intracellular energy levels. Active AMPK directly phosphorylates the Tuberous sclerosis complex 2 (TSC2) tumor suppressor protein. Phosphorylated TSC2 results in the inactivation of Rheb, which directly associates with and activates the mTORC1 kinase. Lee et al. (2010) posited that AMPK modulates mTOR signaling by the direct phosphorylation of raptor at two conserved serine sites, Ser722 and Ser792. The phosphorylation of raptor by AMPK facilitates 14-3-3 binding to raptor, which is essential for mTORC1 inhibition, energy stress-induced cell cycle arrest, and subsequent autophagy activation. Curcumin has been shown to activate AMPK by elevating intracellular AMP/ATP ratios, possibly via moderate suppression of mitochondrial F0F1-ATPase (Lee et al. 2010).

Curcumin's activation of AMPK promotes mitochondrial biogenesis through the induction of PGC-1α and strengthens antioxidant defenses via Nrf2 signaling, thereby reducing oxidative stress and enhancing neuronal survival. Some studies suggest that curcumin's inhibitory effects on mTOR may be partially independent of AMPK, indicating parallel regulation pathways including phosphatase activation or direct manipulation of the mTOR complex. Nevertheless, AMPK activation continues to be a pivotal modulator of curcumin-induced autophagy, energy equilibrium, and neuroprotection (Perrone et al. 2019; Wang et al. 2025).

6.3.3.2 TFEB Nuclear Translocation and Lysosomal Biogenesis

According to recent research, autophagy depends critically on the transcription factor EB (TFEB). By promoting the expression of genes linked to autophagy, TFEB also aids in the synthesis of autophagosomes and their fusion with lysosomes. Phosphorylation processes, particularly at serine residues, are primarily responsible for controlling TFEB activity. Importantly, a key modulator of TFEB activity is mTORC1, a principal negative regulator of autophagy. In the presence of nutrients, mTORC1 phosphorylates TFEB at multiple serine residues on the lysosomal

Table 6.1 Upregulation and downregulation of molecular targets by curcumin

Group of molecular targets	Molecular targets which get upregulate	Molecular target which get downregulate	Ref.
Transcription factors	Aryl hydrocarbon receptor (AhR), activating transcription factor 3 (ATF3), C/EBP homologous protein (CHOP), electrophile response element (EpRE), peroxisome proliferator-activated receptor-gamma (PPAR-γ), and Nrf2	NF-κB, signal transducers and activators of transcription (STATs), activator protein-1 (AP-1), hypoxia-inducible factor-1 (HIF-1), Notch-1, early growth response-1 (Egr-1), and β-catenin	Zhou et al. (2011)
Protein kinases		mTOR, mitogen-activated protein kinases (MAPKs such as ERK, JNK, and p38 MAPK), focal adhesion kinase (FAK), Janus kinase (JAK), protein kinase C (PKC), and serine/threonine kinase Akt (protein kinase B)	
Growth factors and growth factor receptors		Epidermal growth factor (EGF), receptor (EGFR), fibroblast growth factor (FGF), vascular endothelial growth factor (VEGF), transforming growth factor-β (TGF-β), hepatocyte growth factor (HGF), androgen (AR), estrogen receptors (ER-α), and human EGFR 2(HER2/neu)	
Enzymes	Hemeoxygenase-1 (HO-1) and Glutamyl cysteine ligase (GCL)	Cyclooxygenase-2 (COX-2), 5-lipoxygenase (5-LOX), matrix metalloproteinases (MMPs), inducible nitric oxide synthase (iNOS), and telomerase	Fraunberger et al. (2016), Nimiya et al. (2016), and Zhou et al. (2011)
Adhesion molecules		Intracellular adhesion molecule-1 (ICAM-1), vascular cell adhesion molecule-1 (VCAM-1), endothelial leukocyte adhesion molecule-1 (ELAM-1), E-selectin, and P-selectin	
Apoptosis-related proteins	Pro-apoptotic members (Bax, Bak, Bim, Puma, and Noxa), caspases (Barzegarzadeh et al. 2021; Boyao et al. 2019; Çakmak et al. 2022; Chongtham and Agrawal 2016), and death receptors (TRAIL-R1/DR4, TRAIL-R2/DR5, FADD)	Anti-apoptotic proteins (Bcl-2, Bcl-xL, and survivin), cellular FLICE inhibitory protein (c-FLIP), and inhibitor of apoptosis proteins (IAP)	
Other molecular targets		Cyclin D1 (cell cycle regulator), HSP70, p53, P300/CREB-binding protein, multidrug resistance protein (MRP), urokinase-type plasminogen activator (uPA), and its receptor (uPAR)	

surface, which makes it easier for TFEB to bind to 14-3-3 proteins and prevents its nuclear translocation and activation (Medina et al. 2017; Wang et al. 2020).

By reducing mTORC1, curcumin promotes lysosomal gene transcription and biogenesis by facilitating the dephosphorylation and nuclear translocation of TFEB. Studies have revealed a curcumin analogue that, independently of mTOR inhibition, directly binds to and activates TFEB in vitro and in vivo. This suggests a number of complementary mechanisms through which curcumin enhances lysosomal function. Increased lysosomal capacity restores effective autophagic flow, which makes it easier to eliminate clumped proteins and damaged organelles. This is a critical component in preventing neuronal failure in AD, PD, and other neurodegenerative diseases. Despite its intrinsic stability limitations, curcumin's bioavailability and retention are improved by the lipid-rich environment of the brain, increasing its therapeutic potential in the central nervous system. Because TFEB activation is essential for maintaining neuronal homeostasis, curcumin not only initiates autophagy but also guarantees its total breakdown by boosting lysosomal machinery. Curcumin also suppresses the NF-κB pathway, which plays a key role in triggering inflammatory reactions that exacerbate neurodegeneration. By preventing NF-κB activation, curcumin reduces neuroinflammation and increases autophagy to protect neurons. Curcumin's dual mechanism of promoting autophagy through mTOR regulation and preventing NF-κB-mediated inflammation makes it a promising multitarget therapeutic drug for the treatment of neurodegenerative diseases (Medina et al. 2017; Perrone et al. 2019; Settembre et al. 2011; Wang et al. 2020; Zhang et al. 2016).

6.3.3.3 Downregulates the PI3K–AKT–mTOR Pathway

Numerous vital physiological processes, such as stress reactions, cell growth and survival, and metabolism, are regulated by the phosphatidylinositol 3-kinase (PI3K)/Akt (protein kinase B)/mTOR signaling pathway (Glaviano et al. 2023). PtdIns(3,4)P2 (PIP2) and PtdIns(3,4,5)P3 (PIP3) are produced from phosphatidylinositol by activated PI3K. Consequently, Akt and phosphatidylinositol-dependent kinase 1 (PDK1) are drawn to the cell membrane. PDK1 and PDK2 (probably mTORC2) phosphorylate Akt at Thr308 and Ser473, respectively (Manning and Toker 2017). Tuberous sclerosis complexes 1 and 2 (TSC1 and TSC2, also known as hamartin and tuberin) inhibit the action of mTOR. The TSC1/TSC2 complex prevents mTOR activation by separating the brain's abundant Ras homologue from mTOR. In order to activate mTOR and destabilize the TSC1/TSC2 complex, Akt phosphorylates TSC2. On the other hand, an increased AMP/ATP ratio and/or the tumor suppressor LKB1 stimulate active AMPK, which inhibits mTOR activation by activating TSC1/TSC2 (Ihlamur et al. 2024; Tchevkina and Komelkov 2012).

The formation of neurotoxic aggregates, such as amyloid-beta and α-synuclein, which are indicative of AD and PD, respectively, is caused by the inhibition of autophagy by hyperactivated neurons. Curcumin affects this pathway in a number of ways. It reduces PI3K activity, which in turn reduces Akt phosphorylation on Ser473 and Thr308, which in turn limits the activity of mTOR complex 1 (mTORC1). This downregulation eliminates the inhibition on autophagy initiation mechanisms by mitigating mTORC1-mediated phosphorylation of downstream targets, such as p70 S6 kinase (S6K1) and 4E-binding protein 1 (4E-BP1) (Forouzanfar et al. 2020; Perrone et al. 2019). Curcumin therapy improved cognitive impairments, increased the autophagic clearance of amyloidogenic peptides, and significantly decreased phosphorylated Akt and mTOR levels in the hippocampus in mouse models of cerebral ischemia and AD. The PI3K inhibitor LY294002 reduced curcumin's neuroprotective effects, demonstrating the critical role PI3K/Akt/mTOR signaling plays in its mode of action. Furthermore, regardless of upstream PI3K signaling, curcumin directly lowers mTOR kinase activity by interfering with the mTOR-raptor interaction in mTORC1. Curcumin appears to activate serine/threonine phosphatases, such as protein phosphatase 2A (PP2A), which strengthens pathway inhibition and increases the dephosphorylation of Akt and mTOR substrates (Perrone et al. 2019; Wang et al. 2025). Curcumin reduces oxidative stress and proinflammatory cytokine production, two factors that lead to mTOR hyperactivation, by mitigating TLR4/p38 MAPK signaling at the junction of neuroinflammation and mTOR dysregulation. According to Wang et al. (2025), these results highlight curcumin's multifaceted suppression of PI3K/Akt/mTOR signaling, which aids in the restoration of autophagic homeostasis, which is crucial for neuronal survival (Fig. 6.1) (Wang et al. 2025).

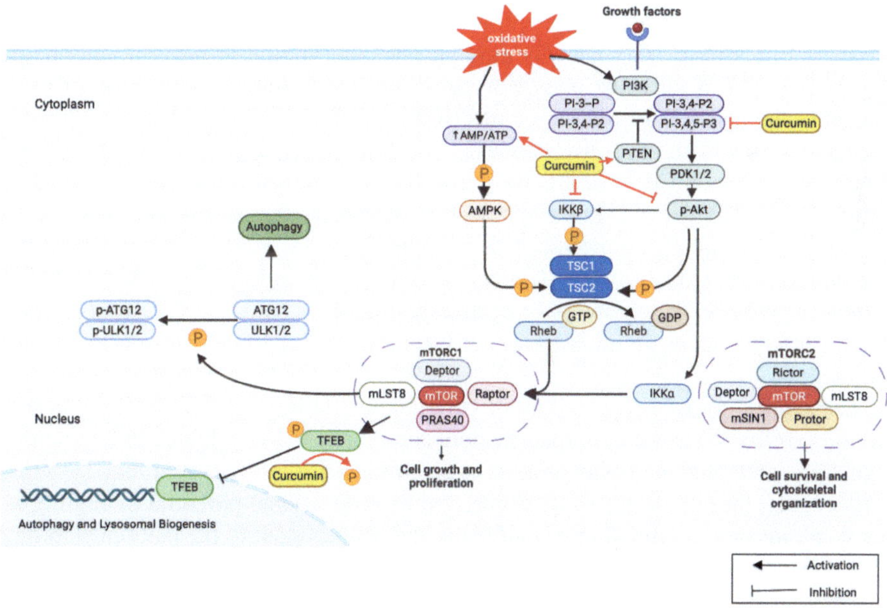

Fig. 6.1 Signal transduction pathways of the PI 3-K, Akt, mTOR, and autophagy cascade

6.3.4 Preclinical, Clinical, and Biomarker-Based Evidence for Curcumin as a Neuroprotective Modulator of mTOR and Autophagy Pathways

Tables 6.2 and 6.3 summarize the preclinical data demonstrating curcumin's neuroprotective potential in the context of neurodegenerative diseases. According to in vivo research summarized in Table 6.2, curcumin has been demonstrated to alter important signaling cascades such as PI3K/Akt/mTOR, NF-κB, AMPK, and BDNF/TrkB in models of multiple sclerosis, Parkinson's disease, Alzheimer's disease, and Huntington's disease. In addition to improving motor coordination, cognitive function, synaptic plasticity, and neuronal survival, these interventions consistently reduced oxidative stress, neuroinflammation, and pathogenic protein aggregation.

During oxidative stress, multiple pathways are affected that involve (i) PI 3-K, Akt, and mTORC1, leading to autophagy ultimately interface with programmed cell death pathways autophagy.

Activation of PI 3-K, such as by tropic factors that include erythropoietin, can promote the production of phosphatidylinositide (3,4)-biphosphate (PI-3,4-P2) and phosphatidylinositide (3,4,5)-triphosphate (PI-3,4,5-P3) that recruits Akt to the plasma membrane. Akt can activate mTORC1 through phosphorylating TSC2 and disrupting the interaction between TSC2 and TSC1. Akt may also activate mTORC1 through I-kappaB kinase (IKK). IKKα associates with Raptor and IKKβ that can phosphorylate TSC1 and suppress TSC1 and its interaction with TSC2. mTORC1 activation also inhibits autophagic proteins, autophagy-related gene 13 (Atg13), and UNC-51 like kinase 1/2(ULK1/2) through phosphorylation to prevent autophagy. (ii) Active AMPK directly phosphorylates the TSC2 tumor suppressor protein. Phosphorylated TSC2 results in the inactivation of Rheb, which directly associates with and activates the mTORC1 kinase. (iii) Autophagy depends critically on the transcription factor EB (TFEB). By promoting the expression of genes linked to autophagy, TFEB also aids in the synthesis of autophagosomes and their fusion with lysosomes. Phosphorylation processes, particularly at serine residues, are primarily responsible for controlling TFEB activity. Importantly, a key modulator of TFEB activity is mTORC1, a principal negative regulator of autophagy. Curcumin as a natural product act as an inhibitor or activator can modulate this process and foster autophagy.

These findings are supported by in vitro data in Table 6.3, which highlights curcumin's ability to reduce apoptosis, restore autophagic activity, lessen inflammatory signaling, and prevent toxic protein aggregation in glial and neuronal systems. Collectively, these findings support curcumin's translational potential by highlighting its pleiotropic modulation of molecular targets pertinent to the pathophysiology of neurodegeneration.

The overall results of clinical trials are intriguing yet ambiguous, as most trials were short term and involved small sample sizes, with significant dose variability. The variability in therapy outcomes across research addressing the same neurodegenerative disorders, such as AD, is particularly significant. Studies utilizing unmodified prototype curcumin formulations demonstrated minimal variations from placebo controls, largely because of the intrinsically low bioavailability and

Table 6.2 In vivo neurodegenerative disease-specific evidence with curcumin

Disease/condition	Regulated pathway/mechanism	Curcumin dosage used/ duration	Key evidence/findings	Ref.
Parkinson's disease	AKT/mTOR	0.5, 1, 2, 5, 10 μM for 6 h	Reduced α-synuclein accumulation, improved motor behavior, and induced autophagy via AKT/mTOR pathway inhibition in PD rats with 6-OHDA-lesioned striatum	He et al. (2022)
	NF-κB/NLRP3	50 mg/kg/day, for 21 days	Blocked NF-κB/NLRP3-driven inflammation, reduced mitochondrial fission, and dysfunction in rotenone-induced PD rats	Xu et al. (2023)
	NF-κB/Glial/NADPH	40 mg/kg for 21 days	Inhibited NF-κB-mediated neuroinflammation, suppressed microglial and astrocyte activation, blocked oxidative stress via NADPH oxidase, and restored dopaminergic function and motor behavior in lipopolysaccharide-induced PD model	Sharma et al. (2017)
	BDNF/TrkB/PI3K	5, 10, and 20 mg/kg/day for 30 days	Activated the BDNF/TrkB/PI3K signaling pathway; enhanced neurotransmitter levels; promoted hippocampal neuron regeneration, repair, and improved mood-related and motor behaviors in 6-OHDA-induced PD rat model	Yang et al. (2014)
	Oxidative stress/mitochondrial complex I inhibition	25, 50 mg/kg/day for 30 days	Improved motor performance, reduced lipid peroxidation, enhanced antioxidant defense, and prevented complex I inhibition in rotenone-induced PD mice	Ramires Júnior et al. (2021)
	Oxidative stress/stress response/neurotransmitter recovery	5, 10, and 20 mg/kg daily for 30 days	Improved cognitive and motor function; protected against oxidative stress and lipid peroxidation; elevated neurotrophic markers (NGF, bFGF, TrkA); reduced stress response protein (HSP70); increased neurotransmitter levels (DA, ACh) in substantia nigra	Song et al. (2016)
	A2A receptor gene expression/α-synuclein, AchE, MDA, Ang-II, CRP, IL-6, Caspase-3, neuroinflammation	80 mg/kg/day for 12 days (combination therapy-curcumin, niacin, ZM241385)	Significantly improved behavioral deficits, oxidative stress, inflammation, and mitochondrial enzyme activity; improved antioxidant status and reduced MDA, IL-6, CRP, caspase-3 showed strongest neuroprotection	Motawi et al. (2020)

(continued)

Table 6.2 (continued)

Disease/ condition	Regulated pathway/ mechanism	Curcumin dosage used/ duration	Key evidence/findings	Ref.
Alzheimer's disease	PI3K/Akt/mTOR	160 ppm, 1000 ppm for 6 months	Curcumin induced autophagy and inhibited Aβ production through downregulation of PI3K/Akt/mTOR pathway in APP/PS1 double transgenic mice	Wang et al. (2014)
	AMPK	150 mg/kg/day for 10 days	Improved cognitive performance, reduced neuronal damage and Aβ buildup, suppressed proinflammatory cytokines (TNF-α, IL-6, IL-1β), and activated AMPK signaling in $A\beta^{1-42}$ -injected mice	Shao et al. (2023)
	PPARγ/NF-κB	150 mg/kg for 1 month	Enhanced spatial memory, reduced microglial/astrocyte activation, lowered cytokine production, inhibited NF-κB, and affected PPARγ activation in APP/PS1 transgenic mice	Liu et al. (2016)
	Wnt/β-catenin, BDNF	100 mg/kg for 14 days	Curcumin upregulated Wnt/β-catenin and BDNF pathways, promoting neurogenesis in Aβ-injected mice	Lou et al. (2024)
	Akt/GSK3β/IL-1β	Not specified	Reduced Aβ accumulation in brain and retina, upregulated Akt, downregulated GSK3β, and lowered IL-1β levels especially effective in early stage AD in APP_SWE/PS1Δ_E9 transgenic mice	Mei et al. (2022)
	Neuroinflammation and oxidative stress mechanisms	200 mg/kg/day for 7 days	Improved cognitive performance, decrease IL-6 and TNF-α levels in hippocampus, oxidative stress (TOS) and increase antioxidant status (TAS) in Wistar Albino rats	Çakmak et al. (2022)
	Inflammatory cytokines and ER stress	40 mg/kg/day for 3 weeks	Improved cognitive function, decrease ApoE4 expression, inflammatory cytokines, and ER stress marker proteins in ApoE4 transgenic mice	Kou et al. (2021)
	Oxidative stress and cholinergic dysfunction	12.5, 25 mg/kg for 10 days	Reduce escape latency and distance in Morris Water Maze (improved learning and memory), lipid peroxidation, and protein carbonyls and increase glutathione and acetylcholine in Wistar male rats	Ahmadi et al. (2021)

Huntington's disease	Central and peripheral dysfunction	25 mg/kg, 1 mg/kg; 3 weeks after offspring birth	Preserved motor function; protected from neuropathological changes and improved gastrointestinal (GI) function, including GI motility and emptying in R6/2 transgenic mice	Elifani et al. (2019)
	Neuronal cell death and toxicity	3, 5, 10, 15, 20 μM; 1, 7, 14 days	Suppressed neuronal cell death and toxicity linked to polyglutamine expanded huntingtin in mutant *Drosophila melanogaster*	Chongtham and Agrawal (2016)
	Nrf2	40 mg/kg/day or 7 days	Restored mitochondrial Complex II activity and cytochrome levels; reduced ROS, lipid peroxidation, protein carbonyls, and mitochondrial swelling; restored glutathione and SOD activity; improved neuromotor coordination and activated Nrf2 pathway in rat model	Sandhir et al. (2014)
	BDNF, TrkB, synaptophysin, PSD-95	100 mg/kg/day; alternate days for 8 weeks	Improved learning and memory (active-avoidance task); preserved medium spiny neuron arborization and spine density; modulated synaptic proteins in YAC128 transgenic mice	Gharaibeh et al. (2020)
Multiple sclerosis	AMPK/SIRT1 and JAK2/STAT3	200 mg/kg/day for 14 days	Neuroprotective effects by: Activating AMPK/SIRT1, enhanced antioxidant response (Nrf2); supporting myelin repair (via CREB/BDNF/MBP) and suppressing inflammation (JAK2/STAT3 inhibition) in autoimmune encephalomyelitis (EAE)-induced mice	Sadek et al. (2024)
	Proinflammatory cytokines: IL-6-17, TNF-α, IFN-γ/anti-inflammatory: TGF-β/antioxidant: GPX-1	20 mg/kg for 21 days	Reduced demyelination; suppressed inflammation and oxidative stress; improved antioxidant enzyme activity in C57BL/6 EAE mouse model	Esmaeilzadeh et al. (2019)
	AKT/mTOR	10 mg/kg/day for 15 days	Modulated autophagy via AKT/mTOR and provided neuroprotective effect in EAE mouse model	Boyao et al. (2019)
	Oxidative stress modulation	Curcumin: 20 and 40 μg/kg for 5 days Conjugated linoleic acid–curcumin: 20 and 40 μg/kg for 5 days	Improved spatial memory both significantly reduced oxidative stress markers	Barzegarzadeh et al. (2021)
	Expression of MBP, NOGO-A, and iNOS genes	100 mg/kg/day from twelfth day for 24 days	Increased MBP expression (remyelination support) and decrease NOGO-A expression (axonal growth inhibition) and iNOS expression (neuroinflammation marker) in Wistar rats	Mavaddatiyan et al. (2021)

Table 6.3 In vitro neurodegenerative diseases-specific evidence

Disease	In vitro system	Regulated pathway/ mechanism	Curcumin dosage used	Key evidence /findings	Ref.
Parkinson's disease	EA.hy926 endothelial cells (vascular model)	Akt/mTOR	5–20 μmol/l for 4 h prior to and for 4 h during exposure to H_2O_2 (200 μmol/l)	Inhibited apoptosis (lowed caspase-3 and Bax, increased Bcl-2), induced autophagy (increased LC3-II), suppressed Akt/mTOR phosphorylation, overall protection against H_2O_2-induced oxidative stress	Guo et al. (2016)
	PC12 cells treated with 6-OHDA	AKT/mTOR	0.5, 1, 2, 5, 10 μM for 6 h	Suppressed 6-OHDA-induced apoptosis, increased autophagy markers (LC3-II/LC3-I, Beclin-1), and inhibited AKT/mTOR phosphorylation	He et al. (2022)
	SH-SY5Y cells overexpressing A53T mutant α-synuclein	mTOR/p70S6K	Cell viability assay: 0, 3, 5, 6, 7, 10, 15, 20, or 30 μM curcumin for 24 h	Suppressed mTOR/p70S6K signaling, restored autophagy, and reduced neurotoxicity associated with α-synuclein accumulation	Jiang et al. (2013)
Alzheimer's disease	N9 microglial cells stimulated with fibrillar Aβ1–42	PGE2–EP2–PKA signaling	Cell treatment: 5, 10, and 20 μM for 30 min	Enhanced Aβ clearance by suppressing PGE2/EP2/PKA signaling and acting through immunomodulation	He et al. (2016)
	Human SH-SY5Y cortical neurons and mouse N2a neuroblastoma cells	CMA/heat shock proteins (HSPs)/macroautophagy markers	1 μM	Restored chaperone function via upregulation of multiple HSPs; revived CMA markers, promoting clearance of Aβ42 and enhanced macroautophagy	Maiti et al. (2017)
	Mouse primary hippocampal neurons treated with Aβ42	Aβ42 reduced NAD^+/NADH ratio, ATP levels, SIRT3 deacetylation, Thrb expression	1, 5, 10 μM	Reversed Aβ42-induced cell apoptosis and metabolic dysfunction; increased NAD^+/NADH ratio, ATP levels, and SIRT3 activity; upregulated Thrb expression	Liu et al. (2021)
	SH-SY5Y cells treated with thapsigargin (TG)	TG induced ER stress, ROS generation,Mfn2 upregulation, and mitochondrial dysfunction	0.625–5 μmol/L; 4 h	Reduced cell apoptosis and improved cell viability; downregulated ER stress markers (GRP78, p-PERK, p-eIF2α); decreased ROS levels; restored mitochondrial function, when Mfn2 was functional	Zhou et al. (2022)

	(i) Primary microglia from neonatal Sprague Dawley rats treated with Aβ25–35 (ii) PC12 cells exposed to microglial conditioned medium	Neuroinflammation in microglia	10 μM 1 h before Aβ 25–35 treatment	Curcumin reduced HMGB1 expression and release; downregulated TLR4 and RAGE; lowered IL-1β and TNF-α levels; and protected neurons from Aβ-induced microglial toxicity	He et al. (2020)
	BV2 microglial cells treated with 1,2-diacetylbenzene (DAB) to induce neuroinflammation	TREM1/DAP12/NLRP3/CASP1/IL1B/TLR4/NF-κB/ROS and AGE accumulation/tau hyperphosphorylation via GSK-3β/Nrf2 activation		Inhibited DAB-induced inflammation via TREM-1/NLRP3 and TLR4/NF-κB pathways; reduced ROS, AGEs, GSK-3β, β-amyloid, and tau hyperphosphorylation; activated Nrf2	Zhou et al. (2022)
	Primary cortical microglia and primary cortical neurons exposed to high molecular weight Aβ42 oligomers (Aβ42Os)	Aβ42 oligomers activate microglia → release of proinflammatory cytokines → neuronal cell death	1–50 μM for 0–30 days	Anti-inflammatory effects on Aβ42O-stimulated microglia; prevented microglia-mediated neuronal death; inhibited Aβ42 aggregation/oligomerization	De Lorenzi et al. (2022)
	In vitro cellular models (oxidative stress and inflammation assays)	Nrf2-ARE phase II antioxidant response/PHF6 peptide aggregation (Tau aggregation)/LPS-induced inflammatory response/oxidative stress from mitochondrial respiratory chain inhibition/tau hyperphosphorylation induced by phosphatase inhibition	0.1–10 μM	Induced Nrf2-ARE antioxidant response, reducing oxidative stress; inhibited PHF6 peptide aggregation, which is linked to Tau pathology; inhibited inflammatory response induced by LPS; neuroprotection against oxidative insults; protected against Tau hyperphosphorylation	Cores et al. (2021)

(continued)

Table 6.3 (continued)

Disease	In vitro system	Regulated pathway/ mechanism	Curcumin dosage used	Key evidence /findings	Ref.
	LPS-stimulated BV-2 microglial cells	Anti-inflammatory action via reduced NO, TNF-α, IL-6 secretion in microglia	10 μM for 24 h	CUR-2GE (curcumin prodrug) more effectively suppressed proinflammatory mediators than curcumin due to enhanced cellular uptake	Jithavech et al. (2022)
Huntington's disease	Primary rat cortical neurons induced with neuronal apoptosis	Heat Shock Protein 70 (HSP70) expression regulation		Curcumin protected neurons from HIV-1 gp120-induced apoptosis by upregulating HSP70	Xia et al. (2015)
	Biochemical + neuron culture	Modulation of aggregation dynamics	At a 0.33 molar ratio for 24 h	Curcumin delayed aggregation and reduced neurotoxicity	Jain et al. (2025)
	PC12 cell model engineered to express exon 1 of mutant huntingtin	Dose-dependent effect on huntingtin aggregate size	555 ppm for 23 weeks	Low doses of curcumin modestly reduced aggregates while high doses increased size	Hickey et al. (2012)
Amyotrophic lateral sclerosis	In vitro assays with premature superoxide dismutase (SOD1)	Inhibition of SOD1 fibrillation and amyloid formation	0.5, 1, 5, 10, 20, 30 μM for 6, 24, 72 h	Curcumin bound amyloidogenic regions of SOD1, inhibited toxic fibril formation, reduced SOD1-mediated toxicity	Bhatia et al. (2015)
	Motoneuron-like cell lines transfected with mutant and wild-type TDP-43	Nav channel function	15 μM for 24 h	Dimethoxy curcumin (DMC) treatment significantly reduced abnormal excitability and restored Nav channel function by alleviating oxidative stress and mitochondrial dysfunction	Dong et al. (2014)
	(i) Motor neurons transfected with mutant TDP-43 (ii) Neuroblastoma-spinal cord-34 (NSC-34) cells with TDP-25	TDP-43 fragment and TDP-25 aggregation/ oxidative stress markers (LDH, MDA)	10 μM for 24 h	Monocarbonyl dimethoxycurcumin) reduced TDP-43/TDP-25 aggregates and oxidative stress (↓ MDA, LDH)	Duan et al. (2014)

Multiple sclerosis	$CD4^+$ T cell subsets (TH1, TH17, Treg) cultured in IMDM medium	Effect on Bcl-2/Bax, active caspase-3 (inducing apoptosis)/$CD4^+$ T cell activation markers	5,10, 20, 40, and 60 μM; 24, 48, 72 h	Inhibited proliferation of TH1, TH17, and Treg cells via apoptosis; reduced colony formation and T cell activation all in a dose-dependent manner	Liu et al. (2019)
	LPS-stimulated BV-2 microglial cells	Inhibition of AXL/JAK2/STAT3 signaling/microglial activation/inflammatory response	1–500 μM for 24 h	Inhibited microglial activation in spinal cord and BV-2 cells	Sun et al. (2022)
	LPS-stimulated C8-D1A astrocyte cells	Modulation of Th1/Th2/Th17 cytokines	10, 20, and 50 μM; 20, 24, 72 h	Enhanced astrocyte proliferation	ELBini-Dhouib et al. (2022)
Epilepsy	Organotypic hippocampal–entorhinal cortex slice cultures from rats	mTOR pathway inhibition/anti-inflammatory effects	10 μmol/L; every day from day 2 culture	Reduced seizure-like events (SLEs); lowered inflammatory marker expression and greater antiepileptogenic efficacy than rapamycin in this model; showed stronger suppression of SLEs despite less pronounced mTOR inhibition	Drion et al. (2019)

insufficient systemic absorption. Tables 6.4 and 6.5 offer a clinical and translational viewpoint on the therapeutic use and mechanistic validation of curcumin. Results from phase II clinical trials in healthy older cohorts, people with Alzheimer's disease, and people with mild cognitive impairment (MCI) are compiled in Table 6.4. Improved forms of curcumin, like Theracurmin®, Longvida®, and CurQfen®, showed quantifiable gains in areas like mood regulation, memory, and executive function, with corroborating biomarker data showing decreases in tau and amyloid burden. Interestingly, studies using traditional curcumin preparations produced relatively mild cognitive effects, underscoring the importance of delivery methods with improved bioavailability. Biomarker-based methods for tracking autophagic flux in preclinical and clinical settings, such as p62 degradation, TFEB nuclear translocation, LC3-II accumulation, and p62–Keap1–Nrf2 pathway modulation, are described in Table 6.5. These biomarkers offer a framework for incorporating molecular readouts into therapeutic evaluation strategies and mechanistic confirmation of curcumin's autophagy-inducing effects.

6.3.5 Modern Therapeutic Paradigms for Efficient Delivery of Curcumin

Curcumin's poor bioavailability, which is characterized by rapid metabolism, low solubility (~11 ng/mL at pH 5.0), and limited blood-brain barrier (BBB) penetration, is the main focus of therapeutic strategies for curcumin delivery (Xie et al. 2022). While surface modifications, such as receptor-targeted ligands, improve BBB traversal for neurological applications, advanced nanocarriers, such as liposomes, polymeric nanoparticles, and cyclodextrins, optimize solubility and stability (Genchi et al. 2024; Urošević et al. 2022). For example, polymeric nanoparticles allow sustained release, extending the therapeutic effects of curcumin, while liposomal formulations encapsulate it to prevent degradation. Because of their hydrophobic cavities, cyclodextrins improve intestinal absorption by forming inclusion complexes that increase curcumin's solubility (Bertoncini-Silva et al. 2024).

Brain-specific distribution is further refined by the best routes of administration (such as intravenous or intranasal). Via olfactory pathways, intranasal delivery circumvents the blood-brain barrier completely, whereas intravenous administration guarantees systemic circulation with controlled release (Jacob et al. 2024; Khayatan et al. 2024). These strategies are crucial for neurological disorders where accurate targeting is necessary.

Furthermore, curcumin provides supplementary advantages through its modulation of the gut-brain axis. Curcumin indirectly reduces neuroinflammation, a major cause of neurodegeneration, by balancing the microbiota and lowering systemic inflammation (Kim et al. 2024). Furthermore the combination of curcumin and vagus nerve stimulation dramatically reduced neurological impairments, infarct volume, neuronal apoptosis, and the release of inflammatory cytokines. The phosphorylation levels of AKT and ERK2 were elevated following the combination therapy (Xu et al. 2018). To realize its full potential, collaborative research in clinical pharmacology, systems biology, and nanotechnology is essential.

Table 6.4 Phase II results in mild cognitive impairment cohorts

Study	Intervention	Cohort	Key phase II findings	Ref.
Theracurmin® (bioavailable curcumin)—18 month trial	90 mg twice daily; double-blind, placebo-controlled trial	Nondemented older adults (MCI-suspected)	Significant improvements in long-term verbal and visual memory, attention; PET imaging (FDDNP) suggested reduced amyloid and tau accumulation	Small et al. (2018)
Theracurmin (retrospective, 6-month follow-up)	180 mg/day oral; observational	Patients with MCI and Alzheimer's disease	Cognitive and Activities of Daily Living (ADLs) scores remained stable in MCI patients on treatment; progression to AD was lower versus untreated	Dost et al. (2021)
Highly absorbable curcumin (curcuRouge)—Japanese trial protocol	Randomized, double-blind, placebo-controlled phase II planned	General public suspected of MCI ($n \approx 60$)	Trial underway; aims to assess safety and cognitive effects via Mini-Mental State Examination, Japanese version (MMSE-J) and dementia biomarkers	Komiyama et al. (2022)
Biocurcumax™ (RCT)	1500 mg/day curcumin, randomized, placebo-controlled, 12 months	Community-dwelling older adults ($n = 96$)	MoCA scores declined in placebo group at 6 months whereas curcumin group remained stable; no other cognitive differences found	Rainey-Smith et al. (2016)
CurQfen®—encapsulated curcumin–galactomannan (CGM) RCT	400 mg sachets twice daily, randomized, double-blind, placebo-controlled, 6 months	Individuals with mild-to-moderate dementia	Significant improvements in MMSE and locomotive function (GLFS); favorable biomarker changes	Das et al. (2023)
Longvida®—Lipidated curcumin	400 mg/day, double-blind RCT, 1 month	Healthy elderly adults ($n = 60$) aged 60–85	Improved computerized mental task performance (~17% increase in treated vs. 3% in placebo)	Cox et al. (2014)
Longvida®	400 mg/day (contains 80 mg curcumin), double-blind, placebo-controlled, parallel-groups trial, 3 months	Healthy older adults aged 50–80 years (n = 80)	Improved working memory at 12 weeks, reduced fatigue, tension, anger, confusion, and overall mood disturbance. Some mood improvements seen as early as 4 weeks	Cox et al. (2020)
Curcumin	1000 to 4000 mg/day double-blind, randomized, placebo-controlled, 6 months, other treatments: Standard treatment of gingko leaf extract	34 subjects with Alzheimer's disease	Curcumin increased vitamin E levels and serum Aβ40 levels; no cognitive improvement due to curcumin No cognitive decline observed in placebo group after 6 months	Baum et al. (2008)

Table 6.5 Curcumin-based biomarker development for autophagy flux monitoring

Biomarker/biomarker assay	Autophagy flux monitoring mechanism	Pros	Cons	Ref.
LC3-II (western blot, GFP-LC3 puncta), p62 degradation (± bafilomycin A1)	Assesses LC3-II increase and p62 decrease upon curcumin treatment; further increases with Bafilomycin A1 indicate enhanced flux	Gold-standard; direct measure of autophagosome formation and degradation; quantitative method	Requires inhibitors such as Baf for interpretation; static, endpoint assays; susceptible to nonspecific effects	Zhang et al. (2016)
LC3-II, p62, Beclin-1, Atg5, Atg16L1; mRFP-GFP-LC3 assay	In AD model (N2a/APP695swe), curcumin increased autophagy-related proteins, enhanced autophagosome formation and lysosome fusion; visualized via mRFP-GFP-LC3	Gold-standard autophagy flux markers; dynamic and visually demonstrable; specific to neurodegenerative model	Requires imaging tools and controls; interpretations must exclude blockade artifacts	Liang et al. (2019)
TFEB activation (via curcumin derivative E4)	E4 translocated TFEB to nucleus via AKT-mTORC1 inhibition, promoting autophagy-lysosome biogenesis in PD cell models	Targets upstream regulator of the autophagy-lysosome pathway; mechanistically insightful	Study used derivative E4, not native curcumin; in vivo bioavailability unresolved	Benameur et al. (2022)
LC3-II, p62, Beclin-1; Akt/mTOR signaling	In 6-OHDA PD models (PC12 cells and rats), curcumin increased LC3, Beclin-1, inhibited Akt/mTOR; neuroprotection reversed by autophagy inhibitors	Validated in both in vitro and in vivo PD models; flux confirmed via inhibitors and functional outcomes	Need multiple markers and pharmacological controls for accurate flux measurement	He et al. (2022)
p62–Keap1–Nrf2 pathway markers	In rotenone-induced PD mice, curcumin modulated p62–Keap1–Nrf2 axis, implying enhanced autophagy and antioxidant defense	Links selective autophagy with neuroprotection via oxidative-stress pathways	Indirect. Nrf2 activity intertwines with autophagy but is not a direct flux measure	Rathore et al. (2023)
LC3-II, Beclin-1 reduction; ROS metrics	In oxidative-stress neural stem cell models, curcumin reduced LC3-II/Beclin-1 and ROS levels via ERK1/2 pathway suppression	May reflect adaptive autophagy suppression under stress; highlights context-dependent role	A decrease in markers could reflect reduced stress rather than true flux modulation	Ruan et al. (2024)

6.3.5.1 Synergistic Approaches with Rapalogs/Autophagy Inducers

Curcumin has the potential to improve neuroprotection through complementary mechanisms when combined with rapalogs (such as rapamycin and everolimus) and autophagy inducers, according to recent research. Although their clinical utility is limited by immunosuppressive effects and difficulties in penetrating the blood-brain barrier, rapalogs suppress mTORC1 activity, directly stimulating autophagy to clear pathogenic protein aggregates (Bové et al. 2011). By altering mTOR-related pathways upstream (PI3K/Akt and AMPK) and downstream (NF-κB and ERK), curcumin overcomes these restrictions and increases autophagic flux while also reducing neuroinflammation (Huang et al. 2018; Su et al. 2020). According to preclinical research, curcumin-rapamycin combinations improve mitochondrial function, reduce oxidative stress, and maintain synaptic integrity in models of PD and AD. These benefits are correlated with better cognitive and motor outcomes (Liu et al. 2020; Rathore et al. 2023). To maximize therapeutic windows and reduce side effects, it is essential to optimize dosing schedules and pharmacodynamic interactions. Thorough trials are necessary for clinical translation in order to determine biomarkers, stratify patient populations, and validate efficacy. AI-driven modeling and systems biology could improve the accuracy of creating these multidrug treatments.

6.3.5.2 Precision Targeting of Neuronal Versus Glial mTOR Pools

In the central nervous system, ongoing research reveals different mTOR signaling dynamics between neuronal and glial cells, necessitating precise therapeutic targeting for curcumin-based interventions. Glial mTOR modulates neuroinflammation and metabolic support, whereas neuronal mTOR mainly controls synaptic plasticity, autophagy, and protein synthesis (Zhang et al. 2023). By differently affecting mTOR complexes (mTORC1/mTORC2) and upstream regulators like AMPK and PI3K/Akt, curcumin exhibits cell-type-specific modulation. Preclinical research demonstrates that curcumin inhibits mTOR-driven NF-κB activation in glia to decrease neuroinflammatory responses (Su et al. 2020) and suppresses neuronal mTORC1 hyperactivity to improve autophagic clearance and synaptic resilience (Huang et al. 2018). Advances in nanotechnology, like ligand-conjugated nanoparticles, allow for cell-specific delivery that maximizes spatial precision and minimizes off-target effects by targeting neuronal receptors or glial endocytosis mechanisms (Liu et al. 2020). Based on molecular profiling and customized delivery systems, these results highlight the need to comprehend the dual functions of mTOR in order to create curcumin treatments that balance neuronal proteostasis with glial anti-inflammatory effects.

6.4 South Asian Perspective

Curcumin has emerged as a promising therapeutic agent for neurological disorders, with South Asia contributing a prominent role in foundational research and clinical applications. India, the leading global producer of turmeric, yields 1.5 million metric tons each year, encompassing high-curcumin "Alleppey Finger" types (Mukherjee et al. 2025). The future outlook relies on the integration of advanced technology with traditional values in South Asia. The comprehensive strategy enables the region to tackle socioeconomic inequalities in rural areas while converting its 4000-year-old turmeric legacy into scientifically validated, globally accessible neurotherapeutics.

Research conducted by South Asian scientists indicates that frequent use of turmeric, a dietary staple including 2–5% curcuminoids, has been consistently associated with a reduction in cognitive decline among aging populations, as seen by epidemiological research from the region (Sarker and Franks 2018). The cultural familiarity with turmeric has facilitated unique longitudinal observations. Research on Kerala's aging population indicates that individuals who consume curry everyday exhibit a 28% reduced incidence of Alzheimer's disease compared to those in the West (Yallapu et al. 2015). Leveraging traditional knowledge, South Asian researchers have created bioavailability-enhanced formulations like CurQfen®, which demonstrated a 40% superior cognitive improvement in Alzheimer's patients compared to standard curcumin, attributed to optimized blood-brain barrier penetration, in a 2023 Indian clinical trial (Das et al. 2023). Concurrent advancements encompass research in Colombo on curcumin-encapsulated solid lipid nanoparticles that enhanced mitochondrial activity in Huntington's disease mice by 65% (Sandhir et al. 2014). A nanoparticle delivery methods by Indian scientist, such as Cur-PLGA-NPs, have been shown to restore hippocampus neurogenesis in animal models by upregulating Wnt/β-catenin signaling pathways (Tiwari et al. 2014).

6.5 Conclusion

Curcumin mainly modulates the mTOR signaling pathway and restores autophagic flux, two mechanisms that are severely compromised in neurodegenerative disorders, to provide its neuroprotective benefits. Curcumin helps remove harmful protein aggregates, lowers oxidative and inflammatory stress, and maintains neural and synaptic function by adjusting these pathways. Its distinct pleiotropic nature, which distinguishes it from traditional single-target therapies, is highlighted by the convergence of data from in vitro, in vivo, and clinical investigations. Curcumin has great potential as a next-generation therapeutic candidate with further development of bioavailable formulations and incorporation of autophagy-related indicators. In addition to addressing the pathogenic characteristics of neurodegeneration, curcumin, as a regulator of the mTOR–autophagy axis, lays the groundwork for novel, mechanism-driven therapies in upcoming neurotherapeutics.

References

Ahmadi N, Hosseini M-J, Rostamizadeh K, Anoush M (2021) Investigation of therapeutic effect of curcumin α and β glucoside anomers against Alzheimer's disease by the nose to brain drug delivery. Brain Res 1766:147517

Baum L, Lam CWK, Cheung SKK, Kwok T, Lui V, Tsoh J, Lam L, Leung V, Hui E, Ng C, Woo J (2008) Six-month randomized, placebo-controlled, double-blind, pilot clinical trial of curcumin in patients with Alzheimer disease. J Clin Psychopharmacol 28(1): 110–113
Ban B-K, Jun M-H, Ryu H-H, Jang D-J, Ahmad ST, Lee J-A (2013) Autophagy negatively regulates early axon growth in cortical neurons. Mol Cell Biol 33(19):3907–3919
Barzegarzadeh B, Hatami H, Dehghan G, Khajehnasiri N, Khoobi M, Sadeghian R (2021) Conjugated linoleic acid-curcumin attenuates cognitive deficits and oxidative stress parameters in the ethidium bromide–induced model of demyelination. Neurotox Res 39(3):815–825
Bertoncini-Silva C, Vlad A, Ricciarelli R, Giacomo Fassini P, Suen VMM, Zingg J-M (2024) Enhancing the bioavailability and bioactivity of curcumin for disease prevention and treatment. Antioxidants 13(3):331
Benameur T, Giacomucci G, Panaro MA, Ruggiero M, Trotta T, Monda V, Pizzolorusso I, Lofrumento DD, Porro C, Messina G (2022) New promising therapeutic avenues of curcumin in brain diseases. Molecules 27(1): 236
Bhatia NK, Srivastava A, Katyal N, Jain N, Khan MAI, Kundu B, Deep S (2015) Curcumin binds to the pre-fibrillar aggregates of cu/Zn superoxide dismutase (SOD1) and alters its amyloidogenic pathway resulting in reduced cytotoxicity. Biochimica et Biophysica Acta (BBA)-Proteins Proteomics 1854(5):426–436
Bové J, Martínez-Vicente M, Vila M (2011) Fighting neurodegeneration with rapamycin: mechanistic insights. Nat Rev Neurosci 12(8):437–452
Boyao Y, Mengjiao S, Caicai B, Xiaoling L, Manxia W (2019) Dynamic expression of autophagy-related factors in autoimmune encephalomyelitis and exploration of curcumin therapy. J Neuroimmunol 337:577067
Çakmak G, Kaplan DS, Yıldırım C, Ulusal H, Tarakçıoğlu M, Öztürk ZA (2022) Improvement of cognitive deficit of curcumin on scopolamine-induced Alzheimer's disease models. Caspian J Intern Med 13(1):16
Chongtham A, Agrawal N (2016) Curcumin modulates cell death and is protective in Huntington's disease model. Sci Rep 6(1):18736
Cores Á, Carmona-Zafra N, Martín-Cámara O, Sánchez JD, Duarte P, Villacampa M, Bermejo-Bescós P, Martín-Aragón S, León R, Menéndez JC (2021) Curcumin-piperlongumine hybrids with a multitarget profile elicit neuroprotection in in vitro models of oxidative stress and hyperphosphorylation. Antioxidants 11(1):28
Cox KH, Pipingas A, Scholey AB (2014) Investigation of the effects of solid lipid curcumin on cognition and mood in a healthy older population. J Psychopharmacol 29(5): 642-651
Cox KH, White DJ, Pipingas A, Poorun K, Scholey A (2020) Further evidence of benefits to mood and working memory from lipidated curcumin in healthy older people: a 12-week, double-blind, placebo-controlled, partial replication study. Nutrients 12(6): 1678
Das SS, Gopal PM, Thomas JV, Mohan MC, Thomas SC, Maliakel BP, Krishnakumar I, Pulikkaparambil Sasidharan BC (2023) Influence of CurQfen®-curcumin on cognitive impairment: a randomized, double-blinded, placebo-controlled, 3-arm, 3-sequence comparative study. Front Dementia 2:1222708
De Lorenzi E, Franceschini D, Contardi C, Di Martino RMC, Seghetti F, Serra M, Bisceglia F, Pagetta A, Zusso M, Belluti F (2022) Modulation of amyloid β-induced microglia activation and neuronal cell death by curcumin and analogues. Int J Mol Sci 23(8):4381
Doherty J, Baehrecke EH (2018) The complex roles and regulation of autophagy during cell death. Nat Cell Biol 20(10):1110
Dong H, Xu L, Wu L, Wang X, Duan W, Li H, Li C (2014) Curcumin abolishes mutant TDP-43 induced excitability in a motoneuron-like cellular model of ALS. Neuroscience 272:141–153
Dost FS, Kaya D, Ontan MS, Erken N, Bulut EA, Aydin AE, Isik AT (2021) Theracurmin supplementation may be a therapeutic option for older patients with Alzheimer's disease: a 6-month retrospective follow-up study. Curr Alzheimer Res 18(14):1087–1092
Drion CM, Kooijman L, Aronica E, van Vliet EA, Wadman WJ, Chameau P, Gorter JA (2019) Curcumin reduces development of seizurelike events in the hippocampal-entorhinal cortex slice culture model for epileptogenesis. Epilepsia 60(4):605–614

Duan W, Guo Y, Xiao J, Chen X, Li Z, Han H, Li C (2014) Neuroprotection by monocarbonyl dimethoxycurcumin C: ameliorating the toxicity of mutant TDP-43 via HO-1. Mol Neurobiol 49(1):368–379

Duarte-Silva S, Silva-Fernandes A, Neves-Carvalho A, Soares-Cunha C, Teixeira-Castro A, Maciel P (2016) Combined therapy with m-TOR-dependent and-independent autophagy inducers causes neurotoxicity in a mouse model of Machado–Joseph disease. Neuroscience 313:162–173

ELBini-Dhouib I, Manai M, Neili N-e, Marzouki S, Sahraoui G, Ben Achour W, Zouaghi S, BenAhmed M, Doghri R, Srairi-Abid N (2022) Dual mechanism of action of curcumin in experimental models of multiple sclerosis. Int J Mol Sci 23(15):8658

Elifani F, Amico E, Pepe G, Capocci L, Castaldo S, Rosa P, Montano E, Pollice A, Madonna M, Filosa S (2019) Curcumin dietary supplementation ameliorates disease phenotype in an animal model of Huntington's disease. Hum Mol Genet 28(23):4012–4021

Esmaeilzadeh E, Soleimani M, Zare-Abdollahi D, Jameie B, Khorram Khorshid HR (2019) Curcumin ameliorates experimental autoimmune encephalomyelitis in a C57BL/6 mouse model. Drug Dev Res 80(5):629–636

Fang C, Gu L, Smerin D, Mao S, Xiong X (2017) The interrelation between reactive oxygen species and autophagy in neurological disorders. Oxidative Med Cell Longev 2017(1):8495160

Forouzanfar F, Read MI, Barreto GE, Sahebkar A (2020) Neuroprotective effects of curcumin through autophagy modulation. IUBMB Life 72(4):652–664

Fraunberger EA, Scola G, Laliberté VL, Duong A, Andreazza AC (2016) Redox modulations, antioxidants, and neuropsychiatric disorders. Oxidative Med Cell Longev 2016(1):4729192

Genchi G, Lauria G, Catalano A, Carocci A, Sinicropi MS (2024) Neuroprotective effects of curcumin in neurodegenerative diseases. Foods 13(11):1774

Gharaibeh A, Maiti P, Culver R, Heileman S, Srinageshwar B, Story D, Spelde K, Paladugu L, Munro N, Muhn N (2020) Solid lipid curcumin particles protect medium spiny neuronal morphology, and reduce learning and memory deficits in the YAC128 mouse model of Huntington's disease. Int J Mol Sci 21(24):9542

Glaviano A, Foo AS, Lam HY, Yap KC, Jacot W, Jones RH, Eng H, Nair MG, Makvandi P, Geoerger B (2023) PI3K/AKT/mTOR signaling transduction pathway and targeted therapies in cancer. Mol Cancer 22(1):138

Gokuladhas S, Miller C, Cooper AA, O'Sullivan JM (2025) Inherited genetic variation in Parkinson's disease: convergence on impaired autophagosome-lysosome fusion through the altered expression of mRNA isoforms. Mol Neurobiol:1–15

Guo S, Long M, Li X, Zhu S, Zhang M, Yang Z (2016) Curcumin activates autophagy and attenuates oxidative damage in EA. hy926 cells via the Akt/mTOR pathway. Mol Med Rep 13(3):2187–2193

He G-L, Luo Z, Yang J, Shen T-t, Chen Y, Yang X-S (2016) Curcumin ameliorates the reduction effect of PGE2 on fibrillar β-amyloid peptide (1-42)-induced microglial phagocytosis through the inhibition of EP2-PKA signaling in N9 microglial cells. PLoS One 11(1):e0147721

He W, Yuan K, Ji B, Han Y, Li J (2020) Protective effects of curcumin against neuroinflammation induced by Aβ25-35 in primary rat microglia: modulation of high-mobility group box 1, toll-like receptor 4 and receptor for advanced glycation end products expression. Ann Trans Med 8(4):88

He H-J, Xiong X, Zhou S, Zhang X-R, Zhao X, Chen L, Xie C-L (2022) Neuroprotective effects of curcumin via autophagy induction in 6-hydroxydopamine Parkinson's models. Neurochem Int 155:105297

Heras-Sandoval D, Pérez-Rojas JM, Pedraza-Chaverri J (2020) Novel compounds for the modulation of mTOR and autophagy to treat neurodegenerative diseases. Cell Signal 65:109442

Hickey MA, Zhu C, Medvedeva V, Lerner RP, Patassini S, Franich NR, Maiti P, Frautschy SA, Zeitlin S, Levine MS (2012) Improvement of neuropathology and transcriptional deficits in CAG 140 knock-in mice supports a beneficial effect of dietary curcumin in Huntington's disease. Mol Neurodegener 7(1):12

Huang L, Chen C, Zhang X, Li X, Chen Z, Yang C, Liang X, Zhu G, Xu Z (2018) Neuroprotective effect of curcumin against cerebral ischemia-reperfusion via mediating autophagy and inflammation. J Mol Neurosci 64(1):129–139

Ihlamur M, Akgül B, Zengin Y, Korkut ŞV, Kelleci K, Abamor EŞ (2024) The mTOR signaling pathway and mTOR inhibitors in cancer: next-generation inhibitors and approaches. Curr Mol Med 24(4):478–494

Jacob S, Kather FS, Morsy MA, Boddu SH, Attimarad M, Shah J, Shinu P, Nair AB (2024) Advances in nanocarrier systems for overcoming formulation challenges of curcumin: current insights. Nano 14(8):672

Jain G, Trombetta-Lima M, Matlahov I, Ribas HT, Chen T, Parlato R, Portale G, Dolga AM, van der Wel PC (2025) Inhibitor-based modulation of huntingtin aggregation mechanisms mitigates fibril-induced cellular stress. Nat Commun 16(1):3588

Jiang T-F, Zhang Y-J, Zhou H-Y, Wang H-M, Tian L-P, Liu J, Ding J-Q, Chen S-D (2013) Curcumin ameliorates the neurodegenerative pathology in A53T α-synuclein cell model of Parkinson's disease through the downregulation of mTOR/p70S6K signaling and the recovery of macroautophagy. J Neuroimmune Pharmacol 8(1):356–369

Jithavech P, Suwattananuruk P, Hasriadi, Muangnoi C, Thitikornpong W, Towiwat P, Vajragupta O, Rojsitthisak P (2022) Physicochemical investigation of a novel curcumin diethyl γ-aminobutyrate, a carbamate ester prodrug of curcumin with enhanced anti-neuroinflammatory activity. PLoS One 17(3):e0265689

Kah G, Chandran R, Abrahamse H (2023) Curcumin a natural phenol and its therapeutic role in cancer and photodynamic therapy: a review. Pharmaceutics 15(2):639

Khayatan D, Razavi SM, Arab ZN, Hosseini Y, Niknejad A, Momtaz S, Abdolghaffari AH, Sathyapalan T, Jamialahmadi T, Kesharwani P (2024) Superoxide dismutase: a key target for the neuroprotective effects of curcumin. Mol Cell Biochem 479(3):693–705

Kim Y, Lim J, Oh J (2024) Taming neuroinflammation in Alzheimer's disease: the protective role of phytochemicals through the gut– brain axis. Biomed Pharmacother 178:117277

Kocaturk NM, Gozuacik D (2018) Crosstalk between mammalian autophagy and the ubiquitin-proteasome system. Front Cell Dev Biol 6:128

Komiyama M, Ozaki Y, Wada H, Yamakage H, Satoh-Asahara N, Kishimoto A, Katsuura Y, Imaizumi A, Hashimoto T, Sunagawa Y (2022) Study protocol to determine the effects of highly absorbable oral curcumin on the indicators of cognitive functioning: a double-blind randomised controlled trial. BMJ Open 12(9):e057936

Kou J, Wang M, Shi J, Zhang H, Pu X, Song S, Yang C, Yan Y, Doring Y, Xie X (2021) Curcumin reduces cognitive deficits by inhibiting neuroinflammation through the endoplasmic reticulum stress pathway in apolipoprotein E4 transgenic mice. Acs Omega 6(10):6654–6662

Kuijpers M, Kochlamazashvili G, Stumpf A, Puchkov D, Swaminathan A, Lucht MT, Krause E, Maritzen T, Schmitz D, Haucke V (2021) Neuronal autophagy regulates presynaptic neurotransmission by controlling the axonal endoplasmic reticulum. Neuron 109(2):299–313. e299

Lee JW, Park S, Takahashi Y, Wang H-G (2010) The association of AMPK with ULK1 regulates autophagy. PLoS One 5(11):e15394

Li Y, Li S, Wu H (2022) Ubiquitination-proteasome system (UPS) and autophagy two main protein degradation machineries in response to cell stress. Cells 11(5):851

Liang J, Zhou F, Xiong X, Zhang X, Li S, Li X, Gao M, Li Y (2019) Enhancing the retrograde axonal transport by curcumin promotes autophagic flux in N2a/APP695swe cells. Aging (Albany NY) 11(17): 7036-7050

Liu Z-J, Li Z-H, Liu L, Tang W-X, Wang Y, Dong M-R, Xiao C (2016) Curcumin attenuates beta-amyloid-induced neuroinflammation via activation of peroxisome proliferator-activated receptor-gamma function in a rat model of Alzheimer's disease. Front Pharmacol 7:261

Liu J, Yan Y, Liu J, Wang Y, Wang X (2019) Curcumin prevents experimental autoimmune encephalomyelitis by inhibiting proliferation and effector CD4+ T cell activation. Eur Rev Med Pharmacol Sci 23(20)

Liu J, Liu C, Zhang J, Zhang Y, Liu K, Song J-X, Sreenivasmurthy SG, Wang Z, Shi Y, Chu C (2020) A self-assembled α-synuclein nanoscavenger for Parkinson's disease. ACS Nano 14(2):1533–1549

Liu M, Zhang X, Wang Y (2021) Curcumin alleviates Aβ42-induced neuronal metabolic dysfunction via the Thrb/SIRT3 axis and improves cognition in APPTG mice. Neurochem Res 46(12):3166–3178

Long J, Luo X, Fang D, Song H, Fang W, Shan H, Liu P, Lu B, Yin X-M, Hong L (2022) Discovery of an autophagy inducer J3 to lower mutant huntingtin and alleviate Huntington's disease-related phenotype. Cell Biosci 12(1):167

Lou S, Gong D, Yang M, Qiu Q, Luo J, Chen T (2024) Curcumin improves neurogenesis in Alzheimer's disease mice via the upregulation of Wnt/β-catenin and BDNF. Int J Mol Sci 25(10):5123

Maiese K (2016) Targeting molecules to medicine with mTOR, autophagy and neurodegenerative disorders. Br J Clin Pharmacol 82(5):1245–1266

Maiti P, Rossignol J, Dunbar G (2017) Curcumin modulates molecular chaperones and autophagy-lysosomal pathways in vitro after exposure to A*β*42. J Alzheimers Dis Parkinsonism 7(1):299

Manning BD, Toker A (2017) AKT/PKB signaling: navigating the network. Cell 169(3):381–405

Mavaddatiyan L, Khezri S, Froushani SMA (2021) Molecular effects of curcumin on the experimental autoimmune encephalomyelitis. Veterinary Research Forum

Medina D, Settembre C, Ballabio A (2017) Methods to monitor and manipulate TFEB activity during autophagy. In: Methods in enzymology, vol 588. Elsevier, pp 61–78

Mei X, Qiu C, Shi L, Li X, Yang M, Hu J, Chen Z, Zou C (2022) The effect of curcumin on Aβ, Akt, and GSK3β on the brain and retina of APP/PS1 mice and in the blood of Alzheimer's patients with early-stage disease. Pharmacogn Mag 18(79)

Monroy A, Lithgow GJ, Alavez S (2013) Curcumin and neurodegenerative diseases. Biofactors 39(1):122–132

Motawi TK, Sadik NA, Hamed MA, Ali SA, Khalil WK, Ahmed YR (2020) Potential therapeutic effects of antagonizing adenosine A2A receptor, curcumin and niacin in rotenone-induced Parkinson's disease mice model. Mol Cell Biochem 465(1):89–102

Mueed Z, Tandon P, Maurya SK, Deval R, Kamal MA, Poddar NK (2019) Tau and mTOR: the hotspots for multifarious diseases in Alzheimer's development. Front Neurosci 12:1017

Mukherjee A, Dutta S, Mukherjee E, Gaikwad K, Khanna T, Sen N (2025) Making India the global hub for turmeric. Academic Foundation/AF Press

Narayanan L, Thekkekkara D, Kondaveeti SN, Babu A, Sailen BA, Chidambaram SB, Manjula SN (2024) Deciphering the intricate role of mTOR signaling and autophagy in Parkinson's disease and therapeutic prospects. J Appl Pharm Sci 14(3):001–010

Nimiya Y, Wang W, Du Z, Sukamtoh E, Zhu J, Decker E, Zhang G (2016) Redox modulation of curcumin stability: redox active antioxidants increase chemical stability of curcumin. Mol Nutr Food Res 60(3):487–494

Panwar V, Singh A, Bhatt M, Tonk RK, Azizov S, Raza AS, Sengupta S, Kumar D, Garg M (2023) Multifaceted role of mTOR (mammalian target of rapamycin) signaling pathway in human health and disease. Signal Transduct Target Ther 8(1):375

Perluigi M, Di Domenico F, Barone E, Butterfield D (2021) mTOR in Alzheimer disease and its earlier stages: links to oxidative damage in the progression of this dementing disorder. Free Radic Biol Med 169:382–396

Perrone L, Squillaro T, Napolitano F, Terracciano C, Sampaolo S, Melone MAB (2019) The autophagy signaling pathway: a potential multifunctional therapeutic target of curcumin in neurological and neuromuscular diseases. Nutrients 11(8):1881

Priyadarsini KI (2014) The chemistry of curcumin: from extraction to therapeutic agent. Molecules 19(12):20091–20112

Rainey-Smith SR, Brown BM, Sohrabi HR, Shah T, Goozee KG, Gupta VB, Martins RN (2016) Curcumin and cognition: a randomised, placebo-controlled, double-blind study of community-dwelling older adults. Br J Nutr 115(12): 2106-2113

Ramires Júnior OV, Alves BdS, Barros PAB, Rodrigues JL, Ferreira SP, Monteiro LKS, Araújo GMS, Fernandes SS, Vaz GR, Dora CL (2021) Nanoemulsion improves the neuroprotective effects of curcumin in an experimental model of Parkinson's disease. Neurotox Res 39(3):787–799

Rathore AS, Singh SS, Birla H, Zahra W, Keshri PK, Dilnashin H, Singh R, Singh S, Singh SP (2023) Curcumin modulates p62–Keap1–Nrf2-mediated autophagy in rotenone-induced Parkinson's disease mouse models. ACS Chem Neurosci 14(8):1412–1423

Ruan Y, Luo H, Tang J, Ji M, Yu D, Yu Q, Cao Z, Mai Y, Zhang B, Chen Y, Liu J, Liao W (2024) Curcumin inhibits oxidative stress and autophagy in C17.2 neural stem cell through ERK1/2 signaling pathways. Aging Medicine 7(5): 559–570

Sadek MA, Rabie MA, El Sayed NS, Sayed HM, Kandil EA (2024) Neuroprotective effect of curcumin against experimental autoimmune encephalomyelitis-induced cognitive and physical impairments in mice: an insight into the role of the AMPK/SIRT1 pathway. Inflammopharmacology 32(2):1499–1518

Sandhir R, Yadav A, Mehrotra A, Sunkaria A, Singh A, Sharma S (2014) Curcumin nanoparticles attenuate neurochemical and neurobehavioral deficits in experimental model of Huntington's disease. NeuroMolecular Med 16(1):106–118

Sarker MR, Franks SF (2018) Efficacy of curcumin for age-associated cognitive decline: a narrative review of preclinical and clinical studies. Geroscience 40(2):73–95

Saxton RA, Sabatini DM (2017) mTOR signaling in growth, metabolism, and disease. Cell 168(6):960–976

Schmeisser K, Parker JA (2019) Pleiotropic effects of mTOR and autophagy during development and aging. Front Cell Dev Biol 7:192

Schmukler E, Pinkas-Kramarski R (2020) Autophagy induction in the treatment of Alzheimer's disease. Drug Dev Res 81(2):184–193

Settembre C, Di Malta C, Polito VA, Arencibia MG, Vetrini F, Erdin S, Erdin SU, Huynh T, Medina D, Colella P (2011) TFEB links autophagy to lysosomal biogenesis. Science 332(6036):1429–1433

Shao S, Ye X, Su W, Wang Y (2023) Curcumin alleviates Alzheimer's disease by inhibiting inflammatory response, oxidative stress and activating the AMPK pathway. J Chem Neuroanat 134:102363

Sharma N, Sharma S, Nehru B (2017) Curcumin protects dopaminergic neurons against inflammation-mediated damage and improves motor dysfunction induced by single intranigral lipopolysaccharide injection. Inflammopharmacology 25(3):351–368

Small GW, Siddarth P, Li Z, Miller KJ, Ercoli L, Emerson ND, Martinez J, Wong K-P, Liu J, Merrill DA (2018) Memory and brain amyloid and tau effects of a bioavailable form of curcumin in non-demented adults: a double-blind, placebo-controlled 18-month trial. Am J Geriatr Psychiatry 26(3):266–277

Song S, Nie Q, Li Z, Du G (2016) Curcumin improves neurofunctions of 6-OHDA-induced parkinsonian rats. Pathol Res Pract 212(4):247–251

Stavrides P, Goulbourne CN, Peddy J, Huo C, Rao M, Khetarpal V, Marchionini DM, Nixon RA, Yang D-S (2025) mTOR inhibition in Q175 Huntington's disease model mice facilitates neuronal autophagy and mutant huntingtin clearance. elife 14:RP104979

Su I-J, Chang H-Y, Wang H-C, Tsai K-J (2020) A curcumin analog exhibits multiple biologic effects on the pathogenesis of Alzheimer's disease and improves behavior, inflammation, and β-amyloid accumulation in a mouse model. Int J Mol Sci 21(15):5459

Sun M, Liu N, Sun J, Li X, Wang H, Zhang W, Xie Q, Wang M (2022) Curcumin regulates anti-inflammatory responses by AXL/JAK2/STAT3 signaling pathway in experimental autoimmune encephalomyelitis. Neurosci Lett 787:136821

Tchevkina E, Komelkov A (2012) Protein phosphorylation as a key mechanism of mTORC1/2 signaling pathways. In: Protein phosphorylation in human health, vol 1. Intech, pp 3–50

Tiwari SK, Agarwal S, Seth B, Yadav A, Nair S, Bhatnagar P, Karmakar M, Kumari M, Chauhan LKS, Patel DK (2014) Curcumin-loaded nanoparticles potently induce adult neurogenesis and reverse cognitive deficits in Alzheimer's disease model via canonical Wnt/β-catenin pathway. ACS Nano 8(1):76–103

Urošević M, Nikolić L, Gajić I, Nikolić V, Dinić A, Miljković V (2022) Curcumin: biological activities and modern pharmaceutical forms. Antibiotics 11(2):135

Valencia M, Kim SR, Jang Y, Lee SH (2021) Neuronal autophagy: characteristic features and roles in neuronal pathophysiology. Biomol Ther 29(6):605

Vo TS, Vo TTBC, Vo TTTN, Lai TNH (2021) Turmeric (Curcuma longa L.): chemical components and their effective clinical applications. J Turk Chem Soc Sect A Chem 8(3):883–898

Wang C, Zhang X, Teng Z, Zhang T, Li Y (2014) Downregulation of PI3K/Akt/mTOR signaling pathway in curcumin-induced autophagy in APP/PS1 double transgenic mice. Eur J Pharmacol 740:312–320

Wang M-M, Feng Y-S, Yang S-D, Xing Y, Zhang J, Dong F, Zhang F (2019) The relationship between autophagy and brain plasticity in neurological diseases. Front Cell Neurosci 13:228

Wang Z, Yang C, Liu J, Chun-Kit Tong B, Zhu Z, Malampati S, Gopalkrishnashetty Sreenivasmurthy S, Cheung K-H, Iyaswamy A, Su C (2020) A curcumin derivative activates TFEB and protects against parkinsonian neurotoxicity in vitro. Int J Mol Sci 21(4):1515

Wang Y, Fung NSK, Lam W-C, Lo ACY (2022) mTOR signalling pathway: a potential therapeutic target for ocular neurodegenerative diseases. Antioxidants 11(7):1304

Wang G, Zhou X, Pang X, Ma K, Li L, Song Y, Hou D, Wang X (2025) Pharmacological effects, molecular mechanisms and strategies to improve bioavailability of curcumin in the treatment of neurodegenerative diseases. Front Pharmacol 16:1625821

Wu H, Chen S, Ammar A-B, Xu J, Wu Q, Pan K, Zhang J, Hong Y (2015) Crosstalk between macroautophagy and chaperone-mediated autophagy: implications for the treatment of neurological diseases. Mol Neurobiol 52(3):1284–1296

Xia C, Cai Y, Li S, Yang J, Xiao G (2015) Curcumin increases HSP70 expression in primary rat cortical neuronal apoptosis induced by gp120 V3 loop peptide. Neurochem Res 40(9):1996–2005

Xie L, Ji X, Zhang Q, Wei Y (2022) Curcumin combined with photodynamic therapy, promising therapies for the treatment of cancer. Biomed Pharmacother 146:112567

Xu J, Kong X, Xiu H, Dou Y, Wu Z, Sun P (2018) Combination of curcumin and vagus nerve stimulation attenuates cerebral ischemia/reperfusion injury-induced behavioral deficits. Biomed Pharmacother 103:614–620

Xu L, Hao L-P, Yu J, Cheng S-Y, Li F, Ding S-M, Zhang R (2023) Curcumin protects against rotenone-induced Parkinson's disease in mice by inhibiting microglial NLRP3 inflammasome activation and alleviating mitochondrial dysfunction. Heliyon 9(5)

Yallapu MM, Nagesh PKB, Jaggi M, Chauhan SC (2015) Therapeutic applications of curcumin nanoformulations. AAPS J 17(6):1341–1356

Yang J, Song S, Li J, Liang T (2014) Neuroprotective effect of curcumin on hippocampal injury in 6-OHDA-induced Parkinson's disease rat. Pathol Res Pract 210(6):357–362

Zhai K, Brockmüller A, Kubatka P, Shakibaei M, Büsselberg D (2020) Curcumin's beneficial effects on neuroblastoma: mechanisms, challenges, and potential solutions. Biomolecules 10(11):1469

Zhang J, Wang J, Xu J, Lu Y, Jiang J, Wang L, Shen H-M, Xia D (2016) Curcumin targets the TFEB-lysosome pathway for induction of autophagy. Oncotarget 7(46):75659

Zhang L, Jia Z, Wu Q, Bai T, Wang B, Hu X, Li T, Liu X, Fu J, Chen Y (2023) Alleviating symptoms of neurodegenerative disorders by astrocyte-specific overexpression of TMEM164 in mice. Nat Metab 5(10):1787–1802

Zhang X, Zhang H, Dong J, Cai H, Le W (2025) Advances in autophagy for Parkinson's disease pathogenesis and treatment. Ageing Neurodegener Dis 5(1):N/A–N/A

Zhou H, Beevers SC, Huang S (2011) The targets of curcumin. Curr Drug Targets 12(3):332–347

Zhou H-Y, Sun Y-Y, Chang P, Huang H-C (2022) Curcumin inhibits cell damage and apoptosis caused by thapsigargin-induced endoplasmic reticulum stress involving the recovery of mitochondrial function mediated by mitofusin-2. Neurotox Res 40(2):449–460

Zhu Z, Yang C, Iyaswamy A, Krishnamoorthi S, Sreenivasmurthy SG, Liu J, Wang Z, Tong BC-K, Song J, Lu J (2019) Balancing mTOR signaling and autophagy in the treatment of Parkinson's disease. Int J Mol Sci 20(3):728

Neuroprotective Potential of Melatonin via ERK1/2 Dependent Nrf2 Activation

7

Nuwan S. Medawaththa, Sachini Raveesha, Lakmal Gonawala, and Nalaka Wijekoon

Abstract

Melatonin is a neurohormone synthesized in the human pineal gland. Melatonin is considered a multifunctional molecule associated with circadian rhythm regulation, impacting metabolism, the immune system, sleep, antioxidant, anti-inflammatory, and anti-apoptotic properties. Melatonin functions as a free radical scavenger, directly neutralizing reactive oxygen species (ROS) and nitrogen-derived species. Dietary patterns rich with fruits and vegetables are reported to enhance melatonin levels, where the gut microbiota plays a crucial role in melatonin production from tryptophan. The neuroprotective role of melatonin is facilitated through Erk1/2 dependent nuclear factor E2-related factor 2 (Nrf2). Erk1/2 plays a paradoxical role where Erk1/2 enhances the generation of mitochondrial ROS and concurrently reduces ROS-induced damage by regulating mitochondrial homeostasis. Intriguingly, this dual activity of Erk1/2 is reported to be orchestrated via Nrf2, promoting neuroprotection. This chapter aims to explore the molecular function of melatonin in the regulation of ERK1/2-dependent Nrf2 activation, Keap1 and Bach1 regulation, and downstream expression of cytoprotective genes, including HO-1, NQO1, and GCLC, as well as the interplay

N. S. Medawaththa
Bioinformatics, Sri Lanka Institute of Biotechnology (SLIBTEC), Homagama, Sri Lanka
e-mail: nuwan.medawaththa@slibtec.gov.lk

S. Raveesha
Quality Assurance, Sri Lanka Institute of Biotechnology (SLIBTEC), Homagama, Sri Lanka
e-mail: sachini.raveesha@slibtec.gov.lk

L. Gonawala · N. Wijekoon (✉)
Molecular Diagnostics and Vaccines, Sri Lanka Institute of Biotechnology (SLIBTEC), Homagama, Sri Lanka
e-mail: lakmal.gonawala@slibtec.gov.lk; nalaka.wijekoon@slibtec.gov.lk

H. Dilnashin, S. P. Singh (eds.), *Phototherapeutic Approaches to Neurodegeneration*,
https://doi.org/10.1007/978-981-95-4751-7_7

between food and gut microbiota in modulating melatonin levels that contribute to neuroprotection.

Keywords

Melatonin · Neuroprotection · Nrf2 · Erk1/2 · Oxidative stress

7.1 Introduction

7.1.1 Melatonin and Neurodegenerative Landscape

Melatonin (N-acetyl-5-methoxytryptamine) stands as one of the most evolutionarily conserved signaling molecules, serving as a pivotal regulator of circadian rhythms and exhibiting remarkable neuroprotective properties that position it as a promising therapeutic candidate for neurodegenerative disorders. This indoleamine hormone has emerged from its traditional role as a sleep-regulating hormone to become recognized as a potent antioxidant and cellular protector with profound implications for neurodegenerative disease pathogenesis and treatment.

The neuroprotective potential of melatonin emerges from its multifaceted mechanisms of action that address the fundamental pathological processes underlying neurodegeneration. These mechanisms encompass antioxidant defense, anti-apoptotic signaling, mitochondrial stabilization, and anti-inflammatory modulation, creating a comprehensive protective profile that targets the complex, interconnected pathways involved in neurodegenerative disease progression (Alghamdi 2018).

A critical feature distinguishing melatonin from many other therapeutic compounds is its exceptional ability to cross the blood-brain barrier efficiently. Studies have demonstrated that melatonin achieves cerebrospinal fluid concentrations that are closely parallel to plasma levels, indicating efficient brain penetration and distribution throughout neural tissues. The clinical significance of melatonin's blood-brain barrier permeability cannot be overstated, as this property enables direct neuroprotective actions within the central nervous system following systemic administration. This pharmacokinetic advantage positions melatonin as an attractive therapeutic candidate for neurodegenerative disorders, where many potentially beneficial compounds fail to achieve therapeutically relevant brain concentrations due to blood-brain barrier limitations (Alluri et al. 2016).

Importantly, melatonin's antioxidant cascade extends beyond the parent molecule, as its metabolites, including cyclic 3-hydroxymelatonin, N^1-acetyl-N^2-formyl-5-methoxykynuramine (AFMK), and N^1-acetyl-5-methoxykynuramine (AMK), retain significant antioxidant capacity, enabling a single melatonin molecule to scavenge up to 10 reactive species (Tan et al. 2007).

Melatonin's antioxidant properties represent perhaps its most extensively characterized neuroprotective mechanism, operating through both direct radical scavenging and indirect enhancement of endogenous antioxidant systems. The direct antioxidant activity involves the molecule's capacity to neutralize various reactive species without forming pro-oxidant intermediates, while the indirect effects include upregulation of antioxidant enzymes such as superoxide dismutase, glutathione peroxidase, and catalase. Additionally, melatonin stimulates the synthesis of reduced glutathione and enhances the activity of glutathione reductase, maintaining optimal cellular redox balance (Ezema et al. 2024).

The anti-apoptotic effects of melatonin involve complex interactions with mitochondrial death pathways and pro-survival signaling cascades. Melatonin inhibits cytochrome c release from mitochondria, prevents activation of initiator caspases, and reduces expression of pro-apoptotic proteins, including Bax and Bad, while promoting anti-apoptotic factors such as Bcl-2. These effects collectively contribute to neuronal survival under pathological stress conditions characteristic of neurodegenerative diseases (Wang 2009; Radogna et al. 2008).

Mitochondrial protection represents a critical aspect of melatonin's neuroprotective profile, given the central role of mitochondrial dysfunction in neurodegeneration. Melatonin enhances mitochondrial respiratory chain efficiency, particularly complexes I, III, and IV, while maintaining mitochondrial membrane potential and promoting mitochondrial biogenesis through PGC-1α activation (Yao et al. 2023). The hormone also protects mitochondrial membrane phospholipids, especially cardiolipin, from oxidative damage, preserving membrane integrity and function (Yao et al. 2023).

Anti-inflammatory mechanisms constitute another fundamental component of melatonin's neuroprotective arsenal. Melatonin inhibits microglial activation, reduces production of pro-inflammatory cytokines, including TNF-α, IL-1β, and IL-6, and promotes anti-inflammatory mediators such as IL-4 and IL-10 (Chahbouni et al. 2010). The hormone also suppresses NF-κB activation and reduces expression of inducible nitric oxide synthase and cyclooxygenase-2, thereby attenuating neuroinflammatory responses that contribute to neurodegenerative pathology (Benedeto-Stojanov et al. 2023).

7.1.2 Mechanistic Hallmarks of Melatonin in Neurodegenerative Disorders (NDDs)

The therapeutic potential of melatonin in neurodegenerative diseases has been extensively validated through preclinical studies demonstrating significant neuroprotective efficacy across diverse experimental models. These investigations have revealed consistent patterns of protection against the pathological hallmarks of major neurodegenerative disorders, including protein aggregation, oxidative stress, mitochondrial dysfunction, and neuroinflammation (M. Chen et al. 2020).

7.1.3 Early Evidence in AD, PD, ALS, and Stroke Models

In Alzheimer's disease models, melatonin has demonstrated remarkable efficacy in addressing both amyloid-β and tau pathologies, the defining neuropathological features of this disorder. Early investigations revealed that melatonin inhibits amyloid-β generation through modulation of amyloid precursor protein processing, promotes non-amyloidogenic pathways, and directly interferes with amyloid fibril formation through structure-dependent interaction (Li et al. 2020). Additionally, melatonin efficiently attenuates tau hyperphosphorylation by modulating the activities of protein kinases, including glycogen synthase kinase-3β and protein kinases, while enhancing protein phosphatase-2A activity to promote tau dephosphorylation (Chen et al. 2020).

Clinical investigations in Alzheimer's disease patients have provided encouraging evidence for melatonin's therapeutic potential, with studies demonstrating improvements in sleep quality, reduction of sundowning behaviors, and slower progression of cognitive impairment. More recent research has indicated that melatonin supplementation may reduce cerebrospinal fluid levels of tau proteins in patients with mild cognitive impairment, suggesting direct anti-tauogenic effects in human subjects (Cardinali et al. 2012).

Parkinson's disease research has yielded compelling evidence for melatonin's TP-induced models of Parkinsonism, melatonin treatment significantly preserved dopaminergic neurons in the substantia nigra, maintained striatal dopamine levels, and improved motor function. The neuroprotective mechanisms involve multiple pathways, including inhibition of α-synuclein aggregation, suppression of neuroinflammation through NLRP3 inflammasome modulation, and enhancement of antioxidant defenses (Zheng et al. 2021). Stereological analyses have demonstrated that melatonin pretreatment can preserve up to 50% of tyrosine hydroxylase-positive neurons and 30% of total neurons in the substantia nigra compared to untreated MPTP-exposed animals (Pérez-Lloret and Cardinali 2021).

Amyotrophic lateral sclerosis studies have provided dramatic evidence for melatonin's therapeutic potential in this devastating motor neuron disease. In SOD1^G93A^ transgenic mice, the most widely validated ALS model, melatonin treatment significantly delayed disease onset by 10.1%, prolonged survival by 7.4%, and reduced motor neuron loss by 51.9%. The protective mechanisms involve inhibition of the Rip2/caspase-1/cytochrome c cell death pathway, prevention of mitochondrial apoptotic factor release, and preservation of melatonin receptor expression in spinal cord tissue (Bald et al. 2021). These findings have translated into clinical observations, where retrospective analyses of ALS patient databases have indicated improved survival outcomes in patients receiving melatonin supplementation (Y. Zhang et al. 2013).

Stroke research has consistently demonstrated melatonin's neuroprotective efficacy in experimental models of cerebral ischemia. In middle cerebral artery occlusion models, melatonin administration significantly reduces infarct volume, brain edema, and oxidative damage while improving long-term functional outcomes. The therapeutic window for melatonin in stroke extends beyond the acute phase, with studies showing continued benefits when treatment is initiated several hours post-ischemia (Lin et al. 2018). Mechanistic investigations have revealed that melatonin enhances synaptic protein expression, preserves dendritic density in affected brain regions, and maintains electrophysiological responsiveness in post-ischemic tissue (Pei et al. 2002).

The convergent evidence from these diverse experimental models establishes melatonin as a versatile neuroprotective agent capable of addressing the fundamental pathological processes underlying major neurodegenerative disorders. The hormone's multifaceted mechanisms of action, combined with its excellent safety profile and ability to cross the blood-brain barrier, position it as a promising therapeutic candidate for translational development in neurodegenerative disease treatment and prevention. The consistent demonstration of neuroprotective efficacy across different disease models and experimental paradigms provides a strong foundation for advancing melatonin-based therapeutic approaches toward clinical applications.

7.2 Circadian Rhythm, Light Biology, and Melatonin

The intricate relationship between circadian rhythms, environmental light exposure, and melatonin synthesis represents a fundamental aspect of chronobiology that significantly impacts neuroprotective mechanisms in neurodegenerative disorders (Fig. 7.1). Understanding this sophisticated biological system is essential for comprehending how melatonin's neuroprotective potential operates through ERK1/2-dependent Nrf2 activation. This section examines the biological foundations of circadian phototransduction, the molecular mechanisms of clock gene regulation, and the pathological consequences of circadian disruption in neurodegeneration.

The fundamental architecture of circadian timing relies on transcriptional-translational feedback loops involving core clock genes, including CLOCK, BMAL1 (aryl hydrocarbon receptor nuclear translocator-like), Period genes (PER1, PER2, PER3), and Cryptochrome genes (CRY1, CRY2) (Takahashi 2017). These molecular oscillators generate approximately 24-h rhythms in gene expression that drive rhythmic outputs in hormone secretion, metabolism, body temperature, and sleep-wake cycles. The suprachiasmatic nucleus (SCN's) role as a master coordinator becomes particularly evident in its regulation of pineal melatonin synthesis, where circadian timing information is converted into a humoral signal that can influence peripheral tissues and provide feedback to the central nervous system (Schwartz et al. 2009).

The tissue-specific nature of melatonin-clock gene interactions adds another layer of complexity to this regulatory system. In the SCN itself, melatonin receptors (MT1 and MT2) are expressed in specific neuronal populations, and melatonin administration can phase-shift behavioral circadian rhythms, suggesting functional clock gene modulation (Gnocchi and Bruscalupi 2017). However, the precise molecular mechanisms by which melatonin receptor activation translates into changes in clock gene expression within SCN neurons remain incompletely understood and represent an important frontier for future investigation.

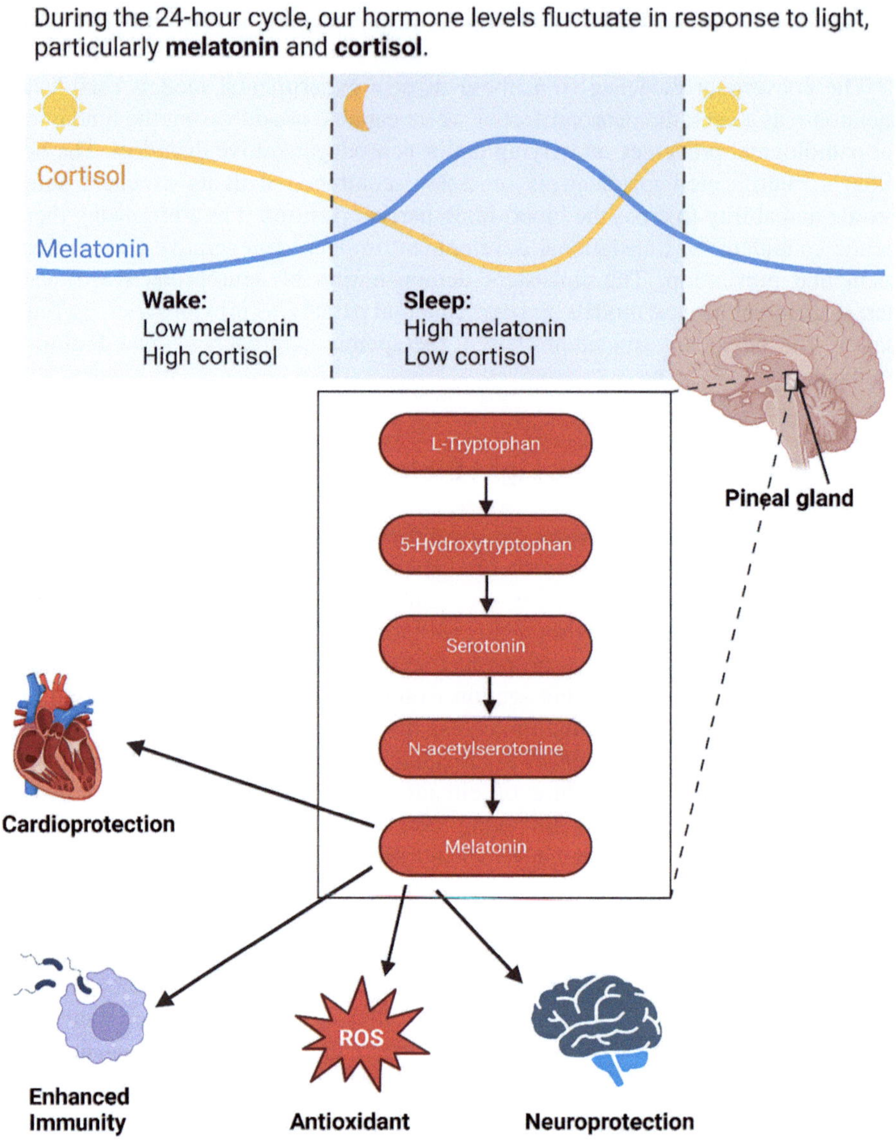

Fig. 7.1 Melatonin biosynthesis in the pineal gland and its protective roles in circadian rhythm regulation, immunity, antioxidant defense, and cardio/neuroprotection

7.3 Chronobiology and Neurodegeneration

The bidirectional relationship between circadian dysfunction and neurodegeneration has emerged as a critical area of investigation, with mounting evidence suggesting that circadian disruption may not merely be a consequence of neurodegenerative pathology but also a contributing factor to disease progression (Musiek 2015). This relationship is particularly evident in major neurodegenerative disorders, where circadian abnormalities often precede or accompany the onset of clinical symptoms, suggesting fundamental interactions between molecular clockwork and neuronal health (Videnovic et al. 2014).

7.4 Circadian Disruption and Neurodegenerative Risk

The epidemiological evidence linking circadian disruption to increased neurodegenerative disease risk has grown substantially in recent years. Particular attention was focused on the consequences of artificial light exposure, shift work, and other forms of chronobiological misalignment. Large-scale longitudinal studies have demonstrated that individuals with disrupted sleep-wake cycles exhibit significantly increased risks for developing Alzheimer's disease and other dementias, with some studies reporting hazard ratios exceeding 1.5 for severely disrupted circadian patterns (S.-J. Chen et al. 2022).

Understanding these complex relationships between circadian biology, light exposure, and neurodegeneration provides essential context for appreciating how melatonin's neuroprotective mechanisms, operating through ERK1/2-dependent Nrf2 activation, may be compromised in modern environments and how therapeutic interventions targeting these pathways might be optimized for maximum benefit. The intricate interplay between environmental light, circadian timing, and molecular neuroprotection represents a fertile area for future research and therapeutic development in the field of neurodegenerative disease prevention and treatment.

7.5 Core Signaling Pathways in Melatonin-Induced Neuroprotection

The molecular mechanisms underlying melatonin's neuroprotective efficacy involve sophisticated signaling networks that coordinate cellular defense responses against oxidative stress, neuroinflammation, and protein aggregation pathologies characteristic of neurodegenerative disorders. Central to these protective mechanisms are two pivotal signaling cascades: the Nuclear Factor Erythroid 2-Related Factor 2 (Nrf2) pathway, which orchestrates cellular antioxidant responses, and the Extracellular Signal-Regulated Kinase 1/2 (ERK1/2) pathway, which mediates cell survival signaling (Sun and Nan 2017). The convergence and crosstalk between these pathways represent a critical nexus through which melatonin exerts its neuroprotective effects,

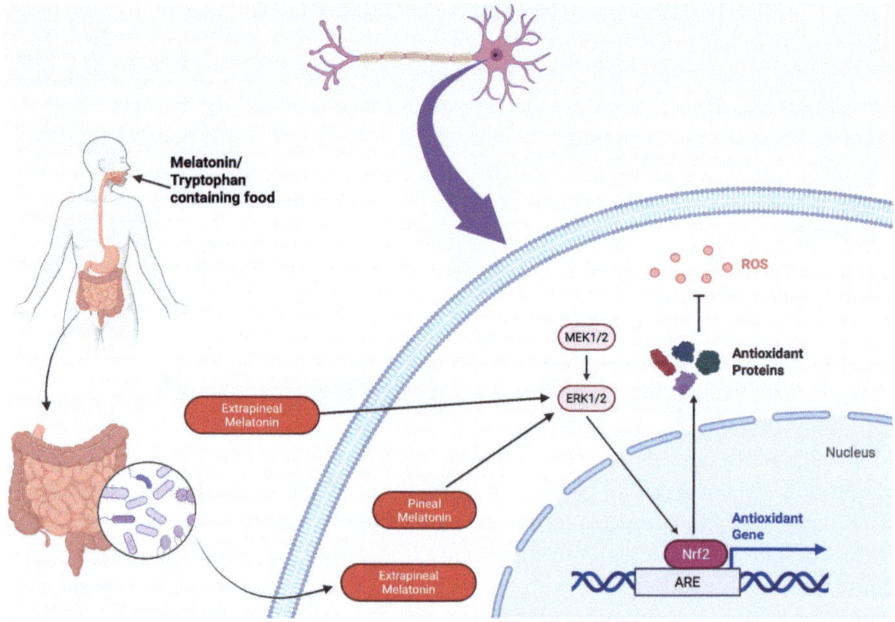

Fig. 7.2 Neuroprotective mechanisms of melatonin via Nrf2-mediated antioxidant response and ERK1/2-mediated cell survival signaling

particularly through ERK1/2-dependent activation of Nrf2-mediated transcriptional programs (Fig. 7.2).

7.5.1 Nrf2 Pathway in Neuronal Redox Homeostasis

The Nrf2 signaling system represents the primary transcriptional defense mechanism against oxidative and electrophilic stress in mammalian cells, functioning as a master regulator of cellular redox homeostasis (Ngo and Duennwald 2022). This basic leucine zipper (bZIP) transcription factor belongs to the cap'n'collar (CNC) family and orchestrates the expression of over 250 cytoprotective genes involved in antioxidant defense, phase II detoxification, and cellular stress adaptation (Vomund et al. 2017). In the central nervous system, Nrf2 expression exhibits cellular heterogeneity, with higher levels observed in astrocytes compared to neurons, reflecting the specialized roles of different cell types in maintaining neuronal redox balance (Vargas et al. 2008).

7.5.2 ERK1/2 Signaling: Role in Cell Survival

The extracellular signal-regulated kinases 1 and 2 (ERK1/2) constitute central components of the mitogen-activated protein kinase (MAPK) superfamily, serving as critical mediators of cell survival, proliferation, and differentiation in response to diverse stimuli, including growth factors, cytokines, and cellular stress (Roux and Blenis 2004). In the context of neurodegeneration, ERK1/2 signaling represents a pivotal pathway through which cells coordinate survival responses against oxidative stress, excitotoxicity, and protein aggregation pathologies characteristic of neurodegenerative disorders (Van Dooren et al. 2014).

7.5.3 ERK1/2 and Nrf2 Crosstalk: Points of Convergence

The functional intersection between ERK1/2 and Nrf2 signaling pathways represents a sophisticated regulatory mechanism that coordinates cellular stress responses and survival signaling. This crosstalk occurs at multiple levels, including direct phosphorylation of Nrf2 by ERK1/2, modulation of upstream regulators, and convergent transcriptional programs that enhance cellular antioxidant capacity (Apopa et al. 2008; Bloom and Jaiswal 2003; Cores et al. 2020; Li and Kong 2009; Niture et al. 2010).

Nuclear ERK1/2 and Nrf2 exhibit transcriptional synergy through several mechanisms that amplify antioxidant gene expression beyond the effects of either pathway alone. ERK1/2 can directly phosphorylate transcriptional co-activators such as CREB-binding protein (CBP) and p300, enhancing their ability to interact with Nrf2 and promote ARE-mediated transcription (Chowdhury et al. 2024; Li and Kong 2009).

This biphasic response allows cells to mount immediate protective responses while establishing longer-term adaptive changes in antioxidant capacity.

7.5.4 ERK1/2-Dependent Nrf2 Activation by Melatonin

The neuroprotective actions of melatonin involve sophisticated molecular mechanisms that integrate ERK1/2 and Nrf2 signaling pathways to coordinate cellular defense responses against oxidative stress and neurodegeneration. Emerging evidence demonstrates that melatonin can activate ERK1/2 through both receptor-dependent and receptor-independent mechanisms, leading to subsequent Nrf2 phosphorylation, nuclear translocation, and antioxidant gene expression (Ahmadi and Ashrafizadeh 2020; Deng et al. 2025; Ikram et al. 2021; Zhang et al. 2024). Experimental and translational evidence for the neuroprotection of melatonin are summarized in Table 7.1 (In vitro studies), Table 7.2 (In vivo studies), and Table 7.3 (Clinical studies).

7.6 Experimental and Translational Evidence

Table 7.1 In vitro studies of neuroprotective effect of melatonin

Model/system	Condition	Melatonin (Dose)	Pathway/mechanism	Key findings	References
Mouse cortical astrocytes	Hemin-induced oxidative stress	50–100 μM (in vitro)	Melatonin activates PKCα, which moves Nrf2 into the nucleus to boost HO-1 production. Blocking PKCα or melatonin receptors stops this effect	Melatonin reduced ROS and apoptosis; protection required PKCα-mediated Nrf2/HO-1 induction	Chen et al. (2019)
Neuronal cells (α-syn PFF PD model)	α-Synuclein fibril toxicity	(MT1 KO/ overexpression)	Removing the MT1 receptor lowers Sirt1, Nrf2, HO-1, GPX4, and ferritin levels. Increasing MT1 has the opposite effect	MT1 signaling (via Nrf2) protects against α-syn-induced ferroptosis; MT1 loss aggravated toxicity	Lv et al. (2024)
SH-SY5Y (human neuroblastoma)	MPP^+ (PD toxin)	10 μM (in vitro)	Melatonin or MT receptor agonists trigger PI3K/Akt, which then activates Nrf2. Blocking PI3K or melatonin receptors prevents this activation	Melatonin had dual antioxidant actions: direct ROS scavenging and indirect MT/PI3K/Akt/Nrf2–mediated protection	Kato et al. (2025)

Table 7.2 In vivo studies of the neuroprotective effect of melatonin

Model/system	Condition	Melatonin (Dose)	Pathway/mechanism	Key findings	References
Rat pups (developing brain)	Acute ethanol neurotoxicity	20 mg/kg i.p., single dose	Upregulated Nrf2 and HO-1 expression reduced ROS levels, suppressed MAPK-p38/JNK phosphorylation, and inhibited NF-κB/IKKβ signaling; Nrf2 silencing abolished these protective effects	Melatonin prevented ethanol-induced ROS, inflammation, and apoptosis in the neonatal brain (in vivo & HT22)	Ali et al. (2018)
Rat (peripheral nerve)	Streptozotocin diabetic neuropathy	3 & 10 mg/kg i.p. daily (weeks 7–8)	Nrf2 activation increased HO-1 expression and reduced NF-κB/IκBα signaling along with TNF-α, IL-6, iNOS, and COX-2 expression in sciatic nerve tissue	Melatonin improved nerve conduction and blood flow; reduced oxidative DNA damage and neuroinflammation	Negi et al. (2011)
Rat (retina)	Streptozotocin diabetic retinopathy	10 mg/kg i.p. daily ×12 weeks	Sustained Nrf2 nuclear retention elevated glutamate–cysteine ligase (GCL) expression and inhibited NF-κB activity and pro-inflammatory cytokine production	Melatonin maintained retinal GSH, prevented ERG decline, and reduced cytokine expression in diabetic retina	Jiang et al. (2016)
Mouse (CNS)	α-Synuclein fibril-induced PD	(MT1 genetic KO vs. OE)	MT1 receptor deletion suppressed Sirt1, Nrf2, HO-1, GPX4, and ferritin (Fth1) expression while promoting ferroptosis; MT1 overexpression reversed these effects	MT1 (melatonin receptor) activation was neuroprotective; its loss aggravated α-synuclein pathology and DA neuron loss	Lv et al. (2024)

Table 7.3 Clinical studies of the neuroprotective effect of melatonin

Study population	Intervention & dosage	Primary outcome	Key findings	Study design & duration	References
Parkinson's Disease patients ($n = 86$)	Melatonin 3 mg/day for 8 weeks vs placebo	Pittsburgh Sleep Quality Index (PSQI); Epworth Sleepiness Scale (ESS)	Significant improvement in sleep quality (PSQI: $p = 0.001$); reduced sleep latency	RCT, double-blind, 8 weeks (2023–2024)	Sugumaran et al. (2024)
Traumatic Brain Injury patients ($n = 33$)	Melatonin 2 mg prolonged-release for 4 weeks vs placebo (crossover)	Pittsburgh Sleep Quality Index (PSQI); Sleep efficiency	Improved sleep quality ($p \leq 0.0001$); reduced anxiety and fatigue	RCT, double-blind crossover, 4 weeks each arm (2011)	Grima et al. (2018)
Mild cognitive impairment patients ($n = 96$)	Melatonin 3-24 mg/day for 15–60 months vs standard care	Mini-mental state exam (MMSE); ADAS-Cog scores	Better performance in all neuropsychological tests; reduced benzodiazepine use	Retrospective cohort, 15–60 months (2007–2012)	Cardinali et al. (2012)
Multiple sclerosis patients ($n = 27$)	Melatonin 3 mg/day for 12 weeks vs placebo	Dynamic postural balance; Fatigue Severity Scale (FSS)	Improved dynamic balance, walking speed, reduced fatigue and pain	RCT, double-blind, 12 weeks (2024)	Jallouli et al. (2025)
Hemorrhagic stroke patients ($n = 40$)	Melatonin 30 mg/day for 3 days vs placebo	Duration of mechanical ventilation; ICU stay	Shorter mechanical ventilation duration; reduced morphine consumption	RCT, double-blind, 3 days (2017)	Dianatkhah et al. (2017)

7.7 Phototherapeutics and Melatonin: A Synergistic Frontier

The convergence of photobiomodulation and melatonin signaling represents an emerging therapeutic paradigm that harnesses complementary mechanisms to address the multifaceted pathophysiology of neurodegenerative disorders. This synergistic relationship operates through overlapping molecular pathways that enhance cellular bioenergetics, modulate oxidative stress responses, and promote neuroprotective signaling cascades. The integration of light-based therapies with endogenous and exogenous melatonin systems offers unprecedented opportunities for optimizing therapeutic outcomes through coordinated activation of ERK1/2-dependent Nrf2 pathways and related cellular defense mechanisms (Abdolmaleki et al. 2024; Brennan et al. 2007; Chen et al. 2017; Hong 2019; Jung and Kim 2024).

7.7.1 Melatonin as a Phototherapeutic Agent

The conceptualization of melatonin as a phototherapeutic agent extends beyond its traditional role as a circadian hormone to encompass its unique properties as a photosensitive molecule that bridges circadian biology with light-responsive cellular mechanisms (Reiter et al. 2010). This perspective recognizes melatonin's evolutionary origins as a photoprotective compound in primitive organisms and its contemporary functions in modulating light-dependent cellular processes in mammalian systems (Okeke et al. 2022; Zhao et al. 2019).

The clinical application of melatonin as a phototherapeutic agent has been explored in various neurodegenerative contexts, where coordinated light exposure and melatonin administration demonstrate synergistic effects on circadian rhythm restoration and neuroprotection (Cremascoli et al. 2022). In Alzheimer's disease patients, combined bright light therapy (2500 lux for 1 h) and evening melatonin administration (5 mg) significantly improved daytime wakefulness, activity levels, and rest-activity rhythm strength compared to either intervention alone (Dowling et al. 2008). This synergistic approach addresses both the circadian disruption characteristic of neurodegenerative diseases and the underlying cellular pathology through complementary mechanisms.

The therapeutic window for melatonin as a phototherapeutic agent is critically dependent on timing relative to circadian phase and light exposure patterns. Studies demonstrate that melatonin's phototherapeutic efficacy is maximized when administered in coordination with specific light exposure protocols that account for individual circadian phase relationships. The phase response curve for light-induced melatonin suppression reveals that therapeutic interventions must be precisely timed to avoid counterproductive effects where exogenous melatonin conflicts with endogenous circadian regulation (Kozaki et al. 2015; Kripke et al. 2007; Lewy 2010; St Hilaire et al. 2012).

7.7.2 Photobiomodulation (PBM) and ERK/Nrf2 Pathways

The mechanistic integration of photobiomodulation with ERK1/2 and Nrf2 signaling pathways represents a sophisticated cellular response system that coordinates multiple aspects of neuroprotection through light-induced molecular cascades (Serrage et al. 2019). This integration occurs at the level of cytochrome c oxidase, the primary photoacceptor for red and near-infrared light, which serves as a critical regulatory node linking photon absorption to downstream signaling events that culminate in ERK1/2 activation and Nrf2-mediated transcriptional responses (dos Cardoso et al. 2022; Hamblin 2017; Hong 2019).

The primary molecular mechanism underlying PBM's effects on ERK/Nrf2 pathways involves the generation of controlled levels of reactive oxygen species through mitochondrial photostimulation. Absorption of photons by cytochrome c oxidase in respiratory chain complex IV leads to enhanced electron transport efficiency and modest ROS production that functions as a redox signal rather than a cytotoxic stress. These ROS, particularly hydrogen peroxide, serve as second messengers that activate upstream kinases in the MAPK-ERK pathway while simultaneously triggering Keap1 oxidation and subsequent Nrf2 nuclear translocation (Hamblin 2017; Serrage et al. 2019).

Experimental validation of PBM's effects on ERK1/2 signaling has been demonstrated across multiple cell types relevant to neurodegeneration. In hippocampal HT-22 cells exposed to 660 nm LED light, significant increases in ERK and CREB phosphorylation were observed within minutes of light exposure, accompanied by enhanced brain-derived neurotrophic factor (BDNF) expression (Bathini et al. 2020; Chou et al. 2020; Salehpour et al. 2018).

7.7.3 PBM-Melatonin Crosstalk

The interaction between photobiomodulation and melatonin systems represents a complex regulatory network that operates at multiple levels to coordinate cellular responses to light exposure and circadian signaling. This crosstalk involves both direct effects of light on melatonin synthesis and metabolism, as well as indirect interactions through shared downstream signaling pathways that influence cellular bioenergetics, oxidative stress responses, and neuroprotective mechanisms (Bonmati-Carrion et al. 2014; Zhao et al. 2019).

The question of whether photobiomodulation can directly induce melatonin biosynthesis represents a critical intersection between circadian biology and phototherapeutic mechanisms that has significant implications for optimizing therapeutic protocols. Current evidence suggests a complex relationship where PBM may influence melatonin production through multiple pathways, including direct effects on pineal gland function, modulation of circadian regulatory mechanisms, and enhancement of extrapineal melatonin synthesis in peripheral tissues (Acuña-Castroviejo et al. 2014).

Experimental investigations into PBM's effects on pineal melatonin production have yielded intriguing but preliminary results that suggest wavelength-specific modulation of biosynthetic pathways. Red light exposure (660–670 nm) at therapeutic intensities appears to have minimal suppressive effects on nocturnal melatonin production compared to blue or white light and may in some contexts actually enhance melatonin synthesis through mechanisms involving improved mitochondrial function in pinealocytes (Markus et al. 2018).

The potential for PBM to induce extrapineal melatonin synthesis represents a particularly promising avenue for therapeutic applications, as numerous tissues throughout the body contain the enzymatic machinery necessary for local melatonin production. Mitochondria in various cell types, including neurons, astrocytes, and immune cells, have been shown to synthesize melatonin independently of pineal gland regulation. PBM-induced enhancement of mitochondrial function could theoretically stimulate local melatonin production in brain tissue, providing neuroprotective benefits without depending on systemic melatonin levels or circadian timing.

Studies investigating the effects of transcranial photobiomodulation on cerebrospinal fluid melatonin levels have provided some support for the hypothesis that PBM can enhance brain melatonin production (Salehpour et al. 2018). Patients with neurodegenerative diseases treated with transcranial Near Infra-Red (NIR) light protocols showed modest increases in cerebrospinal fluid melatonin concentrations compared to baseline, although these changes were not consistently correlated with plasma melatonin levels, suggesting local rather than systemic effects on melatonin biosynthesis. Thus, the molecular mechanisms by which PBM might enhance melatonin synthesis involve several potential pathways that require further investigation.

7.8 Balanced Diet and Melatonin

The gastrointestinal tract contains substantial quantities of melatonin, primarily synthesized locally by enterochromaffin (EC) cells with minimal contribution from systemic sources like the pineal gland. Unlike circulating melatonin that follows circadian patterns, intestinal melatonin concentrations in the ileum and colon fluctuate according to feeding status and digestive processes rather than light exposure or diurnal rhythms (Chen et al. 2011). Melatonin production in the pineal gland exhibits a well-defined circadian pattern, whereas its synthesis in the gastrointestinal tract may be influenced by dietary composition (Liu et al. 2024).

Emerging studies reveal that gut microbiota modulates host melatonin synthesis through specific microbial interactions, offering potential therapeutic avenues for melatonin-associated disorders (Bonmati-Carrion and Tomas-Loba 2021; Liu et al. 2024). Recent research demonstrates that gut microbiota enhances colonic melatonin production by activating the TLR2/4-MyD88-NF-κB pathway in intestinal epithelial cells, upregulating AANAT expression. This finding provides novel mechanistic insights into host melatonin regulation and opens new avenues for targeted therapies (Liu et al. 2024).

Tryptophan metabolism occurs through two distinct pathways, differentiated by their terminal or intermediate products: the serotonin pathway and the kynurenine pathway. In humans, the kynurenine pathway serves as the dominant catabolic route, processing over 90% of dietary tryptophan into primary metabolites such as kynurenine and niacin. Only a minor fraction is channeled through the serotonin pathway for subsequent melatonin synthesis (Barik 2020).

Melatonin is found in significant quantities in whole grains (such as pigmented rice and wheat), nuts (almonds, walnuts), fruits (bananas, cherries, mangoes), and animal products (fish, eggs, and meats). Tryptophan, the precursor to both serotonin and melatonin, is abundant in legumes (chickpeas, soybeans, black beans), dairy (milk, cheese), poultry, and seafood, which are dietary staples in the region. Serotonin, primarily derived from tryptophan, is present in foods like bananas, tomatoes, and pineapples. Notably, traditional South Asian diets which often combine grains, lentils, and vegetables may naturally enhance melatonin and serotonin production. However, food processing methods, such as rice polishing, can reduce melatonin content, suggesting that whole-grain and minimally processed variants may offer greater benefits. This dietary pattern highlights the potential of culturally relevant nutrition in promoting sleep and mental well-being, though further research is needed to quantify its physiological impact.

Table 7.4 outlines a variety of common food sources in South Asia that are rich in melatonin, tryptophan, and serotonin, which are bioactive compounds essential for regulating sleep, mood, and overall health.

7.9 Role of Gut Microbiota in Melatonin Level Modulation

Certain microbes in the human gut have the ability to generate melatonin (Sejbuk et al. 2024). The gastrointestinal tract (GIT) represents the largest extrapineal reservoir of circulating melatonin, with mucosal concentrations surpassing plasma levels by 100 to 400 times. This substantial melatonin synthesis in the GIT occurs predominantly during daylight hours and sustains peripheral blood concentrations, particularly when facilitated by a tryptophan (Trp)-rich diet, as Trp serves as the primary precursor for melatonin biosynthesis. In contrast to pineal melatonin synthesis, which is regulated by photoperiodic signals, gastrointestinal (GIT) melatonin secretion is linked to food intake. Interestingly, this process appears to be modulated through a negative feedback mechanism involving circulating melatonin levels (Konturek et al. 2007).

Gut microbiota play a regulatory role in host melatonin synthesis. Specifically, microbial activation of the TLR2/4/MyD88/NF-κB pathway in intestinal epithelial cells upregulates AANAT expression, thereby increasing colonic melatonin levels. This mechanistic insight opens new avenues for understanding the autonomous regulation of melatonin production (Liu et al. 2024).

The gut microbiota plays a significant role in melatonin production, creating a bidirectional relationship where microbes influence host melatonin levels. Studies demonstrate that injecting the melatonin precursor 5-HTP into the gastrointestinal tract increases melatonin, particularly in the colon, suggesting microbial involvement. Additionally, gut-derived melatonin may enter the bloodstream, as evidenced

Table 7.4 Common South Asian food sources rich in melatonin, tryptophan, and serotonin

Food source	Melatonin/ Tryptophan/ Serotonin	References
Rice	Melatonin	Setyaningsih et al. (2015)
Almonds	Melatonin	Meng et al. (2017), Minich et al. (2022) and Zuraikat et al. (2021)
Walnuts	Melatonin	Minich et al. (2022), Peuhkuri et al. (2012) and Zuraikat et al. (2021)
Wheat	Melatonin	Johns et al. (2013), Meng et al. (2017) and Minich et al. (2022)
Grapes	Melatonin	Iriti et al. (2010), Iriti and Varoni (2015) and Minich et al. (2022)
Cherries	Melatonin	Meng et al. (2017), Minich et al. (2022)
Pineapple	Melatonin	Johns et al. (2013), Pranil et al. (2021) and Sae-Teaw et al. (2013)
Mango	Melatonin	Pranil et al. (2021)
Orange	Melatonin	Sae-Teaw et al. (2013)
Banana	Melatonin	Hu et al. (2017), Johns et al. (2013) and Sae-Teaw et al. (2013)
Pepper	Melatonin	Riga et al. (2014)
Tomato	Melatonin	Riga et al. (2014)
Beer	Melatonin	Iriti et al. (2010), Peuhkuri et al. (2012)
Wine	Melatonin	Iriti and Varoni (2015), Meng et al. (2017)
Coffee	Melatonin	Minich et al. (2022)
Fish	Melatonin	Esposito and Cuzzocrea (2010), Minich et al. (2022)
Eggs	Melatonin	Esposito and Cuzzocrea (2010), Minich et al. (2022)
Chicken	Melatonin	Tan et al. (2014)
Lamb	Melatonin	Esposito and Cuzzocrea (2010), Minich et al. (2022) and Tan et al. (2014)
Pork	Melatonin	Esposito and Cuzzocrea (2010), Minich et al. (2022), Tan et al. (2014)
Beef	Melatonin	Esposito and Cuzzocrea (2010), Minich et al. (2022) and Tan et al. (2014)
Mushrooms (Agaricus bisporus)	Melatonin	Meng et al. (2017)
Ginger	Melatonin	Zuraikat et al. (2021)
Pineapple	Serotonin	Zuraikat et al. (2021)
Banana	Serotonin	Zuraikat et al. (2021)
Plums	Serotonin	Zuraikat et al. (2021)
Tomatoes	Serotonin	Peuhkuri et al. (2012), Zuraikat et al. (2021)
Black walnuts	Serotonin	Zuraikat et al. (2021)
Milk	Tryptophan	Palego et al. (2016), Peuhkuri et al. (2012) and Su et al. (2022)

(continued)

Table 7.4 (continued)

Food source	Melatonin/ Tryptophan/ Serotonin	References
Cheese	Tryptophan	Palego et al. (2016), Su et al. (2022)
Cheddar	Tryptophan	Zuraikat et al. (2021)
Eggs	Tryptophan	Palego et al. (2016), Zuraikat et al. (2021)
Meat (chicken breast, pork)	Tryptophan	Palego et al. (2016), Zuraikat et al. (2021)
Sea food (fish and crustaceous—crab, shrimp)	Tryptophan	Palego et al. (2016), Zuraikat et al. (2021)
Potatoes	Tryptophan	Palego et al. (2016)
Chickpeas	Tryptophan	Palego et al. (2016)
Soybeans	Tryptophan	Palego et al. (2016)
Cocoa beans	Tryptophan	Palego et al. (2016)
Cashew	Tryptophan	Palego et al. (2016)
Wheat bran	Tryptophan	Zuraikat et al. (2021)
Black beans	Tryptophan	Zuraikat et al. (2021)
Kidney beans	Tryptophan	Zuraikat et al. (2021)

by elevated plasma levels in animals after 5-HTP administration. Microbial metabolites like butyrate further enhance melatonin synthesis by activating key enzymes (AANAT and HIOMT) in the gut and immune cells. The microbiota also upregulates melatonin receptor expression in colon cells, potentially mediated by short-chain fatty acids (SCFAs), amplifying melatonin's effects. However, the interaction between pineal and gut-derived melatonin remains unclear, warranting further research to fully elucidate this relationship (Bonmatí-Carrión and Rol 2023).

7.10 Human Clinical Trials and Translational Gaps

The translation of melatonin's ERK1/2-dependent Nrf2 activation mechanisms from preclinical models to clinical applications faces significant challenges related to study design limitations, biomarker validation, outcome measurement standardization, and the identification of appropriate patient populations (Boutin et al. 2023). Current clinical evidence provides encouraging but incomplete support for melatonin's neuroprotective efficacy, with several critical gaps that must be addressed to realize the therapeutic potential of this signaling pathway.

7.11 Intermittent Fasting

The integration of melatonin supplementation with intermittent fasting protocols represents a particularly interesting approach that leverages the natural intersection of circadian biology, metabolic regulation, and cellular stress responses (Anwardeen et al. 2025). Intermittent fasting activates many of the same cellular pathways targeted by melatonin, including AMPK activation, autophagy induction, and Nrf2-mediated antioxidant responses, creating opportunities for synergistic interactions (Charlot et al. 2021; Shabkhizan et al. 2023).

The molecular mechanisms underlying melatonin-intermittent fasting synergies involve the coordinated activation of cellular stress response pathways that promote adaptation and resilience (Charlot et al. 2021). Intermittent fasting induces mild oxidative stress that activates Nrf2-dependent transcriptional programs, while melatonin provides additional antioxidant protection and enhances the efficiency of these adaptive responses. The combination also optimizes autophagy function through dual activation of AMPK (by fasting) and direct effects on autophagy machinery (by melatonin), promoting the clearance of damaged proteins and organelles (Ning et al. 2023).

Circadian alignment represents a critical component of successful melatonin-intermittent fasting combinations, as the timing of fasting periods must be coordinated with natural circadian rhythms to maximize benefits and minimize disruption (Shkorfu et al. 2025). Time-restricted feeding protocols that align with natural circadian patterns of melatonin secretion demonstrate enhanced efficacy in metabolic regulation, weight management, and glycemic control. The addition of exogenous melatonin can potentially optimize these effects by reinforcing circadian signals and enhancing the cellular stress responses activated by fasting (Charlot et al. 2021).

Experimental evidence from animal models demonstrates that melatonin-intermittent fasting combinations produce superior outcomes in neurodegenerative disease models compared to either intervention alone (Zhang et al. 2024). In Alzheimer's disease mouse models, the combination approach reduces amyloid burden, preserves synaptic function, and improves cognitive performance through enhanced autophagy, reduced neuroinflammation, and optimized energy metabolism. These effects are associated with increased expression of brain-derived neurotrophic factor (BDNF), enhanced mitochondrial biogenesis, and improved blood-brain barrier integrity (Cordos et al. 2025).

7.12 Challenges, Limitations, and Future Scope

While melatonin's ERK1/2-dependent Nrf2 activation represents a promising neuroprotective mechanism, several significant challenges and limitations must be carefully addressed to ensure safe and effective clinical translation. This section critically examines the potential risks associated with pathway overactivation, the complex challenges surrounding chronotherapeutic timing and dosing, and the largely unexplored integration of phototherapeutic approaches with melatonin therapy, and it also provides comprehensive recommendations for future research directions.

7.12.1 Chronotherapeutic Mismatches and Dosing Challenges

The optimization of melatonin therapy for neuroprotective applications faces significant challenges related to dosing complexity and circadian timing considerations that are critical for therapeutic efficacy while minimizing adverse effects (Cruz-Sanabria et al. 2023). These challenges are compounded by melatonin's biphasic dose-response characteristics and the potential for chronotherapeutic mistiming to disrupt normal circadian regulation (Arendt and Skene 2005; Emens and Burgess 2015; Hardeland 2013).

Circadian mistiming represents a critical risk factor in melatonin therapy, as inappropriate timing relative to individual circadian phase can diminish therapeutic benefits, disrupt normal sleep-wake cycles, and exacerbate existing circadian dysfunction. The complex relationship between exogenous melatonin administration and endogenous circadian regulation requires precise timing considerations that account for individual chronotype, existing circadian phase, and therapeutic objectives (Baron and Reid 2014; Emens and Burgess 2015).

The physiological basis for circadian mistiming risks stems from melatonin's dual role as both a circadian signal and a therapeutic agent. Endogenous melatonin serves as a primary circadian output signal that conveys timing information throughout the body, with peak secretion typically occurring during the biological night. Exogenous melatonin administration can function as a chronobiotic agent that shifts circadian phase, but the direction and magnitude of phase shifts depend critically on the timing of administration relative to the individual's circadian rhythm (Cruz-Sanabria et al. 2023).

7.13 Recommendations for Future Research

The development of safe and effective melatonin-based neuroprotective therapies requires systematic investigation across multiple research domains, from basic mechanistic studies to large-scale clinical trials. These recommendations address critical knowledge gaps and methodological challenges that must be overcome to translate promising preclinical findings into clinical applications.

Future longitudinal trials should incorporate comprehensive biomarker panels that can monitor pathway activation, safety parameters, and early indicators of therapeutic efficacy. These biomarker assessments should include measures of ERK1/2 phosphorylation, Nrf2 nuclear translocation, downstream target gene expression, oxidative stress markers, inflammatory cytokines, and circadian rhythm parameters. The development of non-invasive biomarker collection methods, such as saliva, tears, or peripheral blood sampling, would facilitate repeated assessments throughout extended study periods.

References

Abdolmaleki A, Karimian A, Khoshnazar SM, Asadi A, Samarein ZA, Smail SW, Bhattacharya D (2024) The role of Nrf2 signaling pathways in nerve damage repair. Toxicol Res 13(3):tfae080. https://doi.org/10.1093/toxres/tfae080

Acuña-Castroviejo D, Escames G, Venegas C, Díaz-Casado ME, Lima-Cabello E, López LC, Rosales-Corral S, Tan D-X, Reiter RJ (2014) Extrapineal melatonin: sources, regulation, and potential functions. Cell Mol Life Sci 71(16):2997–3025. https://doi.org/10.1007/s00018-014-1579-2

Ahmadi Z, Ashrafizadeh M (2020) Melatonin as a potential modulator of Nrf2. Fundam Clin Pharmacol 34(1):11–19. https://doi.org/10.1111/fcp.12498

Alghamdi BS (2018) The neuroprotective role of melatonin in neurological disorders. J Neurosci Res 96(7):1136–1149. https://doi.org/10.1002/jnr.24220

Ali T, Rehman SU, Shah FA, Kim MO (2018) Acute dose of melatonin via Nrf2 dependently prevents acute ethanol-induced neurotoxicity in the developing rodent brain. J Neuroinflammation 15(1):119. https://doi.org/10.1186/s12974-018-1157-x

Alluri H, Wilson RL, Shaji CA, Wiggins-Dohlvik K, Patel S, Liu Y, Peng X, Beeram MR, Davis ML, Huang JH, Tharakan B (2016) Melatonin preserves blood-brain barrier integrity and permeability via matrix Metalloproteinase-9 inhibition. PLoS One 11(5):e0154427. https://doi.org/10.1371/journal.pone.0154427

Anwardeen NR, Naja K, Almuraikhy S, Sellami M, Al-Amri HS, Philip N, Tamimi F, Agil A, Elrayess MA (2025) The influence of circadian rhythm disruption during Ramadan on metabolic responses to physical activity: a pilot study. Front Neurosci 19. https://doi.org/10.3389/fnins.2025.1542016

Apopa PL, He X, Ma Q (2008) Phosphorylation of Nrf2 in the transcription activation domain by casein kinase 2 (CK2) is critical for the nuclear translocation and transcription activation function of Nrf2 in IMR-32 neuroblastoma cells. J Biochem Mol Toxicol 22(1):63–76. https://doi.org/10.1002/jbt.20212

Arendt J, Skene DJ (2005) Melatonin as a chronobiotic. Sleep Med Rev 9(1):25–39. https://doi.org/10.1016/j.smrv.2004.05.002

Bald EM, Nance CS, Schultz JL (2021) Melatonin may slow disease progression in amyotrophic lateral sclerosis: findings from the pooled resource open-access ALS clinic trials database. Muscle Nerve 63(4):572–576. https://doi.org/10.1002/mus.27168

Barik S (2020) The uniqueness of tryptophan in biology: properties, metabolism, interactions and localization in proteins. Int J Mol Sci 21(22):8776. https://doi.org/10.3390/ijms21228776

Baron KG, Reid KJ (2014) Circadian misalignment and health. Int Rev Psychiatry (Abingdon, England) 26(2):139–154. https://doi.org/10.3109/09540261.2014.911149

Bathini M, Raghushaker CR, Mahato KK (2020) The molecular mechanisms of action of Photobiomodulation against neurodegenerative diseases: a systematic review. Cell Mol Neurobiol 42(4):955–971. https://doi.org/10.1007/s10571-020-01016-9

Benedeto-Stojanov D, Ničković VP, Petrović G, Rancić A, Grgov I, Nikolić GR, Marčetić ZP, Popović MR, Lazarević M, Mitić KV, Sokolović D (2023) Melatonin as a promising anti-inflammatory agent in an in vivo animal model of sepsis-induced rat liver damage. Int J Mol Sci 25(1):455. https://doi.org/10.3390/ijms25010455

Bloom DA, Jaiswal AK (2003) Phosphorylation of Nrf2 at Ser40 by Protein Kinase C in Response to Antioxidants Leads to the Release of Nrf2 from INrf2, but Is Not Required for Nrf2 Stabilization/Accumulation in the Nucleus and Transcriptional Activation of Antioxidant Response Element-mediated NAD(P)H:Quinone Oxidoreductase-1 Gene Expression. J Biol Chem 278(45):44675–44682. https://doi.org/10.1074/jbc.M307633200

Bonmatí-Carrión M-Á, Rol M-A (2023) Melatonin as a mediator of the gut microbiota-host interaction: implications for health and disease. Antioxidants (Basel, Switzerland) 13(1):34. https://doi.org/10.3390/antiox13010034

Bonmati-Carrion M-A, Tomas-Loba A (2021) Melatonin and cancer: a polyhedral network where the source matters. Antioxidants (Basel, Switzerland) 10(2):210. https://doi.org/10.3390/antiox10020210

Bonmati-Carrion MA, Arguelles-Prieto R, Martinez-Madrid MJ, Reiter R, Hardeland R, Rol MA, Madrid JA (2014) Protecting the melatonin rhythm through circadian healthy light exposure. Int J Mol Sci 15(12):23448–23500. https://doi.org/10.3390/ijms151223448

Boutin JA, Kennaway DJ, Jockers R (2023) Melatonin: facts, Extrapolations and Clinical Trials. Biomolecules 13(6):943. https://doi.org/10.3390/biom13060943

Brennan R, Jan JE, Lyons CJ (2007) Light, dark, and melatonin: emerging evidence for the importance of melatonin in ocular physiology. Eye 21(7):901–908. https://doi.org/10.1038/sj.eye.6702597

Cardinali DP, Vigo DE, Olivar N, Vidal MF, Furio AM, Brusco LI (2012) Therapeutic application of melatonin in mild cognitive impairment. Am J Neurodegener Dis 1(3):280–291

Chahbouni M, Escames G, Venegas C, Sevilla B, García JA, López LC, Muñoz-Hoyos A, Molina-Carballo A, Acuña-Castroviejo D (2010) Melatonin treatment normalizes plasma pro-inflammatory cytokines and nitrosative/oxidative stress in patients suffering from Duchenne muscular dystrophy. J Pineal Res 48(3):282–289. https://doi.org/10.1111/j.1600-079X.2010.00752.x

Charlot A, Hutt F, Sabatier E, Zoll J (2021) Beneficial effects of early time-restricted feeding on metabolic diseases: importance of aligning food habits with the circadian clock. Nutrients 13(5):1405. https://doi.org/10.3390/nu13051405

Chen C-Q, Fichna J, Bashashati M, Li Y-Y, Storr M (2011) Distribution, function and physiological role of melatonin in the lower gut. World J Gastroenterol 17(34):3888–3898. https://doi.org/10.3748/wjg.v17.i34.3888

Chen N, Zhang R, Konishi T, Wang J (2017) Upregulation of NRF2 through autophagy/ERK 1/2 ameliorates ionizing radiation induced cell death of human osteosarcoma U-2 OS. Mutat Res Genet Toxicol Environ Mutagen 813:10–17. https://doi.org/10.1016/j.mrgentox.2016.11.006

Chen X, Xi Z, Liang H, Sun Y, Zhong Z, Wang B, Bian L, Sun Q (2019) Melatonin prevents mice cortical astrocytes from hemin-induced toxicity through activating PKCα/Nrf2/HO-1 signaling in vitro. Front Neurosci 13. https://doi.org/10.3389/fnins.2019.00760

Chen M, Cecon E, Karamitri A, Gao W, Gerbier R, Ahmad R, Jockers R (2020) Melatonin MT1 and MT2 receptor ERK signaling is differentially dependent on Gi/o and Gq/11 proteins. J Pineal Res 68(4):e12641. https://doi.org/10.1111/jpi.12641

Chen S-J, Deng Y-T, Li Y-Z, Zhang Y-R, Zhang W, Chen S-D, Wu B-S, Yang L, Dong Q, Feng J, Cheng W, Yu J-T (2022) Association of circadian rhythms with brain disorder incidents: a prospective cohort study of 72242 participants. Transl Psychiatry 12:514. https://doi.org/10.1038/s41398-022-02278-1

Chou L-Y, Chao Y-M, Peng Y-C, Lin H-C, Wu Y-L (2020) Glucosamine enhancement of BDNF expression and animal cognitive function. Molecules (Basel, Switzerland) 25(16):3667. https://doi.org/10.3390/molecules25163667

Chowdhury MAR, Haq MM, Lee JH, Jeong S (2024) Multi-faceted regulation of CREB family transcription factors. Front Mol Neurosci 17. https://doi.org/10.3389/fnmol.2024.1408949

Cordos B, Tero-Vescan A, Hampson IN, Oliver AW, Slevin M (2025) Synergistic autophagy-related mechanisms of protection against brain aging and AD: cellular pathways and therapeutic strategies. Pharmaceuticals 18(6):829. https://doi.org/10.3390/ph18060829

Cores Á, Piquero M, Villacampa M, León R, Menéndez JC (2020) NRF2 regulation processes as a source of potential drug targets against neurodegenerative diseases. Biomolecules 10(6):904. https://doi.org/10.3390/biom10060904

Cremascoli R, Sparasci D, Giusti G, Cattaldo S, Prina E, Roveta F, Bruno F, Ghezzi C, Cerri S, Picascia M, Bernini S, Sinforiani E, Terzaghi M, Priano L, Mauro A, Manni R (2022) Effects of circadian phase tailored light therapy on sleep, mood, and cognition in Alzheimer's disease: preliminary findings in a pivotal study. Front Physiol 12:755322. https://doi.org/10.3389/fphys.2021.755322

Cruz-Sanabria F, Carmassi C, Bruno S, Bazzani A, Carli M, Scarselli M, Faraguna U (2023) Melatonin as a Chronobiotic with sleep-promoting properties. Curr Neuropharmacol 21(4):951–987. https://doi.org/10.2174/1570159X20666220217152617

Deng X, Xie A, Ye D, Ma Y, Bao Z, Xie Q, Luo Z, Wang R, Li H, Yu R (2025) Melatonin activates KEAP1/NRF2/PTGS2 pathway to attenuate Hyperoxia-driven Ferroptosis in bronchopulmonary dysplasia. J Inflamm Res 18:7571–7583. https://doi.org/10.2147/JIR.S520404

Dianatkhah M, Najafi A, Sharifzadeh M, Ahmadi A, Sharifnia H, Mojtahedzadeh M, Najmeddin F, Moghaddas A (2017) Melatonin supplementation may improve the outcome of patients with hemorrhagic stroke in the intensive care unit. J Res Pharm Pract 6(3):173–177. https://doi.org/10.4103/jrpp.JRPP_17_49

dos Cardoso FS, Barrett DW, Wade Z, Gomes da Silva S, Gonzalez-Lima F (2022) Photobiomodulation of cytochrome c oxidase by chronic transcranial laser in young and aged brains. Front Neurosci 16. https://doi.org/10.3389/fnins.2022.818005

Dowling GA, Burr RL, Van Someren EJW, Hubbard EM, Luxenberg JS, Mastick J, Cooper BA (2008) Melatonin and Bright-Light Treatment for Rest–Activity Disruption in Institutionalized Patients with Alzheimer's Disease. J Am Geriatr Soc 56(2):239–246. https://doi.org/10.1111/j.1532-5415.2007.01543.x

Emens JS, Burgess HJ (2015) Effect of light and melatonin and other melatonin receptor agonists on human circadian physiology. Sleep Med Clin 10(4):435–453. https://doi.org/10.1016/j.jsmc.2015.08.001

Esposito E, Cuzzocrea S (2010) Antiinflammatory activity of melatonin in central nervous system. Curr Neuropharmacol 8(3):228–242. https://doi.org/10.2174/157015910792246155

Ezema BO, Eze CN, Ayoka TO, Nnadi CO (2024) Antioxidant-enzyme interaction in non-communicable diseases. J Explor Res Pharmacol 9(4):262–275. https://doi.org/10.14218/JERP.2024.00020

Gnocchi D, Bruscalupi G (2017) Circadian rhythms and hormonal homeostasis: pathophysiological implications. Biology 6(1):10. https://doi.org/10.3390/biology6010010

Grima NA, Rajaratnam SMW, Mansfield D, Sletten TL, Spitz G, Ponsford JL (2018) Efficacy of melatonin for sleep disturbance following traumatic brain injury: a randomised controlled trial. BMC Med 16(1):8. https://doi.org/10.1186/s12916-017-0995-1

Hamblin MR (2017) Mechanisms and applications of the anti-inflammatory effects of photobiomodulation. AIMS Biophys 4(3):337–361. https://doi.org/10.3934/biophy.2017.3.337

Hardeland R (2013) Chronobiology of melatonin beyond the feedback to the Suprachiasmatic nucleus—consequences to melatonin dysfunction. Int J Mol Sci 14(3):5817–5841. https://doi.org/10.3390/ijms14035817

Hong N (2019) Photobiomodulation as a treatment for neurodegenerative disorders: current and future trends. Biomed Eng Lett 9(3):359–366. https://doi.org/10.1007/s13534-019-00115-x

Hu W, Yang H, Tie W, Yan Y, Ding Z, Liu Y, Wu C, Wang J, Reiter RJ, Tan D-X, Shi H, Xu B, Jin Z (2017) Natural variation in Banana varieties highlights the role of melatonin in postharvest ripening and quality. J Agric Food Chem 65(46):9987–9994. https://doi.org/10.1021/acs.jafc.7b03354

Ikram M, Park HY, Ali T, Kim MO (2021) Melatonin as a potential regulator of oxidative stress, and Neuroinflammation: mechanisms and implications for the Management of Brain Injury-Induced Neurodegeneration. J Inflamm Res 14:6251–6264. https://doi.org/10.2147/JIR.S334423

Iriti M, Varoni EM (2015) Melatonin in Mediterranean diet, a new perspective. J Sci Food Agric 95(12):2355–2359. https://doi.org/10.1002/jsfa.7051

Iriti M, Varoni EM, Vitalini S (2010) Melatonin in traditional Mediterranean diets. J Pineal Res 49(2):101–105. https://doi.org/10.1111/j.1600-079X.2010.00777.x

Jallouli S, Ghroubi S, Damak M, Sakka S, Elleuch MH, Mhiri C, Yahia A, Driss T, de Marco G, Hammouda O (2025) 12-week melatonin supplementation improved dynamic postural stability and walking performance in persons living with multiple sclerosis: a randomized controlled trial. Behav Brain Res 476:115191. https://doi.org/10.1016/j.bbr.2024.115191

Jiang T, Chang Q, Cai J, Fan J, Zhang X, Xu G (2016) Protective effects of melatonin on retinal inflammation and oxidative stress in experimental diabetic retinopathy. Oxidative Med Cell Longev 2016(1):3528274. https://doi.org/10.1155/2016/3528274

Johns NP, Johns J, Porasuphatana S, Plaimee P, Sae-Teaw M (2013) Dietary intake of melatonin from tropical fruit altered urinary excretion of 6-sulfatoxymelatonin in healthy volunteers. J Agric Food Chem 61(4):913–919. https://doi.org/10.1021/jf300359a

Jung J, Kim T (2024) Photobiomodulation and its therapeutic potential in sleep disturbances. Sleep Med Res 15(4):218–227. https://doi.org/10.17241/smr.2024.02593

Kato Y, Sato K, Nagamine H, Kanatani M, Horikoshi Y, Nakaso K (2025) Cytoprotective effect of melatonin against MPP+ toxicity in SH-SY5Y cells: role sharing of two types of antioxidative activities of melatonin. Biochem Biophys Res Commun 742:151074. https://doi.org/10.1016/j.bbrc.2024.151074

Konturek SJ, Konturek PC, Brzozowska I, Pawlik M, Sliwowski Z, Cześnikiewicz-Guzik M, Kwiecień S, Brzozowski T, Bubenik GA, Pawlik WW (2007) Localization and biological activities of melatonin in intact and diseased gastrointestinal tract (GIT). J Physiol Pharmacol 58(3):381–405

Kozaki T, Kubokawa A, Taketomi R, Hatae K (2015) Effects of day-time exposure to different light intensities on light-induced melatonin suppression at night. J Physiol Anthropol 34(1):27. https://doi.org/10.1186/s40101-015-0067-1

Kripke DF, Elliott JA, Youngstedt SD, Rex KM (2007) Circadian phase response curves to light in older and young women and men. J Circadian Rhythms 5. https://doi.org/10.1186/1740-3391-5-4

Lewy A (2010) Clinical implications of the melatonin phase response curve. J Clin Endocrinol Metab 95(7):3158–3160. https://doi.org/10.1210/jc.2010-1031

Li W, Kong A-N (2009) Molecular mechanisms of Nrf2-mediated antioxidant response. Mol Carcinog 48(2):91–104. https://doi.org/10.1002/mc.20465

Li Y, Zhang J, Wan J, Liu A, Sun J (2020) Melatonin regulates aβ production/clearance balance and aβ neurotoxicity: a potential therapeutic molecule for Alzheimer's disease. Biomed Pharmacother 132:110887. https://doi.org/10.1016/j.biopha.2020.110887

Lin YW, Chen TY, Hung CY, Tai SH, Huang SY, Chang CC, Hung HY, Lee EJ (2018) Melatonin protects brain against ischemia/reperfusion injury by attenuating endoplasmic reticulum stress. Int J Mol Med 42(1):182–192. https://doi.org/10.3892/ijmm.2018.3607

Liu B, Fan L, Wang Y, Wang H, Yan Y, Chen S, Hung I, Liu C, Wei H, Ge L, Ren W (2024) Gut microbiota regulates host melatonin production through epithelial cell MyD88. Gut Microbes 16(1):2313769. https://doi.org/10.1080/19490976.2024.2313769

Lv Q-K, Tao K-X, Yao X-Y, Pang M-Z, Cao B-E, Liu C-F, Wang F (2024) Melatonin MT1 receptors regulate the Sirt1/Nrf2/Ho-1/Gpx4 pathway to prevent α-synuclein-induced ferroptosis in Parkinson's disease. J Pineal Res 76(2):e12948. https://doi.org/10.1111/jpi.12948

Markus RP, Fernandes PA, Kinker GS, da Silveira Cruz-Machado S, Marçola M (2018) Immune-pineal axis – acute inflammatory responses coordinate melatonin synthesis by pinealocytes and phagocytes. Br J Pharmacol 175(16):3239–3250. https://doi.org/10.1111/bph.14083

Meng X, Li Y, Li S, Zhou Y, Gan R-Y, Xu D-P, Li H-B (2017) Dietary sources and bioactivities of melatonin. Nutrients 9(4):367. https://doi.org/10.3390/nu9040367

Minich DM, Henning M, Darley C, Fahoum M, Schuler CB, Frame J (2022) Is melatonin the "next vitamin D"?: a review of emerging science, clinical uses, safety, and dietary supplements. Nutrients 14(19):3934. https://doi.org/10.3390/nu14193934

Musiek ES (2015) Circadian clock disruption in neurodegenerative diseases: cause and effect? Front Pharmacol 6:29. https://doi.org/10.3389/fphar.2015.00029

Negi G, Kumar A, Sharma SS (2011) Melatonin modulates neuroinflammation and oxidative stress in experimental diabetic neuropathy: effects on NF-κB and Nrf2 cascades. J Pineal Res 50(2):124–131. https://doi.org/10.1111/j.1600-079X.2010.00821.x

Ngo V, Duennwald ML (2022) Nrf2 and oxidative stress: a general overview of mechanisms and implications in human disease. Antioxidants 11(12):2345. https://doi.org/10.3390/antiox11122345

Ning B, Hang S, Zhang W, Mao C, Li D (2023) An update on the bridging factors connecting autophagy and Nrf2 antioxidant pathway. Front Cell Dev Biol 11. https://doi.org/10.3389/fcell.2023.1232241

Niture SK, Kaspar JW, Shen J, Jaiswal AK (2010) Nrf2 signaling and cell survival. Toxicol Appl Pharmacol 244(1):37–42. https://doi.org/10.1016/j.taap.2009.06.009

Okeke ES, Ogugofor MO, Nkwoemeka NE, Nweze EJ, Okoye CO (2022) Phytomelatonin: a potential phytotherapeutic intervention on COVID-19-exposed individuals. Microbes Infect 24(1):104886. https://doi.org/10.1016/j.micinf.2021.104886

Palego L, Betti L, Rossi A, Giannaccini G (2016) Tryptophan biochemistry: structural, nutritional, metabolic, and medical aspects in humans. J Amino Acids 2016:8952520. https://doi.org/10.1155/2016/8952520

Pei Z, Ho HTS, Cheung RTF (2002) Pre-treatment with melatonin reduces volume of cerebral infarction in a permanent middle cerebral artery occlusion stroke model in the rat. Neurosci Lett 318(3):141–144. https://doi.org/10.1016/s0304-3940(01)02503-4

Pérez-Lloret S, Cardinali DP (2021) Melatonin as a Chronobiotic and Cytoprotective agent in Parkinson's disease. Front Pharmacol 12. https://doi.org/10.3389/fphar.2021.650597

Peuhkuri K, Sihvola N, Korpela R (2012) Dietary factors and fluctuating levels of melatonin. Food Nutr Res 56. https://doi.org/10.3402/fnr.v56i0.17252

Pranil T, Moongngarm A, Manwiwattanakul G, Loypimai P, Kerr WL (2021) Melatonin and its derivative contents in tropical fruits and fruit tablets. J Food Compos Anal 103:104109. https://doi.org/10.1016/j.jfca.2021.104109

Radogna F, Cristofanon S, Paternoster L, D'Alessio M, De Nicola M, Cerella C, Dicato M, Diederich M, Ghibelli L (2008) Melatonin antagonizes the intrinsic pathway of apoptosis via mitochondrial targeting of Bcl-2. J Pineal Res 44(3):316–325. https://doi.org/10.1111/j.1600-079X.2007.00532.x

Reiter RJ, Tan D-X, Fuentes-Broto L (2010) Melatonin: a multitasking molecule. Prog Brain Res 181:127–151. https://doi.org/10.1016/S0079-6123(08)81008-4

Riga P, Medina S, García-Flores LA, Gil-Izquierdo Á (2014) Melatonin content of pepper and tomato fruits: effects of cultivar and solar radiation. Food Chem 156:347–352. https://doi.org/10.1016/j.foodchem.2014.01.117

Roux PP, Blenis J (2004) ERK and p38 MAPK-activated protein kinases: a family of protein kinases with diverse biological functions. Microbiol Mol Biol Rev 68(2):320–344. https://doi.org/10.1128/MMBR.68.2.320-344.2004

Sae-Teaw M, Johns J, Johns NP, Subongkot S (2013) Serum melatonin levels and antioxidant capacities after consumption of pineapple, orange, or banana by healthy male volunteers. J Pineal Res 55(1):58–64. https://doi.org/10.1111/jpi.12025

Salehpour F, Mahmoudi J, Kamari F, Sadigh-Eteghad S, Rasta SH, Hamblin MR (2018) Brain Photobiomodulation therapy: a narrative review. Mol Neurobiol 55(8):6601–6636. https://doi.org/10.1007/s12035-017-0852-4

Schwartz MD, Wotus C, Liu T, Friesen WO, Borjigin J, Oda GA, de la Iglesia HO (2009) Dissociation of circadian and light inhibition of melatonin release through forced desynchronization in the rat. Proc Natl Acad Sci 106(41):17540–17545. https://doi.org/10.1073/pnas.0906382106

Sejbuk M, Siebieszuk A, Witkowska AM (2024) The role of gut microbiome in sleep quality and health: dietary strategies for microbiota support. Nutrients 16(14):2259. https://doi.org/10.3390/nu16142259

Serrage H, Heiskanen V, Palin W, Cooper PR, Milward MR, Hadis M, Hamblin MR (2019) Under the spotlight: mechanisms of photobiomodulation concentrating on blue and green light. Photochem Photobiol Sci. https://doi.org/10.1039/c9pp00089e

Setyaningsih W, Saputro IE, Barbero GF, Palma M, García Barroso C (2015) Determination of melatonin in Rice (Oryza sativa) grains by pressurized liquid extraction. J Agric Food Chem 63(4):1107–1115. https://doi.org/10.1021/jf505106m

Shabkhizan R, Haiaty S, Moslehian MS, Bazmani A, Sadeghsoltani F, Saghaei Bagheri H, Rahbarghazi R, Sakhinia E (2023) The beneficial and adverse effects of Autophagic response to caloric restriction and fasting. Adv Nutr 14(5):1211–1225. https://doi.org/10.1016/j.advnut.2023.07.006

Shkorfu W, Fadel A, Hamsho M, Ranneh Y, Shahbaz HM (2025) Intermittent fasting and hormonal regulation: pathways to improved metabolic health. Food Sci Nutr 13(8):e70586. https://doi.org/10.1002/fsn3.70586

St Hilaire MA, Gooley JJ, Khalsa SBS, Kronauer RE, Czeisler CA, Lockley SW (2012) Human phase response curve to a 1 h pulse of bright white light. J Physiol 590(Pt 13):3035–3045. https://doi.org/10.1113/jphysiol.2012.227892

Su X, Gao Y, Yang R (2022) Gut microbiota-derived tryptophan metabolites maintain gut and systemic homeostasis. Cells 11(15):2296. https://doi.org/10.3390/cells11152296

Sugumaran R, Sai Krishna KS, Saibaba J, Narayan SK, Sandhiya S, Rajeswari M (2024) Melatonin on sleep in Parkinson's disease: a randomized double blind placebo controlled trial. Sleep Med 124:502–509. https://doi.org/10.1016/j.sleep.2024.10.020

Sun J, Nan G (2017) The extracellular signal-regulated kinase 1/2 pathway in neurological diseases: a potential therapeutic target (Review). Int J Mol Med 39(6):1338–1346. https://doi.org/10.3892/ijmm.2017.2962

Takahashi JS (2017) Transcriptional architecture of the mammalian circadian clock. Nat Rev Genet 18(3):164–179. https://doi.org/10.1038/nrg.2016.150

Tan D-X, Manchester LC, Terron MP, Flores LJ, Reiter RJ (2007) One molecule, many derivatives: a never-ending interaction of melatonin with reactive oxygen and nitrogen species? J Pineal Res 42(1):28–42. https://doi.org/10.1111/j.1600-079X.2006.00407.x

Tan D-X, Zanghi BM, Manchester LC, Reiter RJ (2014) Melatonin identified in meats and other food stuffs: potentially nutritional impact. J Pineal Res 57(2):213–218. https://doi.org/10.1111/jpi.12152

Van Dooren T, Princen K, De Witte K, Griffioen G (2014) Derailed Intraneuronal Signalling drives pathogenesis in sporadic and familial Alzheimer's disease. Biomed Res Int 2014:167024. https://doi.org/10.1155/2014/167024

Vargas MR, Johnson DA, Sirkis DW, Messing A, Johnson JA (2008) Nrf2 activation in astrocytes protects against neurodegeneration in mouse models of familial amyotrophic lateral sclerosis. J Neurosci 28(50):13574–13581. https://doi.org/10.1523/JNEUROSCI.4099-08.2008

Videnovic A, Lazar AS, Barker RA, Overeem S (2014) 'The clocks that time us'—circadian rhythms in neurodegenerative disorders. Nat Rev Neurol 10(12):683–693. https://doi.org/10.1038/nrneurol.2014.206

Vomund S, Schäfer A, Parnham MJ, Brüne B, von Knethen A (2017) Nrf2, the master regulator of anti-oxidative responses. Int J Mol Sci 18(12):2772. https://doi.org/10.3390/ijms18122772

Wang X (2009) The Antiapoptotic activity of melatonin in neurodegenerative diseases. CNS Neurosci Ther 15(4):345–357. https://doi.org/10.1111/j.1755-5949.2009.00105.x

Yao Y, Chen T, Wu H, Yang N, Xu S (2023) Melatonin attenuates bisphenol A-induced colon injury by dual targeting mitochondrial dynamics and Nrf2 antioxidant system via activation of SIRT1/PGC-1α signaling pathway. Free Radic Biol Med 195:13–22. https://doi.org/10.1016/j.freeradbiomed.2022.12.081

Zhang Y, Cook A, Kim J, Baranov SV, Jiang J, Smith K, Cormier K, Bennett E, Browser RP, Day AL, Carlisle D, Ferrante RJ, Wang X, Friedlander RM (2013) Melatonin inhibits the caspase-1/cytochrome c/caspase-3 cell death pathway, inhibits MT1 receptor loss and delays disease progression in a mouse model of amyotrophic lateral sclerosis. Neurobiol Dis 55:26–35. https://doi.org/10.1016/j.nbd.2013.03.008

Zhang C, Ma Y, Zhao Y, Guo N, Han C, Wu Q, Mu C, Zhang Y, Tan S, Zhang J, Liu X (2024) Systematic review of melatonin in cerebral ischemia-reperfusion injury: critical role and therapeutic opportunities. Front Pharmacol 15. https://doi.org/10.3389/fphar.2024.1356112

Zhao D, Yu Y, Shen Y, Liu Q, Zhao Z, Sharma R, Reiter RJ (2019) Melatonin synthesis and function: evolutionary history in animals and plants. Front Endocrinol 10. https://doi.org/10.3389/fendo.2019.00249

Zheng R, Ruan Y, Yan Y, Lin Z, Xue N, Yan Y, Tian J, Yin X, Pu J, Zhang B (2021) Melatonin attenuates Neuroinflammation by Down-regulating NLRP3 Inflammasome via a SIRT1-dependent pathway in MPTP-induced models of Parkinson's disease. J Inflamm Res 14:3063–3075. https://doi.org/10.2147/JIR.S317672

Zuraikat FM, Wood RA, Barragán R, St-Onge M-P (2021) Sleep and diet: mounting evidence of a cyclical relationship. Annu Rev Nutr 41:309–332. https://doi.org/10.1146/annurev-nutr-120420-021719

Role of Berberine in the Modulation of the PI3K/AKT/mTOR Pathway

8

Héctor Eduardo López-Valdés, Hilda Martínez-Coria, and Juan Angel Ortiz-Lara

Abstract

Berberine is a natural compound found in several medicinal plants used for centuries in Asian countries, particularly in China and India. Chemically, berberine is a quaternary benzylisoquinoline alkaloid and its use in preclinical studies has shown a diversity of beneficial effects in models of human diseases including cancer, type 2 diabetes, heart disease, depression, obesity, epilepsy, and neurodegenerative diseases. Berberine exhibits its effects through modulating different cellular mechanisms resulting in decreased oxidative stress, inhibition of proinflammatory cytokines and angiogenesis, increased apoptosis, decreased metastasis, among other effects. Berberine modulates several intracellular signaling cascades including the PI3K/AKT/mTOR pathway, which, under homeostatic conditions, is involved in many processes including angiogenesis, apoptosis, cell migration, and proliferation, but under pathological conditions, overactivation or aberrant activation is associated with several human diseases such as those mentioned above and several neurodegenerative diseases such as Alzheimer's and Parkinson's disease. This chapter aims to analyze the interaction of berberine with the PI3K/AKT/mTOR pathway in two of the most common neurodegenerative illnesses, Alzheimer's disease and Parkinson's disease.

Héctor Eduardo López-Valdés and Hilda Martínez-Coria contributed equally to this work.

H. E. López-Valdés (✉) · H. Martínez-Coria · J. A. Ortiz-Lara
Departamento de Fisiología, Facultad de Medicina, Universidad Nacional Autónoma de México (UNAM), Mexico City, Mexico
e-mail: helopez@comunidad.unam.mx

H. Dilnashin, S. P. Singh (eds.), *Phototherapeutic Approaches to Neurodegeneration*,
https://doi.org/10.1007/978-981-95-4751-7_8

Keywords

Berberine · Neurodegeneration · PI3K/AKT/mTOR pathway · Alzheimer's disease · Parkinson's disease

8.1 Introduction

Berberine (BBR) is a bioactive molecule isolated from plants of different genera, including *Berberis, Annickia, Coptis,* and *Argemone* (Neag et al. 2018). This molecule is found in many commonly known medicinal plants, such as barberry (*Berberis vulgaris*), tree turmeric (*Berberis aristata*), goldenseal (*Hydrastis canadensis*), and goldenthread (*Coptis chinensis*), which have been used in traditional Asian medicine, mainly in China and India, to treat various conditions such as gastric problems, infections, inflammation, and detoxification (Birdsall 1997; Prajwala et al. 2020; Zhu et al. 2022). In modern science, BBR has been tested in a large number of preclinical studies for various human diseases, such as cancer, thrombosis, bacterial infections, metabolic disease, depression, Alzheimer's disease (AD), Parkinson's disease (PD), stroke, Huntington's disease, and respiratory disorders, with positive results (Chaudhari et al. 2025; Shou and Shaw 2022). In addition, there is evidence of its positive effects in various clinical studies, for example, in metabolic diseases (Belwal et al. 2020; X. Xu et al. 2021; Ye et al. 2021). The effects of BBR on this diversity of diseases are due to its multiple cellular effects, resulting from its interactions with different molecules that are part of various intracellular signaling pathways that modify cell physiology. These diverse actions of BBR include inducing apoptosis, inhibiting autophagy, reducing inflammation, decreasing oxidative stress, promoting amyloid beta clearance, and reducing tau protein hyperphosphorylation (He et al. 2017; F. Hu et al. 2024; J. Hu et al. 2012; Huang et al. 2017; Kuo et al. 2004; Lin et al. 2019; Y. Wu et al. 2021). The intracellular signaling pathways with which BBR interacts are diverse, including MAPK, Wnt/β-catenin AMPK, and PI3K/AKT/mTOR (Chaudhari et al. 2025). This chapter will focus on analyzing the interaction between BBR and the PI3K/AKT/mTOR signaling pathway in AD and PD.

8.2 BBR Structure and Pharmacokinetics

The International Union of Pure and Applied Chemistry (IUPAC) name for BBR is 5,6-dihydro-9,10-dimethoxybenzo[g]-1,3-benzodioxole [5,6-a]quinolizinium, whose formula is $C_{20}H_{18}ClNO_4$, with a molecular weight of 371.8, and its structure is that of a heteropentacyclic compound, shown in Fig. 8.1 (PubChem 2025). BBR is yellow, odorless, and bitter tasting, is more soluble in organic solvents, and has low solubility in water (K. Wang et al. 2017). This isoquinoline alkaloid is found in different parts of the plant, such as the rhizomes, bark, roots, and fruits, and its concentration varies between different parts of the plant. For example, in *Berberis*

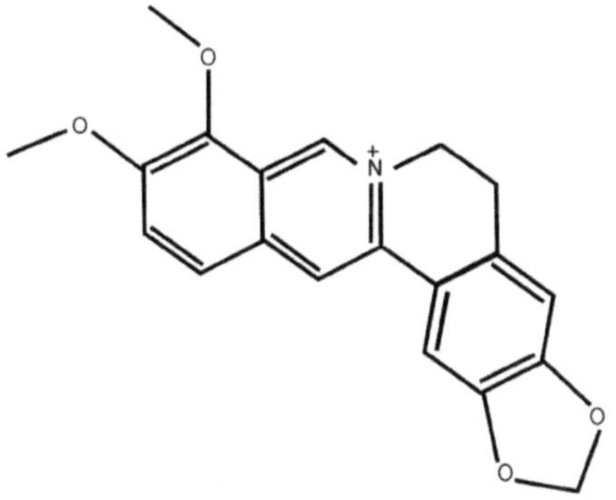

Fig. 8.1 Chemical structure of BBR consists of a system of four fused rings (tetracyclic skeleton), known as the protoberberine moiety, a positively charged nitrogen atom within the C ring, a methylenedioxy group forming a five-membered ring and, two methoxy groups (OCH_3).

darwinii, the roots contain the highest concentration, followed by the bark and leaves, while the roots contain the lowest concentration (Chacón-Fuentes et al. 2025). Some studies on the pharmacokinetics of BBR in humans and rats have reported parameters such as half-life ($t_{1/2}$), maximum concentration (Cmax), and area under the curve (AUC), which provide information on the effects of the organism on the molecule. In healthy male volunteers who took a single 300 mg dose of BBR, the $T_{1/2}$ was 0.87 ± 0.03 h, the Cmax was 394.7 ± 155.4 μg/L, and the AUC was 2799.0 ± 1128.5 μg/L (B. Li et al. 1995). In rats, administration of BBR via the femoral vein (1.0 mg/kg^{-1}) and oral gavage (100 mg/kg^{-1}) had the following values: intravenous administration had a Cmax of 44.9 ± 6. 64 and AUC of 68.7 ± 10.5 ng/h/mL^{-1}, and with oral administration, the Cmax was 9.48 ± 3.40 ng mL^{-1} and the AUC was 46.5 ± 12.8 ng h mL^{-1} (W. Chen et al. 2011), which shows that oral administration of BBR has poor absorption through the intestinal wall and low bioavailability. In rats, after oral administration of 200 mg/kg, BBR shows differential distribution among organs with maximum distribution in the liver, followed by the kidneys, muscle, lungs, brain, heart, pancreas, and fat (Tan et al. 2013). In addition, BBR administered intravenously (3 mg/kg) in rats crosses the blood-brain barrier (BBB) and accumulates in the hippocampus, suggesting that it may interact directly with neurons and neuroglia (X. Wang et al. 2005). After BBR administration, it is metabolized mainly in the liver where phase I (modification) and phase II (conjugation) reactions occur. In humans, rats, and mice, phase I reactions are mediated by cytochrome P450 in the liver and intestine and the main metabolic pathways are demethylation, which produces berberrubine and thalifendine (Y. Li et al. 2011; Spinozzi et al. 2014; Tan et al. 2013), demethylenation produces demethyleneberberine (G. Chen et al. 2013; Y. Guo et al. 2011; Y. Liu et al. 2009), and reduction produces dihydroberberine, jatrorrhizine, and columbamine (Y. Guo et al. 2011; Qiu et al. 2008), while in phase II, BBR and its phase I metabolites are mainly converted into conjugates by glucuronidation (e.g., jatrorrhizine-3-O-glucuronide) and sulfation (e.g., demethyleneberberine-2-O-sulfate) (J.-F. Pan et al. 2002; Shi et al. 2012). In addition, there is evidence to suggest that the intestinal microbiota may metabolize BBR (R. Feng et al. 2015). The excretion of BBR varies depending on the route of administration. For example, in rats, oral administration of BBR results in excretion mainly in feces and minimal excretion in bile and urine, while intravenous administration results in excretion mainly in urine (Alpaslan Ağaçdiken and Göktaş 2025). In summary, BBR is found in many plant species used in traditional medicine, has poor bioavailability, undergoes type I and II metabolism, crosses the BBB, accumulates in various organs including the hippocampus, and is mainly eliminated in feces when administered orally.

8.3 PIP3/AKT/mTOR Pathway

Cells communicate and respond to external stimuli through a series of complex molecular events called cell signaling cascades or molecular pathways. In general, these molecular pathways feature a series of basic elements such as receptors, signal transducers, signal regulators, and effectors. These pathways generally transmit signals from the cell surface to the nucleus, initiating specific responses. Some examples are the Notch, Wnt, and PIP3/AKT/mTOR pathways. The canonical (or classical) activation of the phosphoinositide 3-kinase (PI3K)/protein kinase B (AKT or PKB)/mechanistic target of rapamycin (mTOR) pathway is highly conserved in eukaryotic cells and participates in various cellular events such as cell growth, metabolism, growth, motility, cell proliferation, and differentiation (Omolekan et al. 2024). Its alteration is involved in various diseases such as cancer, AD, PD, Huntington's disease, amyotrophic lateral sclerosis (ALS), and depression (Fakhri et al. 2021; Glaviano et al. 2023; N. Guo et al. 2024). The canonical PIP3/AKT/mTOR pathway can be activated by the interaction of different agonists, mainly with members of two families of transmembrane receptors, receptor tyrosine kinases (RTKs) and G protein-coupled receptors (GPCRs). The former are activated by growth factors such as epidermal growth factor (EGF), insulin-like growth factor 1 (IGF-1) and insulin, and the latter by a wide variety of agonists such as neurotransmitters, lipids, hormones, peptides, and odorants (Du and Lovly 2018; Vanhaesebroeck et al. 2010). The activation of the PI3P/AKT/mTOR pathway through the activation of RTKs begins when the agonist interacts with its binding site on the receptor and causes the dimerization or oligomerization of these receptors, which have intrinsic tyrosine kinase activity and autophosphorylate their own tyrosine residues or those of other proteins to initiate the activation of this pathway or, alternatively, could activate another pathway such as JAK2/STAT or RAS/MAPK (Du and Lovly 2018). Activated RTK can phosphorylate and activate PI3K either directly or indirectly through phosphorylating a scaffolding adaptor such as IRS1, which binds and activates PI3K (Manning and Cantley 2007). The activation of any of the members of the three different classes of PI3K catalyzes the phosphorylation of a serine ring in one or more of the different types of phosphatidylinositol (PtdIns) that are in the cell membrane, and depending on the activated PI3K, there will be a preference for the catalyzed PtdIns (Evangelisti and Martelli 2023). Class I PI3K preferentially phosphorylates PtdIns-4.5-diphosphate (PtdIns-4,5P2 or PIP2) and converts it to PtdIns-3-4,5-triphosphate (PtdIns-3-4,5P3 or PIP3), which recruits downstream some of the AKT isoforms, which belong to the serine/threonine kinase family and have a pleckstrin homology (PH) domain, a sequence of 120 amino acids, which binds to PtdIns-3-4, 5P3, causing phosphorylation and partial activation of AKT (Evangelisti and Martelli 2023; Hanada et al. 2004; Powis et al. 2023). Three genes coding for AKT1, AKT2, and AKT3 have been reported in mammals. AKT1 and AKT2 are expressed in all tissues, while AKT3 is expressed in kidney, heart, and brain (Manning and Toker 2017). To achieve maximum activation, AKT must be phosphorylated at two different residues, which are T308 and S473 in AKT1, T309 and S474 in AKT2, and T305 and S472 in AKT3 (Manning and Toker 2017). The binding of AKT1 to PIP3 induces conformational changes in the AKT1 molecule, exposing residue T308, which is phosphorylated by phosphoinositide-dependent protein kinase 1 (PDK1), and this event promotes the phosphorylation of residue S473 by the mTOR complex 2 (mTORC2), which is activated by RTK through an unknown pathway (Manning and Cantley 2007; Miricescu et al. 2021). In addition to mTORC2, there is also mTORC1, and both protein complexes have mTOR as their catalytic subunit but differ in the other regulatory proteins of the complex and sensibility to rapamycin (G. Y. Liu and Sabatini 2020). Once any of the AKT isoforms have been activated, they translocate to different subcellular compartments such as the endoplasmic reticulum, mitochondria, Golgi apparatus, and nucleus, where they phosphorylate various downstream effectors that regulate a variety of essential processes such as the cell cycle, apoptosis, RNA translation, glycolysis, and angiogenesis (Evangelisti and Martelli 2023). Activated AKT1 plays a very important role in many cellular functions, including cell growth and survival, glucose, protein and fatty acid synthesis, and inhibition of apoptosis (Manning and Cantley 2007), and this diversity of functions occurs because AKT1 (like the other isoforms) has a variety of substrates that share a common characteristic: they have a specific amino acid sequence (consensus motif) that the AKT isoforms recognize and bind to in order to regulate their function, which may be activation or inhibition, altering their subcellular localization, or modifying their stability (Evangelisti and Martelli 2023; Manning and Cantley 2007). The consensus motif of substrates of the AGC kinase family, to which AKT belongs, has a minimal structure for recognition consisting of an R-X-R-X-S/T-B sequence, where X represents any amino acid, B represents bulky hydrophobic residues, S and T are Ser and Thr residues at the phospho-accepting site at position 0, and R represents residues at positions −5 and −3 that appear to confer specificity to the different kinases, and in the case of AKT to Arg (Balasuriya et al. 2020; Evangelisti and Martelli 2023). The diversity of cellular functions that AKT1 modulates is due to its ability to control a large number and diversity of substrates; currently, ~200 have been catalogued (Balasuriya et al. 2020). AKT1 can modulate a cellular process by phosphorylating various substrates involved in a particular cellular function, such as apoptosis. AKT1 inhibits the activity of the transcription factor forkhead box class O (FoxO3a), which reduces the expression of various pro-apoptotic genes (e.g., caspase-3, caspase-9, and Noxa), and, on the other hand, phosphorylates cyclic AMP response element binding protein (CREB), another transcription factor that promotes the expression of anti-apoptotic genes (bcl-2 and mcl-1) (Evangelisti and Martelli 2023). Similarly, AKT1 controls other cellular processes by inhibiting or activating its substrates, which, in turn, can activate or inhibit downstream signaling that will produce a specific action on the final effector (e.g., transcription factor). To illustrate this, we will briefly analyze the modulation of AKT1 in protein synthesis. Activated AKT phosphorylates tuberous sclerosis complex 2 (TSC2), causing its separation from TSC1 and thereby inhibiting the TSC1/TSC2 complex, since this separation also releases the small G protein Ras homolog enriched in the brain (Rheb) complex, which activates mTORC1. Activated mTOR in turn phosphorylates other key molecules in protein synthesis, such as ribosomal protein S6 kinase 1 (S6K1), which activates protein synthesis by phosphorylating ribosomal protein S6. On the other hand, the phosphorylation of eukaryotic translation initiation factor 4E-binding proteins (4E-BPs) by activated mTOR causes the separation and activation of eukaryotic translation initiation factor 4E (eIF4E), a regulator of translation (Amorim et al. 2018; Evangelisti and Martelli 2023; X. Wu et al. 2022).

Another way to activate the PIP3/AKT/mTOR pathway is through the interaction of agonists of the GPCR family, which consists of hundreds of different receptors and ligands (Wacker et al. 2017). Ligands include photons, ions, vitamins, odorants, hormones, and neurotransmitters, but not all activate the PIP3/AKT/mTOR pathway. Some examples that activate this pathway are receptors such as angiotensin II type 1, thyroid-stimulating hormone (TSH), and serotonin (Wacker et al. 2017). Activated GPCRs can coordinate a wide variety of downstream signaling responses, but the most common is through G proteins (Inoue et al. 2019). GPCRs activate four families of heterotrimeric G proteins (G_s, $G_{i/o}$, $G_{q/11}$, and $G_{12/13}$), and the members of these families are composed of α, β, and γ subunits. Each G protein is referred to by the type of α subunit it contains; for example, G_s protein contains the αs subunit, and G_q protein contains the α_q subunit. Each family of G proteins couples to a signaling pathway (Pierce et al. 2002). In their resting state, GPCRs are not associated with G proteins, but when the agonist binds to the GPCRs, it undergoes conformational changes that induce interaction with G proteins, particularly with the α subunit, which changes its conformation and releases GTP and binds GTP, causing the G protein to separate from the GPCRs. Subsequently, the protein separates into a monomer (Gα) and a heterodimer (Gβγ), and both become active and can interact with different signaling elements such as calcium channels, phospholipase C (PLC), adenylyl cyclase (AC), depending on the types of G proteins (New et al. 2007; M. Zhang et al. 2024). The activation of IP3K occurs when Gβγ binds directly to IP3K and activates it, while it is known that Gα can activate this IP3K indirectly by transactivating other receptors such as integrin and RTKs (Fig. 8.2) (New et al. 2007).

8.4 Neurodegeneration and IP3/AKT/mTOR Pathway in AD and PD

8.4.1 Neurodegeneration

Neurodegeneration is mainly characterized by the degeneration or progressive loss of neurons in the central nervous system (CNS) or peripheral nervous system (PNS), which leads to the alteration of functions such as balance, memory, movement,

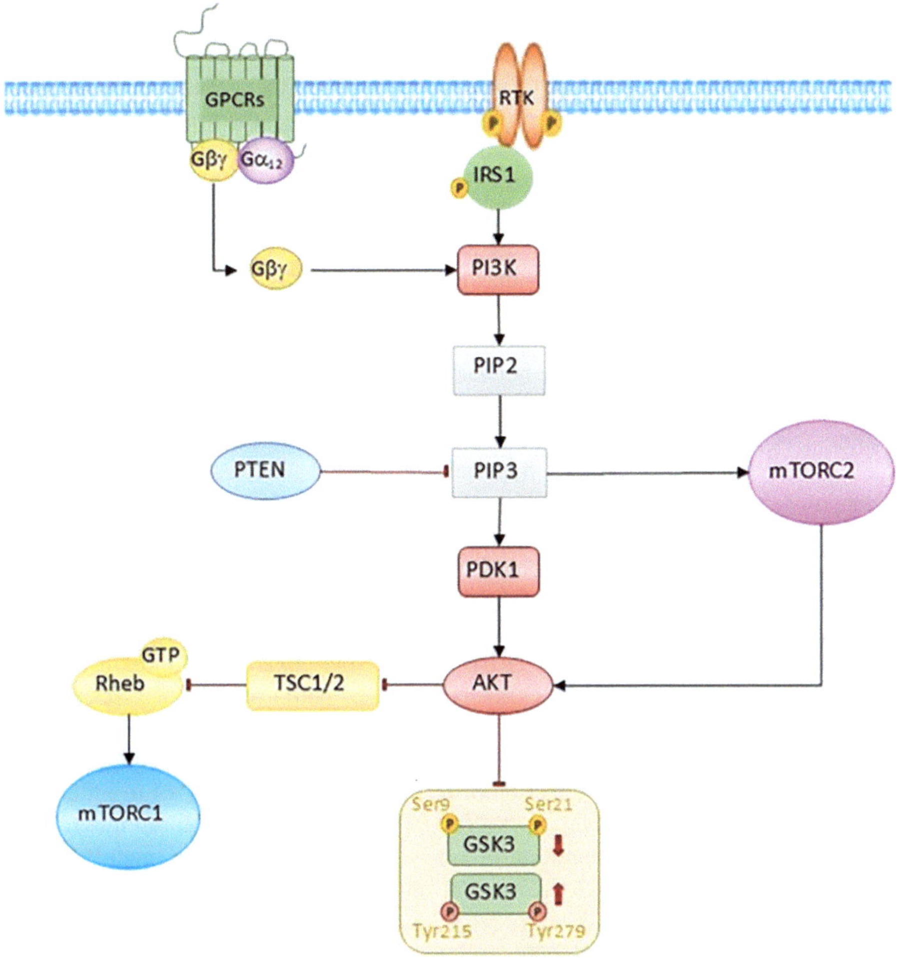

Fig. 8.2 Activation of the IP3K/AKT/mTOR pathway. Activated RTKs and GPCRs can activate this pathway. Activated RTKs can activate phosphoinositide 3-kinase (IP3K) either directly or indirectly through phosphorylation of PtdIns-3-4,5-diphosphate (PIP2) and its conversion to PIP3. Activated GPCRs interact with G proteins and induce their cleavage into the active Gα and Gβγ subunits. The Gβγ subunit directly activates IP3K. Active PIP3 phosphorylates phosphoinositide-dependent protein kinase 1 (PDK1), which partially phosphorylates AKT, which completes its activation by being phosphorylated by the mTORC2. Fully activated AKT phosphorylates tuberous sclerosis complex 2 (TSC2) and causes the separation of the TSC1/TSC2 complex and the release of the small G protein Ras homolog enriched in brain (Rheb) complex, which activates mTORC1. PTEN is short for phosphatase and tensin homolog

language, and other cognitive functions, gradually leading to the loss of independent living. Neurodegenerative diseases include a wide variety of disorders such as ALS, prion disease, multiple sclerosis, PD, Huntington's disease, and AD, with the latter two being the most common (Matilla-Dueñas et al. 2017; Wilson et al. 2023).

8.4.2 Modulation of the IP3/AKT/mTOR Pathway in AD by BBR

AD is an incurable and progressive neurodegenerative illness that commonly presents difficulties with communication, problem solving, abnormal behavior, and memory loss (resulting from the loss of neurons and brain atrophy in areas such as the hippocampus and cerebral cortex), loss of neuronal synapses, and eventually death (Martínez-Coria et al. 2023). AD is the leading cause of neurodegeneration and dementia and is generally divided into familial Alzheimer's disease (FAD) and idiopathic or sporadic Alzheimer's disease (SAD). FAD is caused by genetic mutations in presenilin 1, presenilin 2, and the amyloid precursor protein A4 (APP) and accounts for around 5% of cases, while SAD accounts for 95% of reported cases and its cause is unknown, but it is believed to be a combination of genetic and environmental factors (Alzheimer's Association 2019; Pons and Rivest 2022). There are various risk factors for this disease, but the most important are aging and diseases such as cerebrovascular disease, hypertension, type 2 diabetes, brain trauma, and poor physical activity (Edwards III et al. 2019). AD is related to the accumulation of insoluble forms of amyloid β (Aβ) in plaques and the intraneuronal deposition of neurofibrillary tangles (NFTs) formed by hyperphosphorylated tau protein. These two alterations are the hallmarks of the disease but are accompanied by other alterations such as loss of neurons and synapses, reactive astrocytes and microglia, alterations of the blood-brain barrier (BBB), neuroinflammation, and brain atrophy (Breijyeh and Karaman 2020). AD involves various pathological processes such as excitotoxicity (mainly due to excessive glutamate secretion and excessive activation of NMDA-type glutamate receptors), oxidative stress (mainly due to mitochondrial dysfunction and excessive production of reactive oxygen species), neuroinflammation (mainly due to the activation of NLRP1 and NLRP3 inflammasomes), BBB disruption, dysbiosis, and alteration of various signaling pathways (mainly due to the downregulation of PI3K/AKT/mTOR and the upregulation of Notch, among others) (Das et al. 2023).

Several studies suggest that there is a link between type 2 diabetes and AD (Beeri et al. 2008; Peila et al. 2002), and some research shows that insulin-triggered signaling and glucose metabolic dysfunction in the brain may be related to the cognitive problems that occur in AD, which suggests that there is an alteration in the activity of the IP3/AKT/mTOR pathway. In non-diabetic patients with AD, type 1 and 2 insulin receptors (IRs), a member of the RTK family, and IP3K are reduced, and this is associated with cognitive decline (Talbot et al. 2012). In animal models of AD, double transgenic mice (APPswe/PS1dE9) and rats treated with scopolamine showed that IRs, phosphorylated IP3K, and phosphorylated AKT were significantly reduced (Das et al. 2023; H. Feng et al. 2016). In healthy brains, active AKT promotes the expression of GLUT3 and GLUT4 glucose transporters, inhibits glycogen synthase kinase 3 beta (GSK3β), a key threonine/serine kinase in insulin-dependent glycogen synthesis, insulin signaling, and participates in the regulation of signaling pathways involved in protein synthesis, proliferation, differentiation, and survival; its inactivation depends on active AKT (Doble and Woodgett 2003). In AD, the reduction of active AKT causes GSK3β to be phosphorylated and to interact with the tau protein, which causes abnormal phosphorylation of tau, thereby promoting the formation of NFT. Furthermore, it also interacts with presenilin I, which is the catalytic core of γ-secretase that cleaves the amyloid precursor protein (APP), so the degradation of the C-terminal of presenilin I caused by GSK3β alters the activity of γ-secretase and promotes the accumulation of toxic Aβ (Hanger et al. 1992; Kirschenbaum et al. 2001). The deregulation of mTOR in relation to autophagy in AD is well documented, and there is experimental evidence supporting mTOR hyperactivity caused by Aβ (Caccamo et al. 2011), but it is now known that mTORC1 plays a very important role in blocking the early stages of autophagy, and this protein complex is activated by a small GTPase (Rheb) and is not considered part of the IP3/AKT/mTOR pathway (Deleyto-Seldas and Efeyan 2021). However, mTORC1 can activate S6K1, which in turn inhibits mTORC2 (Szwed et al. 2021), thereby altering the regulatory role of the latter. Furthermore, a decrease in active AKT would reduce the expression of glucose transporters, which could aggravate neurodegeneration.

In a recent meta-analysis, the authors concluded that regardless of the animal model used (rats and mice), BBR improved memory through a mechanism that includes the reduction of tau phosphorylation, Aβ, APP, and BACE1, a β-secretase involved in the cleavage of the APP protein (J.-Y. Liu et al. 2024). Although many studies on the effect of BBR in preclinical studies have been published, the molecular mechanisms of its beneficial effects on AD are incomplete, but it is known that BBR decreases acetylcholinesterase, anti-inflammatory cytokines IL-6 and TNF-α, the production of oxygen free radicals, BACE1 and APP, which are the result of modulating signaling pathways such as AMPK and PI3K/AKT pathway (Akbar et al. 2021). The modulation of BBR in the IP3/AKT/mTOR pathway in AD is poorly understood, but some effects of BBR on some of the key molecules in this pathway are known. For example, in an in vitro study using mouse neuroblastoma N2a cells expressing the human mutant APP695, it was shown that BBR decreases tau hyperphosphorylation through modulation of the AKT/glycogen synthase kinase 3 (GSK3) pathway, increasing AKT activity and, as a result, inhibiting GSK3 (Durairajan et al. 2012). It is known that GSK3 stimulates neurodegeneration by promoting hyperphosphorylation of tau protein, the production and accumulation of Aβ, which cause neuronal damage, microglial activation, and other alterations in structures such as the hippocampus that alter memory formation (Limantoro et al. 2023). In another in vitro study, BBR increased Aβ clearance in a primary culture of hippocampal neurons from 3xTg-AD transgenic mice, a model of AD. BBR reduced intracellular Aβ levels through autophagy stimulated by IP3K activation (Huang et al. 2017). Using a cellular model of excitotoxicity with an N18TG2 neuroblastoma culture, treatment with BBR produces an increase in the expression of IRs, phosphorylated AKT, and the transcription factor Nrf2 (Hsu et al. 2012). In a study using HEK293 cells expressing tau treated with calyculin-A, a compound that inhibits PP2A and PP1 phosphatases, and used as a cellular model to study cell viability, tau phosphorylation, and oxidative stress, BBR restored PP2A activity and reversed GSK3 activation (Yu et al. 2011), suggesting increased AKT activity.

8.4.3 Modulation of the IP3/AKT/mTOR Pathway in PD by BBR

PD is the second most common neurodegenerative disease, the main characteristic of which is the degeneration of dopaminergic neurons in the substantia nigra pars compacta (SNpc), part of the basal ganglia, and intraneuronal aggregates of the α-synuclein protein called Lewy bodies. Patients with this disease present motor impairments, particularly resting tremor and bradykinesia (progressive decrease in the speed and amplitude of voluntary and automatic movements) and non-motor impairments such as sleep disturbances, fatigue, depression, and dementia (Martínez-Coria et al. 2023; Poewe et al. 2017). Although PD is generally considered an idiopathic disease, it has been suggested that certain genes may be involved (e.g., Park1, Park4), along with various environmental factors such as stress, brain damage, physical inactivity, and exposure to pesticides (M. S. Khan et al. 2025). Although the exact cause of PD is unknown, research suggests that it is a combination of genetic predisposition and environmental factors that contribute to the development of the disease, and it is this combination of factors that triggers the aggregation of α-synuclein. Excitotoxicity, oxidative stress, neuroinflammation, apoptosis, mitochondrial dysfunction, and dysbiosis also occur (M. S. Khan et al. 2025). There is evidence of the beneficial effects of BBR administration in animal models of PD that include the restoration of dopaminergic neurotransmission, decreased mitochondrial dysfunction, improved motor coordination, reduced apoptosis, reduced neurodegeneration of neurons in SNpc, improved short-term memory, among other beneficial effects (Kim et al. 2014; Kwon et al. 2010; Tseng et al. 2024; L. Wang et al. 2021a). The pathogenesis of PD is unknown, but there is evidence of the involvement of mTORC1 as a neuroprotective or neurotoxic factor. Some studies have been conducted using different toxins (e.g., 1-methyl-4-phenyl-1,2,3,6-tetrahydropyridine (MPTP) and rotenone) to produce animal models of PD and in vitro studies with neurons, and in these models, it was observed that inhibition of mTORC1 signaling reduces the accumulation of α-synuclein, protects against cell damage, reduces oxidative stress, and reduces apoptosis (Dijkstra et al. 2015; Jiang et al. 2013; Malagelada et al. 2010; T. Pan et al. 2009; Tain et al. 2009). However, similar studies show opposite effects (L. Chen et al. 2010; Rodríguez-Blanco et al. 2012; Y. Xu et al. 2014; Zhou et al. 2015). In a recent article, Y. Wang et al. (2021b) point out that the discrepancy between the neuroprotective and neurotoxic effects of BBR could be due to the route of administration. With oral administration of BBR, an increase in bacterial production of Dopa and dopamine is observed, which is reflected in the blood and brain, as well as an improvement in brain function, while intraperitoneal administration has very poor absorption. However, there is no evidence of the mechanism by which this could regulate mTORC1. On the other hand, the combination of the mTORC1 inhibitor rapamycin and L-Dopa (a dopamine precursor) used in an animal model of PD did not cause dyskinesia (one of the adverse effects of L-Dopa) and prevented the loss of bidirectional communication in the striatum (Calabrese et al. 2020; Santini et al. 2009). These studies support the idea that the beneficial effects of BBR are related to the inhibition of mTORC1, together with other events in addition to the increase in dopamine, such as the modulation of autophagy, the decrease in the accumulation of α-synuclein, and neuroinflammation. Moreover, it is known that α-synuclein suppresses TSC1/2, an inhibitor of mTORC, which causes an increase in its activity and increase in protein synthesis, both in animal models and in postmortem studies of brains from patients with PD (M. R. Khan et al. 2023). On the other hand, phosphatase and tensin homolog (PTEN) acts as a suppressor of PtdIns-3-4,5P3 signaling and reduces IR signaling and PI3K/AKT pathway activity. It has been found in PD patients that this protein is elevated in SNpc neurons and this correlates with energy metabolism dysfunction, which occurs in high percentages (50–80%). This alteration is related to cognitive decline and motor and non-motor alterations (Sekar and Taghibiglou 2018). In in vitro and in vivo studies, BBR significantly decreases PTEN (El-Horany et al. 2020; G. Li et al. 2021; C. Zhang et al. 2015). However, this effect has not been explored in the context of PD. Figure 8.3 shows the main modulatory effects of BBR on AD and PD.

8.5 Conclusions and Future Perspectives

Experimental evidence shows that BBR counteracts several important alterations in AD and PD pathology, suggesting a potential therapeutic agent for these conditions, but more detailed research is needed on the modulation of BBR on the IP3K/AKT/mTOR pathway and other signaling pathways to better understand the mechanisms of damage involved. In addition, the therapeutic potential of BBR cannot be fully exploited due to some significant limitations, such as poor gastrointestinal absorption, low water solubility, low bioavailability, and rapid metabolism. Currently, many laboratories are working to improve the pharmacokinetics of BBR using different approaches, including various types of nanocarriers such as lipid-based nanoparticles (e.g., liposomes), polymeric nanoparticles, metallic nanoparticles (e.g., gold, silver, zinc oxide), nanoemulsions, nanogels, and others (Baidoo et al. 2025; Chaudhari et al. 2025), which generally show significant improvements. However, their clinical application is still premature, as more research is needed to determine their biocompatibility, biosafety, stability, and toxicity.

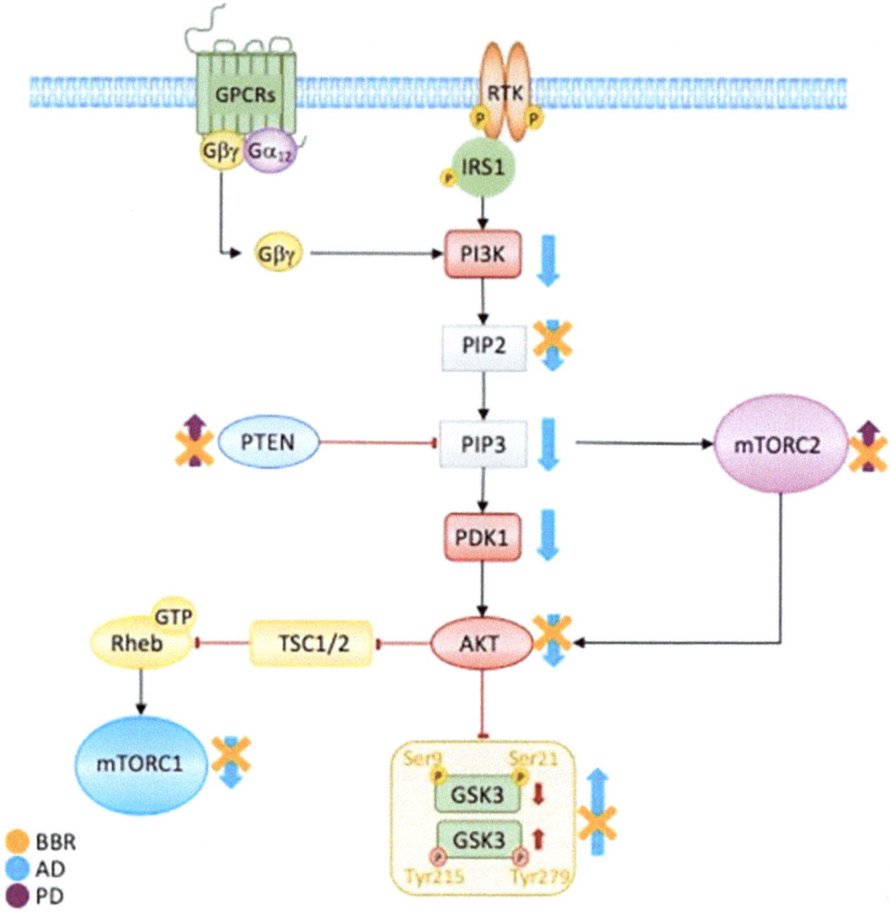

Fig. 8.3 Dysregulation of the IP3K/AKT/mTOR pathway in AD and PD, and modulation by BBR. The blue arrow indicates alterations in AD, the purple arrow indicates alterations in PD, and the orange cross indicates regulation by BBR. RTKs tyrosine kinases, GPCRs G protein-coupled receptors, IP3K phosphoinositide 3-kinase, PIP2 PtdIns-3-4,5-diphosphate, PIP3 PtdIns-3-4,5-triphosphate, Gα alpha subunit of the G protein, Gβγ beta gamma subunit of the G protein, PDK1 phosphoinositide-dependent protein kinase 1, mTORC2 mechanistic target of rapamycin complex 2, mTORC1 mechanistic target of rapamycin complex 1, PTEN phosphatase and tensin homolog, and Ras small G protein

References

Akbar M, Shabbir A, Rehman K, Akash MSH, Shah MA (2021) Neuroprotective potential of berberine in modulating Alzheimer's disease via multiple signaling pathways. J Food Biochem 45(10):e13936. https://doi.org/10.1111/jfbc.13936

Alpaslan Ağaçdiken A, Göktaş Z (2025) Berberine-induced browning and energy metabolism: mechanisms and implications. PeerJ 13:e18924. https://doi.org/10.7717/peerj.18924

Alzheimer's Association (2019) 2019 Alzheimer's disease facts and figures. Alzheimers Dement 15(3):321–387. https://doi.org/10.1016/j.jalz.2019.01.010

Amorim IS, Lach G, Gkogkas CG (2018) The role of the eukaryotic translation initiation factor 4E (eIF4E) in neuropsychiatric disorders. Front Genet 9:561. https://doi.org/10.3389/fgene.2018.00561

Baidoo I, Sarbadhikary P, Abrahamse H, George BP (2025) Metal-based nanoplatforms for enhancing the biomedical applications of berberine: current progress and future directions. Nanomedicine (Lond) 20(8):851–868. https://doi.org/10.1080/17435889.2025.2480051

Balasuriya N, Davey NE, Johnson JL, Liu H, Biggar KK, Cantley LC, Li SS-C, O'Donoghue P (2020) Phosphorylation-dependent substrate selectivity of protein kinase B (AKT1). J Biol Chem 295(24):8120–8134. https://doi.org/10.1074/jbc.RA119.012425

Beeri MS, Schmeidler J, Silverman JM, Gandy S, Wysocki M, Hannigan CM, Purohit DP, Lesser G, Grossman HT, Haroutunian V (2008) Insulin in combination with other diabetes medication is associated with less Alzheimer neuropathology. Neurology 71(10):750–757. https://doi.org/10.1212/01.wnl.0000324925.95210.6d

Belwal T, Bisht A, Devkota HP, Ullah H, Khan H, Pandey A, Bhatt ID, Echeverría J (2020) Phytopharmacology and clinical updates of Berberis species against diabetes and other metabolic diseases. Front Pharmacol 11:41. https://doi.org/10.3389/fphar.2020.00041

Birdsall TC (1997) Berberine: therapeutic potential of an alkaloid found in several medicinal plants. Altern Med Rev 2(2)

Breijyeh Z, Karaman R (2020) Comprehensive review on Alzheimer's disease: causes and treatment. Molecules (Basel, Switzerland) 25(24):5789. https://doi.org/10.3390/molecules25245789

Caccamo A, Maldonado MA, Majumder S, Medina DX, Holbein W, Magrí A, Oddo S (2011) Naturally secreted amyloid-β increases mammalian target of rapamycin (mTOR) activity via a PRAS40-mediated mechanism*. J Biol Chem 286(11):8924–8932. https://doi.org/10.1074/jbc.M110.180638

Calabrese V, Di Maio A, Marino G, Cardinale A, Natale G, De Rosa A, Campanelli F, Mancini M, Napolitano F, Avallone L, Calabresi P, Usiello A, Ghiglieri V, Picconi B (2020) Rapamycin, by inhibiting mTORC1 signaling, prevents the loss of striatal bidirectional synaptic plasticity in a rat model of L-DOPA-induced dyskinesia. Front Aging Neurosci 12:230. https://doi.org/10.3389/fnagi.2020.00230

Chacón-Fuentes M, Burgos-Díaz C, Opazo-Navarrete M, Mercado A, Westermeyer F (2025) Berberine and Palmatine distribution across plant organs in Berberis darwinii: basis for selecting superior-producing accessions. Molecules 30(8):8. https://doi.org/10.3390/molecules30081849

Chaudhari S, Dalabehera M, Subudhi RN, Dua K, Kaur M, Paudel KR, Kumar J (2025) From nature to nanotech: unlocking Berberine's therapeutic approaches. J Drug Deliv Sci Technol 108:106924. https://doi.org/10.1016/j.jddst.2025.106924

Chen G, Lu F, Xu L, Dong H, Yi P, Wang F, Huang Z, Zou X (2013) The anti-diabetic effects and pharmacokinetic profiles of berberine in mice treated with Jiao-Tai-Wan and its compatibility. Phytomedicine 20(10):780–786. https://doi.org/10.1016/j.phymed.2013.03.004

Chen L, Xu B, Liu L, Luo Y, Yin J, Zhou H, Chen W, Shen T, Han X, Huang S (2010) Hydrogen peroxide inhibits mTOR signaling by activation of AMPKα leading to apoptosis of neuronal cells. Lab Investig 90(5):762–773. https://doi.org/10.1038/labinvest.2010.36

Chen W, Miao Y-Q, Fan D-J, Yang S-S, Lin X, Meng L-K, Tang X (2011) Bioavailability study of berberine and the enhancing effects of TPGS on intestinal absorption in rats. AAPS PharmSciTech 12(2):705–711. https://doi.org/10.1208/s12249-011-9632-z

Das TK, Ganesh BP, Fatima-Shad K (2023) Common signaling pathways involved in Alzheimer's disease and stroke: two faces of the same coin. J Alzheimers Dis Rep 7(1):381–398. https://doi.org/10.3233/ADR-220108

Deleyto-Seldas N, Efeyan A (2021) The mTOR–autophagy axis and the control of metabolism. Front Cell Dev Biol 9. https://doi.org/10.3389/fcell.2021.655731

Dijkstra AA, Ingrassia A, de Menezes RX, van Kesteren RE, Rozemuller AJM, Heutink P, van de Berg WDJ (2015) Evidence for immune response, axonal dysfunction and reduced endocytosis in the substantia Nigra in early stage Parkinson's disease. PLoS One 10(6):e0128651. https://doi.org/10.1371/journal.pone.0128651

Doble B, Woodgett JR (2003) GSK-3: tricks of the trade for a multi-tasking kinase. J Cell Sci 116(Pt 7):1175. https://doi.org/10.1242/jcs.00384

Du Z, Lovly CM (2018) Mechanisms of receptor tyrosine kinase activation in cancer. Mol Cancer 17(1):58. https://doi.org/10.1186/s12943-018-0782-4

Durairajan SSK, Liu L-F, Lu J-H, Chen L-L, Yuan Q, Chung SK, Huang L, Li X-S, Huang J-D, Li M (2012) Berberine ameliorates β-amyloid pathology, gliosis, and cognitive impairment in an Alzheimer's disease transgenic mouse model. Neurobiol Aging 33(12):2903–2919. https://doi.org/10.1016/j.neurobiolaging.2012.02.016

Edwards GA III, Gamez N, Escobedo G Jr, Calderon O, Moreno-Gonzalez I (2019) Modifiable risk factors for Alzheimer's disease. Front Aging Neurosci 11. https://doi.org/10.3389/fnagi.2019.00146

El-Horany HE-S, Gaballah HH, Helal DS (2020) Berberine ameliorates renal injury in a rat model of D-galactose-induced aging through a PTEN/Akt-dependent mechanism. Arch Physiol Biochem 126(2):157–165. https://doi.org/10.1080/13813455.2018.1499117

Evangelisti C, Martelli AM (2023) The PI3K/Akt/mTOR pathway. In: Bradshaw RA, Hart GW, Stahl PD (eds) Encyclopedia of cell biology, 2nd edn. Academic Press, pp 153–161. https://doi.org/10.1016/B978-0-12-821618-7.00081-X

Fakhri S, Iranpanah A, Gravandi MM, Moradi SZ, Ranjbari M, Majnooni MB, Echeverría J, Qi Y, Wang M, Liao P, Farzaei MH, Xiao J (2021) Natural products attenuate PI3K/Akt/mTOR signaling pathway: a promising strategy in regulating neurodegeneration. Phytomedicine 91:153664. https://doi.org/10.1016/j.phymed.2021.153664

Feng H, Dang H, Fan H, Chen X, Rao Y, Ren Y, Yang J, Shi J, Wang P, Tian J (2016) Curcumin ameliorates insulin signalling pathway in brain of Alzheimer's disease transgenic mice. Int J Immunopathol Pharmacol 29(4):734–741. https://doi.org/10.1177/0394632016659494

Feng R, Shou J-W, Zhao Z-X, He C-Y, Ma C, Huang M, Fu J, Tan X-S, Li X-Y, Wen B-Y, Chen X, Yang X-Y, Ren G, Lin Y, Chen Y, You X-F, Wang Y, Jiang J-D (2015) Transforming berberine into its intestine-absorbable form by the gut microbiota. Sci Rep 5(1):12155. https://doi.org/10.1038/srep12155

Glaviano A, Foo ASC, Lam HY, Yap KCH, Jacot W, Jones RH, Eng H, Nair MG, Makvandi P, Geoerger B, Kulke MH, Baird RD, Prabhu JS, Carbone D, Pecoraro C, Teh DBL, Sethi G, Cavalieri V, Lin KH et al (2023) PI3K/AKT/mTOR signaling transduction pathway and targeted therapies in cancer. Mol Cancer 22(1):138. https://doi.org/10.1186/s12943-023-01827-6

Guo N, Wang X, Xu M, Bai J, Yu H, Le Zhang N (2024) PI3K/AKT signaling pathway: molecular mechanisms and therapeutic potential in depression. Pharmacol Res 206:107300. https://doi.org/10.1016/j.phrs.2024.107300

Guo Y, Li F, Ma X, Cheng X, Zhou H, Klaassen CD (2011) CYP2D plays a major role in berberine metabolism in liver of mice and humans. Xenobiotica 41(11):996–1005. https://doi.org/10.3109/00498254.2011.597456

Hanada M, Feng J, Hemmings BA (2004) Structure, regulation and function of PKB/AKT—a major therapeutic target. Biochim Biophys Acta Proteins Proteom 1697(1):3–16. https://doi.org/10.1016/j.bbapap.2003.11.009

Hanger DP, Hughes K, Woodgett JR, Brion JP, Anderton BH (1992) Glycogen synthase kinase-3 induces Alzheimer's disease-like phosphorylation of tau: generation of paired helical filament epitopes and neuronal localisation of the kinase. Neurosci Lett 147(1):58–62. https://doi.org/10.1016/0304-3940(92)90774-2

He W, Wang C, Chen Y, He Y, Cai Z (2017) Berberine attenuates cognitive impairment and ameliorates tau hyperphosphorylation by limiting the self-perpetuating pathogenic cycle between NF-κB signaling, oxidative stress and neuroinflammation. Pharmacol Rep 69(6):1341–1348. https://doi.org/10.1016/j.pharep.2017.06.006

Hsu Y-Y, Chen C-S, Wu S-N, Jong Y-J, Lo Y-C (2012) Berberine activates Nrf2 nuclear translocation and protects against oxidative damage via a phosphatidylinositol 3-kinase/Akt-dependent mechanism in NSC34 motor neuron-like cells. Eur J Pharm Sci 46(5):415–425. https://doi.org/10.1016/j.ejps.2012.03.004

Hu F, Hu T, Qiao Y, Huang H, Zhang Z, Huang W, Liu J, Lai S (2024) Berberine inhibits excessive autophagy and protects myocardium against ischemia/reperfusion injury via the RhoE/AMPK pathway. Int J Mol Med 53(5):49. https://doi.org/10.3892/ijmm.2024.5373

Hu J, Chai Y, Wang Y, Kheir MM, Li H, Yuan Z, Wan H, Xing D, Lei F, Du L (2012) PI3K p55γ promoter activity enhancement is involved in the anti-apoptotic effect of berberine against cerebral ischemia-reperfusion. Eur J Pharmacol 674(2–3):132–142. https://doi.org/10.1016/j.ejphar.2011.11.014

Huang M, Jiang X, Liang Y, Liu Q, Chen S, Guo Y (2017) Berberine improves cognitive impairment by promoting autophagic clearance and inhibiting production of β-amyloid in APP/tau/PS1 mouse model of Alzheimer's disease. Exp Gerontol 91:25–33. https://doi.org/10.1016/j.exger.2017.02.004

Inoue A, Raimondi F, Kadji FMN, Singh G, Kishi T, Uwamizu A, Ono Y, Shinjo Y, Ishida S, Arang N, Kawakami K, Gutkind JS, Aoki J, Russell RB (2019) Illuminating G-protein-coupling selectivity of GPCRs. Cell 177(7):1933–1947.e25. https://doi.org/10.1016/j.cell.2019.04.044

Jiang J, Jiang J, Zuo Y, Gu Z (2013) Rapamycin protects the mitochondria against oxidative stress and apoptosis in a rat model of Parkinson's disease. Int J Mol Med 31(4):825–832. https://doi.org/10.3892/ijmm.2013.1280

Khan MR, Yin X, Kang S-U, Mitra J, Wang H, Ryu T, Brahmachari S, Karuppagounder SS, Kimura Y, Jhaldiyal A, Kim HH, Gu H, Chen R, Redding-Ochoa J, Troncoso J, Na CH, Ha T, Dawson VL, Dawson TM (2023) Enhanced mTORC1 signaling and protein synthesis in pathologic α-synuclein cellular and animal models of Parkinson's disease. Sci Transl Med 15(724):eadd0499. https://doi.org/10.1126/scitranslmed.add0499

Khan MS, Nasiripour S, Bopassa JC (2025) Parkinson disease signaling pathways, molecular mechanisms, and potential therapeutic strategies: a comprehensive review. Int J Mol Sci 26(13):13. https://doi.org/10.3390/ijms26136416

Kim M, Cho K-H, Shin M-S, Lee J-M, Cho H-S, Kim C-J, Shin D-H, Yang HJ (2014) Berberine prevents nigrostriatal dopaminergic neuronal loss and suppresses hippocampal apoptosis in mice with Parkinson's disease. Int J Mol Med 33(4):870–878. https://doi.org/10.3892/ijmm.2014.1656

Kirschenbaum F, Hsu SC, Cordell B, McCarthy JV (2001) Glycogen synthase kinase-3beta regulates presenilin 1 C-terminal fragment levels. J Biol Chem 276(33):30701–30707. https://doi.org/10.1074/jbc.M102849200

Kuo C-L, Chi C-W, Liu T-Y (2004) The anti-inflammatory potential of berberine in vitro and in vivo. Cancer Lett 203(2):127–137. https://doi.org/10.1016/j.canlet.2003.09.002

Kwon IH, Choi HS, Shin KS, Lee BK, Lee CK, Hwang BY, Lim SC, Lee MK (2010) Effects of berberine on 6-hydroxydopamine-induced neurotoxicity in PC12 cells and a rat model of Parkinson's disease. Neurosci Lett 486(1):29–33. https://doi.org/10.1016/j.neulet.2010.09.038

Li B, Zhang M, Bao L (1995) Study on the pharmacokinetics of berberine after oral administration in human bein. Ha'erbin yi ke da xue xue bao = Journal of Harbin Medical University = Harbin yike daxue xuebao

Li G, Zhang C, Liang W, Zhang Y, Shen Y, Tian X (2021) Berberine regulates the Notch1/PTEN/PI3K/AKT/mTOR pathway and acts synergistically with 17-AAG and SAHA in SW480 colon cancer cells. Pharm Biol 59(1):21–30. https://doi.org/10.1080/13880209.2020.1865407

Li Y, Ren G, Wang Y-X, Kong W-J, Yang P, Wang Y-M, Li Y-H, Yi H, Li Z-R, Song D-Q, Jiang J-D (2011) Bioactivities of berberine metabolites after transformation through CYP450 isoenzymes. J Transl Med 9:62. https://doi.org/10.1186/1479-5876-9-62

Limantoro J, de Liyis BG, Sutedja JC (2023) Akt signaling pathway: a potential therapy for Alzheimer's disease through glycogen synthase kinase 3 beta inhibition. Egypt J Neurol Psychiatry Neurosurg 59(1):147. https://doi.org/10.1186/s41983-023-00751-2

Lin J, Cai Q, Liang B, Wu L, Zhuang Y, He Y, Lin W (2019) Berberine, a traditional Chinese medicine, reduces inflammation in adipose tissue, polarizes M2 macrophages, and increases energy expenditure in mice fed a high-fat diet. Med Sci Monit 25:87–97. https://doi.org/10.12659/MSM.911849

Liu GY, Sabatini DM (2020) mTOR at the nexus of nutrition, growth, ageing and disease. Nat Rev Mol Cell Biol 21(4):183–203. https://doi.org/10.1038/s41580-019-0199-y

Liu J-Y, Dai Y, He Y-X, Lin L (2024) Effect of berberine on cognitive function and β-amyloid precursor protein in Alzheimer's disease models: a systematic review and meta-analysis. Front Pharmacol 14:1301102. https://doi.org/10.3389/fphar.2023.1301102

Liu Y, Hao H, Xie H, Lv H, Liu C, Wang G (2009) Oxidative demethylenation and subsequent glucuronidation are the major metabolic pathways of berberine in rats. J Pharm Sci 98(11):4391–4401. https://doi.org/10.1002/jps.21721

Malagelada C, Jin ZH, Jackson-Lewis V, Przedborski S, Greene LA (2010) Rapamycin protects against neuron death in in vitro and in vivo models of Parkinson's disease. J Neurosci Off J Soc Neurosci 30(3):1166–1175. https://doi.org/10.1523/JNEUROSCI.3944-09.2010

Manning BD, Cantley LC (2007) AKT/PKB signaling: navigating downstream. Cell 129(7):1261–1274. https://doi.org/10.1016/j.cell.2007.06.009

Manning BD, Toker A (2017) AKT/PKB signaling: navigating the network. Cell 169(3):381–405. https://doi.org/10.1016/j.cell.2017.04.001

Martínez-Coria H, Arrieta-Cruz I, Gutiérrez-Juárez R, López-Valdés HE (2023) Anti-inflammatory effects of flavonoids in common neurological disorders associated with aging. Int J Mol Sci 24(5):4297. https://doi.org/10.3390/ijms24054297

Matilla-Dueñas A, Corral-Juan M, Rodríguez-Palmero Seuma A, Vilas D, Ispierto L, Morais S, Sequeiros J, Alonso I, Volpini V, Serrano-Munuera C, Pintos-Morell G, Álvarez R, Sánchez I (2017) Rare neurodegenerative diseases: clinical and genetic update. Adv Exp Med Biol 1031:443–496. https://doi.org/10.1007/978-3-319-67144-4_25

Miricescu D, Totan A, Stanescu-Spinu I-I, Badoiu SC, Stefani C, Greabu M (2021) PI3K/AKT/mTOR signaling pathway in breast cancer: from molecular landscape to clinical aspects. Int J Mol Sci 22(1):1. https://doi.org/10.3390/ijms22010173

Neag MA, Mocan A, Echeverría J, Pop RM, Bocsan CI, Crişan G, Buzoianu AD (2018) Berberine: botanical occurrence, traditional uses, extraction methods, and relevance in cardiovascular, metabolic, hepatic, and renal disorders. Front Pharmacol 9:557. https://doi.org/10.3389/fphar.2018.00557

New DC, Wu K, Kwok AWS, Wong YH (2007) G protein-coupled receptor-induced Akt activity in cellular proliferation and apoptosis. FEBS J 274(23):6025–6036. https://doi.org/10.1111/j.1742-4658.2007.06116.x

Omolekan TO, Chamcheu JC, Buerger C, Huang S (2024) PI3K/AKT/mTOR signaling network in human health and diseases. Cells 13(17):1500. https://doi.org/10.3390/cells13171500

Pan J-F, Yu C, Zhu D-Y, Zhang H, Zeng J-F, Jiang S-H, Ren J-Y (2002) Identification of three sulfate-conjugated metabolites of berberine chloride in healthy volunteers' urine after oral administration. Acta Pharmacol Sin 23(1):77–82

Pan T, Rawal P, Wu Y, Xie W, Jankovic J, Le W (2009) Rapamycin protects against rotenone-induced apoptosis through autophagy induction. Neuroscience 164(2):541–551. https://doi.org/10.1016/j.neuroscience.2009.08.014

Peila R, Rodriguez BL, Launer LJ, Honolulu-Asia Aging Study (2002) Type 2 diabetes, APOE gene, and the risk for dementia and related pathologies: the Honolulu-Asia aging study. Diabetes 51(4):1256–1262. https://doi.org/10.2337/diabetes.51.4.1256

Pierce KL, Premont RT, Lefkowitz RJ (2002) Seven-transmembrane receptors. Nat Rev Mol Cell Biol 3(9):639–650. https://doi.org/10.1038/nrm908

Poewe W, Seppi K, Tanner CM, Halliday GM, Brundin P, Volkmann J, Schrag A-E, Lang AE (2017) Parkinson disease. Nat Rev Dis Primers 3(1):1–21. https://doi.org/10.1038/nrdp.2017.13

Pons V, Rivest S (2022) Targeting systemic innate immune cells as a therapeutic avenue for Alzheimer disease. Pharmacol Rev 74(1):1–17. https://doi.org/10.1124/pharmrev.121.000400

Powis G, Meuillet EJ, Indarte M, Booher G, Kirkpatrick L (2023) Pleckstrin Homology [PH] domain, structure, mechanism, and contribution to human disease. Biomed Pharmacother 165:115024. https://doi.org/10.1016/j.biopha.2023.115024

Prajwala B, Raghu N, Gopenath T, Shanmukhappa B, Kaginelli, Karthikeyan M, Ashok G, Ranjith M, Srinivasan V, Basalingappa K (2020) Berberine and its pharmacology potential: a review. Eur J Biomed Pharm Sci 7(5):115–123

PubChem (2025) Berberine. https://pubchem.ncbi.nlm.nih.gov/compound/2353

Qiu F, Zhu Z, Kang N, Piao S, Qin G, Yao X (2008) Isolation and identification of urinary metabolites of berberine in rats and humans. Drug Metab Dispos 36(11):2159–2165. https://doi.org/10.1124/dmd.108.021659

Rodríguez-Blanco J, Martín V, García-Santos G, Herrera F, Casado-Zapico S, Antolín I, Rodriguez C (2012) Cooperative action of JNK and AKT/mTOR in 1-methyl-4-phenylpyridinium-induced autophagy of neuronal PC12 cells. J Neurosci Res 90(9):1850–1860. https://doi.org/10.1002/jnr.23066

Santini E, Heiman M, Greengard P, Valjent E, Fisone G (2009) Inhibition of mTOR signaling in Parkinson's disease prevents L-DOPA-induced dyskinesia. Sci Signal 2(80):ra36. https://doi.org/10.1126/scisignal.2000308

Sekar S, Taghibiglou C (2018) Elevated nuclear phosphatase and tensin homolog (PTEN) and altered insulin signaling in substantia nigral region of patients with Parkinson's disease. Neurosci Lett 666:139–143. https://doi.org/10.1016/j.neulet.2017.12.049

Shi R, Zhou H, Ma B, Ma Y, Wu D, Wang X, Luo H, Cheng N (2012) Pharmacokinetics and metabolism of jatrorrhizine, a gastric prokinetic drug candidate. Biopharm Drug Dispos 33(3):135–145. https://doi.org/10.1002/bdd.1779

Shou J-W, Shaw P-C (2022) Therapeutic efficacies of Berberine against neurological disorders: an update of pharmacological effects and mechanisms. Cells 11(5):796. https://doi.org/10.3390/cells11050796

Spinozzi S, Colliva C, Camborata C, Roberti M, Ianni C, Neri F, Calvarese C, Lisotti A, Mazzella G, Roda A (2014) Berberine and its metabolites: relationship between physicochemical properties and plasma levels after administration to human subjects. J Nat Prod 77(4):766–772. https://doi.org/10.1021/np400607k

Szwed A, Kim E, Jacinto E (2021) Regulation and metabolic functions of mTORC1 and mTORC2. Physiol Rev 101(3):1371–1426. https://doi.org/10.1152/physrev.00026.2020

Tain LS, Mortiboys H, Tao RN, Ziviani E, Bandmann O, Whitworth AJ (2009) Rapamycin activation of 4E-BP prevents parkinsonian dopaminergic neuron loss. Nat Neurosci 12(9):1129–1135. https://doi.org/10.1038/nn.2372

Talbot K, Wang H-Y, Kazi H, Han L-Y, Bakshi KP, Stucky A, Fuino RL, Kawaguchi KR, Samoyedny AJ, Wilson RS, Arvanitakis Z, Schneider JA, Wolf BA, Bennett DA, Trojanowski JQ, Arnold SE (2012) Demonstrated brain insulin resistance in Alzheimer's disease patients is associated with IGF-1 resistance, IRS-1 dysregulation, and cognitive decline. J Clin Invest 122(4):1316–1338. https://doi.org/10.1172/JCI59903

Tan X-S, Ma J-Y, Feng R, Ma C, Chen W-J, Sun Y-P, Fu J, Huang M, He C-Y, Shou J-W, He W-Y, Wang Y, Jiang J-D (2013) Tissue distribution of berberine and its metabolites after oral administration in rats. PLoS One 8(10):e77969. https://doi.org/10.1371/journal.pone.0077969

Tseng H-C, Wang M-H, Fang C-H, Lin Y-W, Soung H-S (2024) Neuroprotective potentials of Berberine in rotenone-induced Parkinson's disease-like motor symptoms in rats. Brain Sci 14(6):6. https://doi.org/10.3390/brainsci14060596

Vanhaesebroeck B, Guillermet-Guibert J, Graupera M, Bilanges B (2010) The emerging mechanisms of isoform-specific PI3K signalling. Nat Rev Mol Cell Biol 11(5):329–341. https://doi.org/10.1038/nrm2882

Wacker D, Stevens RC, Roth BL (2017) How ligands illuminate GPCR molecular pharmacology. Cell 170(3):414–427. https://doi.org/10.1016/j.cell.2017.07.009

Wang K, Feng X, Chai L, Cao S, Qiu F (2017) The metabolism of berberine and its contribution to the pharmacological effects. Drug Metab Rev 49(2):139–157. https://doi.org/10.1080/03602532.2017.1306544

Wang L, Sheng W, Tan Z, Ren Q, Wang R, Stoika R, Liu X, Liu K, Shang X, Jin M (2021a) Treatment of Parkinson's disease in zebrafish model with a berberine derivative capable of crossing blood brain barrier, targeting mitochondria, and convenient for bioimaging experi-

ments. Comp Biochem Physiol C Toxicol Pharmacol 249:109151. https://doi.org/10.1016/j.cbpc.2021.109151
Wang X, Wang R, Xing D, Su H, Ma C, Ding Y, Du L (2005) Kinetic difference of berberine between hippocampus and plasma in rat after intravenous administration of Coptidis rhizoma extract. Life Sci 77(24):3058–3067. https://doi.org/10.1016/j.lfs.2005.02.033
Wang Y, Tong Q, Ma S-R, Zhao Z-X, Pan L-B, Cong L, Han P, Peng R, Yu H, Lin Y, Gao T-L, Shou J-W, Li X-Y, Zhang X-F, Zhang Z-W, Fu J, Wen B-Y, Yu J-B, Cao X, Jiang J-D (2021b) Oral berberine improves brain dopa/dopamine levels to ameliorate Parkinson's disease by regulating gut microbiota. Signal Transduct Target Ther 6(1):77. https://doi.org/10.1038/s41392-020-00456-5
Wilson DM, Cookson MR, Van Den Bosch L, Zetterberg H, Holtzman DM, Dewachter I (2023) Hallmarks of neurodegenerative diseases. Cell 186(4):693–714. https://doi.org/10.1016/j.cell.2022.12.032
Wu X, Xie W, Xie W, Wei W, Guo J (2022) Beyond controlling cell size: functional analyses of S6K in tumorigenesis. Cell Death Dis 13(7):646. https://doi.org/10.1038/s41419-022-05081-4
Wu Y, Chen Q, Wen B, Wu N, He B, Chen J (2021) Berberine reduces Aβ42 deposition and tau hyperphosphorylation via ameliorating endoplasmic reticulum stress. Front Pharmacol 12:640758. https://doi.org/10.3389/fphar.2021.640758
Xu X, Yi H, Wu J, Kuang T, Zhang J, Li Q, Du H, Xu T, Jiang G, Fan G (2021) Therapeutic effect of berberine on metabolic diseases: both pharmacological data and clinical evidence. Biomed Pharmacother 133:110984. https://doi.org/10.1016/j.biopha.2020.110984
Xu Y, Liu C, Chen S, Ye Y, Guo M, Ren Q, Liu L, Zhang H, Xu C, Zhou Q, Huang S, Chen L (2014) Activation of AMPK and inactivation of Akt result in suppression of mTOR-mediated S6K1 and 4E-BP1 pathways leading to neuronal cell death in in vitro models of Parkinson's disease. Cell Signal 26(8):1680–1689. https://doi.org/10.1016/j.cellsig.2014.04.009
Ye Y, Liu X, Wu N, Han Y, Wang J, Yu Y, Chen Q (2021) Efficacy and safety of Berberine alone for several metabolic disorders: a systematic review and meta-analysis of randomized clinical trials. Front Pharmacol 12:653887. https://doi.org/10.3389/fphar.2021.653887
Yu G, Li Y, Tian Q, Liu R, Wang Q, Wang J-Z, Wang X (2011) Berberine attenuates Calyculin A-induced cytotoxicity and tau hyperphosphorylation in HEK293 cells. J Alzheimer's Dis 24(3):525–535. https://doi.org/10.3233/JAD-2011-101779
Zhang C, Shi Y-R, Liu X-R, Cao Y-C, Zhen D, Jia Z-Y, Jiang J-Q, Tian J-H, Gao J-M (2015) The anti-apoptotic role of Berberine in preimplantation embryo in vitro development through regulation of miRNA-21. PLoS One 10(6):e0129527. https://doi.org/10.1371/journal.pone.0129527
Zhang M, Chen T, Lu X, Lan X, Chen Z, Lu S (2024) G protein-coupled receptors (GPCRs): advances in structures, mechanisms and drug discovery. Signal Transduct Target Ther 9(1):1–43. https://doi.org/10.1038/s41392-024-01803-6
Zhou Q, Liu C, Liu W, Zhang H, Zhang R, Liu J, Zhang J, Xu C, Liu L, Huang S, Chen L (2015) Rotenone induction of hydrogen peroxide inhibits mTOR-mediated S6K1 and 4E-BP1/eIF4E pathways, leading to neuronal apoptosis. Toxicol Sci Off J Soc Toxicol 143(1):81–96. https://doi.org/10.1093/toxsci/kfu211
Zhu C, Li K, Peng X-X, Yao T-J, Wang Z-Y, Hu P, Cai D, Liu H-Y (2022) Berberine a traditional Chinese drug repurposing: its actions in inflammation-associated ulcerative colitis and cancer therapy. Front Immunol 13:1083788. https://doi.org/10.3389/fimmu.2022.1083788

Neuroprotective Efficacy of Photomedicine in Mitigating Aging-Related Neurodegeneration

9

Adnan Khan, Malakpogu Ravindra Babu,
Shriyansh Srivastava, Rakesh Sahu, Mohd. Tariq,
and Ghulam Md Ashraf

Abstract

Neurodegenerative illnesses like Alzheimer's and Parkinson's, which are affected by accumulated oxidative stress, mitochondrial dysfunction, synaptic degradation, and neuroinflammation, have age as a critical risk factor. The recent research has spotlighted photomedicine, which includes Photo-Bio-Modulation treatment (PBM) and Photo-Dynamic Therapy (PDT), as a promising noninvasive approach to mitigate these pathogenic processes. Photomedicine makes use of certain wavelengths of light, primarily red and near-infrared spectrum, to revive intracellular pathways that promote neuronal survival. It lessens the inflammatory cytokines, enriches mitochondrial function, and supports synaptic plasticity. It has been indicated by recent clinical and preclinical studies that light-based therapy can improve cognitive function, slow neuronal degeneration, and stimulate neurogenesis in aging animals. These abstracts analyze the molecular support of photomedicine's neuroprotective benefits, evaluate existing experimental and

A. Khan · M. R. Babu (✉) · S. Srivastava (✉)
School of Medical and Allied Sciences, Galgotias University, Greater Noida, Uttar Pradesh, India

R. Sahu
School of Pharmacy, Sharda University, Greater Noida, Uttar Pradesh, India

M. Tariq
Department of Academics, Sumandeep Vidyapeeth (Deemed to be University), Vadodara, Gujarat, India

Department of Biotechnology, Graphic Era (Deemed to be University), Dehradun, Uttarakhand, India

G. M. Ashraf (✉)
Department of Biomedical Sciences, College of Medicine, Gulf Medical University, Ajman, United Arab Emirates

H. Dilnashin, S. P. Singh (eds.), *Phototherapeutic Approaches to Neurodegeneration*,
https://doi.org/10.1007/978-981-95-4751-7_9

clinical evidence, and study translational barriers and potential opportunities for integrating photomedicine into the treatment of neurodegenerative disorders.

Keywords

Neurodegenerative · Photomedicine · Photo-Bio-Modulation Treatment · Photodynamic Therapy · Neurogenesis · Translational Barrier

9.1 Introduction

The term neurodegeneration is a popular term, and its meaning is thought to be familiar to all (Dugger and Dickson 2017). However, it is far more difficult to come up with an accurate definition of neurodegeneration than one would think (Jellinger 2010). In many cases, neurodegeneration is mentioned casually and barely discussed in large medical textbooks, and is not fully defined in the most extensive dictionaries (Przedborski et al. 2003). Etymologically, the term consists of the prefix neuro, meaning nerve cell (i.e., neuron), and degeneration, meaning, in the case of tissues or organs, a process of structural or functional loss (Brown and Al-Chalabi 2017). In the strict sense of the word, therefore, neurodegeneration can be equated with any pathological process that, first of all, involves neurons (Soto and Pritzkow 2018). Practically, neurodegenerative diseases constitute a huge category of neurological disorders that have a heterogencous clinical and pathological manifestation that target particular subsets of neurons of particular functional anatomical systems (Goedert et al. 2017); they have an unknown etiology, and their progression is inexorable (Dauer and Przedborski 2003). On the other hand, neoplasm, edema, hemorrhage, and nervous system trauma, not primary neuronal disorders, are not regarded as neurodegenerative disorders (Troppan et al. 2016). Neurodegenerative diseases (NDs) constitute a very heterogeneous and broad category of disorders that affect the central nervous system (CNS) and the peripheral nervous system (Erkkinen et al. 2018). They are defined as irreversible, progressive deterioration of the neurological functioning that was intact in the past and which gets worse with age (Soto and Pritzkow 2018). Such diseases are Alzheimer's, Parkinson's, Huntington's, motor neuron disease, etc. (Kalia and Lang 2015) NDs pathogenesis has not been elucidated and might differ depending on the particular condition (Gitler et al. 2017). There are currently no curative measures in spite of their immense morbidity and mortality impact in the global arena (Mehta et al. 2017). NDs seriously diminish the quality of life of patients and their families (Schrag and Quinn 2000). Currently, the use of pharmaceutical therapy is the first choice, yet in most cases, it is symptomatic and sometimes is associated with side effects, including diarrhea, nausea, and headache (Birks et al. 2015). So to prevent these side effects, light therapy, also called heliotherapy, is exposure to daylight or artificial light under controlled conditions and has been identified as a cost-effective, convenient, safe, and effective way of reducing symptoms associated with sleep disorders, depression, and cognitive decline (Vandewalle et al. 2009). There is some research on the efficiency of light therapy in the treatment of neurodegenerative diseases. The review article presents the existing evidence on the mechanisms of action, therapeutic uses, and success of light therapy in neurodegenerative diseases.

9.2 Mechanism of Aging-Related Neurodegeneration

Aging is a progressive process of functional impairment, which is time-dependent and causes death (López-Otín et al. 2013). The most outstanding attributes of the aging processes are gradual diminution of physiological capacity, diminished responsiveness to environmental stimuli, diminished adaptive capacity, predisposition to diseases, and elevated mortality (Kirkwood 2005). Numerous theories have been proposed on how to explain aging, but the biological mechanism(s) underlying the aging process remain unknown. Major theories of aging are changes in proteins, DNA damage and reduced DNA repair, inappropriate cross-linking between proteins, DNA, and other structural molecules, failure of neuroendocrine secretion, cellular senescence in the cell culture system, elevated free radical-mediated oxidative stress, and alteration of the sequence of gene expression (Harman 2001). One of the most common ideas is that the trend of oncogenic development of every species is hereditary. Thus, one of the possible causes of aging can be genetic: a gradual loss of molecular components (e.g., loss of code, loss of gene expression devices, loss of conditions of gene expression, inappropriate regulation of genes) (Kenyon 2010), a coordinated action of which is critical to cell viability and proliferation. High-density oligonucleotide microarrays have been used to identify several classes of genes that are differentially expressed in the monkey corpus callosum during aging (Farkas et al. 2011). The genes majorly regulate the rise in stress factors and the decline of cell functionality. The cell functional changes are enhanced cell cycle inhibition, proteolysis, and a reduction in mitochondrial activity, signal transduction, and protein translation (Tahara et al. 2009). Although most of these categories have been reported already in functional brain aging, this was the first time they were directly related to white matter. Microarray analysis has also been used in identifying age-activated neuroprotective response pathways in white matter and several genes that are involved in lifespan (Lu et al. 2004). The discovery of Klotho, a multifunctional protein that controls phosphate and calcium metabolism, insulin resistance, and which is recognized to protect against oxidative stress and apoptosis (Kuro-o et al. 1997), was of specific interest. The other processes that regulate aging include a reduction in the effectiveness of DNA repair, end-replication issues linked to telomeres, and mutation of mitochondrial DNA (Vijg and Suh 2013). The reduction in these elements can also be connected with cellular senescence, apoptosis, and age-related diseases (Campisi 2013). Aging causes a number of effects on the human brain (Peters 2006). The brain of human beings decreases in size as one grows old (Fjell and Walhovd 2010). The assumption is that the loss of weight and volume in the aging brain is as a result of loss of neurons and myelinated axons (Peters 2002). The aging brain is characterized by changes in white matter of the brain (Bennett and Madden 2014). The overexpression and proteolytic fragmentation of 2′,3′-cyclic nucleotide 3′-phosphodiesterase due to the calpain-mediated proteolysis are accompanied by those changes and lead to myelin and axonal pathology in the aging brain (Tran et al. 2008). The activation of microglia also increases in multiple brain parts like the hippocampus with age (Streit et al. 2004). The potential mechanisms can involve microglial response to advanced glycation end products (AGEPs), which leads to activation of nuclear transcription factor-kappaB (NFκB) and transcription of pro-inflammatory cytokines with the process of aging (Symposium announcement 2001). In this way, age-related changes in gene expression lead to the violation of metabolic homeostasis (Zahn et al. 2006).

One of the memory processes is learning and it involves the adaptation of the brain to the input in the environment (Kandel 2001). Memory can be described as the continuity of learning over time by means of the storage and retrieval of information, and it explains all the knowledge that is acquired in the course of experience (Squire and Wixted 2011). Memory comprises: learning or encoding, short-term or long-term memory, and recall or retrieval (Baddeley 2003). Hence, when we test recall, it will inform us what we have learned (Kirkwood 2005). We cannot think of learning without memory, memory without learning (Tulving 2002). Memory has a number of forms. The capacity to directly and consciously recall or recognize information that has been recently processed is one type of memory; impairment of this type of memory is characteristic of global amnesia (Kopelman 2002). It is no secret that with age, the ability to recall and recognize new facts and events decreases (Nyberg et al. 2012). The age-related deterioration of memory performance is strongly connected to age-related structural and functional alterations in the hippocampus (Small et al. 2011).

9.3 Introduction to Photomedicine

Photomedicine uses light for the diagnosis, treatment, and prevention of diseases (Hamblin and Demidova 2006). Its principles apply from physics, biology, engineering, and medicine to obtain an accurate wavelength of light which can be used in diagnosis, treatment, or prevention (Brazzelli et al. 2007) (Table 9.1).

Near-infrared (NIR) light has been demonstrated to facilitate healing and better motor, cognitive, behavioral, and metabolic functions (Wang et al. 2017). A combination of the two methods with functional and integrative biomedical interventions, such as stem cell therapy, gene therapy, and optogenetically delivered chemotherapy (localized optical modulation of selected genes via photoexcitable ion channels), forms a systemic intervention strategy that can enhance the effective delivery of precision medicine (Ririe et al. 2021).

PBM is a technique that uses nonthermal light of the laser or light-emitting diodes (LEDs) in the red to near-infrared spectrum (650–1200 nm) to modulate biological activity or produce therapeutic effects (Rahman et al. 2016). In simple terms, it is a light-based treatment of different diseases. The amount of energy that this mode of treatment provides is minimal, and therefore, there is no thermal effect.

Table 9.1 Different photomedicine descriptions

Concept	Description
Therapeutic light	Use of specific wavelengths (e.g., red, blue, NIR) to promote healing or kill cells
Photobiomodulation (PBM)	Low-level light therapy that stimulates or inhibits cellular function
Photodynamic therapy (PDT)	Combines light-sensitive drugs with light to destroy abnormal cells (e.g., cancer)
Laser surgery	Use of focused laser beams for cutting or removing tissue precisely
Photoacoustic imaging	Combines laser-induced ultrasound with imaging to visualize tissues noninvasively

Several past works showed that PBM has a beneficial effect on neurodegenerative diseases, such as AD, PD, and various forms of epilepsy (Hamieh and Hamieh 2020). Although the animal studies have had successful results, there have been limited clinical studies that have depicted positive results of PBM against the neurodegenerative disease. With the complicated nature of the pathogenesis of neurodegenerative diseases, the broad-acting methods, like PBM, are promising. In the recent past, PBM used on the nose and the brain has proven to be effective in delaying the progression of neurodegenerative diseases, including AD and PD (Kim et al. 2017). Investigations have indicated that PBM is used to apply to target organs at the right wavelengths and optimal fluences or energy density to generate therapeutic effects, and with no harmful effects. PBM therapy increases brain blood flow, brain energy metabolism, and antioxidant defense (Yang et al. 2021). Also, a second group of studies analyzed the impact of PBM and the use of red to near-infrared light on body tissues, neuroinflammatory, and oxidative stress in animal models of neurodegenerative diseases (Quatman-Yates et al. 2013). Neurodegenerative diseases are becoming common in the aging populations across the globe, but a cure to the diseases has not been found (Fisher 2019). Consequently, the creation of new treatment approaches that are safer and more efficient in managing the symptoms of neurodegenerative diseases is a topical issue of the current research (Karami et al. 2021).

The use of natural light in the treatment of ailments dates back to 3000 years. An example is the use of sunlight by ancient Egyptians and the Chinese in treating psoriasis, rickets, and vitiligo (Romand et al. 2005). Phototherapy was introduced by Niels Finsen in the late nineteenth century, and he discovered that exposure to red light concealed the development of smallpox pustules. He later found out that the ultraviolet light of the sun was also useful in the treatment of cutaneous tuberculosis. These discoveries earned him the Nobel Prize in 1903 (Anías-Calderón et al. 2004). Later in 1913, a German scientist, Friedrich Meyer-Betz tested the effects of hematoporphyrin on his skin, and he felt swelling and pain when parts of the body were exposed to light. The main processes of photodynamic reactions gained a lot of scientific interest in the 1960s (Hart and Belotto 2010). Photodynamic therapy (PDT) is currently a rapidly developing method of detecting and treating a wide range of pathologies, not only precancerous and cancerous lesions, such as brain tumors, but also a variety of other disorders of the nervous system, such as Alzheimer's disease (AD) and Parkinson's disease (PD) (Zhou et al. 2020). It is capable of treating meningeal lymphatic vessels and is increasingly becoming common with its good antimicrobial characteristics (Wu et al. 2014). The majority of the existing cases of PDT applications are based on the usage of photosensitizers (PS). PS are physiochemical substances that can absorb light of certain wavelength, transform, and transfer it as useful energy to an oxygen molecule in the production of reactive oxygen species (ROS) (Dolmans et al. 2003), which are oxygen-based free radicals, non-radical species, and electronically excited singlet oxygen. The PDT is beneficial because it is efficient, has less side effects, and is less invasive than chemotherapy and radiotherapy (Wang et al. 2009). PDT also possesses the most important feature of using the integrity of the blood-brain barrier (BBB) of the brain in biomedical applications, and nowadays is regarded as a potent technique targeting malignant cells and other brain abnormalities (Henderson and Dougherty 1992).

9.4 Principle of Photomedicine

The mechanism of action of photomedicine is the interaction of light with biological tissues that is described by complex cascade of photophysical, photochemical, and photobiological processes (Vijayavenkataraman et al. 2017). Endogenous chromophores in the visible and near-infrared wavelengths can be absorbed especially by cytochrome c oxidase in the mitochondria, which acts as a primary photoacceptor (Karu and Kolyakov 2005). When absorbed, the mitochondrial activity is enhanced leading to an increased electron transport, high production of ATP, and regulation of reactive oxygen species (ROS) which are intracellular messengers (Harpavat et al. 2016). Such effects help in improving cell proliferation, migration, anti-inflammatory effects, and tissue regeneration (Shin et al. 2017). The secondary cellular processes include alterations in the homeostatic parameters of intracellular calcium levels, redox state, and cyclic AMP, and create an amplification cascade that propagates the initial photonic signal across the biochemical networks (Wong-Riley et al. 2005). The biphasic dose-response relationship is one of the most important concepts to learn, in which low doses of light may cause a biological effect, and excessive light may inhibit or kill the tissue (Hamblin et al. 2011). The processes were used in photodynamic therapy, where photosensitizers are activated by light to generate cytotoxic species to destroy target cells and in photobiomodulation, which has therapeutic effects by promoting healing without causing thermal damage (Agostinis et al. 2011). Together, these principles provide the scientific rationale of the use of light in therapeutic uses in dermatology, oncology, neurology, and regenerative medicine (Posten et al. 2006).

9.5 Application of Photomedicine in Aging Brain Disorder

Photomedicine has come out as a potential noninvasive approach in the treatment of age-related brain diseases like Alzheimer's disease (AD), Parkinson's disease (PD), and age-related cognitive impairment (Shulyatnikova and Verkhratsky 2020). Transcranial photobiomodulation (tPBM) is one of the most actively studied applications where the low-level red or near-infrared light penetrates the scalp and skull to stimulate the brain tissues (Dai et al. 2016). Such an approach acts on mitochondrial chromophores, mainly cytochrome c oxidase, which increases ATP production and regulates oxidative stress, which is severely impaired in neurodegenerative disorders (Wong-Riley et al. 2005). In addition, the stimulation of light has also been proved to boost brain-derived neurotrophic factor (BDNF) and support synaptic plasticity, which is essential to memory and cognitive performance in the aging brain (Chang and Meyerhoefer 2019). Preclinical and small-scale clinical trials have demonstrated tPBM to be effective in working memory, attention, and mood, and therefore, it may have neuroprotective and neurorestorative effects (Eisfeld and Kawaoka 2017). Also, photobiomodulation seems to inhibit neuroinflammation by reducing microglial activation and pro-inflammatory cytokines (Wang et al. 2025). The results encourage photomedicine in multimodal therapy to decelerate or promote cognitive aging in populations (Johnson et al. 2017). Even though further large-scale studies are necessary, the available evidence points to photomedicine as a potentially effective means of enhancing brain health and performance in the elderly that is associated with minimal risks.

Transcranial photobiomodulation (tPBM) and photomedicine, in particular, have been receiving increasing attention as a noninvasive means of neuromodulation that potentially can have therapeutic effects on aging brain disorders such as Alzheimer's disease (AD), Parkinson's disease (PD), vascular dementia, and mild cognitive impairment (Salehpour et al. 2018). The basic process is that low-level red or near-infrared (NIR) light (usually 600–1100 nm) is delivered to the cerebral cortex and penetrates tissues where it is absorbed by mitochondrial photo acceptors, primarily cytochrome c oxidase (Wang et al. 2025). This stimulation promotes oxidative phosphorylation, leading to the production of ATP that is of great importance in the survival of neurons and synaptic transmission (Hamblin 2016). In addition, it was demonstrated that photobiomodulation has the ability to decrease neuroinflammation, which is one of the features of neurodegeneration, by inhibiting microglial activation and preventing pro-inflammatory cytokines, such as TNF-α and IL-6 (Salehpour and Hamblin 2020).

The recent preclinical studies have shown that NIR can decrease the burden of amyloid-beta in AD models, increase its clearance by promoting glymphatic drainage, and increase neurogenesis and angiogenesis (Padovani et al. 2010). PBM has been demonstrated to protect against dopaminergic cells, mitochondrial dynamics, and locomotion in animal models of Parkinson's disease (Johnstone et al. 2016). Pilot clinical trials on elderly patients and AD patients show that photobiomodulation positively affects cognitive parameters such as working memory, attention span, and emotional stability with regular sessions (Chao 2019). These benefits can be partially explained by the fact that it causes an expansion of cerebral blood flow and an elevation of neurotrophic factors such as brain-derived neurotrophic factor (BDNF) and nerve growth factor (NGF) (Meng et al. 2013).

Notably, photomedicine is the only technology that can affect brain functioning with no side effects, which makes it especially appealing to frail or polypharmacy-prone groups of people (Nairuz et al. 2024). Its neuroprotective, neurorestorative, and metabolic-regulating effects can be seen as additional properties that make it a synergistic intervention in combination with cognitive training or pharmacotherapy (Berman and Nichols 2019). Ongoing randomized controlled trials are exploring the best parameters, such as pulse modulation, fluence, and frequency of treatment, in order to achieve the best long-term effects of neurodegenerative and age-related cognitive disorders (Berman et al. 2019). With the maturation of the field, photomedicine is an emerging, precision-targeted, and low-risk approach to the goal of delaying or ameliorating the effects of cognitive aging and dementia (Premi et al. 2022).

9.5.1 Alzheimer's Disease Cognitive Enhancement

Treatment of cognitive impairments and neuropathology of Alzheimer's disease (AD) is one of the most researched photomedicine applications (Padovani et al. 2010). Transcranial photobiomodulation (tPBM) is a method where near-infrared light (usually 810–1064 nm) is applied to the skull and activates cortical and subcortical brain areas (Hamblin 2016). There is an uptake of light energy by cytochrome c oxidase in the mitochondria, resulting in an increase in ATP production, cellular metabolism, and decreased oxidative stress (Hamblin 2016). Such procedures have been associated with decreased deposition of amyloid-beta and increased clearance of the same because of improved glymphatic flow (Iliff et al. 2012). Moreover, PBM promotes dendritic development and stabilization of dendritic systems, which is essential in memory storage (Purushothuman et al. 2014). PBM therapy has been shown to enhance spatial learning and memory in AD animal models and exhibit measurable cognitive performance, verbal fluency, and executive functioning in human pilot studies (Saleh et al. 2017). The positive results suggest that PBM can be a noninvasive therapy that can prevent or even abolish the initial phase of cognitive decline in AD (Salehpour et al. 2018).

9.5.2 Parkinson's Disease and Neuroprotection

Parkinson's disease (PD) has also been treated using photomedicine in order to enhance the survival of the neurons, especially those located in the substantia nigra where the dopaminergic neurons are damaged (Tang et al. 2014). PBM enhances the mitochondrial bioenergetics and inhibits pro-inflammatory microglia, which seroprotects the neurons under oxidative stress and apoptosis (Lattanzi et al. 2018). According to preclinical studies on PD animal models, PBM treatment was found to maintain motor functions and striatum dopamine levels in a very good manner (Gao et al. 2017). It is notable that tPBM has been found to modulate the expression of neuroprotective-related genes, including the Bcl-2, and prevent the apoptotic pathways (Amaroli et al. 2019). In certain clinical observations, patients with PD treated with the use of tPBM have exhibited less rigidity, tremor, and bradykinesia. These results justify PBM as an add-on or monotherapeutic treatment of neurodegenerative motor diseases (Hamblin 2016).

9.5.3 Prevention of Age-Related Deterioration of Cognition

In addition to pathological aging, photomedicine has shown improvement of the normal age-related cognitive decline (Shi et al. 2020). It has been shown that repeated administration of PBM in old rodents can enhance the neurogenesis of hippocampus, increase the cerebral blood flow, and enhance the functional connectivity between the brain regions (Sun et al. 2023). This is in part due to the upregulation of neurotrophic factors like brain-derived neurotrophic factor (BDNF), vascular endothelial growth factor (VEGF), and insulin-like growth factor-1 (IGF-1) which are normally downregulated in aging (Hamblin 2016). Short-term PBM has been found to have positive effects in older adults in the areas of attention span, working memory, and processing speed (Cercignani et al. 2017). Further, PBM can regulate the circadian gene expression, and this is helpful to restore the sleep-wake patterns, which has an indirect effect on cognition and mood regulation in the elderly (Daly and Leahy 2013).

9.5.4 Behavioral Symptom Neuromodulation

Neuropsychiatric symptoms of old age brain diseases are depression, anxiety, irritability, and emotional lability (Chen et al. 2020). Photomedicine has been found to be effective in the mitigation of these behavioral disturbances, by exerting neuromodulatory effects (Barrett and Gonzalez-Lima 2013). PBM causes mood stabilization through the alteration of the cortical excitability and neurotransmitter levels (serotonin and dopamine pathways, in particular) (Carey and Preston 2019). In clinical trials, anxiety and depressive symptoms have been reduced in elderly people who have been subjected to regular tPBM sessions (López-García et al. 2019). Anxiolytic property of PBM is also evidenced by animal studies, which showed the reduction in stress-induced behavior change along with normalization of corticosterone levels (Schneider et al. 2019). PBM has been especially attractive since it is a noninvasive and well-tolerated alternative to pharmacological interventions in patients who are resistant to or experience side effects of pharmacological interventions (Non-neuronal evoked and spontaneous hemodynamic 2025).

9.5.5 Targets of Central and Peripheral Pain Modulation

Neurodegenerative disorders in old patients are a comorbidity of chronic neuropathic and musculoskeletal pain. The new approach is the use of photomedicine, which acts centrally and peripherally on the pain processing pathways (Nwazota et al. 2019). Recent advances in wireless optoelectronic devices have enabled the targeted delivery of photonic energy to the dorsal root ganglia and the spinal cord regions in an accurate way. They are able to prevent nociceptive transmission and reduce perception of pain without use of systemic analgesics (Gundersen 2016). The experimental study has proved that PBM therapy has reduced the mechanical allodynia and thermal hyperalgesia in animal models. In addition to that, PBM improves local circulation and decreases inflammation, which also adds to the analgesic effect (Qin et al. 2011). Wearable or implantable PBM devices would be a huge step forward in the quality of life of the aging population with chronic pain, especially in case of failure or contraindication of the standard treatment (Shirkavand et al. 2021).

9.6 Clinical Studies

Photomedicine or rather photobiomodulation (PBM) therapy is gaining a lot of interest as it has neuroprotective effects against age-related neurodegeneration (Chung et al. 2012). It is a noninvasive treatment, implying that it is a low-intensity red or near-infrared light applied to the tissues of the body, including the brain, and has demonstrated positive outcomes in the prevention of neurodegenerative processes (Hamblin 2016). PBM can influence cellular metabolism by improving the activity of the mitochondrion and the synthesis of adenosine triphosphate (ATP) (Vähävihu et al. 2010) which is of great importance to the health of the neurons. The mitochondrial dysfunction, oxidative stress, and chronic inflammation that occur in aging brains are the major causes of neurodegenerative disorders like Alzheimer's, Parkinson's, and cognitive impairment in general (Mattson and Arumugam 2018). Animal models as well as human studies have shown that PBM reduces markers of oxidative stress and pro-inflammatory cytokines and thus creates a more conducive environment in relation to neuronal survival and repair (Mouton-Liger et al. 2014).

A number of clinical trials confirmed that PBM is able to modify some of the key biological functions including neuroinflammation, regulation of apoptosis, and synaptic plasticity (Lasselin et al. 2016). In an example, transcranial PBM using near-infrared light (700-1400 nm) could not only preserve the integrity of brain tissue of neurodegenerative models, but also improve memory, cognition, and motor coordination in old-age categories to an appreciable extent (Karu and Kolyakov 2005). PBM also has been linked to augmented expression of neuroprotective proteins, and diminished expression of proteins associated with neural deterioration in models of Alzheimer's disease such as amyloid-beta and phosphorylated tau (Basurto-Islas et al. 2014). Even distant PBM, in which light is applied to the distant regions of the body, has shown neuroprotective effects, so there may be systemic effects, such as improved circulation or immune modulation (Zhao et al. 2014).

PBM has been claimed to be able to repeatedly change the gene expression pattern that is involved in neuronal survival, induce neurogenesis, and activate synapses in both clinical and preclinical reviews (Lanzafame et al. 2013). It has been established that the treatment also alters the brain metabolomic markers of aged rodents to boost their metabolic plasticity and neurochemical resilience. It is interesting that some systematic reviews and meta-analyses showed similar results that long-term PBM therapy leads to reduced hippocampal degeneration, enhanced neuronal connectivity, and enhanced behavioral performance, particularly in aged animals and aged human beings with mild cognitive impairment (Hamblin 2019).

In addition to its direct effects on the neurons, PBM was also shown to possibly have an effect on glial activity, especially that of astrocytes and microglia, which are essential to neuroinflammation and brain homeostasis (Kholyavin and Polyakov 2022). The result of the modulation of these cells by PBM treatment is the reduction of the release of the destructive inflammatory mediators and the promotion of the release of neurotrophic factors. It has also been recently shown that PBM may enhance cerebral blood flow and oxygenation, which is otherwise impaired by age and in the majority of neurodegenerative diseases (Tedford et al. 2015). PBM vascular effects are an additional neuroprotective advantage, as aging is also a factor that is associated with impaired vascular wellbeing and decreased delivery of nutrients to the brain (Henderson and Morries 2015).

One of the most compelling clinical studies that prove these benefits was published in PLOS ONE (2025), and the group of researchers applied 808-nm transcranial photobiomodulation therapy (tPBMT) to healthy BALB/c mice during a 30-day period. One hour of NIR light exposure per day was done on mice, and a complete set of molecular assays (complete transcriptomic profiling [RNA-seq] and Western blot protein quantification) was performed (Tian et al. 2016). It was disclosed that PBM regulated more than 1400 genes in cortex and hippocampus, such as oxidative response (NF-kBIa, JUND, PKC family), inflammation (TNFaIP6, IL-1RAPL1), and apoptosis (Caspase-3, CDKN1A). Notably, the genes that are of vital importance in terms of Alzheimer's disease pathology, including BACE1, BACE2, PSEN2, and others involved in the processing of amyloid precursor protein, were downregulated, which points to a prophylactic effect on the development of amyloid plaques (Fisher et al. 2013). It should be mentioned that the level of APP was reduced, but the level of protective proteins like MANF and PKC was increased. Histological results also indicated that there were more anti-inflammatory microglia and less pro-inflammatory cytokines such as TNF-alpha and IL-17 (Shahapal et al. 2019). The results suggest that tPBMT is not only neuroprotective but also has a part in the brain homeostasis via the systemic transcriptional reprogramming. This mechanistic depth and multi-layered approach are a very good reason to translate PBM therapy into clinical practice in the field of age-related cognitive decline (Xu et al. 2020).

Moreover, the clinical applicability of PBM is improved due to the new technologies that enable the implementation of PBM with more precision and consistency (Chung et al. 2012). The available body of evidence is in favor of the notion that PBM, with its multimodal actions, such as anti-inflammatory, anti-apoptotic, neurogenic, and metabolic regulation, is a very promising therapeutic approach to slowing down, mitigating, or even reversing part of the aging-related neurodegeneration (Feng et al. 2019). It is noninvasive, safe, and has wide-ranging physiological implication and thus an attractive add-on or alternative to pharmacological intervention in the setting of neurodegenerative aging (Kothapalli et al. 2019).

9.7 New Technology

Photomedicine has traditionally used externally applied light sources, such as lasers, LEDs, and broadband UV systems, to deliver therapeutic effects in dermatology, oncology, neurology, and antimicrobial applications (Rad et al. 2023). Nonetheless, there is currently a major paradigm shift in the field, due to the emergence of new materials and light-engineering technologies that make autonomous, precise, and long-term phototherapy possible (Ramabadran et al. 2023). The technological advancement of external energy-driven devices to self-sustainable light-emitting systems is redefining the scope of operation of photomedicine (Jia et al. 2022). In 2024–2025, research is especially targeted at combining long-duration photonic activity with spectral selectivity to enhance effectiveness of treatment, comfort of users, and device independence (Thammathiwat et al. 2023).

Among the most outstanding developments is the creation of persistent luminescent materials or what is commonly referred to as the afterglow phosphors (Seow 2013). They are able to store energy that is external—usually the sunlight, UV, or LED light and then gradually discharge it over long durations as visible or ultraviolet light (Heoa et al. 2013). As opposed to traditional light-emitting systems, these phosphors do not need a constant electrical current to sustain the emission, a revolutionary idea since it implies autonomous light delivery (Seidel et al. 2019). This has the prospect of transforming how light therapy can be administered, with off-grid, wearable, and even implantable phototherapeutic systems that do not need any external light sources (Kumthekar and Ogdie 2020). The developments do not represent incremental improvements but a total redesign of the phototherapy paradigm that makes it possible to perform extended and spatially limited light exposure with minimal discomfort and risk to the patient (Shaikh et al. 2021).

One of the brightest examples of such innovation is the development of a $LiCaPO_4:Gd^{3+}$ phosphor with narrowband ultraviolet-B (NB-UVB) emission, which can be used in continuous operation to produce its luminescent effect in more than 100 hours after a single excitation (Fong et al. 2014). NB-UVB radiation is well recognized to be clinically effective in the treatment of diseases of the skin including psoriasis, vitiligo, eczema, and some types of skin cancer (Ali 2018). The specificity of narrowband UV-B is a benefit since it minimizes the side effects that are usually attributed to broadband UV treatment, which include photoaging and DNA damage (Bertrand et al. 2015). The long-lasting emission of NB-UVB by this phosphor allows patients to be exposed to therapeutic doses continuously over a period of days without the use of large external light systems (Saftoiu 2015). It also creates the prospects of night-time therapy, when the treatment may proceed even during sleep, which also enhances the patient compliance and clinical outcomes greatly (Watanabe et al. 2020).

Practically, these long-lasting phosphors may be incorporated into biocompatible patches, flexible films, or implantable matrices that fit a particular anatomical location. These systems would enable local and programmable light therapy, which is of great value in chronic cases or cases where continuous low-dose stimulation is needed (Wen et al. 2021). Take the example of chronic wound, the controlled release of UV-B light may assist in avoiding colonization of bacteria and stimulating tissue regeneration without daily dressing change or exposure to hospital-grade equipment. Likewise, implants placed after surgery and including light-emitting substances might speed healing or forestall infection without the need to intervene. This is a major step in the direction of patient-controlled decentralized healthcare (Browne 2023).

Besides dermatology, these technologies have an impact in other areas that are high impact. In oncology, light sources that remain on can activate photosensitizers that are used in photodynamic therapy (PDT), allowing the targeted destruction of cancerous cells without harming healthy tissue (Lucky et al. 2015). Given that tumors are typically deep in tissues and in low-oxygen conditions where conventional light can hardly penetrate, it would be particularly helpful to have implantable materials that would continue emitting light even after excitation. In addition, it minimizes the logistical challenge on the clinical systems as well as patients, and it is possible that outpatient or home-based cancer phototherapy may be possible in some instances (Liu et al. 2021).

Persistent luminescence may be used in neurology to address the invasive drawbacks of the traditional fiber-optic delivery of light, as in optogenetics, where light is applied to neurons by activating genetically expressed light-sensitive proteins (Matsubara and Yamashita 2021). The new generation of phosphor-based implants may enable scientists and clinicians to activate or inhibit neural activity over long periods of time with low invasiveness and high spatial resolution. It has the potential to transform research into brain circuitry and treatment of diseases like epilepsy, Parkinson's, and depression (Fan et al. 2017).

Photoactivated drug delivery systems are another field that is growing at a fast rate. Smart materials are being developed to deliver drugs upon exposure to certain light wavelengths so that medicine is only released at the intended location and only when necessary (Lucky et al. 2015). Such systems could be turned on in a sustained and programmable manner by persistent light sources embedded near or within tissues, and would increase the precision of medicine approaches and reduce off-target effects. As an example, NB-UVB phosphors could be used to activate photocleavable linkers in chemotherapeutic drugs to locally release drugs with greatly reduced systemic toxicity (Jiang et al. 2019).

9.8 Challenges

Although photomedicine has shown therapeutic potential in a broad spectrum of neurological diseases, including Alzheimer's disease, Parkinson's disease, traumatic brain injury (TBI), stroke, and depression, clinical translation of photomedicine is fraught with many technical, biological, and logistical hurdles. Among the most basic and long-standing barriers is a low penetration of light into brain tissues, particularly through the intact human skull. Photons are scattered and absorbed by the scalp, skull bones, cerebrospinal fluid, and brain tissue, which is a major contributor to the light intensity attenuation before it reaches the deep brain structures (Salehpour et al. 2019). Despite the relative advantages of the penetration depth of near-infrared (NIR) light (700-1400 nm) compared to visible or ultraviolet light, it is not sufficient to penetrate into the human brains to therapeutic depths. The optical window is also very sensitive to the personal differences in skull thickness, skin pigmentation, hair, and cerebral blood volume which causes variable therapeutic results. Therefore, disorders that require deep brain targets, such as the hippocampus in Alzheimer's or basal ganglia in Parkinson's disease, are challenging to treat with noninvasive transcranial phototherapy (Salehpour et al. 2018). The other major challenge is the absence of standard dosimetry protocols. Although preclinical studies are promising, no consensus has been reached on the best parameters including wavelength, fluence (light dose), exposure time, pulse frequency, and duration of treatment. Photothermal or photochemical damage can be produced by overexposure, and there can be no therapeutic effect with under-exposure. There is a wide variation in the methods of study, some using continuous wave lasers, others pulsed light-emitting diodes (LEDs), and it is hard to compare one with another (Algorri et al. 2021). In addition, there is no real-time monitoring of the distribution of light in the brain currently, and therefore, clinicians cannot adjust the dose to changing patient conditions or treatment efficacy. In the absence of strong biofeedback mechanisms, treatment has to be based on approximated models of light propagation, or animal models, which are both uncertain (Alekseeva et al. 2024).

In addition, there is a paucity of long-term safety and efficacy data in humans. Human trials have been mostly of small scale and in most cases restricted to case series or early-phase studies with rather short follow-up durations. As an example, although photobiomodulation (PBM) has demonstrated neuroprotective activity in rodent models of Alzheimer's disease and stroke, its application in clinical practice in humans remains in its early stages (Godaert and Dramé 2024). It is also feared that there may be cumulative effects of repeated exposure to light on the neural tissue especially in vulnerable groups like the aged or people with comorbidities. Moreover, since the progression of brain diseases is usually slow and has a complex, multifactorial mechanism of etiology, the window of therapeutic light-based interventions has not been determined. Whether PBM can be applied as a preventive, acute, or chronic treatment, and at what point it can be most advantageous, remains unknown (Cassano et al. 2019).

One of the scientific challenges that face photomedicine is the incomplete picture of the molecular mechanisms of photomedicine in the neural contexts. Though there are studies indicating that light can bind to the cytochrome c oxidase of the mitochondrion to increase ATP production, decrease oxidative stress, and regulate inflammatory signaling, the downstream biological processes are still largely hypothetical. The therapeutic effects also appear to differ between brain areas, neuron types, and disease models, and it is challenging to come up with universally applicable treatment paradigms (Bathini et al. 2020). In addition, the immune reaction to light stimulation in the central nervous system (CNS) is not completely known. Although PBM potentially decreases neuroinflammation in some models, it also has the potential to induce unwanted immune responses or disrupt the blood-brain barrier in other situations (Xie et al. 2023).

Also, implantable or wearable light delivery systems are not yet developed in an effective and safe way. Although the limitations of skull penetration are being circumvented using implantable optical fibers and persistent luminescent nanoparticles, these methods give rise to significant issues of biocompatibility, tissue heating, device stability, and risk of infection (Wu and Guo 2021). The noninvasive alternative is wearable helmets and LED caps intended to be used at home during PBM, which also have issues regarding the homogeneous distribution of light to the brain and patient compliance. Such systems can be customized to individual brain anatomy with MRI-based targeting or adaptive photonics, but this increases cost and complexity, and so is not in wide clinical use (Fernandes et al. 2024).

Finally, regulatory and ethical issues are becoming more and more topical as photomedicine tools move out of the laboratory and into the clinic. Because all these interventions are combinations of medical devices and light sources and may include photosensitizing agents (in the case of photodynamic therapy), their approval tends to reside in regulatory gray areas (National Academies of Sciences, Engineering, and Medicine, et al. 2017). There is also an ethical issue of PBM usage in vulnerable groups, including cognitively impaired patients who cannot give full consent. The possibility of manipulating the brain with light also poses philosophical issues of identity, agency, and enhancement, especially when such technologies are applied to cognitive enhancement or mood control in the absence of any therapeutic need (Zuk et al. 2018).

9.9 Conclusion

The evidence in this chapter taken together highlights the potential of photomedicine, photobiomodulation (PBM) and photodynamic therapy (PDT), as a new, noninvasive form of treatment in the fight against age-related neurodegeneration. Alzheimer's disease (AD), Parkinson's disease (PD), and cognitive impairment are multifactorial neurodegenerative disorders that are driven by oxidative stress, mitochondrial dysfunction, loss of synapses, neuroinflammation, and cerebral blood flow impairment. By addressing these pathophysiological processes at the cellular and molecular level, photomedicine provides a neuroprotection strategy that works outside the confines of symptomatic pharmacotherapies. Mechanistically, PBM activates the mitochondrial cytochrome c oxidase to increase the production of ATP, regulate oxidative stress responses, optimize redox homeostasis, and decrease neuroinflammation. It maintains synaptic plasticity, upregulates neurotrophic factors such as BDNF and NGF, and enhances neurogenesis—restoring neural integrity and functionality. PDT, with its light-induced production of reactive oxygen species, has the potential to selectively destroy abnormal cells and has become an effective adjuvant to the treatment of neuro-oncological and inflammatory brain disorders. The creation of long-lasting luminescent materials, like narrowband UV-B-emitting phosphors, also indicates a transition in phototherapeutic design, between externally driven and self-powered, wearable, or implantable light sources, which increases patient compliance and broadens therapeutic windows. The preclinical and early clinical trials show promising results: enhanced memory, motor coordination, and mood stabilization, decreased neuroinflammation, and amyloid load. In addition, PBM has been shown to have protective effects on dopaminergic neurons, improve cerebral perfusion, and regulate important gene expression pathways involved in aging and neurodegeneration. The results confirm photomedicine as a disease-modifying intervention and as a modality that can restore neurorestoration and functional improvement of the aging brain. But there are still considerable obstacles. Its wide clinical use is hampered by the penetration of light through the skull, the anatomical variability of individuals, the lack of standard treatment regimes, the incomplete understanding of dose-response relationships, and insufficient long-term clinical data. In addition, there are ethical and regulatory issues pertaining to neuromodulation in cognitively vulnerable or impaired individuals, which should be dealt with by properly designed policies and informed consent procedures. To sum up, photomedicine is on the cutting edge of a paradigm shift in the neuromodulator treatment of aging brain disorders. It offers a paradigm shift in the manner cognitive health and neurodegeneration can be preserved due to its multi-mechanism, safety profile, and capability to synergize with the ongoing pharmacological and behavioral therapy. Its full potential in neurotherapeutics should be sought in future research by optimizing delivery systems, large-scale randomized clinical trials, and expanding mechanistic knowledge.

References

(2001) Symposium announcement. Exp Neurol 170(1):226 https://doi.org/10.1006/EXNR.2001.7724

Agostinis P et al (2011) Photodynamic therapy of cancer: an update. CA Cancer J Clin 61(4):250–281. https://doi.org/10.3322/CAAC.20114

Alekseeva P, Makarov V, Efendiev K, Shiryaev A, Reshetov I, Loschenov V (2024) Devices and methods for dosimetry of personalized photodynamic therapy of tumors: a review on recent trends. Cancers (Basel) 16(13). https://doi.org/10.3390/CANCERS16132484

Algorri JF, Ochoa M, Roldán-Varona P, Rodríguez-Cobo L, López-Higuera JM (2021) Light technology for efficient and effective photodynamic therapy: a critical review. Cancers (Basel) 13(14):3484. https://doi.org/10.3390/CANCERS13143484

Ali M (2018) Abdominal compartment syndrome: the importance of urinary catheter placement in measuring intra-abdominal pressure. BMJ Case Rep 2018:bcr2018226786. https://doi.org/10.1136/BCR-2018-226786

Amaroli A, Ravera S, Baldini F, Benedicenti S, Panfoli I, Vergani L (2019) Photobiomodulation with 808-nm diode laser light promotes wound healing of human endothelial cells through increased reactive oxygen species production stimulating mitochondrial oxidative phosphorylation. Lasers Med Sci 34(3):495–504. https://doi.org/10.1007/S10103-018-2623-5/METRICS

Anías-Calderón J, Verdugo-Díaz L, Drucker-Colín R (2004) Adrenalectomy and dexamethasone replacement on yawning behavior. Behav Brain Res 154(1):255–259. https://doi.org/10.1016/j.bbr.2004.02.013

Baddeley A (2003) Working memory: looking back and looking forward. Nat Rev Neurosci 4(10):829–839. https://doi.org/10.1038/nrn1201

Barrett DW, Gonzalez-Lima F (2013) Transcranial infrared laser stimulation produces beneficial cognitive and emotional effects in humans. Neuroscience 230:13–23. https://doi.org/10.1016/J.NEUROSCIENCE.2012.11.016/ASSET/FD0C435C-26CC-41F0-9944-CD27B64C8002/MAIN.ASSETS/GR5.SML

Basurto-Islas G et al (2014) Therapeutic benefits of a component of coffee in a rat model of Alzheimer's disease. Neurobiol Aging 35(12):2701–2712. https://doi.org/10.1016/J.NEUROBIOLAGING.2014.06.012

Bathini M, Raghushaker CR, Mahato KK (2020) The molecular mechanisms of action of photobiomodulation against neurodegenerative diseases: a systematic review. Cell Mol Neurobiol 42(4):955. https://doi.org/10.1007/S10571-020-01016-9

Bennett IJ, Madden DJ (2014) Disconnected aging: cerebral white matter integrity and age-related differences in cognition. Neuroscience 276:187–205. https://doi.org/10.1016/J.NEUROSCIENCE.2013.11.026

Berman MH, Nichols TW (2019) Treatment of neurodegeneration: integrating photobiomodulation and neurofeedback in Alzheimer's dementia and Parkinson's: a review. Photobiomodul Photomed Laser Surg 37(10):623–634. https://doi.org/10.1089/PHOTOB.2019.4685;PAGE:STRING:ARTICLE/CHAPTER

Berman MH, Nichols T, Huang J, Nizamutdinov D (2019) Noninvasive neurotherapeutic treatment of neurodegeneration: integrating photobiomodulation and neurofeedback training. In: Photobiomodulation in the brain: low-level laser (light) therapy in neurology and neuroscience, pp 447–462. https://doi.org/10.1016/B978-0-12-815305-5.00032-4

Bertrand TE, Cruz A, Binitie O, Cheong D, Letson GD (2015) Do surgical margins affect local recurrence and survival in extremity, nonmetastatic, high-grade osteosarcoma? Clin Orthop Relat Res 474(3):677. https://doi.org/10.1007/S11999-015-4359-X

Birks JS, Chong LY, Grimley Evans J (2015) Rivastigmine for Alzheimer's disease. Cochrane Database Syst Rev 2015(9). https://doi.org/10.1002/14651858.CD001191.PUB4/MEDIA/CDSR/CD001191/IMAGE_N/NCD001191-CMP-002-17.PNG

Brazzelli V et al (2007) A long-term time course of colorimetric assessment of the effects of imatinib mesylate on skin pigmentation: a study of five patients. J Eur Acad Dermatol Venereol 21(3):384–387. https://doi.org/10.1111/J.1468-3083.2006.01981.X

Brown RH, Al-Chalabi A (2017) Amyotrophic lateral sclerosis. N Engl J Med 377(2):162–172. https://doi.org/10.1056/NEJMRA1603471

Browne K (2023) Novel antimicrobial strategies to prevent healthcare-associated infections, pp. 2024–2036. https://doi.org/10.26190/unsworks/25254

Campisi J (2013) Aging, cellular senescence, and cancer. Annu Rev Physiol 75:685–705. https://doi.org/10.1146/ANNUREV-PHYSIOL-030212-183653/CITE/REFWORKS

Carey M, Preston C (2019) Investigating the components of body image disturbance within eating disorders. Front Psychol 10:474709. https://doi.org/10.3389/FPSYT.2019.00635/BIBTEX

Cassano P et al (2019) Reported side effects, weight and blood pressure, after repeated sessions of transcranial photobiomodulation. Photobiomodul Photomed Laser Surg 37(10):651–656. https://doi.org/10.1089/PHOTOB.2019.4678

Cercignani M et al (2017) Characterizing axonal myelination within the healthy population: a tract-by-tract mapping of effects of age and gender on the fiber g-ratio. Neurobiol Aging 49:109–118. https://doi.org/10.1016/J.NEUROBIOLAGING.2016.09.016

Chang HH, Meyerhoefer CD (2019) Inter-brand competition in the convenience store industry, store density and healthcare utilization. J Health Econ 65:117–132. https://doi.org/10.1016/J.JHEALECO.2019.03.001

Chao LL (2019) Effects of home photobiomodulation treatments on cognitive and behavioral function, cerebral perfusion, and resting-state functional connectivity in patients with dementia: a pilot trial. Photobiomodul Photomed Laser Surg 37(3):133–141. https://doi.org/10.1089/PHOTOB.2018.4555. https://homeliebertpubcom/photob

Chen J et al (2020) Anterior communicating artery aneurysms: anatomical considerations and microsurgical strategies. Front Neurol 11:528998. https://doi.org/10.3389/FNEUR.2020.01020/BIBTEX

Chung H, Dai T, Sharma SK, Huang YY, Carroll JD, Hamblin MR (2012) The nuts and bolts of low-level laser (light) therapy. Ann Biomed Eng 40(2):516–533. https://doi.org/10.1007/S10439-011-0454-7/METRICS

Dai L, Huang Y, Chen Y, er Long Z (2016) Cloning and characterization of filamentous temperature-sensitive protein Z from Xanthomonas oryzae pv. Oryzae. Springerplus 5(1):1–7. https://doi.org/10.1186/S40064-016-1876-3

Daly SM, Leahy MJ (2013) Go with the flow ': a review of methods and advancements in blood flow imaging. J Biophotonics 6(3):217–255. https://doi.org/10.1002/JBIO.201200071;WGROUP:STRING:PUBLICATION

Dauer W, Przedborski S (2003) Parkinson's disease: mechanisms and models. Neuron 39(6):889–909. https://doi.org/10.1016/S0896-6273(03)00568-3/ASSET/F42A31EC-01FD-4FE1-B9F2-DE67FDF35ABA/MAIN.ASSETS/GR4.JPG

Dolmans DEJGJ, Fukumura D, Jain RK (2003) Photodynamic therapy for cancer. Nat Rev Cancer 3(5):380–387. https://doi.org/10.1038/NRC1071

Dugger BN, Dickson DW (2017) Pathology of neurodegenerative diseases. Cold Spring Harb Perspect Biol 9(7):a028035. https://doi.org/10.1101/CSHPERSPECT.A028035

Eisfeld AJ, Kawaoka Y (2017) Calculated risk: a new single-nucleotide polymorphism linked to severe influenza disease. Nat Med 23(8):911–912. https://doi.org/10.1038/NM.4383

Erkkinen MG, Kim MO, Geschwind MD (2018) Clinical neurology and epidemiology of the major neurodegenerative diseases. Cold Spring Harb Perspect Biol 10(4):a033118. https://doi.org/10.1101/CSHPERSPECT.A033118

Fan W et al (2017) Enhanced afterglow performance of persistent luminescence implants for efficient repeatable photodynamic therapy. ACS Nano 11(6):5864–5872. https://doi.org/10.1021/ACSNANO.7B01505/SUPPL_FILE/NN7B01505_SI_001.PDF

Farkas E, Obrenovitch TP, Institóris Á, Bari F (2011) Effects of early aging and cerebral hypoperfusion on spreading depression in rats. Neurobiol Aging 32(9):1707–1715. https://doi.org/10.1016/J.NEUROBIOLAGING.2009.10.002

Feng X et al (2019) 2-deoxy-D-glucose augments photodynamic therapy induced mitochondrial caspase-independent apoptosis and energy-mediated autophagy. Lasers Surg Med 51(4):352–362. https://doi.org/10.1002/LSM.23020

Fernandes F et al (2024) Devices used for photobiomodulation of the brain—a comprehensive and systematic review. J Neuroeng Rehabil 21(1). https://doi.org/10.1186/S12984-024-01351-8

Fisher RA (2019) Anonymous live liver donor: the good Samaritan stranger. Nat Rev Gastroenterol Hepatol 16(11):650–651. https://doi.org/10.1038/S41575-019-0207-X

Fisher CJ, Niu CJ, Lai B, Chen Y, Kuta V, Lilge LD (2013) Modulation of PPIX synthesis and accumulation in various normal and glioma cell lines by modification of the cellular signaling and temperature. Lasers Surg Med 45(7):460–468. https://doi.org/10.1002/LSM.22161;WGROUP:STRING:PUBLICATION

Fjell AM, Walhovd KB (2010) Structural brain changes in aging: courses, causes and cognitive consequences. Rev Neurosci 21(3):187–221. https://doi.org/10.1515/REVNEURO.2010.21.3.187/PDF/FIRSTPAGE

Fong DL et al (2014) Transmission of Chagas disease via blood transfusions in 2 immunosuppressed pigtailed macaques (Macaca nemestrina). Comp Med 64(1):63. Accessed: August 10, 2025. [Online]. Available: https://pmc.ncbi.nlm.nih.gov/articles/PMC3929221/

Gao L, Li X, Yang W, Sun X (2017) Modulation of azimuth tuning plasticity in rat primary auditory cortex by medial prefrontal cortex. Neuroscience 347:36–47. https://doi.org/10.1016/J.NEUROSCIENCE.2017.01.046/ASSET/65BF5529-95B5-4B47-9A0A-9A5883F55A99/MAIN.ASSETS/GR5.SML

Gitler AD, Dhillon P, Shorter J (2017) Neurodegenerative disease: models, mechanisms, and a new hope. Dis Model Mech 10(5):499–502. https://doi.org/10.1242/DMM.030205

Godaert L, Dramé M (2024) Efficacy of photobiomodulation therapy in older adults: a systematic review. Biomedicine 12(7). https://doi.org/10.3390/BIOMEDICINES12071409

Goedert M, Eisenberg DS, Crowther RA (2017) Propagation of tau aggregates and neurodegeneration. Annu Rev Neurosci 40:189–210. https://doi.org/10.1146/ANNUREV-NEURO-072116-031153/CITE/REFWORKS

Gundersen BB (2016) Automating brain mapping. Nat Methods 13(9):719–719. https://doi.org/10.1038/nmeth.3990

Hamblin MR (2016) Shining light on the head: photobiomodulation for brain disorders. BBA Clin 6:113–124. https://doi.org/10.1016/J.BBACLI.2016.09.002

Hamblin MR (2019) Photobiomodulation for Alzheimer's disease: has the light dawned? Photonics 6(3):77. https://doi.org/10.3390/PHOTONICS6030077

Hamblin MR, Demidova TN (2006) Mechanisms of low level light therapy. Proc SPIE Int Soc Opt Eng 6140:614001. https://doi.org/10.1117/12.646294

Hamblin MR, Huang YY, Sharma SK, Carroll J (2011) Biphasic dose response in low level light therapy - an update. Dose-Response 9(4):602–618. https://doi.org/10.2203/DOSE-RESPONSE.11-009.HAMBLIN

Hamieh A, Hamieh M (2020) Brucella prosthetic valve endocarditis with septic and cardiogenic shock. IDCases 21. https://doi.org/10.1016/j.idcr.2020.e00881

Harman D (2001) Aging: overview. Ann N Y Acad Sci 928(1):1–21. https://doi.org/10.1111/J.1749-6632.2001.TB05631.X

Harpavat S et al (2016) Newborn direct or conjugated bilirubin measurements as a potential screen for biliary atresia. J Pediatr Gastroenterol Nutr 62(6):799–803. https://doi.org/10.1097/MPG.0000000000001097

Hart R, Belotto M (2010) The institutional review board. Semin Nucl Med 40(5):385–392. https://doi.org/10.1053/J.SEMNUCLMED.2010.03.007

Henderson BW, Dougherty TJ (1992) How does photodynamic therapy work? Photochem Photobiol 55(1):145–157. https://doi.org/10.1111/J.1751-1097.1992.TB04222.X

Henderson TA, Morries LD (2015) Near-infrared photonic energy penetration: can infrared phototherapy effectively reach the human brain? Neuropsychiatr Dis Treat 11:2191–2208. https://doi.org/10.2147/NDT.S78182

Heoa SM, Ruhla S, Scannapieco FA (2013) Implications of salivary protein binding to commensal and pathogenic bacteria. J Oral Biosci 55(4):169. https://doi.org/10.1016/J.JOB.2013.06.004

Iliff JJ et al (2012) A paravascular pathway facilitates CSF flow through the brain parenchyma and the clearance of interstitial solutes, including amyloid β. Sci Transl Med 4(147). https://doi.org/10.1126/SCITRANSLMED.3003748/SUPPL_FILE/MOVIE_2.MOV

Jellinger KA (2010) Basic mechanisms of neurodegeneration: a critical update. J Cell Mol Med 14(3):457–487. https://doi.org/10.1111/J.1582-4934.2010.01010.X

Jia H et al (2022) Analysis of factors affecting medical personnel seeking employment at primary health care institutions: developing human resources for primary health care. Int J Equity Health 21(1):37. https://doi.org/10.1186/S12939-022-01638-Z

Jiang W, Huang L, Mo F, Zhong Y, Xu L, Fu F (2019) Persistent luminescent multifunctional drug delivery nano-platform based on nanomaterial ZnGa2O4:Cr3+,Sn4+ for imaging-guided cancer chemotherapy. J Mater Chem B 7(18):3019–3026. https://doi.org/10.1039/C9TB00109C

Johnson JA, Jakubovski E, Reed MO, Bloch MH (2017) Predictors of long-term risky driving behavior in the multimodal treatment study of children with attention-deficit/hyperactivity disorder. J Child Adolesc Psychopharmacol 27(8):747–754. https://doi.org/10.1089/CAP.2017.0025

Johnstone DM, Moro C, Stone J, Benabid AL, Mitrofanis J (2016) Turning on lights to stop neurodegeneration: the potential of near infrared light therapy in Alzheimer's and Parkinson's disease. Front Neurosci 9(JAN):170021. https://doi.org/10.3389/FNINS.2015.00500/XML

Kalia LV, Lang AE (2015) Parkinson's disease. Lancet 386(9996):896–912. https://doi.org/10.1016/S0140-6736(14)61393-3

Kandel ER (2001) The molecular biology of memory storage: a dialogue between genes and synapses. Science (1979) 294(5544):1030–1038. https://doi.org/10.1126/SCIENCE.1067020;WEBSITE:WEBSITE:AAAS-SITE;JOURNAL:JOURNAL:SCIENCE;WGROUP:STRING:PUBLICATION

Karami H, Vahidi S, Abouie S, Gholizadeh H (2021) Percutaneous extraction of big kidney stone without fragmentation: the novel technique in percutaneous nephrolithotomy (EN BLOC PCNL). Urol Case Rep 39. https://doi.org/10.1016/j.eucr.2021.101749

Karu TI, Kolyakov SF (2005) Exact action spectra for cellular responses relevant to phototherapy. Photomed Laser Surg 23(4):355–361. https://doi.org/10.1089/PHO.2005.23.355

Kenyon CJ (2010) The genetics of ageing. Nature 464(7288):504–512. https://doi.org/10.1038/NATURE08980

Kholyavin AI, Polyakov YI (2022) Stereotactic cingulotomy and capsulotomy for obsessive-compulsive disorders: indications and comparative results. Prog Brain Res 272(1):1–21. https://doi.org/10.1016/BS.PBR.2022.03.010

Kim HJ, Lee GS, Kim H, Kim BJ (2017) Hwangryunhaedok-tang induces the depolarization of pacemaker potentials through 5-HT3 and 5-HT4 receptors in cultured murine small intestine interstitial cells of Cajal. World J Gastroenterol 23(29):5313–5323. https://doi.org/10.3748/WJG.V23.I29.5313

Kirkwood TBL (2005) Understanding the odd science of aging. Cell 120(4):437–447. https://doi.org/10.1016/J.CELL.2005.01.027

Kopelman MD (2002) Disorders of memory. Brain 125(10):2152–2190. https://doi.org/10.1093/BRAIN/AWF229

Kothapalli PR, Wyler von Ballmoos MC, Chinnadurai P, Lumsden AB, Ramchandani MK (2019) Value of the hybrid operating theater for an integrated approach to diagnosis and treatment of pulmonary nodules in 2019. Front Surg 6:36. https://doi.org/10.3389/FSURG.2019.00036

Kumthekar A, Ogdie A (2020) Obesity and psoriatic arthritis: a narrative review. Rheumatol Ther 7(3):447. https://doi.org/10.1007/S40744-020-00215-6

Kuro-o M et al (1997) Mutation of the mouse klotho gene leads to a syndrome resembling ageing. Nature 390(6655):45–51. https://doi.org/10.1038/36285

Lanzafame RJ, Blanche RR, Bodian AB, Chiacchierini RP, Fernandez-Obregon A, Kazmirek ER (2013) The growth of human scalp hair mediated by visible red light laser and LED sources in males. Lasers Surg Med 45(8):487–495. https://doi.org/10.1002/LSM.22173;REQUESTEDJOURNAL:JOURNAL:10969101;WGROUP:STRING:PUBLICATION

Lasselin J, Alvarez-Salas E, Jan-Sebastian G (2016) Well-being and immune response: a multi-system perspective. Curr Opin Pharmacol 29:34–41. https://doi.org/10.1016/J.COPH.2016.05.003

Lattanzi GM, Buzzanca A, Frascarelli M, Di Fabio F (2018) Genetic and clinical features of social cognition in 22q11.2 deletion syndrome. J Neurosci Res 96(10):1631–1640. https://doi.org/10.1002/JNR.24265;WGROUP:STRING:PUBLICATION

Liu N, Chen X, Sun X, Sun X, Shi J (2021) Persistent luminescence nanoparticles for cancer theranostics application. J Nanobiotechnology 19(1):1–24. https://doi.org/10.1186/S12951-021-00862-Z

López-García S et al (2019) A rare PSEN1 (Leu85Pro) mutation causing Alzheimer's disease in a 29-year-old woman presenting as corticobasal syndrome. J Alzheimer's Dis 70(3):655–658. https://doi.org/10.3233/JAD-190107

López-Otín C, Blasco MA, Partridge L, Serrano M, Kroemer G (2013) The hallmarks of aging. Cell 153(6):1194. https://doi.org/10.1016/J.CELL.2013.05.039/ASSET/4B4DCA34-3BF7-4A23-90DA-524E4B221C55/MAIN.ASSETS/GR2.JPG

Lu T et al (2004) Gene regulation and DNA damage in the ageing human brain. Nature 429(6994):883–891. https://doi.org/10.1038/NATURE02661

Lucky SS, Soo KC, Zhang Y (2015) Nanoparticles in photodynamic therapy. Chem Rev 115(4):1990–2042. https://doi.org/10.1021/CR5004198/ASSET/CR5004198.FP.PNG_V03

Matsubara T, Yamashita T (2021) Remote optogenetics using up/down-conversion phosphors. Front Mol Biosci 8:771717. https://doi.org/10.3389/FMOLB.2021.771717/XML

Mattson MP, Arumugam TV (2018) Hallmarks of brain aging: adaptive and pathological modification by metabolic states. Cell Metab 27(6):1176–1199. https://doi.org/10.1016/J.CMET.2018.05.011/ASSET/34582356-CB61-47EB-AE90-DED8A29AB8D0/MAIN.ASSETS/GR1.JPG

Mehta D, Jackson R, Paul G, Shi J, Sabbagh M (2017) Why do trials for Alzheimer's disease drugs keep failing? A discontinued drug perspective for 2010-2015. Expert Opin Investig Drugs 26(6):735–739. https://doi.org/10.1080/13543784.2017.1323868;SUBPAGE:STRING:ACCESS

Meng C, He Z, Xing D (2013) Low-level laser therapy rescues dendrite atrophy via upregulating BDNF expression: implications for Alzheimer's disease. J Neurosci 33(33):13505–13517. https://doi.org/10.1523/JNEUROSCI.0918-13.2013

Mouton-Liger F et al (2014) Developmental molecular and functional cerebellar alterations induced by PCP4/PEP19 overexpression: implications for down syndrome. Neurobiol Dis 63:92–106. https://doi.org/10.1016/J.NBD.2013.11.016

Nairuz T, Sangwoo-Cho, Lee JH (2024) Photobiomodulation therapy on brain: pioneering an innovative approach to revolutionize ognitive dynamics. *Cells* 13(11):966. https://doi.org/10.3390/CELLS13110966

National Academies of Sciences, Engineering, and Medicine, Health and Medicine Division; Board on Health Sciences Policy, Forum on Neuroscience and Nervous System Disorders (2017) International perspectives on integrating ethical, legal, and social considerations into the development of non-invasive neuromodulation devices. National Academies Press (US). https://doi.org/10.17226/24643

Non-neuronal evoked and spontaneous hemodynamic changes in the anterior temporal region of the human head may lead to misinterpretations of functional near-infrared spectroscopy signals. Accessed: August 10, 2025. [Online]. Available: https://www.spiedigitallibrary.org/journals/neurophotonics/volume-5/issue-01/011002/Non-neuronal-evoked-and-spontaneous-hemodynamic-changes-in-the-anterior/10.1117/1.NPh.5.1.011002.full

Nwazota N, Pyati S, Fisher K, Roy L (2019) Device review: Pulsante™ sphenopalatine ganglion microstimulator. Pain Manag 9(6):535–541. https://doi.org/10.2217/PMT-2018-0082

Nyberg L, Lövdén M, Riklund K, Lindenberger U, Bäckman L (2012) Memory aging and brain maintenance. Trends Cogn Sci 16(5):292–305. https://doi.org/10.1016/J.TICS.2012.04.005/ASSET/108C8514-DE3C-4F30-B636-44FC46CCF407/MAIN.ASSETS/GR8.JPG

Padovani A et al (2010) The speech and language FOXP2 gene modulates the phenotype of frontotemporal lobar degeneration. J Alzheimer's Dis 22(3):923–931. https://doi.org/10.3233/JAD-2010-101206

Peters A (2002) The effects of normal aging on myelin and nerve fibers: a review. J Neurocytol 31(8–9):581–593. https://doi.org/10.1023/A:1025731309829/METRICS

Peters R (2006) Ageing and the brain: this article is part of a series on ageing edited by Professor Chris Bulpitt. Postgrad Med J 82(964):84–88. https://doi.org/10.1136/PGMJ.2005.036665

Posten W, Wrone DA, Dover JS, Arndt KA, Silapunt S, Alam M (2006) Low-level laser therapy for wound healing: mechanism and efficacy. Dermatologic Surg 31(3):334–340. https://doi.org/10.1111/J.1524-4725.2005.31086

Premi E et al (2022) An automated toolbox to predict single subject atrophy in Presymptomatic Granulin mutation carriers. J Alzheimers Dis 86(1):205–218. https://doi.org/10.3233/JAD-215447/SUPPL_FILE/SJ-PDF-1-ALZ-10.3233_JAD-215447.PDF

Przedborski S, Vila M, Jackson-Lewis V (2003) Series introduction: neurodegeneration: what is it and where are we? J Clin Invest 111(1):3–10. https://doi.org/10.1172/JCI17522

Purushothuman S, Johnstone DM, Nandasena C, Mitrofanis J, Stone J (2014) Photobiomodulation with near infrared light mitigates Alzheimer's disease-related pathology in cerebral cortex - evidence from two transgenic mouse models. Alzheimers Res Ther 6(1):1–13. https://doi.org/10.1186/ALZRT232/FIGURES/7

Qin C et al (2011) Comparison of permissible source region and multispectral data using efficient bioluminescence tomography method. J Biophotonics 4(11–12):824–839. https://doi.org/10.1002/JBIO.201100049

Quatman-Yates CC, Myer GD, Ford KR, Hewett TE (2013) A longitudinal evaluation of maturational effects on lower extremity strength in female adolescent athletes. Pediatr Phys Ther 25(3):271–276. https://doi.org/10.1097/PEP.0B013E31828E1E9D

Rad EM, Ilali HM, Majnoon MT, Zeinaloo A (2023) Mechanical QT and JT intervals by M-mode echocardiography: an extrapolation from the concurrent electrocardiographic tracings. Ann Pediatr Cardiol 15(4):364. https://doi.org/10.4103/APC.APC_169_21

Rahman A, Hoque MM, Khan MAK, Sarwar MG, Halim MA (2016) Non-covalent interactions involving halogenated derivatives of capecitabine and thymidylate synthase: a computational approach. Springerplus 5(1):1–18. https://doi.org/10.1186/S40064-016-1844-Y

Ramabadran R et al (2023) DNMT3A-coordinated splicing governs the stem state switch toward differentiation in embryonic and hematopoietic stem cells. Nat Cell Biol 25(4):528. https://doi.org/10.1038/S41556-023-01109-9

Ririe DG, Eisenach JC, Martin TJ (2021) A painful beginning: early life surgery produces long-term behavioral disruption in the rat. Front Behav Neurosci 15. https://doi.org/10.3389/FNBEH.2021.630889

Romand R, Gourgou S, Sancho-Garnier H (2005) A randomized trial assessing the five-day plan for smoking cessation. Addiction 100(10):1546–1554. https://doi.org/10.1111/J.1360-0443.2005.01215.X

Saftoiu A (2015) Editorial. Ultrasound Int Open 1(2):E33. https://doi.org/10.1055/S-0035-1565206

Saleh S, Fluet G, Qiu Q, Merians A, Adamovich SV, Tunik E (2017) Neural patterns of reorganization after intensive robot-assisted virtual reality therapy and repetitive task practice in patients with chronic stroke. Front Neurol 8(SEP):254385. https://doi.org/10.3389/FNEUR.2017.00452/BIBTEX

Salehpour F, Hamblin MR (2020) Photobiomodulation for Parkinson's disease in animal models: a systematic review. Biomolecules 10(4):610. https://doi.org/10.3390/BIOM10040610

Salehpour F, Mahmoudi J, Kamari F, Sadigh-Eteghad S, Rasta SH, Hamblin MR (2018) Brain photobiomodulation therapy: a narrative review. Mol Neurobiol 55(8):6601–6636. https://doi.org/10.1007/S12035-017-0852-4

Salehpour F et al (2019) Penetration profiles of visible and near-infrared lasers and light-emitting diode light through the head tissues in animal and human species: a review of literature. Photobiomodul Photomed Laser Surg 37(10):581–595. https://doi.org/10.1089/PHOTOB.2019.4676

Schneider A et al (2019) Growth hormone-mediated reprogramming of macrophage transcriptome and effector functions. Sci Rep 9(1):1–13. https://doi.org/10.1038/s41598-019-56017-6

Schrag A, Quinn N (2000) Dyskinesias and motor fluctuations in Parkinson's disease: a community-based study. Brain 123(11):2297–2305. https://doi.org/10.1093/BRAIN/123.11.2297

Seidel S, Aigner M, Wildner B, Sycha T, Pablik E (2019) Antipsychotics for the treatment of neuropathic pain in adults. Cochrane Database Syst Rev 2019(6):CD012916. https://doi.org/10.1002/14651858.CD012916.PUB2

Seow CY (2013) Hill's equation of muscle performance and its hidden insight on molecular mechanisms. J Gen Physiol 142(6):561. https://doi.org/10.1085/JGP.201311107

Shahapal A et al (2019) FAM19A5 expression during embryogenesis and in the adult traumatic brain of FAM19A5-LacZ Knock-in mice. Front Neurosci 13(AUG):471299. https://doi.org/10.3389/FNINS.2019.00917/BIBTEX

Shaikh OH, Kumbhar US, Jain A, Chakkalakkoombil SV (2021) Klippel-Trenaunay syndrome in a young patient with the involvement of gastrointestinal and genitourinary tracts: an unusual and rare presentation. BMJ Case Rep 14(3):e239420. https://doi.org/10.1136/BCR-2020-239420

Shi L et al (2020) Dickkopf-1 overexpression in vitro nominates candidate blood biomarkers relating to Alzheimer's disease pathology. J Alzheimers Dis 77(3):1353–1368. https://doi.org/10.3233/JAD-200208/SUPPL_FILE/SJ-PDF-1-ALZ-10.3233_JAD-200208.PDF

Shin YS, Sim HB, Yoon JR (2017) Tibial nerve neuropathy following medial opening-wedge high tibial osteotomy—case report of a rare technical complication. Eur J Orthop Surg Traumatol 27(4):563–567. https://doi.org/10.1007/S00590-017-1944-6

Shirkavand A, Mohajerani E, Farivar S, Ataie-Fashtami L, Ghazimoradi MH (2021) Monitoring the response of skin melanoma cell line (A375) to treatment with Vemurafenib: a pilot in vitro optical spectroscopic study. Photobiomodul Photomed Laser Surg 39(3):164–177. https://doi.org/10.1089/PHOTOB.2020.4887;PAGE:STRING:ARTICLE/CHAPTER

Shulyatnikova T, Verkhratsky A (2020) Astroglia in sepsis associated encephalopathy. Neurochem Res 45(1):83–99. https://doi.org/10.1007/S11064-019-02743-2

Small SA, Schobel SA, Buxton RB, Witter MP, Barnes CA (2011) A pathophysiological framework of hippocampal dysfunction in ageing and disease. Nat Rev Neurosci 12(10):585–601. https://doi.org/10.1038/nrn3085

Soto C, Pritzkow S (2018) Protein misfolding, aggregation, and conformational strains in neurodegenerative diseases. Nat Neurosci 21(10):1332–1340. https://doi.org/10.1038/s41593-018-0235-9

Squire LR, Wixted JT (2011) The cognitive neuroscience of human memory since H.M. Annu Rev Neurosci 34:259–288. https://doi.org/10.1146/ANNUREV-NEURO-061010-113720/CITE/REFWORKS

Streit WJ, Sammons NW, Kuhns AJ, Sparks DL (2004) Dystrophic microglia in the aging human brain. Glia 45(2):208–212. https://doi.org/10.1002/GLIA.10319;WGROUP:STRING:PUBLICATION

Sun Q et al (2023) MiR-181a-5p promotes neural stem cell proliferation and enhances the learning and memory of aged mice. Aging Cell 22(4):e13794. https://doi.org/10.1111/ACEL.13794

Tahara EB, Navarete FDT, Kowaltowski AJ (2009) Tissue-, substrate-, and site-specific characteristics of mitochondrial reactive oxygen species generation. Free Radic Biol Med 46(9):1283–1297. https://doi.org/10.1016/J.FREERADBIOMED.2009.02.008

Tang Y et al (2014) Time-specific microRNA changes during spinal motoneuron degeneration in adult rats following unilateral brachial plexus root avulsion: ipsilateral vs. contralateral changes. BMC Neurosci 15(1):1–14. https://doi.org/10.1186/1471-2202-15-92/FIGURES/6

Tedford CE, Delapp S, Jacques S, Anders JJ (2015) Quantitative analysis of transcranial and intraparenchymal light penetration in human cadaver brain tissue. Lasers Surg Med 47(4):312–322. https://doi.org/10.1002/LSM.22343

Thammathiwat T, Banjongjit A, Iampenkhae K, Townamchai N, Kanjanabuch T (2023) ANCA associated glomerulonephritis following SARS-CoV-2 vaccination: a case series and systematic review. Vaccines (Basel) 11(5):983. https://doi.org/10.3390/VACCINES11050983/S1

Tian F, Hase SN, Gonzalez-Lima F, Liu H (2016) Transcranial laser stimulation improves human cerebral oxygenation. Lasers Surg Med 48(4):343–349. https://doi.org/10.1002/LSM.22471;WEBSITE:WEBSITE:PERICLES;REQUESTEDJOURNAL:JOURNAL:10969101;WGROUP:STRING:PUBLICATION

Tran MD, Wanner IB, Neary JT (2008) Purinergic receptor signaling regulates N-cadherin expression in primary astrocyte cultures. J Neurochem 105(1):272–286. https://doi.org/10.1111/J.1471-4159.2008.05214.X;PAGEGROUP:STRING:PUBLICATION

Troppan KT et al (2016) The clinical significance of fibrinogen plasma levels in patients with diffuse large B cell lymphoma. J Clin Pathol 69(4):326–330. https://doi.org/10.1136/JCLINPATH-2015-203356

Tulving E (2002) Episodic memory: from mind to brain. Annu Rev Psychol 53:1–25. https://doi.org/10.1146/ANNUREV.PSYCH.53.100901.135114/CITE/REFWORKS

Vähävihu K, Ylianttila L, Kautiainen H, Tuohimaa P, Reunala T, Snellman E (2010) Spore film dosimeters are feasible for UV dose monitoring during heliotherapy. Photochem Photobiol 86(5):1174–1178. https://doi.org/10.1111/J.1751-1097.2010.00769.X;REQUESTEDJOURNAL:JOURNAL:17511097

Vandewalle G, Maquet P, Dijk DJ (2009) Light as a modulator of cognitive brain function. Trends Cogn Sci 13(10):429–438. https://doi.org/10.1016/J.TICS.2009.07.004/ASSET/D60D626A-ED17-49E6-AF98-853F904D8574/MAIN.ASSETS/FX7.SML

Vijayavenkataraman S, Shuo Z, Fuh JYH, Lu WF (2017) Design of three-dimensional scaffolds with tunable matrix stiffness for directing stem cell lineage specification: an in silico study. Bioengineering 4(3). https://doi.org/10.3390/BIOENGINEERING4030066

Vijg J, Suh Y (2013) Genome instability and aging. Annu Rev Physiol 75:645–668. https://doi.org/10.1146/ANNUREV-PHYSIOL-030212-183715/CITE/REFWORKS

Wang JH et al (2009) FibroScan and ultrasonography in the prediction of hepatic fibrosis in patients with chronic viral hepatitis. J Gastroenterol 44(5):439–446. https://doi.org/10.1007/S00535-009-0017-Y

Wang X et al (2017) Elevated expression of EBV and TLRs in the brain is associated with Rasmussen's encephalitis. Virol Sin 32(5):423–430. https://doi.org/10.1007/S12250-017-4058-8

Wang L, Mao L, Huang Z, Switzer JA, Hess DC, Zhang Q (2025) Photobiomodulation: shining a light on depression. Theranostics 15(2):362. https://doi.org/10.7150/THNO.104502

Watanabe K, Hasegawa G, Kashima K, Sato Y, Hara N, Nishiyama T (2020) Primary malignant melanoma of the female urethra: a case report. Urol Case Rep 34:101493. https://doi.org/10.1016/J.EUCR.2020.101493

Wen Y et al (2021) Cross-linked thermosensitive nanohydrogels for ocular drug delivery with a prolonged residence time and enhanced bioavailability. Mater Sci Eng C 119:111445. https://doi.org/10.1016/J.MSEC.2020.111445

Wong-Riley MTT et al (2005) Photobiomodulation directly benefits primary neurons functionally inactivated by toxins: role of cytochrome c oxidase. J Biol Chem 280(6):4761–4771. https://doi.org/10.1074/JBC.M409650200

Wu X, Guo Q (2021) Bioresorbable photonics: materials, devices and applications. Photonics *8*(7):235. https://doi.org/10.3390/PHOTONICS8070235

Wu CH et al (2014) Does non-central nervous system tuberculosis increase the risk of ischemic stroke? A population-based propensity score-matched follow-up study. PLoS One 9(7). https://doi.org/10.1371/JOURNAL.PONE.0098158

Xie K, El Khoury H, Mitrofanis J, Austin PJ (2023) A systematic review of the effect of photobiomodulation on the neuroinflammatory response in animal models of neurodegenerative diseases. Rev Neurosci 34(4):459–481. https://doi.org/10.1515/REVNEURO-2022-0109

Xu JT et al (2020) Effect of stromal cell-derived factor-1/CXCR4 axis in neural stem cell transplantation for Parkinson's disease. Neural Regen Res 15(1):112–119. https://doi.org/10.4103/1673-5374.264470

Yang Z et al (2021) Downregulation of CDH11 promotes metastasis and resistance to paclitaxel in gastric cancer cells. J Cancer 12(1):65–75. https://doi.org/10.7150/JCA.48193

Zahn JM et al (2006) Transcriptional profiling of aging in human muscle reveals a common aging signature. PLoS Genet 2(7):e115. https://doi.org/10.1371/JOURNAL.PGEN.0020115

Zhao Q, Wei H, He Y, Ren Q, Zhou C (2014) Evaluation of ultrasound and glucose synergy effect on the optical clearing and light penetration for human colon tissue using SD-OCT. J Biophotonics 7(11–12):938–947. https://doi.org/10.1002/JBIO.201300141

Zhou L et al (2020) DHZCP modulates microglial M1/M2 polarization via the p38 and TLR4/NF-κB signaling pathways in LPS-stimulated microglial cells. Front Pharmacol 11. https://doi.org/10.3389/FPHAR.2020.01126

Zuk P, Torgerson L, Sierra-Mercado D, Lázaro-Muñoz G (2018) Neuroethics of neuromodulation: an update. Curr Opin Biomed Eng 8:45. https://doi.org/10.1016/J.COBME.2018.10.003

Photo-Secondary Metabolites as a Potential Therapeutic Gateway for Alzheimer's Disease Management

10

Divya Sri Kathiresan, Vinayaga Moorthi Puthamohan, Sivamani Ganesan, and Arun Nagendran N.

Abstract

Waking up to the scientific world's advancements through its own miracles, it still relies on the ancient methods for many therapeutic studies and uses the medicinal plants for their impeccable results, where the World Health Organization (WHO) makes a statement that up to 80% of the developing countries use traditional medicines for the treatment of many diseases. The drugs and health products are mostly from the secondary metabolites (majorly polyphenols, alkaloids, and terpenoids) of the medicinal plants. Photobiomodulation and light therapy are the therapy techniques used non-invasively to treat many diseases, including Alzheimer. The light is not only used in treatments of humans, but it is also a major component of the medicinal plant's growth, marking the photoperiod (consisting of the quality, intensity, and duration of light), which marks the differential secondary metabolites. The flavonoids, plant polyphenols, are the major secondary metabolites used in Alzheimer's disease (AD) treatment, which are found to be increased using UV-B in species like *Artemisia annua*, *Prunella vulgaris*, *Withania somnifera*, and *Ixeris dentata* and through blue light in species like *Ocimum basilicum* and *Rodiola imbricata*. Curcumin, a phenolic compound from *Curcuma longa* supplementation, is found to increase the neuroprotective action and is used in AD treatment. Some of the secondary metabolites

D. S. Kathiresan · V. M. Puthamohan (✉)
Department of Human Genetics and Molecular Biology, Bharathiar University, Coimbatore, Tamil Nadu, India
e-mail: pvmhgmb@buc.edu.in

S. Ganesan
PG & Research Department of Zoology and Biotechnology, A.V.V.M. Sri Pushpam College, Poondi, Thanjavur, Tamil Nadu, India

N. Arun Nagendran
Department of Zoology, Thiagarajar College, Madurai, Coimbatore, Tamil Nadu, India

H. Dilnashin, S. P. Singh (eds.), *Phototherapeutic Approaches to Neurodegeneration*,
https://doi.org/10.1007/978-981-95-4751-7_10

from plants, phenolic acids like gallic acid, rosmarinic acid, and salvianolic acid B; flavonoids like luteolin, quercetin, myricetin, dihydromyricetin, fisetin, and silibinin; and curcumin are found to inhibit the amyloid beta fibrillogenesis and are used in the treatment of AD through thioflavin T assays showing the comparisons with thioflavin T and amyloid beta intensities, thus making the use of inhibiting properties of the secondary metabolites to function as a perfect therapy for AD and related diseases. This chapter will delve deeper into the role of medicinal plant-derived secondary metabolites and the influence of light-based modulation in enhancing their therapeutic potential, with a specific focus on their application in the treatment of AD.

Keywords

Alzheimer disease · Phototherapy · Secondary metabolites · Therapeutic potential · Treatment

10.1 Introduction

Dementia in its most common form which is present in almost 60–80% of the cases around the world is Alzheimer's disease (AD), causing the senile dementia owing to the chronic and progressive neurodegenerative disorder that affects the cognitive decline, causing the decline in reasoning, thinking, memory, problem-solving ability, and also language skills (Llibre-Guerra et al. 2025; Knopman et al. 2021). Although the pathophysiology of AD is complex, most of the AD complexities are investigated microscopically, showing the amyloid beta (Aβ) accumulation in the neuronal region and also the neurofibrillary tangles (NFT) caused by the tau proteins (Karaduran et al. 2025). The Global Burden of Diseases, injuries, and risk factors study (GBD), 2021 database, showed AD to be the major contributor for the global burden of disability-adjusted life years (DALYs) among the ones affecting the nervous system, accounting for 36.3 million, which is more than 150% of 1990, and it ranks eighth in the estimation of the global death rate within the age standardized mortality rate (ASMR), contributing to 1.96 million people (Alzheimer's Association 2024; GBD 2021 2024; GBD DALY 2024; He et al. 2025). The AD treatment conventionally done is using the choline esterase (ChE) inhibitors, which help in the removal or reduction of the amyloid plaques and increase the cognitive ability, where some of the ChE inhibitors are donepezil, rivastigmine, and galantamine (Rabins et al. 2007; Pathan 2023; Swanwick and Lawlor 1999). Many other drugs apart from this were employed, but it had a shorter half life, cholinergic side effects, and also caused hepatotoxicity in AD patients (Lublin and Gandy 2010); it was also found to have milder side effects with facial flushing, nausea, vomiting, and diarrhea (Benninghoff and Perneczky 2022; Mimica and Presečki 2009). Many new techniques for handling the disease and its progression have started involving the development of science and technology, where one such kind is phototherapy (Gao et al. 2025; Zeng et al. 2021). Most of the compounds used in the treatment of the disease lose their efficiency before reaching the precise region of the action of the drug; here, the photodynamic therapy using the light as a major component of action and the different forms of photoemmission make it precise to the tissues of interest, which also achieves deep tissue penetration (Öztürk et al. 2025; Ma et al. 2023). There are many plant compounds that act as a natural source of treatment strategies for the AD progression, which, when combined with phototherapy, lead to the application of precision medicine and less invasive procedures for effective management of the disease (Sulaiman et al. 2022; Meng-Zhen et al. 2022). Many studies use the thioflavin T assay (ThT) for the anti-aggregating studies of the Aβ, concentrating on the secondary metabolites produced from plant compounds and observations done on animal models to study the efficacy of them (Lee et al. 2015; Saha and Ahmad 2024; Witter et al. 2018). In this chapter, we will discuss the photo-secondary metabolites as potential therapeutic gateways for AD management.

10.2 AD Pathophysiology and Targets for Phototherapy

The major pathophysiological aspects during the diagnosis and treatment of AD (Fig. 10.1) are the Aβ aggregation and the plaque formation at the neuronal regions of the brain (Dubois et al. 2014; Hampel et al. 2021a, b), the tau hyperphosphorylation and the presence of NFT (Price et al. 2009; Bejanin et al. 2017), oxidative stress and mitochondrial dysfunction (Kathiresan et al. 2025; Chauhan and Chauhan 2006; Christen 2000), neuroinflammation and glial activation (Meda et al. 2001; Nordengen et al. 2019), and metal dyshomeostasis and redox imbalance (Hu and Wang 2025; Li et al. 2025). According to this, there are many biomarkers necessary for the AD pathology, namely the amyloid biomarker, tau biomarkers, neural damage biomarkers, neuroinflammation biomarkers, synaptic biomarkers, and metabolic biomarkers (Hampel et al. 2021a, b; Khan et al. 2020). The amyloid biomarkers consist of the APP, where it cleaves and forms the alpha and beta secretases and forms the Aβ, and the ratio of the Aβ42/Aβ40 is found to be the best biomarker, as the levels of them are found to be reduced in AD patients (Tiwari et al. 2019; Khan et al. 2020). There are two tau biomarkers, namely the T-tau and P-tau, where both proteins are found to be high in cases of AD at prodromal cases and AD with dementia; the P-tau proteins are found in high concentration at the CSF, which is a unique marker for AD, and the hyperphosphorylation of this P-tau protein leads to the formation of NFT in AD patients (Tiwari et al. 2019). There are neuronal biomarkers that are available, like neurofilament light, which detects the damage in the brain and shows the neurodegeneration but is also not specific for AD. Another neuronal marker is the S100β, which indicates the atrophy of the brain during its higher expression (Sheng et al. 1994; Nehra et al. 2025; Saavedra et al. 2025). The neuroinflammation biomarkers in AD are glial cell line-derived neurotrophic factor, which shows the activation of astrocytes, found in higher concentration in patients with preclinical AD (Gupta et al. 2025; Sil et al. 2025; Sirimaharaj et al. 2025); the elevated levels of pro-inflammatory cytokines like TNF-α in the CSF and blood

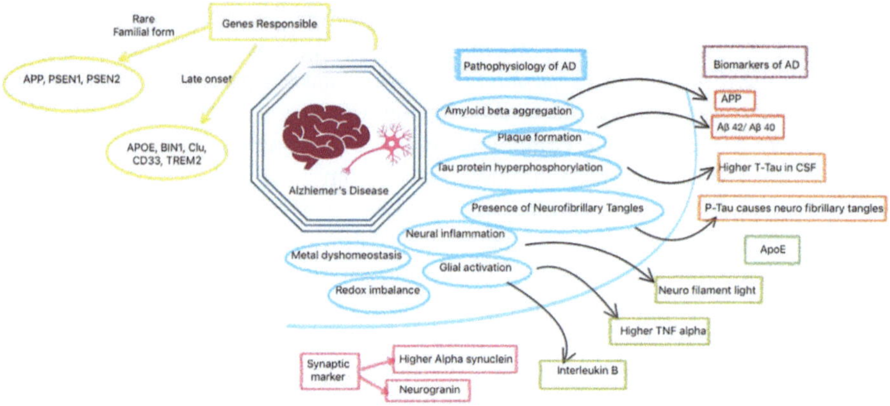

Fig. 10.1 Diagrammatic representation of Alzhiemer's Disease, its pathophysiology, the genes responsible and the biomarkers for diagnosis of the disease

plasma of AD patients (Azargoonjahrom et al. 2025; Tobinick et al. 2006); and IL-β which is the cause for Aβ and NFT formation in AD (Weisman et al. 2006; Hasegawa 2020). The synaptic biomarkers like α-synuclein, which is found elevated in patients with mild cognitive impairment in the CSF, and another synaptic marker, neurogranin, are found to be elevated in AD when dendritic degeneration occurs (Williams et al. 2013; Jaworski et al. 2011). One of the major biomarkers in the analysis of AD progression is the APOE, which is the major lipid transporter of the brain and leads to the synaptic defects and also the cognitive impairment (Farlow et al. 2004; Risacher et al. 2013). There are many genes responsible for the pathogenesis of AD, which have diverse functions and locations on the chromosome (Fan et al. 2019; Oveisgharan et al. 2018). The rarely occurring familial form of AD, which has less than a 1% occurrence rate, is caused by mutations in the genes *APP, PSEN1,* and *PSEN2;* the sporadic form of AD, which is the late onset AD that occurs in most of the population and is reported to be the most common type of AD affected (Cruchaga et al. 2012; Kabir et al. 2020), is due to the gene mutations in APOE, BIN1, Clu, CD33, and TREM2 (Chapuis et al. 2013; Gratuze et al. 2018; Tini et al. 2020). The major pathophysiologies of AD, such as Aβ aggregation, can cause the cytotoxicity of the neurons (Peng et al. 2022; Koffie et al. 2009; Mucke et al. 2000), which causes the identifiable changes in the synaptic number, shape, size, and composition, leading to the deficiency of the cognitive function (Koffie et al. 2009; Kopeikina et al. 2013). The synapse loss and dysfunction are inherited through the ApoE4 gene, causing the abnormality in glial cells, where their normal function of providing a neuroprotective role fails, leading to the neuroinflammation to rise, causing the AD pathology (Peng et al. 2022; Tiwari et al. 2019; Breijyeh and Karaman 2020).

The approach of targeting the treatment strategies through the application of photoactivation and the use of photo compounds is called phototherapy, which is also photobiomodulation, which uses the diverse lights to its use, like the artificial/

natural light sources, comprising the infrared light, laser light source, and ultraviolet light (Meduri et al. 2007; Zhu et al. 2022; Powell and Gach 2015). They are classified according to their source of lighting and their penetration, namely the full-spectrum irradiation, bright light irradiation, and monochromatic light irradiation (Barros et al. 2021; Kurz et al. 2023). Here the laser light, when given at a low level for laser therapy (100–200 mW), has the ability to produce low amounts of biological stimulation and causes changes in the functions of body parts, which is employed in the AD treatment for the reducing of plaque formation and also improving the cognitive development (Wang et al. 2017; Mei et al. 2024). Many investigations have found phototherapy to be effective in treating the conditions of the disease through low-intensity red or infrared light radiation, which is found to have increased the metabolism of the cell and reduced the oxidative stress, toxicity, and inflammation, which is of greater use for the glial cells in the case of AD treatment (Dyson 2006; Grossweiner et al. 2005). As phototherapy has effects on targeting the exact parts of the affected AD brain, through deep penetration, they are a promising strategy for the treatment and management of AD (Henderson and Morries 2015; Mallidi et al. 2016).

10.3 Photo-Secondary Metabolites

The organic compounds produced by the plants with low molecular weight that are responsible for the plant defense, protection, and regulation but not involved in the major fundamental growth are the secondary metabolites of the plant, some of them being the compounds of flavonoids, alkaloids, terpenoids, and phenolics (Bennett and Wallsgrove 1994; Bhatla and Lal 2023; Agostini-Costa et al. 2012). Light is the key factor for the activation, regulation, and expression of these compounds, which, upon varying the intensity, changes in the activation and suppression of different metabolic pathways, which is now employed in the treatment strategies and the development of therapies for many major diseases, including cancer (Hashim et al. 2021; Thoma et al. 2020). The photo-secondary metabolites are the metabolites that absorb the photons in the different types of light radiations and generate the ROS, sometimes undergo photoisomerization, or simply act as natural photosensitizers (Berni et al. 2019; Marslin et al. 2017; Chobot et al. 2014). Based on their chemical structure, photo physical properties and the biological mode of action, they are broadly classified as, (i) polyphenols and flavanoids, which contain the conjugated aromatic systems which absorb the UV light and act as the modulators of amyloid plaques (Bakhtiari et al. 2017; Baptista et al. 2014; Zhang and Yan 2023; Matos et al. 2017) and tau proteins (Kim et al. 2025; Rasouli et al. 2022), which are the antioxidants in the treatment of AD (eg. Resveratrol, quercetin, epigallocatechin gallate and luteolin); (ii) curcuminoids, which are the forms of curcumin that contains the strong beta diketone chromophore that are known to chelation of metals, and involved in the anti-inflammatory and photo dynamic therapy (Hamaguchi et al. 2010; Mishra and Palanivelu 2008), whose absorption range falls between

400-500 nm; (iii) the anthraquinones and naphthodianthrones, which is capable of releasing the singlet oxygen and has the natural photoactive property and mostly mentioned as the natural photosensitizers in the photodynmaic therapy (Wink 2015; Campora et al. 2021; Peng et al. 2023) (e.g., hypericin, emodin, and rhein); (iv) the porphyrinoids and chlorophyll derivatives, which are used in the deep tissue penetration as they are the major absorbers of visible and NIR, thus increasing the ROS (Woodburn et al. 1992; Ma et al. 2023) (e.g., Pheophorbide-a, chlorins, and bacteriochlorins); (v) coumarins and xanthones, which have the ability to adsorb the UV visible light range and undergo photoisomerization and cross-linking, which are responsible for the role of neuroprotection and antioxidant property (Vigueras et al. 2025; Munan et al. 2024; Bento et al. 2017) (e.g., psoralens and mangiferin), and (vi) alkaloids and terpenoids with photo-modifiable potential, which are the compounds that are not strongly photoactive in nature but are incorporated into nanocarriers that are photoresponsive or chemically modified by the activation of light (Lai Shi Min et al. 2022; Yu et al. 2018; Potbhare et al. 2023) (Fig. 10.2) (e.g., huperzine A, ginkgolides, and bilobalide).

Exposure to UV-B radiation triggers excessive ROS generation in plants, largely through NADPH oxidase activity, leading to oxidative stress, lipid peroxidation, and cellular damage (Sharma et al. 2017; Hideg et al. 2013). Studies in Arabidopsis, maize, soybean, Brassica, and Vigna species revealed that UV-B reduces photosynthetic efficiency, pollen germination, and leaf expansion while inducing ROS-mediated membrane damage and cell death (Nawkar et al. 2013; Zhu et al. 2017). To counteract this, plants activate enzymatic antioxidants such as superoxide

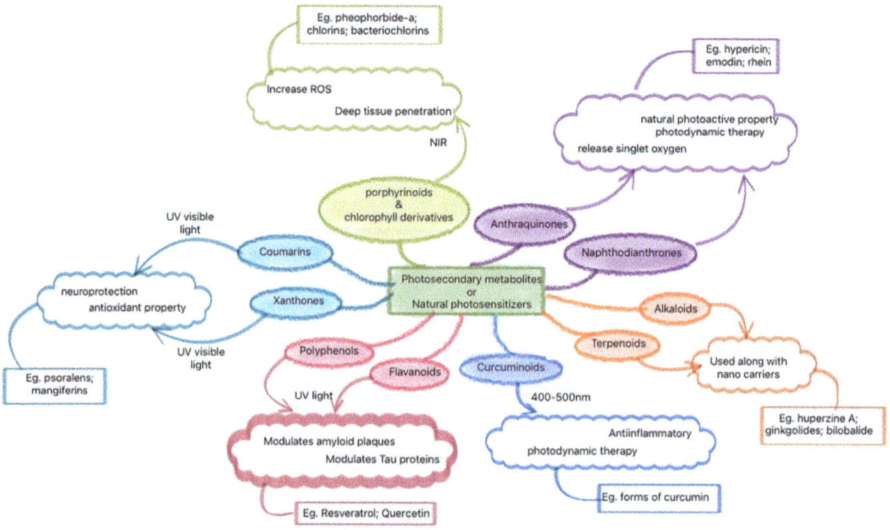

Fig. 10.2 The flow chart diagram showing the different photosecondary metabolities or the photosensitizers and its action over exposure of different light for the management and treatment of Alzhiemer's disease

dismutase, catalase, and ascorbate peroxidase, along with non-enzymatic defenses like flavonoids, carotenoids, proline, ascorbic acid, glutathione, and tocopherols. This balance between ROS production and scavenging is crucial for plant stress adaptation under UV-B exposure, which can also be used in AD management (Maugeri et al. 2022).

Elevated ROS, such as hydrogen peroxide (H_2O_2) and superoxide (O_2^-), can impair photosynthesis, membrane integrity, and pollen viability, leading to lipid peroxidation, DNA lesions, and growth inhibition (Dumont and Beal 2011; Mittler et al. 2022). To counteract this, plants upregulate enzymatic antioxidants, including catalase (CAT), superoxide dismutase (SOD), and ascorbate peroxidase (APX), as well as non-enzymatic scavengers such as glutathione, flavonoids, carotenoids, proline, and ascorbic acid. Interestingly, this balance between ROS-mediated cellular damage and adaptive antioxidant defense mirrors pathological and therapeutic processes in human neurodegenerative diseases such as AD (Delibas 2002). In AD, oxidative stress accelerates Aβ aggregation, synaptic dysfunction, and neuronal death, paralleling UV-B–induced ROS damage in plants. However, just as plants harness both enzymatic and non-enzymatic antioxidants to alleviate oxidative injury, therapeutic strategies for AD also emphasize ROS modulation. Phytochemicals like gallic acid, rosmarinic acid, salvianolic acid B, and luteolin have been shown to inhibit Aβ fibrillogenesis, reduce plaque burden, and enhance memory by acting as antioxidants and anti-aggregation agents (Hajipour et al. 2016; Andrade et al. 2022). Additionally, phototherapeutic approaches such as near-infrared light and photoactive nanoparticles exploit controlled ROS generation or modulation to disaggregate Aβ fibrils and restore neuronal function, representing a parallel to UV-driven stress signaling in plants that triggers protective pathways. Thus, both plant responses to UV-B radiation and AD treatments converge on the delicate balance between ROS toxicity and ROS signaling, where excessive accumulation drives damage but regulated production supports adaptation and therapeutic intervention. Understanding the shared oxidative framework not only deepens our insight into plant resilience but also reinforces the potential of antioxidant and phototherapy-based strategies in combating neurodegeneration.

10.4 Polyphenols and Flavonoids as Photo-Secondary Metabolites in AD Management

Commonly used methods to compare aggregation and Aβ fibril formation under varying conditions include the ThT assay and electron and atomic force microscopy (Khurana et al. 2005; Krebs et al. 2005; Freire et al. 2014). Other methods, such as sandwich ELISA and UV spectroscopy, quantify aggregation levels but cannot differentiate between oligomer and fibril forms (Klaver et al. 2011; Englund et al. 2007). ThT binds to Aβ fibrils and oligomers, but not monomers. Aβ fibrils, including Aβ(1–28) and Aβ(1–40), cause ThT to show a low-intensity absorption maximum at 450 nm and a high emission maximum at 482 nm. The ThT interacts with the beta-sheets of Aβ(1–42) molecules at amino acid positions 17–42. The natural plant-derived compounds, through the ThT assay, show the inhibitory activity of the chemical compounds for Aβ fibrillogenesis, which is found to aid in the treatment of AD (Hashimoto et al. 2010; Bruggink et al. 2012).

Some of the secondary metabolites obtained or extracted from plants, majorly medicinal plants, that are specifically identified to be the inhibitors of Aβ fibrillogenesis are gallic acid, rosmarinic acid, salvianolic acid B, luteolin, quercetin, fisetin, myricetin, dihydromyricetin, epigallocatechin-3-gallate (EGCG), silibinin, oleuropein, rutin, curcumin, crocin, cryptotanshinone, and tabersonine (Stefanescu et al. 2020). Gallic acid from Vitaceae, *Vitis vinifera,* which is a common grapevine, has the property of diminishing or blocking Aβ fibril formation and has the property of disaggregating the already formed fibrils, as demonstrated through ThT assays; in vivo it reduces Aβ(1–42) plaque size, improves spatial reference and working memory in a study on transgenic AD mice, and alleviates cognitive deficits (Liu et al. 2013; Yu et al. 2019). Rosmarinic acid from *Rosmarinus officinalis* from the family Lamiaceae, commonly known as mint, similarly has the property to block or reduce the formation of Aβ fibrils in a dose-dependent manner, as shown by the ThT assay, significantly reduces Aβ plaque numbers, improves memory, decreases A11-positive oligomers, and increases TBS-soluble Aβ levels in vivo (Porzoor et al. 2015; Sun et al. 2019). Salvianolic acid B from *Salvia miltiorrhiza* from the family of Lamiaceae, which is found profoundly in the roots of the plants, also diminishes Aβ fibril formation in a dose-dependent manner, decreases Aβ(1–42) and Aβ(1–40) levels in the hippocampus, and reduces the abilities of orientation, distance estimation, and route finding, which are often referred to as the spatial cognitive impairments, as studied in vivo (Durairajan et al. 2008; Xu et al. 2018). Likewise, luteolin from *Daucus carota* (Apiaceae) diminishes Aβ fibril formation as shown by the ThT assay, inhibits soluble Aβ(1–40) and Aβ(1–42) generation by 25% and 49%, respectively, and attenuates cognitive impairments in vivo (Akaishi et al. 2008; Churches et al. 2014; Rezai-Zadeh et al. 2005). Quercetin from *Malus domestica* (Rosaceae) has been shown to diminish Aβ fibril formation in ThT assays and to reduce both tauopathy and extracellular amyloidosis (Khan et al. 2019; Perret and Luo 2009; Sabogal-Guáqueta et al. 2015). Similarly, fisetin from *Fragaria moschata* from the family Rosaceae, which is a strawberry, has the ability to diminish the Aβ fibril formation and prevent the characteristic feature of progressive memory loss and learning impairments in AD patients (Akaishi et al. 2008; Currais et al. 2014). Myricetin, derived from *Vitis vinifera* (Vitaceae), exhibits dose-dependent inhibition of Aβ fibril formation in ThT assays, reduces A11-positive oligomers, and shows a tendency to decline the amount of Aβ plaque settling on the neurons (Hamaguchi et al. 2009; Akaishi et al. 2008; Shimmyo et al. 2008). Dihydromyricetin from the plant *Ampelopsis grossedentata* of the family Vitaceae, a popular vine tea plant of China, not only diminishes and blocks fibril formation in a dose-dependent manner but also disaggregates preformed fibrils, reduces spatial cognitive impairments, ameliorates behavioral deficits, and lowers hippocampal Aβ(1–42) and Aβ(1–40) levels in vivo (Jia et al. 2019; Liang et al. 2014; Ma et al. 2020).

Epigallocatechin gallate (EGCG) from *Thea sinensis* (Theaceae) also displays potent anti-amyloidogenic effects, diminishing and blocking fibril formation in a dose-dependent manner, disaggregating preformed fibrils as shown by ThT assays, reducing plaque formation, decreasing both soluble and insoluble Aβ(1–40) and Aβ(1–42) levels, and improving working memory performance (Adler et al. 2014; Li et al. 2006; Jia et al. 2013). Silibinin from *Silybum marianum* (Asteraceae) likewise diminishes fibril formation in a dose-dependent manner, markedly reduces the surface area of Aβ plaques, lowers acetylcholinesterase activity and levels, enhances synaptic protection, gliogenesis, and neurogenesis, and ameliorates cognitive deficits in vivo (Yin et al. 2011; Duan et al. 2015).

Oleuropein from *Olea europaea* (Oleaceae) diminishes Aβ fibril formation in a dose-dependent manner and reduces Aβ levels and plaque areas in the cortex and hippocampus (Omar et al. 2018; Grossi et al. 2013). Curcumin from *Curcuma longa* (Zingiberaceae) similarly diminishes and blocks fibril formation, disaggregates preformed fibrils, reduces soluble and insoluble Aβ levels, decreases plaque burden and astrocytic marker, a protein, glial fibrillary acidic protein expression at a lower gradient fraction, and decreases the expression of the A11 antibody that recognizes the prefibrillar or the fibrillary tangle proteins (Jiang et al. 2012; Lim et al. 2001). Rutin from *Malus domestica* (Rosaceae) shows comparable effects, diminishing fibril formation in a gradient way, reducing Aβ oligomer levels, attenuating memory deficits, decreasing microgliosis, astrocytosis, oxidative stress markers (glutathione peroxidase, malondialdehyde), and inflammatory mediators (interleukin-6, interleukin-1β), and increasing the cellular redox state by the measuring of glutathione and its reduced form as glutathione disulfide in its ratio (Wang et al. 2012; Xu et al. 2014; Hu et al. 2015).

In addition, crocin from *Crocus sativus* (Iridaceae) has been reported to diminish the grouping of fibrils and lower the levels of Aβ(1–40) and Aβ(1–42) by 25% and 29%, respectively (Ghahghaei et al. 2012; Batarseh et al. 2017). Cryptotanshinone from *Salvia miltiorrhiza* (Lamiaceae) further contributes by diminishing fibril formation, attenuating Aβ deposits, and ameliorating spatial learning and memory deficits in vivo (Mei et al. 2012). Collectively, these findings highlight that a diverse spectrum of natural polyphenols and related photo-secondary metabolites not only diminish or disaggregate Aβ fibrils but also reduce plaque burden, tauopathy, oxidative stress, and neuroinflammation, thereby offering strong preclinical evidence for their potential as multifaceted therapeutic agents in AD management.

10.5 Conclusion

Polyphenolic compounds derived from medicinal plants hold significant promise in modulating AD pathology through their ability to diminish or block Aβ fibril formation, disaggregate preformed fibrils, and reduce amyloid burden, as demonstrated by ThT assays and in vivo studies. These agents not only alleviate amyloid deposition but also exert broader neuroprotective effects, including suppression of oxidative stress, attenuation of neuroinflammatory cytokines, and promotion of synaptic resilience, neurogenesis, and cognitive performance. Compounds such as curcumin, EGCG, silibinin, and rutin highlight the potential of natural polyphenols to act as multi-target therapeutics against both amyloidogenesis and neuroinflammation. In parallel, the phototherapeutic strategies, including photodynamic therapy, photothermal modulation, and nanoparticle-assisted light activation, have emerged as innovative non-invasive approaches to disrupt Aβ aggregates, enhance clearance, and restore neuronal function. Recent advances combining phototherapy with nanomaterials further improve precision targeting, reduce fibrillar load, and minimize off-target effects, thereby complementing the biochemical actions of polyphenols. Thus, the integration of polyphenolic interventions and phototherapeutic approaches provides a synergistic platform for preventing and attenuating AD progression. While preclinical evidence is compelling, further clinical translation is needed to optimize dosage, bioavailability, light delivery parameters, and long-term efficacy. This combinatorial strategy could ultimately pave the way for safe, multi-modal therapies that slow or even reverse neurodegeneration in AD.

References

Adler BL, Yarchoan M, Hwang HM, Louneva N, Blair JA, Palm R et al (2014) Neuroprotective effects of the amylin analogue pramlintide on Alzheimer's disease pathogenesis and cognition. Neurobiol Aging 35(4):793–801

Agostini-Costa TDS, Vieira RF, Bizzo HR, Silveira D, Gimenes MA (2012) Secondary metabolites. In: Chromatography and its applications, vol 1, pp 131–164

Akaishi T, Morimoto T, Shibao M, Watanabe S, Sakai-Kato K, Utsunomiya-Tate N, Abe K (2008) Structural requirements for the flavonoid fisetin in inhibiting fibril formation of amyloid β protein. Neurosci Lett 444(3):280–285

Alzheimer's Association (2024) 2024 Alzheimer's disease facts and figures. Alzheimers Dement 20:3708–3821. https://doi.org/10.1002/alz.13809

Andrade S, Loureiro JA, Pereira MC (2022) Transferrin-functionalized liposomes for the delivery of gallic acid: a therapeutic approach for Alzheimer's disease. Pharmaceutics 14(10):2163

Azargoonjahromi A, Alzheimer's Disease Neuroimaging Initiative (2025) Plasma and CSF Amyloid-β42 predict plasma sortilin, which influences cognitive impairment via mediation of whole-brain volume: a 12-month longitudinal study across the Alzheimer's Disease spectrum. J Mol Neurosci 75(3):108

Bakhtiari M, Panahi Y, Ameli J, Darvishi B (2017) Protective effects of flavonoids against Alzheimer's disease-related neural dysfunctions. Biomed Pharmacother 93:218–229

Baptista FI, Henriques AG, Silva AM, Wiltfang J, da Cruz e Silva, O. A. (2014) Flavonoids as therapeutic compounds targeting key proteins involved in Alzheimer's disease. ACS Chem Neurosci 5(2):83–92

Barros NDM, Sbroglio LL, Buffara MDO, Baka JLCES, Pessoa ADS, Azulay-Abulafia L (2021) Phototherapy. An Bras Dermatol 96(4):397–407

Batarseh YS, Bharate SS, Kumar V, Kumar A, Vishwakarma RA, Bharate SB, Kaddoumi A (2017) Crocus sativus extract tightens the blood-brain barrier, reduces amyloid β load and related toxicity in 5XFAD mice. ACS Chem Neurosci 8(8):1756–1766

Bejanin A, Schonhaut DR, La Joie R, Kramer JH, Baker SL, Sosa N, Ayakta N, Cantwell A, Janabi M, Lauriola M et al (2017) Tau pathology and neurodegeneration contribute to cognitive impairment in Alzheimer's disease. Brain 140:3286–3300

Bennett RN, Wallsgrove RM (1994) Secondary metabolites in plant defence mechanisms. New Phytol 127(4):617–633

Benninghoff J, Perneczky R (2022) Anti-dementia medications and anti-alzheimer's disease drugs: side effects, contraindications, and interactions. In: NeuroPsychopharmacotherapy. Springer International Publishing, Cham, pp 2779–2788

Bento C, Gonçalves AC, Jesus F, Simões M, Silva LR (2017) Phenolic compounds: sources, properties and applications. In: Porter R, Parker N (eds) Bioactive compounds: sources, properties and applications, pp 271–299

Berni R, Luyckx M, Xu X, Legay S, Sergeant K, Hausman JF et al (2019) Reactive oxygen species and heavy metal stress in plants: impact on the cell wall and secondary metabolism. Environ Exp Bot 161:98–106

Bhatla SC, Lal MA (2023) Secondary metabolites. In: Plant physiology, development and metabolism. Springer Nature Singapore, Singapore, pp 765–808

Breijyeh Z, Karaman R (2020) Comprehensive review on Alzheimer's disease: causes and treatment. Molecules 25(24):5789

Bruggink KA, Müller M, Kuiperij HB, Verbeek MM (2012) Methods for analysis of amyloid-β aggregates. J Alzheimer's Dis 28(4):735–758

Campora M, Canale C, Gatta E, Tasso B, Laurini E, Relini A et al (2021) Multitarget biological profiling of new naphthoquinone and anthraquinone-based derivatives for the treatment of Alzheimer's disease. ACS Chem Neurosci 12(3):447–461

Chapuis J, Hansmannel F, Gistelinck M, Mounier A, Van Cauwenberghe C, Kolen KV et al (2013) Increased expression of BIN1 mediates Alzheimer genetic risk by modulating tau pathology. Mol Psychiatry 18(11):1225–1234

Chauhan V, Chauhan A (2006) Oxidative stress in Alzheimer's disease. Pathophysiology 13(3):195–208

Chobot V, Hadacek F, Kubicova L, Battino M (2014) Effects of selected dietary secondary metabolites on reactive oxygen species production caused by iron (II) autoxidation. Molecules 19(12):20023–20033

Christen Y (2000) Oxidative stress and Alzheimer disease. Am J Clin Nutr 71(2):621S–629S

Churches QI, Caine J, Cavanagh K, Epa VC, Waddington L, Tranberg CE et al (2014) Naturally occurring polyphenolic inhibitors of amyloid beta aggregation. Bioorg Med Chem Lett 24(14):3108–3112

Cruchaga C, Chakraverty S, Mayo K, Vallania FL, Mitra RD, Faber K et al (2012) Rare variants in APP, PSEN1 and PSEN2 increase risk for AD in late-onset Alzheimer's disease families. PLoS One 7(2):e31039

Currais A, Prior M, Dargusch R, Armando A, Ehren J, Schubert D et al (2014) Modulation of p25 and inflammatory pathways by fisetin maintains cognitive function in Alzheimer's disease transgenic mice. Aging Cell 13(2):379–390

Delibas N (2002) Malondialdehyde, superoxide dismutase, melatonin, iron, copper, and zinc blood concentrations in patients with Alzheimer disease: cross-sectional study. PubMed

Duan S, Guan X, Lin R, Liu X, Yan Y, Lin R et al (2015) Silibinin inhibits acetylcholinesterase activity and amyloid β peptide aggregation: a dual-target drug for the treatment of Alzheimer's disease. Neurobiol Aging 36(5):1792–1807

Dubois B, Feldman HH, Jacova C, Hampel H, Molinuevo JL, Blennow K et al (2014) Advancing research diagnostic criteria for Alzheimer's disease: the IWG-2 criteria. Lancet Neurol 13(6):614–629

Dumont M, Beal MF (2011) Neuroprotective strategies involving ROS in Alzheimer disease. Free Radic Biol Med 51(5):1014–1026

Durairajan SSK, Yuan Q, Xie L, Chan WS, Kum WF, Koo I et al (2008) Salvianolic acid B inhibits Aβ fibril formation and disaggregates preformed fibrils and protects against Aβ-induced cytotoxicty. Neurochem Int 52(4–5):741–750

Dyson M (2006) Primary, secondary, and tertiary effects of phototherapy: a review. In: Mechanisms for low-light therapy, vol 6140, pp 29–40

Englund H, Sehlin D, Johansson AS, Nilsson LN, Gellerfors P, Paulie S et al (2007) Sensitive ELISA detection of amyloid-β protofibrils in biological samples. J Neurochem 103(1):334–345

Fan J, Tao W, Li X, Li H, Zhang J, Wei D et al (2019) The contribution of genetic factors to cognitive impairment and dementia: apolipoprotein E gene, gene interactions, and polygenic risk. Int J Mol Sci 20(5):1177

Farlow MR, He Y, Tekin S, Xu J, Lane R, Charles HC (2004) Impact of APOE in mild cognitive impairment. Neurology 63(10):1898–1901

Freire S, de Araujo MH, Al-Soufi W, Novo M (2014) Photophysical study of Thioflavin T as fluorescence marker of amyloid fibrils. Dyes Pigments 110:97–105

Gao F, Hou Y, Wang Y, Liu L, Yi X, Xia N (2025) Photothermal and photodynamic strategies for diagnosis and therapy of Alzheimer's disease by modulating Amyloid-β aggregation. Biosensors 15(8):480

GBD 2021 Diseases and Injuries Collaborators (2024) Global incidence, prevalence, years lived with disability (YLDs), disability-adjusted life-years (DALYs), and healthy life expectancy (HALE) for 371 diseases and injuries in 204 countries and territories and 811 subnational locations, 1990–2021: a systematic analysis for the global burden of disease study 2021. Lancet. 403:2133–2161. https://doi.org/10.1016/S0140-6736(24)00757-8

GBD 2021 Nervous System Disorders Collaborators (2024) Global, regional, and national burden of disorders affecting the nervous system, 1990–2021: a systematic analysis for the global burden of disease study 2021. Lancet Neurol 23:344–381. https://doi.org/10.1016/S1474-4422(24)00038-3

Ghahghaei A, Bathaie SZ, Bahraminejad E (2012) Mechanisms of the effects of crocin on aggregation and deposition of Aβ1–40 fibrils in Alzheimer's disease. Int J Pept Res Ther 18(4):347–351

Gratuze M, Leyns CE, Holtzman DM (2018) New insights into the role of TREM2 in Alzheimer's disease. Mol Neurodegener 13(1):66

Grossi C, Rigacci S, Ambrosini S, Ed Dami T, Luccarini I, Traini C et al (2013) The polyphenol oleuropein aglycone protects TgCRND8 mice against Aß plaque pathology. PLoS One 8(8):e71702

Grossweiner LI, Grossweiner JB, Gerald Rogers BH (2005) The science of phototherapy: an introduction. Springer Netherlands, Dordrecht

Gupta A, Ghosh A, Sharma N, Gorain B (2025) Advancements in stem cell research for effective therapies against Alzheimer's disease: current investigation and future insight. Mol Neurobiol 62:1–25

Hajipour S, Sarkaki A, Farbood Y, Eidi A, Mortazavi P, Valizadeh Z (2016) Effect of gallic acid on dementia type of Alzheimer disease in rats: electrophysiological and histological studies. Basic Clin Neurosci 7(2):97

Hamaguchi T, Ono K, Yamada M (2010) Curcumin and Alzheimer's disease. CNS Neurosci Ther 16(5):285–297

Hamaguchi T, Ono K, Murase A, Yamada M (2009) Phenolic compounds prevent Alzheimer's pathology through different effects on the amyloid-β aggregation pathway. Am J Pathol 175(6):2557–2565

Hampel H, Cummings J, Blennow K, Gao P, Jack CR Jr, Vergallo A (2021a) Developing the ATX (N) classification for use across the Alzheimer disease continuum. Nat Rev Neurol 17(9):580–589

Hampel H, Hardy J, Blennow K, Chen C, Perry G, Kim SH et al (2021b) The amyloid-β pathway in Alzheimer's disease. Mol Psychiatry 26(10):5481–5503

Hasegawa M (2020) Structure of NFT: biochemical approach. In: Tau biology. Springer, pp 23–34

Hashim M, Ahmad B, Drouet S, Hano C, Abbasi BH, Anjum S (2021) Comparative effects of different light sources on the production of key secondary metabolites in plants in vitro cultures. Plants 10(8):1521

Hashimoto M, Bogdanovic N, Volkmann I, Aoki M, Winblad B, Tjernberg LO (2010) Analysis of microdissected human neurons by a sensitive ELISA reveals a correlation between elevated intracellular concentrations of Aβ42 and Alzheimer's disease neuropathology. Acta Neuropathol 119(5):543–554

He Z, Zhang H, Hu G, Qiao Y, Yin C, Li J et al (2025) The current status, trends, and challenges of Alzheimer's disease and other dementias in Asia (1990–2036). Front Public Health 13:1583339

Henderson TA, Morries LD (2015) Near-infrared photonic energy penetration: can infrared phototherapy effectively reach the human brain? Neuropsychiatr Dis Treat 11:2191–2208

Hideg É, Jansen MAK, Strid Å (2013) UV-B exposure, ROS, and stress: inseparable companions or loosely linked associates? Trends Plant Sci 18(2):107–115

Hu B, Dai F, Fan Z, Ma G, Tang Q, Zhang X (2015) Nanotheranostics: Congo red/Rutin-MNPs with enhanced magnetic resonance imaging and H2O2-responsive therapy of Alzheimer's disease in APPswe/PS1dE9 transgenic mice. Adv Mater 27(37):5499–5505

Hu J, Wang WX (2025) Lead induced intracellular copper redox imbalance with aggravated cellular dysfunction. Environ Pollut 382:126710

Jaworski T, Lechat B, Demedts D, Gielis L, Devijver H, Borghgraef P et al (2011) Dendritic degeneration, neurovascular defects, and inflammation precede neuronal loss in a mouse model for tau-mediated neurodegeneration. Am J Pathol 179(4):2001–2015

Jia L, Zhao W, Sang J, Wang W, Wei W, Wang Y et al (2019) Inhibitory effect of a flavonoid dihydromyricetin against Aβ40 amyloidogenesis and its associated cytotoxicity. ACS Chem Neurosci 10(11):4696–4703

Jia N, Han K, Kong JJ, Zhang XM, Sha S, Ren GR, Cao YP (2013) (−)-Epigallocatechin-3-gallate alleviates spatial memory impairment in APP/PS1 mice by restoring IRS-1 signaling defects in the hippocampus. Mol Cell Biochem 380(1):211–218

Jiang T, Zhi XL, Zhang YH, Pan LF, Zhou P (2012) Inhibitory effect of curcumin on the Al (III)-induced Aβ42 aggregation and neurotoxicity in vitro. Biochimica et Biophysica Acta (BBA) 1822(8):1207–1215

Kabir MT, Uddin MS, Setu JR, Ashraf GM, Bin-Jumah MN, Abdel-Daim MM (2020) Exploring the role of PSEN mutations in the pathogenesis of Alzheimer's disease. Neurotox Res 38(4):833–849

Karaduran K, Aydogdu A, Gelisin O, Gunpinar S (2025) Predictive value of complete blood count parameters for Alzheimer's disease in relation to periodontal status. Curr Alzheimer Res 22:302

Kathiresan DS, Balasubramani R, Marudhachalam K, Jaiswal P, Ramesh N, Sureshbabu SG et al (2025) Role of mitochondrial dysfunctions in neurodegenerative disorders: advances in mitochondrial biology. Mol Neurobiol 62(6):6827–6855

Khan H, Ullah H, Aschner M, Cheang WS, Akkol EK (2019) Neuroprotective effects of quercetin in Alzheimer's disease. Biomolecules 10(1):59

Khan S, Barve KH, Kumar MS (2020) Recent advancements in pathogenesis, diagnostics and treatment of Alzheimer's disease. Curr Neuropharmacol 18(11):1106–1125

Khurana R, Coleman C, Ionescu-Zanetti C, Carter SA, Krishna V, Grover RK et al (2005) Mechanism of thioflavin T binding to amyloid fibrils. J Struct Biol 151(3):229–238

Kim YS, Park TE, Eom DB, Ko MS, Lee CH, Park SY (2025) Flavonoids inhibiting Tau hyperphosphorylation: implications for Alzheimer's disease. Nat Prod Sci 31(2):75–83

Klaver AC, Coffey MP, Smith LM, Bennett DA, Finke JM, Dang L, Loeffler DA (2011) ELISA measurement of specific non-antigen-bound antibodies to Aβ1-42 monomer and soluble oligomers in sera from Alzheimer's disease, mild cognitively impaired, and noncognitively impaired subjects. J Neuroinflammation 8(1):93

Knopman DS, Amieva H, Petersen RC, Chételat G, Holtzman DM, Hyman BT et al (2021) Alzheimer disease. Nat Rev Dis Prim 7(1):33

Koffie RM, Meyer-Luehmann M, Hashimoto T, Adams KW, Mielke ML, Garcia-Alloza M et al (2009) Oligomeric amyloid β associates with postsynaptic densities and correlates with excitatory synapse loss near senile plaques. Proc Natl Acad Sci 106(10):4012–4017

Kopeikina KJ, Polydoro M, Tai HC, Yaeger E, Carlson GA, Pitstick R et al (2013) Synaptic alterations in the rTg4510 mouse model of tauopathy. J Comp Neurol 521(6):1334–1353

Krebs MR, Bromley EH, Donald AM (2005) The binding of thioflavin-T to amyloid fibrils: localisation and implications. J Struct Biol 149(1):30–37

Kurz B, Berneburg M, Bäumler W, Karrer S (2023) Phototherapy: theory and practice. JDDG J Dtsch Dermatol Ges 21(8):882–897

Lai Shi Min S, Liew SY, Chear NJY, Goh BH, Tan WN, Khaw KY (2022) Plant terpenoids as the promising source of cholinesterase inhibitors for anti-AD therapy. Biology 11(2):307

Lee JS, Lee BI, Park CB (2015) Photo-induced inhibition of Alzheimer's β-amyloid aggregation in vitro by rose bengal. Biomaterials 38:43–49

Li Q, Gordon M, Tan J, Morgan D (2006) Oral administration of green tea epigallocatechin-3-gallate (EGCG) reduces amyloid beta deposition in transgenic mouse model of Alzheimer's disease. Exp Neurol 198(2):576

Li SY, Gong XY, Ndikuryayo F, Yang WC (2025) The emerging role of mitochondria oxygen redox in pathological progression of neurodegenerative disorders. Ageing Res Rev 104:102660

Liang J, Kerstin Lindemeyer A, Shen Y, López-Valdés HE, Martínez-Coria H, Shao XM, Olsen RW (2014) Dihydromyricetin ameliorates behavioral deficits and reverses neuropathology of transgenic mouse models of Alzheimer's disease. Neurochem Res 39(6):1171–1181

Lim GP, Chu T, Yang F, Beech W, Frautschy SA, Cole GM (2001) The curry spice curcumin reduces oxidative damage and amyloid pathology in an Alzheimer transgenic mouse. J Neurosci 21(21):8370–8377

Liu Y, Pukala TL, Musgrave IF, Williams DM, Dehle FC, Carver JA (2013) Gallic acid is the major component of grape seed extract that inhibits amyloid fibril formation. Bioorg Med Chem Lett 23(23):6336–6340

Llibre-Guerra JJ, Fernandez MV, Joseph-Mathurin N, Bian S, Carter K, Li Y et al (2025) Longitudinal analysis of a dominantly inherited Alzheimer disease mutation carrier protected from dementia. Nat Med 31(4):1267–1275

Lublin AL, Gandy S (2010) Amyloid-β oligomers: possible roles as key neurotoxins in Alzheimer's disease. Mt Sinai J Med 77(1):43–49

Ma M, Wang J, Jiang H, Chen Q, Xiao Y, Yang H, Lin L (2023) Transcranial deep-tissue phototherapy for Alzheimer's disease using low-dose X-ray-activated long-afterglow scintillators. Acta Biomater 155:635–643

Ma Q, Cai S, Jia Y, Sun X, Yi J, Du J (2020) Effects of hot-water extract from vine tea (Ampelopsis grossedentata) on acrylamide formation, quality and consumer acceptability of bread. Foods 9(3):373

Mallidi S, Anbil S, Bulin AL, Obaid G, Ichikawa M, Hasan T (2016) Beyond the barriers of light penetration: strategies, perspectives and possibilities for photodynamic therapy. Theranostics 6(13):2458

Marslin G, Sheeba CJ, Franklin G (2017) Nanoparticles alter secondary metabolism in plants via ROS burst. Front Plant Sci 8:832

Matos AM, Cristovao JS, Yashunsky DV, Nifantiev NE, Viana AS, Gomes CM, Rauter AP (2017) Synthesis and effects of flavonoid structure variation on amyloid-β aggregation. Pure Appl Chem 89(9):1305–1320

Maugeri G, D'Amico AG, Giunta S, Giallongo C, Tibullo D, Bucolo C et al (2022) Activity-dependent neuroprotective protein (ADNP)-derived peptide (NAP) counteracts UV-B radiation-induced ROS formation in corneal epithelium. Antioxidants 11(1):128

Meda L, Baron P, Scarlato G (2001) Glial activation in Alzheimer's disease: the role of Aβ and its associated proteins. Neurobiol Aging 22(6):885–893

Meduri NB, Vandergriff T, Rasmussen H, Jacobe H (2007) Phototherapy in the management of atopic dermatitis: a systematic review. Photodermatol Photoimmunol Photomed 23(4):106–112

Mei X, Zou CJ, Zheng CY, Hu J, Zhou DS (2024) Effect of bright-light therapy on depression and anxiety of a patient with Alzheimer's disease combined with sleep disorder: a case report. World J Psychiatry 14(12):1982

Mei Z, Yan P, Situ B, Mou Y, Liu P (2012) Cryptotanshinione inhibits β-amyloid aggregation and protects damage from β-amyloid in SH-SY5Y cells. Neurochem Res 37(3):622–628

Meng-Zhen S, Ju L, Lan-Chun Z, Cai-Feng D, Shu-Da Y, Hao-Fei Y, Wei-Yan H (2022) Potential therapeutic use of plant flavonoids in AD and PD. Heliyon 8(11):e11440

Mimica N, Presečki P (2009) Side effects of approved antidementives. Psychiatr Danub 21(1):108–113

Mishra S, Palanivelu K (2008) The effect of curcumin (turmeric) on Alzheimer's disease: an overview. Ann Indian Acad Neurol 11(1):13–19

Mittler R, Zandalinas SI, Fichman Y, Van Breusegem F (2022) Reactive oxygen species signalling in plant stress responses. Nat Rev Mol Cell Biol 23(10):663–679
Mucke L, Masliah E, Yu GQ, Mallory M, Rockenstein EM, Tatsuno G et al (2000) High-level neuronal expression of Aβ1–42 in wild-type human amyloid protein precursor transgenic mice: synaptotoxicity without plaque formation. J Neurosci 20(11):4050–4058
Munan S, Chang YT, Samanta A (2024) Chronological development of functional fluorophores for bio-imaging. Chem Commun 60(5):501–521
Nawkar GM, Maibam P, Park JH, Sahi VP, Lee SY, Kang CH (2013) UV-induced cell death in plants. Int J Mol Sci 14(1):1608–1628
Nehra G, Maloney BJ, Smith RR, Chumboatong W, Abner EL, Nelson PT et al (2025) Plasma S100β is a predictor for pathology and cognitive decline in Alzheimer's disease. Fluids Barriers CNS 22(1):4
Nordengen K, Kirsebom BE, Henjum K, Selnes P, Gísladóttir B, Wettergreen M et al (2019) Glial activation and inflammation along the Alzheimer's disease continuum. J Neuroinflammation 16(1):46
Omar SH, Scott CJ, Hamlin AS, Obied HK (2018) Olive biophenols reduces alzheimer's pathology in SH-SY5Y cells and APPswe mice. Int J Mol Sci 20(1):125
Oveisgharan S, Buchman AS, Yu L, Farfel J, Hachinski V, Gaiteri C et al (2018) APOE ε2ε4 genotype, incident AD and MCI, cognitive decline, and AD pathology in older adults. Neurology 90(24):e2127–e2134
Öztürk B, Demir H, Günay MS, Akdag Y, Sahin S, Gulsun T (2025) Photodynamic and photothermal therapies using nanotechnology approach in Alzheimer's disease. Curr Neuropharmacol 23:1985
Pathan A (2023) Limitations of Alzheimer's disease medications. NeuroPharmac J:11–17
Peng L, Bestard-Lorigados I, Song W (2022) The synapse as a treatment avenue for Alzheimer's disease. Mol Psychiatry 27(7):2940–2949
Peng Z, Lu J, Liu K, Xie L, Wang Y, Cai C et al (2023) Hypericin as a promising natural bioactive naphthodianthrone: a review of its pharmacology, pharmacokinetics, toxicity, and safety. Phytother Res 37(12):5639–5656
Perret D, Luo ZD (2009) Targeting voltage-gated calcium channels for neuropathic pain management. Neurotherapeutics 6(4):679–692
Porzoor A, Alford B, Hügel HM, Grando D, Caine J, Macreadie I (2015) Anti-amyloidogenic properties of some phenolic compounds. Biomolecules 5(2):505–527
Potbhare MS, Barik R, Khobragade DS (2023) Management of alzheimer's disease: a review of herbal drugs having potential pharmacological and therapeutic activity. J Young Pharm 15(1):13–30
Powell JB, Gach JE (2015) Phototherapy in the elderly. Clin Exp Dermatol 40(6):605–610
Price JL, McKeel DW Jr, Buckles VD, Roe CM, Xiong C, Grundman M et al (2009) Neuropathology of nondemented aging: presumptive evidence for preclinical Alzheimer disease. Neurobiol Aging 30(7):1026–1036
Rabins PV, Blacker D, Rovner BW, Rummans T, Schneider LS, Tariot PN et al (2007) American Psychiatric Association practice guideline for the treatment of patients with Alzheimer's disease and other dementias. Am J Psychiatry 164(12 Suppl):5–56
Rasouli H, Hosseini Ghazvini SMB, Yarani R, Altıntaş A, Jooneghani SGN, Ramalho TC (2022) Deciphering inhibitory activity of flavonoids against tau protein kinases: a coupled molecular docking and quantum chemical study. J Biomol Struct Dyn 40(1):411–424
Rezai-Zadeh K, Shytle D, Sun N, Mori T, Hou H, Jeanniton D et al (2005) Green tea epigallocatechin-3-gallate (EGCG) modulates amyloid precursor protein cleavage and reduces cerebral amyloidosis in Alzheimer transgenic mice. J Neurosci 25(38):8807–8814
Risacher SL, Kim S, Shen L, Nho K, Foroud T, Green RC et al (2013) The role of apolipoprotein E (APOE) genotype in early mild cognitive impairment (E-MCI). Front Aging Neurosci 5:11
Saavedra J, Nascimento M, Figueira AJ, Oliveira da Silva MI, Gião T, Oliveira J et al (2025) Neuronal expression of S100B triggered by oligomeric Aβ peptide contributes to protection against cytoskeletal damage and synaptic loss. Front Mol Neurosci 18:1636365

Sabogal-Guáqueta AM, Muñoz-Manco JI, Ramírez-Pineda JR, Lamprea-Rodriguez M, Osorio E, Cardona-Gómez GP (2015) The flavonoid quercetin ameliorates Alzheimer's disease pathology and protects cognitive and emotional function in aged triple transgenic Alzheimer's disease model mice. Neuropharmacology 93:134–145
Saha P, Ahmad F (2024) Neuroprotective, anti-inflammatory and antifibrillogenic offerings by Emodin against Alzheimer's dementia: a systematic review. ACS Omega 9(7):7296–7309
Sharma S, Chatterjee S, Kataria S, Joshi J, Datta S, Vairale MG, Veer V (2017) A review on responses of plants to UV-B radiation related stress. In: UV-B radiation: from environmental stressor to regulator of plant growth. Wiley, pp 75–97
Sheng JG, Mrak RE, Griffin WST (1994) S100β protein expression in Alzheimer disease: potential role in the pathogenesis of neuritic plaques. J Neurosci Res 39(4):398–404
Shimmyo Y, Kihara T, Akaike A, Niidome T, Sugimoto H (2008) Multifunction of myricetin on Aβ: neuroprotection via a conformational change of Aβ and reduction of Aβ via the interference of secretases. J Neurosci Res 86(2):368–377
Sil M, Mukherjee N, Chatterjee I, Ghosh A, Goswami A (2025) Glial cells in Alzheimer's disease: pathogenic mechanisms and therapeutic frontiers. J Mol Neurosci 75(3):87
Sirimaharaj N, Thiankhaw K, Chattipakorn N, Chattipakorn SC (2025) Unveiling the protective roles of melatonin on glial cells in the battle against Alzheimer's disease—insights from in vivo and in vitro studies. Mol Neurobiol:1–13
Stefanescu R, Stanciu GD, Luca A, Paduraru L, Tamba BI (2020) Secondary metabolites from plants possessing inhibitory properties against beta-amyloid aggregation as revealed by thioflavin-T assay and correlations with investigations on transgenic mouse models of Alzheimer's disease. Biomolecules 10(6):870
Sulaiman C, George BP, Balachandran I, Abrahamse H (2022) Photoactive herbal compounds: a green approach to photodynamic therapy. Molecules 27(16):5084
Sun J, Jiang G, Shigemori H (2019) Inhibitory activity on amyloid aggregation of rosmarinic acid and its substructures from Isodon japonicus. Nat Prod Commun 14(5):1934578X19843039
Swanwick GR, Lawlor BA (1999) Initiating and monitoring cholinesterase inhibitor treatment for Alzheimer's disease. Int J Geriatr Psychiatry 14(4):244–248
Thoma F, Somborn-Schulz A, Schlehuber D, Keuter V, Deerberg G (2020) Effects of light on secondary metabolites in selected leafy greens: a review. Front Plant Sci 11:497
Tini G, Scagliola R, Monacelli F, La Malfa G, Porto I, Brunelli C, Rosa GM (2020) Alzheimer's disease and cardiovascular disease: a particular association. Cardiol Res Pract 2020(1):2617970
Tiwari S, Atluri V, Kaushik A, Yndart A, Nair M (2019) Alzheimer's disease: pathogenesis, diagnostics, and therapeutics. Int J Nanomedicine 14:5541–5554
Tobinick E, Gross H, Weinberger A, Cohen H (2006) TNF-alpha modulation for treatment of Alzheimer's disease: a 6-month pilot study. Medscape Gen Med 8(2):25
Vigueras G, Izquierdo-García E, de la Torre-Rubio E, Abad-Montero D, Santana MD, Marchán V, Ruiz J (2025) Metal–coumarin derivatives as promising photosensitizers: unlocking their cancer phototherapy potential. Inorg Chem Front 12(14):4355–4375
Wang SW, Wang YJ, Su YJ, Zhou WW, Yang SG, Zhang R et al (2012) Rutin inhibits β-amyloid aggregation and cytotoxicity, attenuates oxidative stress, and decreases the production of nitric oxide and proinflammatory cytokines. Neurotoxicology 33(3):482–490
Wang X, Han M, Wang Q, Zeng Y, Meng Q, Zhang J, Wei X (2017) A light therapy for treating Alzheimer's disease. In: Biophotonics and immune responses XII, vol 10065. SPIE, pp 129–134
Weisman D, Hakimian E, Ho GJ (2006) Interleukins, inflammation, and mechanisms of Alzheimer's disease. Vitam Horm 74:505–530
Williams PA, Thirgood RA, Oliphant H, Frizzati A, Littlewood E, Votruba M et al (2013) Retinal ganglion cell dendritic degeneration in a mouse model of Alzheimer's disease. Neurobiol Aging 34(7):1799–1806
Wink M (2015) Modes of action of herbal medicines and plant secondary metabolites. Medicine 2(3):251–286

Witter S, Witter R, Vilu R, Samoson A (2018) Medical plants and nutraceuticals for amyloid-β fibrillation inhibition. J Alzheimer's Dis Rep 2(1):239–252

Woodburn KW, Stylli S, Hill JS, Kaye AH, Reiss JA, Phillips DR (1992) Evaluation of tumour and tissue distribution of porphyrins for use in photodynamic therapy. Br J Cancer 65(3):321–328

Xu J, Wei K, Zhang G, Lei L, Yang D, Wang W et al (2018) Ethnopharmacology, phytochemistry, and pharmacology of Chinese Salvia species: a review. J Ethnopharmacol 225:18–30

Xu PX, Wang SW, Yu XL, Su YJ, Wang T, Zhou WW et al (2014) Rutin improves spatial memory in Alzheimer's disease transgenic mice by reducing Aβ oligomer level and attenuating oxidative stress and neuroinflammation. Behav Brain Res 264:173–180

Yin F, Liu J, Ji X, Wang Y, Zidichouski J, Zhang J (2011) Silibinin: a novel inhibitor of aβ aggregation. Neurochem Int 58(3):399–403

Yu B, Liu Y, Pan Y, Liu J, Wang H, Tang Z (2018) Light enhanced the biosynthesis of terpenoid indole alkaloids to meet the opening of cotyledons in process of photomorphogenesis of Catharanthus roseus. Plant Growth Regul 84(3):617–626

Yu M, Chen X, Liu J, Ma Q, Zhuo Z, Chen H et al (2019) Gallic acid disruption of Aβ1–42 aggregation rescues cognitive decline of APP/PS1 double transgenic mouse. Neurobiol Dis 124:67–80

Zeng F, Peng K, Han L, Yang J (2021) Photothermal and photodynamic therapies via NIR-activated nanoagents in combating Alzheimer's disease. ACS Biomater Sci Eng 7(8):3573–3585

Zhang Q, Yan Y (2023) The role of natural flavonoids on neuroinflammation as a therapeutic target for Alzheimer's disease: a narrative review. Neural Regen Res 18(12):2582–2591

Zhu G, Tong Q, Ye X, Li J, Zhou L, Sun P et al (2022) Phototherapy for cognitive function in patients with dementia: a systematic review and meta-analysis. Front Aging Neurosci 14:936489

Zhu X, Li N, Wang Y, Ding L, Chen H, Yu Y, Shi X (2017) Protective effects of quercetin on UVB irradiation induced cytotoxicity through ROS clearance in keratinocyte cells. Oncol Rep 37(1):209–218

Terpenoids and Alkaloids: Potential Therapeutic Agents in Neurology

11

Sumit Durgapal, Bhuwan Chandra Joshi, Sushmita Uniyal, and Aanchal Loshali

Abstract

Natural products like terpenoids and alkaloids that originate from plants are widely distributed in nature. In neurology, these phytoconstituents have great potential as therapeutic agents, especially when it concerns managing neurological conditions like epilepsy, Parkinson's disease, Alzheimer's disease and other neurodegenerative conditions. These compounds exhibit neuroprotective and other beneficial effects, making them valuable potential drug candidates. Terpenoids like eugenol, linalool, phytol and pinene have shown potential in modulation of inflammation and oxidative stress that are known as the major contributors to neurodegenerative disease progression. Many preclinical studies have shown potential of terpenoids in improving cognition, memory and mood. Traditional medicine has utilized alkaloids to treat a variety of diseases including neurological conditions. A number of alkaloids are also being studied for their neuroprotective potential such as ergotamine and galantamine, which are used therapeutically to treat postpartum haemorrhage and Alzheimer's disease, respectively. It can act as neuroprotective agents by increasing the neurotransmitter

S. Durgapal
Department of Pharmaceutics, Uttaranchal Institute of Pharmaceutical Sciences, Uttaranchal University, Dehradun, Uttarakhand, India

B. C. Joshi (✉)
Faculty of Pharmacy, Maharaja Agrasen Himalayan Garhwal University, Pokhra, Pauri Garhwal, Uttarakhand, India

S. Uniyal
Department of Pharmacology, School of Pharmaceutical Sciences, Shri Guru Ram Rai University, Dehradun, Uttarakhand, India

A. Loshali
Department of Pharmacognosy and Phytochemistry, School of Pharmaceutical Education and Research, Jamia Hamdard, New Delhi, India

H. Dilnashin, S. P. Singh (eds.), *Phototherapeutic Approaches to Neurodegeneration*,
https://doi.org/10.1007/978-981-95-4751-7_11

GABA, inhibiting acetylcholinesterase (AChE) enzyme and promoting cellular autophagy. Alkaloids and terpenoids can reduce inflammation by suppression of pro-inflammatory cytokines and alteration of inflammatory pathways. Moreover, they are known to modify the signalling route of the brain-derived neurotropic factor (BNDF) which is crucial for neuronal survival, neurogenesis and plasticity. The purpose of this chapter is to explore the role of terpenoids and alkaloids as phytochemicals in the investigation of various neurological disorders, evaluate their efficacy and look into their importance in preclinical research for the development of novel medications. Terpenoids and alkaloids present significant opportunities for the development of new therapeutic approaches in neurology, especially for neurodegenerative diseases because of their various modes of action and possible advantages.

Keywords

Neurodegeneration · Oxidative stress · Mechanisms · Therapeutic agents · Alkaloids · Terpenoids

11.1 Introduction

Neurodegeneration affects millions of people worldwide with limited effective treatments. Progressive neuronal death and cognitive decline are hallmarks of neurodegenerative illnesses including Alzheimer's (AD), Parkinson's (PD) and Huntington's disease (HD) (Battaglia et al. 2024). Most neurodegenerative diseases demonstrate intracellular protein aggregation or extracellular deposition (plaques). For example, HD has a huntingtin (Htt) protein with glutamine repeats, PD has α-synuclein (α-Syn), AD has amyloid-beta (Aβ)/tau aggregates, and amyotrophic lateral sclerosis (ALS) has superoxide dismutase (SOD) (Sirin et al. 2023). Table 11.1 outlines the primary cellular processes linked to the principal neurodegenerative diseases, including HD, PD, AD and ALS. A feature of numerous diseases is the aggregation of certain proteins in neurons which results in cellular dysfunction and ultimately neuronal death.

Table 11.1 Cellular mechanisms underlying major neurodegenerative disorders

Disease	Mechanisms	Aggregated proteins	Reference
HD	Proteostatic dysfunction and neuroinflammation	Mutant Htt	Ross and Tabrizi (2011), McColgan and Tabrizi (2018)
PD	Oxidative stress and mitochondrial dysfunction	α-Syn	Poewe et al. (2017), Zhan et al. (2024)
AD	Synaptic dysfunction and inflammation	Beta-amyloid and tau	Selkoe and Hardy (2016), Henstridge et al. (2019)
ALS	Oxidative stress, mitochondrial dysfunction and excitotoxicity	Various proteins, i.e. TDP-43 and SOD1	Taylor et al. (2016), Brown and Al-Chalabi (2017)

Millions of people worldwide suffer from neurodegeneration, which damages their social and psychological well-being. This illness specifically relates to symptoms that affect social skills, memory and logic (Agrawal 2020; Tanaka 2024). Numerous molecular and pathogenic processes are shared by neurological diseases at various phases of the disease's course (Muddapu et al. 2020). Neurological diseases are primarily caused by oxidative stress, mitochondrial dysfunction, protein misfolding/aggregation, synaptic dysfunction, calcium dysregulation, glutamate excitotoxicity, neuroinflammation, cell cycle dysregulation, dysfunction of the mitogen-activated protein kinase (MAPK) signalling pathways and, ultimately, neuronal degradation (Park et al. 2020; Muddapu et al. 2020; Agrawal 2020; Joseph et al. 2020; Harper and Wilkie 2003; Rekatsina et al. 2020) (Fig. 11.1).

The oxidative stress caused by an excess of reactive oxygen species (ROS) and reactive nitrogen species (RNS) leads to cellular damage, including neurological disorders (Tanaka and Vécsei 2024). The disabilities caused by these disorders have led to increasing attention on more translational research, i.e. from laboratory research to clinical practice. Natural compounds offer multi-target strategies with fewer side effects. Plant-derived bioactive chemicals, including terpenoids and alkaloids, have emerged as prospective neuroprotective agents due to their antioxidant, anti-inflammatory and signalling-modulatory activities. They have the ability to scavenge free radicals, which reduces oxidative stress while also suppressing

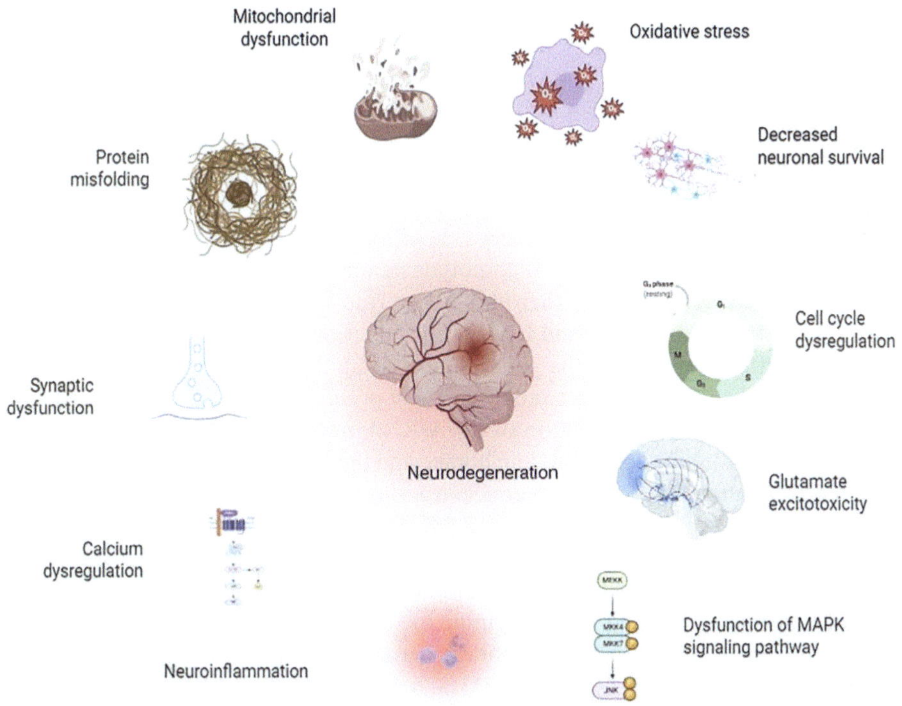

Fig. 11.1 Mechanistic insights into neurodegenerative disorders

pro-inflammatory cytokines. Their mechanism involves enhancement of synaptic plasticity and memory formation. Terpenoids and alkaloids are abundant in medicinal plants and have shown neuroprotective effects in preclinical studies. Several phytochemicals like ginkgolides (*Ginkgo biloba*), curcuminoids (*Curcuma longa*), limonene found in citrus oils have shown neuroprotective effects (de Lima et al. 2025; Utpal et al. 2025). There is evidence to show the clinical relevance of Huperzine A obtained from *Huperzia serrata* for Alzheimer's treatment by reversal inhibition of acetylcholinesterase (Wang et al., 2005). Also, another alkaloid Berberine obtained from rhizomes of *Coptis chinensis* shows promise in reducing amyloid-beta accumulation (Cheng et al., 2022). The pathogenesis of PD and AD is significantly influenced by transcription factors like nuclear factor κB (NF-κB), nuclear factor erythroid-2-related factor 2 (Nrf2) protein, peroxisome proliferator–activated receptor γ (PPARγ) (Kumar 2023; Jing et al. 2016). Chemical compounds isolated from natural plant sources target these transcription factors, thus playing an important role in the treatment of neurodegenerative diseases. There are several studies that prove that oxidative stress and mitochondrial dysfunction are also involved in progression of AD and PD. Some terpenoids and alkaloids have been considered in treating neurodegenerative diseases due to their ability to reduce inflammation and oxidative stress. This book chapter explores their mechanisms of action, therapeutic potential and future directions in neurological research.

11.2 Oxidative Stress

Oxidative stress causes a major imbalance in the body's antioxidant defence mechanisms and quantities of free radicals or their by-products. It affects the structure and function of biomolecules, which leads to the degeneration of neurone cells (Dasuri et al. 2013; Gandhi and Abramov 2012). The consequences of oxidative stress are particularly likely to affect neurons because of their high oxygen demand, low antioxidant (low glutathione) levels, high polyunsaturated fatty acid content and availability of redox active metals that play a key role in the generation of oxidative stress (Lee et al. 2020). ROS and RNS overproduction is associated with a number of adverse effects, including the peroxidation of lipids in the neuronal membrane, disruption of the interaction between lipids and proteins, inhibition of the activities of specific enzymes and oxidation of DNA/RNA molecules. These effects collectively lead to the inhibition of calcium uptake and the induction of glutamate release, which raises intracellular calcium concentration, triggers neuroinflammatory reactions and promotes apoptosis, neuronal injury or necrosis. This implies that the biology and progression of several neurological disorders are significantly influenced by oxidative stress (Xiao et al. 2008; Gan and Johnson 2014). Thus, a lot of research has been done so far to determine the exact part that oxidative stress plays in neurodegeneration. It has been reported that abnormal protein aggregates including superoxide dismutase 1 (SOD1) in ALS, α-Syn in PD, Aβ with hyperphosphorylated tau in AD and Htt in HD are the established marker of neurodegeneration. Moreover, transactivation response (TAR) DNA binding protein-43 (TDP-43) is associated with neuroinflammation, which in turn promotes ROS generation and oxidative stress (Schmidt et al. 2021). Oxidative stress studies in AD patients have shown elevated levels of several biomarkers. These include lipid peroxidation products (malondialdehyde [MDA], 4-hydroxynonenal [4-HNE], thiobarbituric acid reactive substances [TBARs], acrolein, and iso- and neuroprostanes), protein oxidation products (3-nitrotyrosine [3-NT] and protein carbonyl [PC]) and DNA/RNA oxidation products (8-hydroxy-2-deoxyguanosine [8-OHdG] and 8-hydroxyguanosine [8-OHG]). Similarly, PD patients have elevated levels of acrolein, MDA, HNE, TBARs, PC, 3-NT, 8-OHdG and 8-OHG, with the exception of iso- and neuroprostanes (Niedzielska et al. 2016; Barnham et al. 2004). In ALS patients, elevated biomarkers include MDA, HNE, TBARs, 3-NT and 8-OHdG, whereas HD patient is associated with increased levels of isoprostane, TBARs, 3-NT, 8-OHdG and neuron-specific enolases (Ciancarelli et al. 2015; Kumar and Ratan 2016).

11.3 Phytoconstituents Prevent Oxidative Stress in Neurodegenerative Diseases

11.3.1 Alkaloids

Alkaloids are low molecular weight nitrogen-containing chemical molecules that are mostly present in plants and are generated from amino acids. There are about 21,000 reported plant-derived alkaloids of about 27,000 alkaloids so far. Alkaloids possess the potential to treat neurodegenerative diseases by targeting the oxidative stress pathway due to their antioxidant potential (Peng et al. 2019; Dalimunthe et al. 2018). A wide range of alkaloids have been explored for their potential neuroprotective action. These phytoconstituents are primarily obtained from plants and serve as promising lead constituents for the development of neurological therapeutics. Piperine obtained from *Piper longum* L. has demonstrated neuroprotective action in the PD model (Li et al. 2022). Caffeine is a natural alkaloid from *Coffea arabica* (Rubiaceae) enriched with neuromodulatory effects (Yelanchezian et al. 2022). Huperzine A isolated from *Huperzia serrata* shows significant neuroprotective action and is considered a potential candidate for the treatment of AD (Friedli and Inestrosa 2021). Trigonelline obtained from *Trigonella foenum-graecum* L. has demonstrated the neuroprotective potential against various neurological diseases (Liang et al. 2023).

11.3.2 Terpenoids

Terpenes, a class of secondary metabolites, comprise about 30,000 molecules derived from isoprene (C_5H_8). Terpenoids are formed by the addition of functional groups to the hydrocarbon chains of isoprene derivatives. These compounds are classified according to the number of carbons in each molecule with the most prevalent in medicinal plants being monoterpenes (C_{10}), sesquiterpenes (C_{15}) and diterpenes (C_{20}) (Pham et al. 2024; Guesmi et al. 2017). Preclinical research has investigated terpenes for their therapeutic potential in neurodegenerative diseases and other disorders involving neuronal degeneration. Animal studies have demonstrated the strong neuroprotective actions for compounds such as borneol, camphene and limonene which primarily act through anti-inflammatory and antioxidant mechanisms both crucial in counteracting chronic inflammation and oxidative stress implicated in various disease development (Mony et al. 2022; Eddin et al. 2023). Terpenes from *Cannabis sativa* L. (*C. sativa*) are particularly valued for their numerous neuroprotective actions and are considered promising candidates for the development of drug formulations targeting neurological diseases. Immunomodulatory activity has been demonstrated by α-pinene, linalool, phytol and trans-nerolidol isolated from the *C. sativa* indicating their potential as anti-inflammatory agents for the treatment of neurodegenerative diseases (Blevins et al. 2022). Moreover, studies using cannabis-derived terpenoids have shown that α-pinene and β-pinene significantly protect neurons from Aβ toxicity. This effect is linked to Aβ1–42 fibrillization and density inhibition (Laws III and Smid 2024).

11.3.3 General Neuroprotective Mechanisms of Action of Terpenoids and Alkaloids

11.3.3.1 Antioxidant Activity and Inhibition of Inflammation

Oxidative stress is one of the principal reasons for neurodegenerative disorders such as AD and PD, which results from an imbalance between antioxidant defence and ROS. Numerous terpenoids are potent free radical scavengers, such as ginkgolides, carnosol and thymol. These compounds boost endogenous antioxidant enzymes like glutathione peroxidase (GPx) and SOD, scavenge ROS and inhibit lipid peroxidation (Ali et al. 2015). One of the features shared by most CNS diseases is chronic neuroinflammation induced by the activation of astrocytes and microglia. Terpenoids such as ursolic acid have been shown to reduce the secretion of pro-inflammatory cytokines (viz. TNF-α, IL-6, IL-1β) and enzymes (COX-2, iNOS) through inhibition of the NF-κB and MAPK signalling pathways (Wang et al. 2022). Such anti-inflammatory action is a primary mechanism in the protection of neurons. Alkaloids such as berberine and tetrandrine suppress neuroinflammation by inhibiting NF-κB and MAPK pathways. These compounds reduce the production of pro-inflammatory mediators and counteract oxidative stress, key drivers of neuronal damage (Kamila et al. 2025).

11.3.3.2 Modulation of Neurotransmitters

Some terpenoids can modulate neurotransmitter systems, such as GABAergic, cholinergic and dopaminergic systems. Linalool is an example that has been shown to enhance GABAergic transmission, which may be utilized to relieve anxiety and epilepsy (Milanos et al. 2017). It has been shown that ginkgolides from *Ginkgo biloba* can modulate cholinergic neurotransmission which is relevant for the treatment of AD (Singh et al. 2019). Several alkaloids have also been reported to stimulate neurotrophic factors such as brain-derived neurotrophic factor (BDNF), thereby promoting neuronal survival and synaptic plasticity. Phytochemicals have been reported to upregulate BDNF expression and synaptic density in models of depression and AD (Singh et al. 2025).

11.3.3.3 Regulation of Apoptosis and Protection Against Mitochondria

One of the most important factors for neuronal degeneration is mitochondrial dysfunction. Terpenoids like oleanolic acid can inhibit apoptosis by maintaining mitochondrial membrane potential, reducing excessive calcium and restricting cytochrome c release (Sapkota and Choi 2021). Excitotoxicity, primarily due to overactivation of glutamate receptors, plays a major role in acute CNS injuries and chronic neurodegeneration (Verma et al. 2022). Alkaloids such as huperzine A protect neurons by stabilizing mitochondrial function and preventing calcium-induced excitotoxic cascades (da Rosa et al. 2022).

11.3.4 Neuropharmacological Potential of Terpenoids

Ginkgolide B, which is derived from *Ginkgo biloba L.*, is widely known for its ability to reduce neuroinflammation by blocking the activation of the NLRP3 inflammasome, a crucial component of the inflammatory cascade. Moreover, it encourages microglia to polarize into the M2 state which is linked to neuroprotection and inflammation resolution (Zhang et al. 2021). Figure 11.2 illustrates the primary impacts of limonene and ginkgolides. It has been shown that ginkgolides effectively prevent neurodegeneration by blocking a number of signalling pathways linked to neuroinflammation and synaptic dysfunctions. Furthermore, limonene has positive effects on neuronal dysfunction and protein deposition.

The potential of some phytocompounds to lower oxidative stress and inflammation has led to their consideration in the treatment of neurodegenerative disorders. Mechanisms are summarized in Table 11.2. A number of phytocompounds are listed in the table together with their corresponding neuroprotective qualities showing how important these substances can be in the management and prevention of neurological disorders.

11.3.5 Clinical Insights on Phytoconstituents in Neurodegeneration

According to clinical investigations, the use of these phytoconstituents is associated with notable improvements in cognitive functions, memory, reduced disinhibition and irritability, and reduced aberrant behaviour, hallucinations and depression. Various clinical trials that assessed the effect of phytoconstituents on neurodegenerative diseases are displayed in Table 11.3.

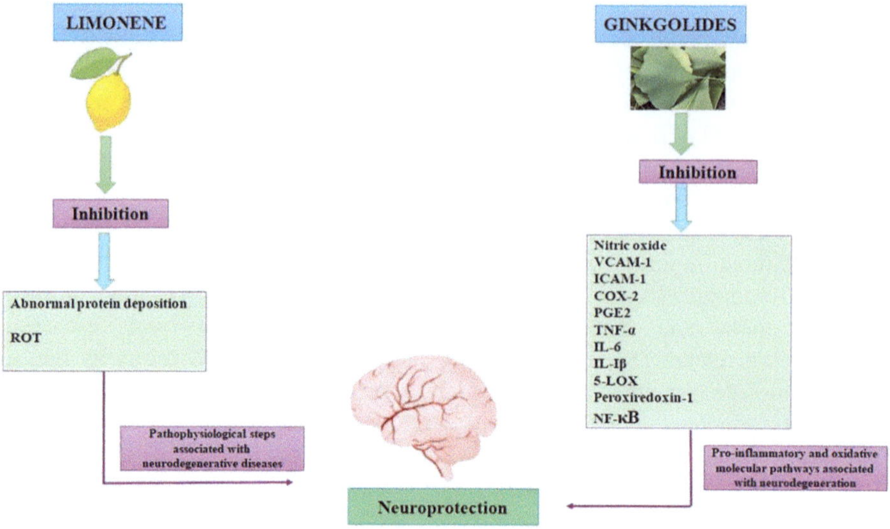

Fig. 11.2 Important oxidative and pro-inflammatory molecular pathways have a role in the development of neurodegenerative disorders during neurodegeneration. However, neurofunction may be enhanced by antioxidant and protective anti-inflammatory pathways. Brain disorders and protein deposition may be predisposed by imbalances between these two sets of processes. Ginkgolides play a vital role in eliminating a number of inflammatory chemicals, including nitric oxide, prostaglandin E2 (PGE2), cyclooxygenase-2 (COX-2), nuclear factor kappa-light-chain-enhancer of activated B cells (NF-κB), interleukin-6 (IL-6), interleukin-1beta (IL-1β) and cyclooxygenase-2 (COX-2). Additionally, it inhibits adhesion factors that are essential for defence cell activity such as vascular cell adhesion molecule 1 (VCAM-1) and intercellular adhesion molecule 1 (ICAM-1), hence reducing inflammatory processes in the brain. Additionally, limonene lessens locomotor deficits in PD by inhibiting rotenone (ROT), which is responsible for inhibition of complex-1 of the mitochondrial electron transport chain. Also, it prevents mitochondrial dysfunction and triggers the death of abnormal cells by inhibiting abnormal protein deposition

11.4 Future Perspectives

The traditional use of terpenoids and alkaloids for memory and cognitive enhancement and their diverse structures and function have gained them importance for investigation in alternative medicines for neurodegenerative disorders. The primary mechanisms by which these phytocompounds prevent or treat dementia include their anti-inflammatory, antioxidant and inhibitory effects on specific neurotransmitter-degrading enzymes, which elevate the level of and activity of other neurotransmitters, decrease neuroinflammation, protect defence cells like microglia, slow the progression of dementia and enhance pro- and anti-apoptotic abilities. The preparation of nanoparticles combined with phytochemicals that are capable of crossing BBB can prove to be effective in personalized therapies against neurodegenerative diseases.

Table 11.2 Neuroprotection mechanisms of phytoconstituents

Class of constituents	Phytoconstituents	Mechanisms	References
Terpenoids	Ginkgolide B	Mitigates neuroinflammation by inhibiting the activation of the NOD-like receptor protein 3 (NLRP3) inflammasome and facilitating the polarization of microglia towards the M2 phenotype	Zhang et al. (2021)
	1,8-cineole; α-terpineol	TNFα and IL-6 suppression for anti-inflammatory action; oxidative stress biomarker regulation	Ayala-Ruiz et al. (2022)
	Limonene	Reduces PD locomotor impairments by rotenone (ROT) inhibition	Eddin et al. (2023)
	β-Caryophyllene	TGF-β1 and nuclear factor kappa B (NF-κB) are negatively modulated; also has antioxidant and anti-inflammatory action	Rajab et al. (2022)
	α-Bisabolol and β-caryophyllene	Shown to have neuroprotective activity on NSC-34 cells from Aβ-induced neurotoxicity	Laws III et al. (2022)
Alkaloids	Berberine	NLRP3 (nucleotide-binding oligomerization domain–containing protein 3) inflammasome inactivation	Sun et al. (2024), Coppinger et al. (2024)
	Caffeine	Inhibits the A1R, A2AR, A2BR and A3R adenosine receptors (ARs)	Merighi et al. (2023)
	Allocryptopine	Modulation of the Akt/GSK-3β signalling cascade; reduction in OS-induced tau hyperphosphorylation	Dolanbay et al. (2021), Dolanbay and Aslim (2021)
	Huperzine A	Inhibits pyroptosis of endothelial cells caused by subarachnoid haemorrhage and thus ameliorates neurological deficits	Tripathi et al. (2024), Hu et al. (2024)
	Galantamine	Improves cognitive behaviour	Saito et al. (2019)

11.5 Conclusion

Terpenoids and alkaloids are abundant in medicinal plants, and their multifaceted actions make them ideal candidates for developing therapies against neurodegenerative diseases. Continued research is essential to unlock their full therapeutic potential. The effectiveness and safety of terpenoids and alkaloids in treating neurological disorders in humans require more investigation. Combining terpenoids and alkaloids may offer enhanced neuroprotection by showing synergistic effects. Translational research and clinical trials to validate efficacy of these constituents are needed. Future avenues for their use to treat human illnesses are also provided by the investigation of innovative delivery technologies (such as nanocarriers) to

Table 11.3 Phytoconstituent-based clinical trials in neurodegenerative disorders

Phytoconstituents	Clinical trial details	References
Cannabidiol	*NCT06542445*—a randomized phase 2 clinical trial was carried out to ascertain the effectiveness and safety of cannabidiol oil in children with ADHD	ClinicalTrials.gov. U.S. National Library of Medicine (2020). ClinicalTrials.gov.
	NCT05233761—a randomized phase 2 clinical trial was conducted to evaluate the effectiveness of oral cannabidiol-terpene formulation on people suffering from insomnia	
Caffeine	*NCT01190735*—a randomized phase 2 clinical trial aimed to access both the safety and effectiveness of caffeine in managing patients suffering from PD	
Huperzine A	*Double-blind study*—huperzine A capsules and tablets for treatment in AD patients	Xu et al. (1999)

increase the formulation's bioavailability. In conclusion, because of their various modes of action and their advantages, terpenoids and alkaloids present encouraging opportunities for creating innovative therapeutic treatments in neurology, especially for neurodegenerative illnesses and mental health conditions.

References

Agrawal M (2020) Molecular basis of chronic neurodegeneration. In: Clinical molecular medicine. Academic Press, pp 447–460

Ali T, Badshah H, Kim TH, Kim MO (2015) Melatonin attenuates D-galactose-induced memory impairment, neuroinflammation and neurodegeneration via RAGE/NF-KB/JNK signaling pathway in an aging mouse model. J Pineal Res 58(1):71–85

Ayala-Ruiz LA, Ortega-Pérez LG, Piñón-Simental JS, Magana-Rodriguez OR, Meléndez-Herrera E, Rios-Chavez P (2022) Role of the major terpenes of Callistemon citrinus against the oxidative stress during a hypercaloric diet in rats. Biomed Pharmacother 153:113505

Barnham KJ, Masters CL, Bush AI (2004) Neurodegenerative diseases and oxidative stress. Nat Rev Drug Discov 3(3):205–214

Battaglia S, Avenanti A, Vécsei L, Tanaka M (2024) Neurodegeneration in cognitive impairment and mood disorders for experimental, clinical and translational neuropsychiatry. Biomedicine 12(3):574

Blevins LK, Bach AP, Crawford RB, Zhou J, Henriquez JE, Rizzo MD et al (2022) Evaluation of the anti-inflammatory effects of selected cannabinoids and terpenes from cannabis sativa employing human primary leukocytes. Food Chem Toxicol 170:113458

Brown RH, Al-Chalabi A (2017) Amyotrophic lateral sclerosis. N Engl J Med 377(2):162–172

Cheng Z, Kang C, Che S, Su J, Sun Q, Ge T, Cui R (2022) Berberine: a promising treatment for neurodegenerative diseases. Frontiers in pharmacology, 13:845591. https://doi.org/10.3389/fphar.2022.845591

Ciancarelli I, De Amicis D, Di Massimo C, Sandrini G, Pistarini C, Carolei A, Ciancarelli MGT (2015) Influence of intensive multifunctional neurorehabilitation on neuronal oxidative damage in patients with Huntington's disease. Funct Neurol 30(1):47

ClinicalTrials.gov (2020) U.S. National Library of Medicine. Https://ClinicalTrials.gov

Coppinger C, Pomales B, Movahed MR, Marefat M, Hashemzadeh M (2024) Berberine: a multi-target natural PCSK9 inhibitor with the potential to treat diabetes, Alzheimer's, cancer and cardiovascular disease. Curr Rev Clin Exp Pharmacol 19(4):312–326

Dalimunthe A, Hasibuan PAZ, Silalahi J, Sinaga SF, Satria D (2018) Antioxidant activity of alkaloid compounds from Litsea cubeba Lour. Orient J Chem 34(2):1149

Dasuri K, Zhang L, Keller JN (2013) Oxidative stress, neurodegeneration, and the balance of protein degradation and protein synthesis. Free Radic Biol Med 62:170–185

Dolanbay SN, Aslim B (2021) Alkaloid extracts from Glaucium corniculatum attenuates oxidative stress-induced tau hyperphosphorylation via regulating AKT/GSK-3β signaling pathway in differentiated PC12 cells. In: International Marmara Sciences Congress (Spring 2021), p 77

Dolanbay SN, Kocanci FG, Aslim B (2021) Neuroprotective effects of allocryptopine-rich alkaloid extracts against oxidative stress-induced neuronal damage. Biomed Pharmacother 140:111690

Eddin LB, Azimullah S, Jha NK, Nagoor Meeran MF, Beiram R, Ojha S (2023) Limonene, a monoterpene, mitigates rotenone-induced dopaminergic neurodegeneration by modulating neuroinflammation, hippo signaling and apoptosis in rats. Int J Mol Sci 24(6):5222

Friedli MJ, Inestrosa NC (2021) Huperzine A and its neuroprotective molecular signaling in Alzheimer's disease. Molecules 26(21):6531

Gan L, Johnson JA (2014) Oxidative damage and the Nrf2-ARE pathway in neurodegenerative diseases. Biochim Biophys Acta (BBA) 1842(8):1208–1218

Gandhi S, Abramov AY (2012) Mechanism of oxidative stress in neurodegeneration. Oxid Med Cell Longev 2012(1):428010

Guesmi F, Prasad S, Tyagi AK, Landoulsi A (2017) Antinflammatory and anticancer effects of terpenes from oily fractions of Teucruim alopecurus, blocker of IκBα kinase, through downregulation of NF-κB activation, potentiation of apoptosis and suppression of NF-κB-regulated gene expression. Biomed Pharmacother 95:1876–1885

Harper SJ, Wilkie N (2003) MAPKs: new targets for neurodegeneration. Expert Opin Ther Targets 7(2):187–200

Henstridge CM, Hyman BT, Spires-Jones TL (2019) Beyond the neuron–cellular interactions early in Alzheimer disease pathogenesis. Nat Rev Neurosci 20(2):94–108

Hu Q, Zhang R, Dong X, Yang D, Yu W, Du Q (2024) Huperzine A ameliorates neurological deficits after spontaneous subarachnoid hemorrhage through endothelial cell pyroptosis inhibition: role of Huperzine A in neurological deficit. Acta Biochim Biophys Sin 56(4):645

Jing X, Wei X, Ren M, Wang L, Zhang X, Lou HJNR (2016) Neuroprotective effects of tanshinone I against 6-OHDA-induced oxidative stress in cellular and mouse model of Parkinson's disease through upregulating Nrf2. Neurochem Res 41:779–786. https://doi.org/10.1007/s11064-015-1751-6

Joseph C, Mangani AS, Gupta V, Chitranshi N, Shen T, Dheer Y et al (2020) Cell cycle deficits in neurodegenerative disorders: uncovering molecular mechanisms to drive innovative therapeutic development. Aging Dis 11(4):946

Kamila P, Kar K, Chowdhury S, Chakraborty P, Dutta R, Singh A, Prajapati BG (2025) Effect of neuroinflammation on the progression of Alzheimer's disease and its significant ramifications for novel anti-inflammatory treatments. IBRO Neurosci Rep 18:771–782

Kumar G (2023) Editorial: neuroprotective mechanisms by phytochemicals in neurological disorders. Front Neurosci 17:1149639. https://doi.org/10.3389/fnins.2023.1149639

Kumar A, Ratan RR (2016) Oxidative stress and Huntington's disease: the good, the bad, and the ugly. J Huntington's Dis 5(3):217–237

Laws JS III, Smid SD (2024) Characterizing cannabis-prevalent terpenes for neuroprotection reveal a role for α and β-pinenes in mitigating amyloid β-evoked neurotoxicity and aggregation in vitro. Neurotoxicology 100:16–24

Laws JS III, Shrestha S, Smid SD (2022) Cannabis terpenes display variable protective and anti-aggregatory actions against neurotoxic β amyloid in vitro: highlighting the protective bioactivity of α-bisabolol in motorneuronal-like NSC-34 cells. Neurotoxicology 90:81–87

Lee KH, Cha M, Lee BH (2020) Neuroprotective effect of antioxidants in the brain. Int J Mol Sci 21(19):7152

Li R, Lu Y, Zhang Q, Liu W, Yang R, Jiao J et al (2022) Piperine promotes autophagy flux by P2RX4 activation in SNCA/α-synuclein-induced Parkinson disease model. Autophagy 18(3):559–575

Liang Y, Dai X, Cao Y, Wang X, Lu J, Xie L et al (2023) The neuroprotective and antidiabetic effects of trigonelline: a review of signaling pathways and molecular mechanisms. Biochimie 206:93–104

de Lima EP, Laurindo LF, Catharin VCS, Direito R, Tanaka M, Jasmin Santos German I, Lamas CB, Guiguer EL, Araújo AC, Fiorini AMR, Barbalho SM (2025) Polyphenols, alkaloids, and terpenoids against neurodegeneration: evaluating the neuroprotective effects of phytocompounds through a comprehensive review of the current evidence. Metabolites 15(2):124. https://doi.org/10.3390/metabo15020124

McColgan P, Tabrizi SJ (2018) Huntington's disease: a clinical review. Eur J Neurol 25(1):24–34

Merighi S, Travagli A, Nigro M, Pasquini S, Cappello M, Contri C et al (2023) Caffeine for prevention of Alzheimer's disease: is the A2A adenosine receptor its target? Biomolecules 13(6):967

Milanos S, Elsharif SA, Janzen D, Buettner A, Villmann C (2017) Metabolic products of linalool and modulation of GABAA receptors. Front Chem 5:46

Mony TJ, Elahi F, Choi JW, Park SJ (2022) Neuropharmacological effects of terpenoids on preclinical animal models of psychiatric disorders: a review. Antioxidants 11(9):1834

Muddapu VR, Dharshini SAP, Chakravarthy VS, Gromiha MM (2020) Neurodegenerative diseases–is metabolic deficiency the root cause? Front Neurosci 14:213

Niedzielska E, Smaga I, Gawlik M, Moniczewski A, Stankowicz P, Pera J, Filip M (2016) Oxidative stress in neurodegenerative diseases. Mol Neurobiol 53(6):4094–4125

Park G, Nhan HS, Tyan SH, Kawakatsu Y, Zhang C, Navarro M, Koo EH (2020) Caspase activation and caspase-mediated cleavage of APP is associated with amyloid β-protein-induced synapse loss in Alzheimer's disease. Cell Rep 31(13):107839

Peng J, Zheng TT, Li X, Liang Y, Wang LJ, Huang YC, Xiao HT (2019) Plant-derived alkaloids: the promising disease-modifying agents for inflammatory bowel disease. Front Pharmacol 10:351

Pham TH, Ganeshalingam M, Thomas R (2024) A multimodal analytical approach is important in accurately assessing terpene composition in edible essential oils. Food Chem 454:139792

Poewe W, Seppi K, Tanner CM, Halliday GM, Brundin P, Volkmann J et al (2017) Parkinson disease. Nat Rev Dis Prim 3(1):1–21

Rajab BS, Albukhari TA, Khan AA, Refaat B, Almehmadi SJ, Nasreldin N et al (2022) Antioxidative and anti-inflammatory protective effects of β-Caryophyllene against amikacin-induced nephrotoxicity in rat by regulating the Nrf2/AMPK/AKT and NF-κB/TGF-β/KIM-1 molecular pathways. Oxidative Med Cell Longev 2022(1):4212331

Rekatsina M, Paladini A, Piroli A, Zis P, Pergolizzi JV, Varrassi G (2020) Pathophysiology and therapeutic perspectives of oxidative stress and neurodegenerative diseases: a narrative review. Adv Ther 37(1):113–139

da Rosa MM, de Amorim LC, de Oliveira Alves JV, da Silva Aguiar IF, da Silva Oliveira FG, da Silva MV, dos Santos MTC (2022) The promising role of natural products in Alzheimer's disease. Brain Disorders 7:100049

Ross CA, Tabrizi SJ (2011) Huntington's disease: from molecular pathogenesis to clinical treatment. Lancet Neurol 10(1):83–98

Saito T, Hisahara S, Iwahara N, Emoto MC, Yokokawa K, Suzuki H et al (2019) Early administration of galantamine from preplaque phase suppresses oxidative stress and improves cognitive behavior in APPswe/PS1dE9 mouse model of Alzheimer's disease. Free Radic Biol Med 145:20–32

Sapkota A, Choi JW (2021) Oleanolic acid provides neuroprotection against ischemic stroke through the inhibition of microglial activation and NLRP3 inflammasome activation. Biomol Ther 30(1):55

Schmidt MF, Gan ZY, Komander D, Dewson G (2021) Ubiquitin signalling in neurodegeneration: mechanisms and therapeutic opportunities. Cell Death Differ 28(2):570–590

Selkoe DJ, Hardy J (2016) The amyloid hypothesis of Alzheimer's disease at 25 years. EMBO Mol Med 8(6):595–608

Singh SK, Srivastav S, Castellani RJ, Plascencia-Villa G, Perry G (2019) Neuroprotective and antioxidant effect of Ginkgo biloba extract against AD and other neurological disorders. Neurotherapeutics 16(3):666–674
Singh AA, Katiyar S, Song M (2025) Phytochemicals targeting BDNF signaling for treating neurological disorders. Brain Sci 15(3):252
Sirin S, Dolanbay SN, Aslim B (2023) Role of plant derived alkaloids as antioxidant agents for neurodegenerative diseases. Health Sci Rev 6:100071
Sun J, Zeng Q, Wu Z, Huang L, Sun T, Ling C et al (2024) Berberine inhibits NLRP3 inflammasome activation and proinflammatory macrophage M1 polarization to accelerate peripheral nerve regeneration. Neurotherapeutics 21(4):e00347
Tanaka M (2024) Beyond the boundaries: transitioning from categorical to dimensional paradigms in mental health diagnostics. Adv Clin Exp Med 33(12):1295–1301
Tanaka M, Vécsei L (2024) A decade of dedication: pioneering perspectives on neurological diseases and mental illnesses. Biomedicine 12(5):1083
Taylor JP, Brown RH Jr, Cleveland DW (2016) Decoding ALS: from genes to mechanism. Nature 539(7628):197–206
Tripathi PN, Lodhi A, Rai SN, Nandi NK, Dumoga S, Yadav P et al (2024) Review of pharmacotherapeutic targets in Alzheimer's disease and its management using traditional medicinal plants. Degener Neurol Neuromuscul Dis 14:47–74
Utpal BK, Al Amin M, Zehravi M et al (2025) Alkaloids as neuroprotectors: targeting signaling pathways in neurodegenerative diseases. Mol Cell Biochem 480:4587–4612. https://doi.org/10.1007/s11010-025-05258-3
Verma M, Lizama BN, Chu CT (2022) Excitotoxicity, calcium and mitochondria: a triad in synaptic neurodegeneration. Transl Neurodegenerat 11(1):3
Wang R, & Tang XC (2005) Neuroprotective effects of huperzine A: a natural cholinesterase inhibitor for the treatment of Alzheimer's disease. Neurosignals, 14(1–2), 71–82. https://doi.org/10.1159/000085387
Wang N, Wang E, Wang R, Muhammad F, Li T, Yue J et al (2022) Ursolic acid ameliorates amyloid β-induced pathological symptoms in Caenorhabditis elegans by activating the proteasome. Neurotoxicology 88:231–240
Xiao X, Liu J, Hu J, Zhu X, Yang H, Wang C, Zhang Y (2008) Protective effects of protopine on hydrogen peroxide-induced oxidative injury of PC12 cells via Ca2+ antagonism and antioxidant mechanisms. Eur J Pharmacol 591(1–3):21–27
Xu SS, Cai ZY, Qu ZW, Yang RM, Cai YL, Wang GQ et al (1999) Huperzine-A in capsules and tablets for treating patients with Alzheimer disease. Zhongguo Yao li xue bao= Acta Pharmacol Sin 20(6):486–490
Yelanchezian MYM, Waldvogel HJ, Faull RL, Kwakowsky A (2022) Neuroprotective effect of caffeine in Alzheimer's disease. Molecules 27(12):3737
Zhan F, Lin G, Su L, Xue L, Duan K, Chen L, Ni J (2024) The association between methylmalonic acid, a biomarker of mitochondrial dysfunction, and cause-specific mortality in Alzheimer's disease and Parkinson's disease. Heliyon 10(8):e29357
Zhang Y, Zhao Y, Zhang J, Gao Y, Li S, Chang C et al (2021) Ginkgolide B inhibits NLRP3 inflammasome activation and promotes microglial M2 polarization in Aβ1-42-induced microglia cells. Neurosci Lett 764:136206

Part III

Photomedicine: Traditional Knowledge to Modern Scientific Assessment

12 Envisioning the Neuroprotective Effect of *Ocimum sanctum Linn* (Tulsi) in Neurodegenerative Diseases: A Portrait of Molecular Crosstalk

Disha Arora, Rupa Joshi, and Anshu Gupta

Abstract

Through a complex molecular crosstalk, *Ocimum sanctum* Linn (tulsi) has shown notable neuroprotective effects in a variety of neurodegenerative disease models. The ability of tulsi extract to maintain dendritic branching in substantia nigra neurons—a region that is significantly impacted in models of Parkinson's disease—indicates that it can prevent and mitigate behavioral abnormalities and neuronal damage brought on by stress. Tulsi extract may help with memory and cognitive processes because it stabilizes the expression of choline acetyltransferase (ChAT), a crucial enzyme in cholinergic neurotransmission, and increases the viability of neuronal cells exposed to neurotoxic agents in vitro. Tulsi's strong anti-inflammatory and antioxidant qualities are responsible for its neuroprotective effects, as they aid in reducing inflammation and oxidative stress, two key factors in the pathophysiology of neurodegenerative diseases. Furthermore, tulsi has been demonstrated to modify the lipid profile of the brain after ischemic injury, changing important lipid species involved in neuronal repair and neurotransmission, thus promoting recovery from stroke-like damage. Tulsi and its bioactive compounds may reduce β-amyloid oligomerization and inhibit neuronal apoptosis by downregulating caspase-3 and -9 at the molecular level, providing protection against neurotoxicity similar to Alzheimer's disease. All of these results point to tulsi's neuroprotective potential, which is mediated by anti-inflammatory, antioxidant, lipidomic, and anti-apoptotic pathways. As such, it is

D. Arora
Faculty of Pharmaceutical Sciences, The ICFAI University, Baddi, Himachal Pradesh, India

R. Joshi (✉) · A. Gupta
Department of Pharmacology, Maharishi Markandeshwar Institute of Medical Sciences and Research, Mullana, Ambala, India

H. Dilnashin, S. P. Singh (eds.), *Phototherapeutic Approaches to Neurodegeneration*,
https://doi.org/10.1007/978-981-95-4751-7_12

positioned as a promising adjunctive therapy option for neurodegenerative diseases.

Keywords

Ocimum sanctum · Tulsi · Neuroprotection · Neurodegenerative diseases · Molecular crosstalk · Neuroinflammation · Phytochemicals

12.1 Introduction

Neurodegenerative diseases are defined as chronic neurological disorders marked by the gradual degeneration of neurons and synaptic connections with advancing age. They are sharing a complex aetiology and instigating a huge burden on healthcare centers globally. Broadly, neurodegenerative diseases include Alzheimer's disease (AD), Parkinson's disease (PD), Huntington's disease (HD), and amyotrophic lateral sclerosis (ALS), which mainly affect cognition and motor functioning of the body. The complexity of the underlying molecular mechanisms of these diseases imposes a devastating challenge to the scientific community as well as healthcare centers. Moreover, the constrained availability of drugs against these neurodegenerative diseases and their associated side effects is augmenting the complexity of management of these diseases, which in turn proves to be an economic burden for both developing and developed countries across the globe (Gadhave et al. 2024).

Approximately 15% of the world population has been reported to be suffering from neurological diseases, and it is estimated that the burden of neurodegenerative diseases will be doubled in the coming two decades. The risk of common neurodegenerative diseases like Alzheimer's and Parkinson's disease is affirmatively linked to age, and risk increases in the elderly above the age of 65 years. However, ALS and Huntington's disease can also affect people of young age groups (Van Schependom and D'haeseleer 2023).

In India, the prevalence of AD is approximately 7.4% among the age group of 60 years and older (Lee et al. 2023), and the prevalence of Parkinson's disease is 67.71 per 1 lakh persons and approximately 10% of the global burden of Parkinson's disease (Verma et al. 2017).

12.1.1 Importance of Exploring Alternative Therapeutic Approaches

Treatment of neurodegenerative diseases is itself a complex phenomenon employing multiple pathophysiological targets, which poses significant challenges for their management. Currently, all the therapies primarily focus on symptomatic treatment of the disease. These include drugs that increase acetylcholine levels in Alzheimer's disease, like donepezil, rivastigmine, galantamine, and memantine, and drugs that increase levels of dopamine and decrease levels of acetylcholine in Parkinson's disease, like levodopa-carbidopa, pramipexole, ropinirole, rasagiline, entacapone, opicapone, amantadine, trihexyphenidyl, benztropine, etc. Merely, therapies are targeting or modifying the disease pathophysiology, thus emerging the need for advanced alternative multitargeted therapies for its management. Among them are immunotherapies, including monoclonal antibodies; gene therapies employing gene editing technologies, like CRISPR-Cas9, to modify underlying gene defects; stem cell therapies to replace the damaged or lost neurons with stem cell-derived healthy neurons; and nanotechnology, enriching the efficacy of drugs by using novel drug delivery systems directly into the brain. Additionally, many non-pharmacological interventions are also being investigated to treat these neurodegenerative diseases. Cognitive stimulation therapy, virtual reality therapy, physical exercise, and psychological support significantly improve cognitive function and quality of life. Even lifestyle modifications like intake of the Mediterranean diet, social engagement, and adequate sleep are some of the options to manage the neurodegenerative diseases symptomatically (Krsek and Baticic 2024; Marchiş and Iorga 2025).

Despite the availability of different options for the treatment of neurodegenerative diseases, their cure is still a major challenge for the health care community as well as the patients who are the real sufferers. The pharmacological treatment therapies provide palliative treatment and have numerous side effects associated with their long-term use, which in turn increases the burden of the disease both for the patient and their caregivers. Additionally, these treatment options become inadequate and inefficient with the progression of the disease. Thus arises the need for more efficacious and safer treatment options for the management of neurodegenerative diseases. Currently, researchers are focusing on early detection of the disease, personalized medicines, and combination therapies with a multitargeted approach employing both allopathic and herbal treatment options.

12.1.2 Significance of *Ocimum sanctum* Linn (Tulsi) in Traditional Medicine

The pathophysiological complexity of the neurodegenerative diseases necessitated the need for a comprehensive approach for their treatment. Traditional systems of medicine like Ayurveda and herbal medicines offer a holistic perspective for the management of neurodegenerative diseases. These systems are considered natural and safer than the current treatment options and also employ a multitargeted approach, addressing different aspects of the disease pathology and addressing all the aspects of the individual, like the physical, mental, and spiritual well-being of the patients. Among them, Ocimum sanctum, commonly known as tulsi, has multiple neuroprotective activities and has been used in the treatment of various diseases in the Ayurvedic system of medicine for centuries (Mukherjee and Wahile 2006; Patwardhan and Mashelkar 2009). It is rich in medicinal compounds possessing antioxidant, anti-inflammatory, antimicrobial, immunomodulatory, neurogenic, and antidepressant activity. It reduces chronic inflammation in disease models of Alzheimer's and Parkinson's disease. These constituents' target the underlying molecular mechanisms of the disease (Arya et al. 2024). Thus, in this chapter, understanding the molecular mechanisms of the neurodegenerative diseases and their treatment strategies with traditional herbal medicines like *Ocimum sanctum* would be a boon for patients suffering from these disorders as well as for the health care professionals to adopt a holistic approach for their patients' well-being.

12.2 Neurodegenerative Diseases: A Molecular Perspective

Progressive degeneration of the neurons in the central nervous system is the hallmark of neurodegenerative diseases like AD, PD, HD, and ALS. The underlying molecular mechanisms of these diseases differ from each other, and thus their symptoms do too. Therefore, it is very important to study the molecular targets of these diseases and the potential treatment options.

12.2.1 Alzheimer's Disease

It is the most common cause of dementia in the elderly population, and its pathophysiology is complex and involves multiple factors. The imperative factor is the accumulation of amyloid β plaques in the brain, which is also considered a hallmark of AD. Amyloid-β peptides aggregated in the brain to form insoluble fibrils, which deposit in the form of plaques in the brains of patients suffering from AD. These deposits of amyloid β plaques lead to neuronal damage and the ultimate death of the neurons (Zhang et al. 2023). The second important characteristic of AD is the formation of neurofibrillary tangles. These tangles are formed by the hyperphosphorylation of the tau protein, which is actually required for the stability of the microtubules in the brain, and thus causes instability of the microtubules and leads to further dysfunction of the neurons (Alonso et al. 2018). Thirdly, inflammation and oxidative stress also cause activation of microglia and astrocytes, which release various proinflammatory cytokines and reactive oxygen species, which ultimately lead to neuronal damage (Fanlo-Ucar et al. 2024).

Aβ aggregation initiates due to the formation of small oligomers, which increases the leakage of calcium ions into the cell and thus destabilizes the neurons and affects synaptic plasticity, which is the main regulator of memory and learning in the body. These oligomers also overactivate the NMDA receptors and thus cause excitotoxicity, and ultimately, enhanced calcium influx will lead to cell death. With age, these oligomers join to form larger fibrils, which leads to the formation of amyloid plaques. Thus, it activates immune responses like microglia and astrocytes. Various ions, like zinc and copper, also accelerate plaque formation. Blocking the activity of beta and gamma secretase enzymes, chelating metal ions, and using antibodies helped to block the formation of A beta oligomers and thus prevent the formation of plaques (Viola and Klein 2015).

Normally, the tau protein acts like a highway in neurons, which helps in the transport of nutrients and signals. In AD, these tau proteins get hyperphosphorylated and destabilize the microtubules by losing affinity to them, and these detached tau proteins create a twisted filament, which further forms neurofibrillary tangles within the neurons. Kinase inhibitors can prevent the excessive tau phosphorylation and can slow down the progression of the disease (Kammula et al. 2023). There are various enzymes involved in the formation of amyloid beta plaques and hyperphosphorylation of tau, whose dysfunction can lead to neuronal damage. Even the altered levels of various neurotransmitters, like acetylcholine, also play an important role in causing dementia and cognitive impairment in these patients.

12.2.2 Parkinson's Disease

The misfolding of alpha-synuclein and its accumulation in the form of Lewy bodies is the hallmark of PD, as it regulates dopamine release and SNCA mutations, and various toxins trigger this misfolding. The formation of highly toxic oligomers of alpha-synuclein disrupts the integrity of cellular membranes by forming pores, which increases calcium influx and disturbs the cellular homeostasis, excessive free radical production, and apoptosis. Dopaminergic neurons are most susceptible to these damages. Alpha-synuclein spreads rapidly like prions and mimics the clinical progression of PD. Thus, targeting stabilization of alpha-synuclein, antibodies against misfolded proteins, enhancement of autophagy, and degradation of aggregation of alpha-synuclein can be the potential therapies for treatment of PD (Siwecka et al. 2023).

12.2.3 Huntington's Disease

A mutation in the HTT gene is the main cause of Huntington's disease, and aggregation of mHTT specifically harms the striatum, which in turn affects motor incoordination and other cellular processes. Therefore, therapeutic approaches like silencing of mHTT via interference RNA and antisense oligonucleotides can be the potential targets for the treatment of Huntington's disease (Srinivasan et al. 2022). Researchers are mainly focusing on both early detection and early treatment by targeting the primary stages of misfolded and aggregated proteins, thus interrupting the whole cascade of cellular dysfunction, which is the hallmark of the pathophysiology of all neurodegenerative diseases. Major molecular mechanisms underlying neurodegeneration and their potential therapeutic targets for early treatment have been shown in Fig. 12.1.

Potential therapeutic targets for the treatment of neurodegenerative diseases have been summarized in Table 12.1 (Toader et al. 2024).

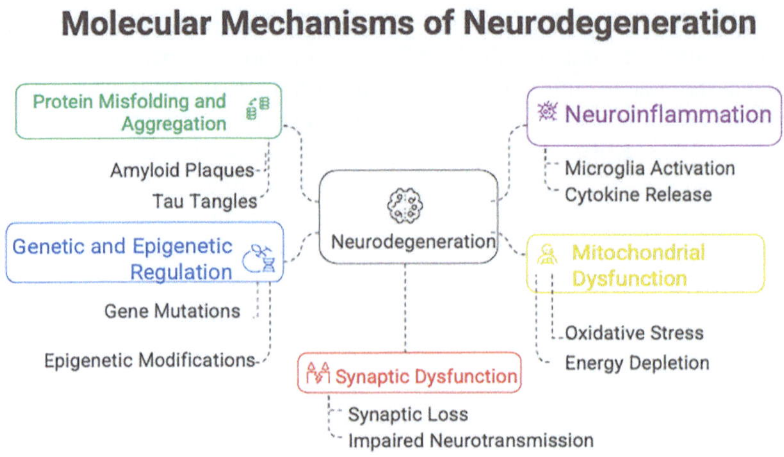

Fig. 12.1 Different molecular mechanisms of neurodegenerative diseases

12.3 *Ocimum sanctum* Linn (Tulsi): Phytochemical Profile and Pharmacological Properties

The varied phytochemical profile (Fig. 12.2) of *Ocimum sanctum* Linn, commonly referred to as holy basil or tulsi, adds to its therapeutic qualities (Sen et al. 2022; Srivastav et al. 2024). Polyphenols, flavonoids, terpenoids, alkaloids, phenolics, essential oils, tannins, and saponins are some examples of these phytochemicals (Fantasma et al. 2024; Sinoriya and Singh 2024).

12.3.1 Important Phytochemicals and Their Importance

Flavonoids and polyphenols are two important and necessary ingredients in tulsi extracts that have a variety of pharmacological properties, such as antioxidant and anticancer effects (Fantasma et al. 2024). Different basil species and extraction solvents have varying levels of total phenolic content (TPC) and total flavonoid content (TFC) (Pathak and Borah 2022). For instance, in a study that extracted phytochemicals from two varieties of *Ocimum basilicum* using various solvents, water typically produced the highest phenolic content, followed by ethanol and then ethyl acetate (Pathak and Borah 2022).

Eugenol, Ocimum sanctum, has a high concentration of this active ingredient (Monga et al. 2017). To increase the amount of eugenol produced from tulsi, supercritical carbon dioxide (SC-CO_2), extraction can be used (Monga et al. 2017).

Two essential chemical components of tulsi are *ursolic acid and rosmarinic acid* (Prasongdee et al. 2024; Ghosh et al. 2013). Particularly, rosmarinic acid is well-known for its antibacterial, anti-inflammatory, and antioxidant qualities (Venuprasad et al. 2013).

Table 12.1 Potential therapeutic targets for the treatment of neurodegenerative diseases

Main pathways	Targets at the molecular levels
Enhancing protein clearance pathways	Enhancing the ubiquitin-proteasome system
	Stimulating autophagy and the lysosomal pathway
Enhancing chaperone levels	In AD: HSP-70 and HSP-90 In PD—HSP70 and HSP27 In HD—HSP40 and HSP70
Pattern recognition receptors	NLRs, TLRs, RAGE
Damage-associated molecular patterns (DAMPs)	HMGB1, extracellular ATP, and mtDNA
Microglial activation states	M1 vs. M2 phenotypes
Signaling pathways in activation	NF-κB, MAPK pathways, PPAR-γ, STAT6
Cytokine networks	Proinflammatory cytokines (TNF-α, IL-1β, and IL-6)
	Anti-inflammatory mediators (IL-10 and TGF-β)
Molecular targets (colony-stimulating factor 1 Receptor—CSF1R inhibitors)	Inhibitors of the NF-κB pathway
	Modulators of the JAK/STAT pathway
Modulation of microglial activation	PPAR-γ agonists
	Colony-stimulating factor 1 Receptor—CSF1R inhibitors
	Enhancement of IL-10 and TGF-β levels
	HDAC inhibitors
Genetic epigenetic mechanisms	Mitochondrial dysfunction
	Oxidative stress
	Defects in the electron transport chain
	Mitochondrial DNA mutations
	ROS generation
	Role of compartment-specific antioxidants
	Gene therapy for endogenous antioxidants like SOD, GPx, and Nrf2

Fig. 12.2 Phytochemical composition of tulsi

One of the active chemical components of *Ocimum tenuiflorum* is *terpenoids* (Sen et al. 2022). The plant's therapeutic qualities are attributed to terpenoids, phenols, flavonoids, and phenylpropanoids (Sen et al. 2022).

Aromatic Oils: It has been discovered that *Ocimum sanctum* leaves cultivated in Nigeria contain essential oils along with a variety of phytochemicals, such as alkaloids, phlobatannins, tannins, terpernoids, cardiac glycosides, steroids, and saponins (Mandal et al. 2022).

Apigenin, carvacrol (Sinoriya and Singh 2024), caryophyllene (Prasongdee et al. 2024; Sinoriya and Singh 2024), linalool, myretenal, and β-sitosterol are some other noteworthy phytochemicals (Prasongdee et al. 2024; Sinoriya and Singh 2024).

12.3.2 Methods of Phytochemical Analysis

The phytochemicals found in *Ocimum sanctum* are identified and measured using a variety of techniques:

Spectrophotometry, nuclear magnetic resonance, thin-layer chromatography, and Fourier transform infrared spectroscopy are methods used to screen for phytochemical components in an ethanolic extract of *Ocimum sanctum* (Sahoo et al. 2022).

The Soxhlet extraction technique is used to extract bioactive substances like proteins, carbohydrates, terpenoids, flavonoids, saponins, and tannins using various solvents like ethanol and hexane (Singh et al. 2022).

Secondary phytochemical metabolites are identified using gas chromatography-mass spectrometry, or GC-MS. For instance, nine secondary phytochemical metabolites were identified by GC-MS analysis of krishna tulsi (*Ocimum tenuiflorum*) (Kumar et al. 2022).

High-performance liquid chromatography, or HPLC, is used to profile secondary metabolites in various tulsi varieties, such as flavonoids and polyphenols (Neto et al. 2013).

12.3.3 Pharmacological Effects

Holy basil, also known as tulsi, is a valuable medicinal plant with a variety of pharmacological qualities (Bawankule et al. 2015). These characteristics (Fig. 12.3) include immunomodulatory, antidiabetic, hepatoprotective, anti-inflammatory, antimicrobial, antistress, antioxidant, and anticancer effects (Almatroodi et al. 2020; Lukianenko et al. 2021).

Antioxidant Properties: Tulsi's numerous phytochemicals, which aid in scavenging free radicals and lowering oxidative stress, are responsible for its antioxidant activity (Nita and Grzybowski, 2016). According to studies, tulsi aqueous extracts can strengthen the body's natural antioxidant defenses. In particular, tulsi raises the body's levels of glutathione, a vital antioxidant (Sinoriya and Singh 2024).

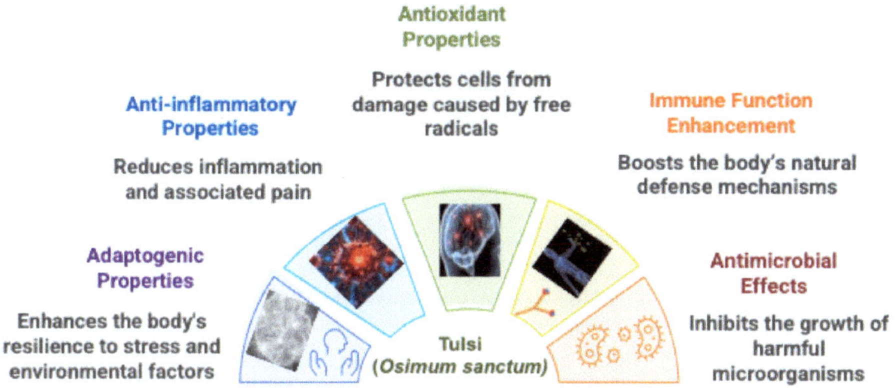

Fig. 12.3 Factors contributing to tulsi's medicinal value

Anti-inflammatory Properties: Numerous in vitro and in vivo investigations have shown that tulsi has anti-inflammatory properties (Singh and Chaudhuri 2018; Preeti et al. 2021). Tulsi leaf extracts have the capacity to suppress the transcription factor NF-κB, which is essential for triggering the expression of genes linked to inflammation (Garg and Sardana 2016). One of tulsi's main ingredients, rosmarinic acid, greatly enhances its anti-inflammatory qualities (Ravindra and Murugesh Babu 2016; Oliveira et al. 2024).

Antimicrobial Properties: Tulsi shows antimicrobial activity against a variety of viruses, fungi, and bacteria (Chaudhari et al. 2022; Wang et al. 2024). Research has demonstrated that it works well against bacteria like *Candida albicans*, *Staphylococcus aureus*, *Escherichia coli*, and *Bacillus subtilis* (Wang et al. 2024). Tulsi may be used to treat a number of infections because of its antimicrobial properties (Santwani 2023).

Antistress Properties: As an adaptogen, tulsi helps the body adjust to stress by balancing its systems (Van Hoyweghen et al. 2013). Tulsi has been shown to lessen stress on the body, mind, and metabolism (Richard et al. 2016; Pooja and Kumar 2023). It has an impact on the stress response-related hypothalamic-pituitary-adrenal (HPA) axis (Kumar and Patel 2022).

Immunomodulatory Properties: Tulsi can alter the immune system because of its immunomodulatory properties (Lukianenko et al. 2021). When necessary, it has been shown to either strengthen or weaken immune responses (Sarfraz and Ahmad 2023).

Antidiabetic Properties: Tulsi's antidiabetic qualities are widely known (Almatroodi et al. 2020). It is advantageous for managing diabetes and can assist in controlling blood sugar levels (Sarfraz and Ahmad 2023).

Hepatoprotective Properties: Because of its hepatoprotective properties, tulsi shields the liver from harm brought on by illnesses or poisons (Baliga et al. 2013; Sarfraz and Ahmad 2023).

Neuroprotective Effects: The neuroprotective properties of tulsi have been the subject of much research. The main ways that tulsi's active ingredients prevent neuronal damage in stress models are through adaptogenic, anti-inflammatory, and antioxidant processes (Rodrigues et al. 2022).

12.4 Molecular Crosstalk: Tulsi's Bioactive Compounds and Neurodegenerative Pathways

Tulsi's (*Ocimum sanctum*) bioactive compounds have the ability to alter a number of signaling pathways (Fig. 12.4), such as the PI3K/Akt and Mitogen-Activated Protein Kinase (MAPK) pathways, which can affect different cellular functions (Guo et al. 2020; Dahms-Verster et al. 2020).

One essential intracellular signaling cascade that controls cell growth, proliferation, survival, and metabolism is the PI3K/Akt pathway (Savova et al. 2023; Shaohui et al. 2025). Numerous illnesses, such as obesity and cancer, commonly exhibit dysregulation of this pathway (Zhang et al. 2022). Eugenol, one of the bioactive substances found in tulsi, has shown promise in blocking the PI3K/Akt pathway, which lowers cancer cell survival and proliferation. Celastrol affects immune function by modulating the PI3K-Akt signaling pathway, although it is not specific to tulsi (Shaohui et al. 2025). According to research, the PI3K/Akt pathway is essential for maintaining metabolic homeostasis, and addressing this pathway may help treat obesity (Savova et al. 2023). Cancers such as melanoma and triple-negative breast cancer frequently exhibit hyperactivation of the PI3K/Akt/mTOR pathway (Zhang et al. 2022).

Another crucial signaling cascade that controls cell division, growth, inflammation, and apoptosis is the Mitogen-Activated Protein Kinase (MAPK) pathway (De Luca et al. 2012; Pappalardo et al. 2016). This pathway has multiple sub-pathways, including p38 MAPK, JNK, and ERK, each of which mediates unique cellular responses in response to various stimuli (Macedo et al. 2024).

According to a number of studies, tulsi's bioactive substances may be able to alter the MAPK pathway. For example, ERK1/2 and p38 MAPK can increase RGS4 expression, which is involved in the MAPK pathway in the context of inflammation (Hu et al. 2009). A crucial part of the MAPK cascade, the RAS/RAF/MEK/ERK pathway, is commonly dysregulated in human cancers (De Luca et al. 2012).

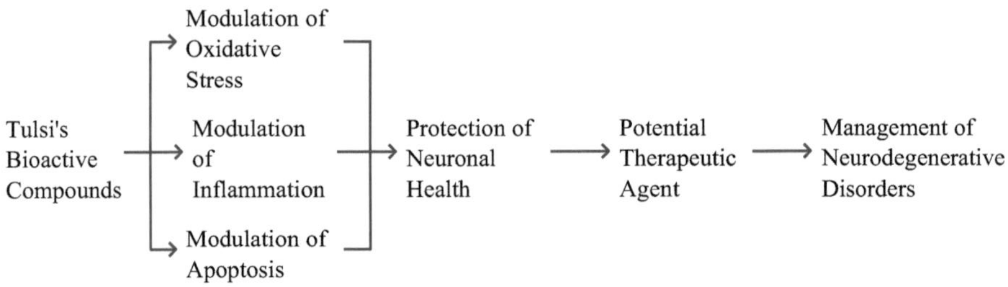

Fig. 12.4 Tulsi's neuroprotective mechanisms

Additionally, the role of MAPK pathways in melanocytic transformation is supported by computational modeling (Pappalardo et al. 2016).

Tulsi contains ursolic acid, which has been shown to modulate the PI3K/Akt and MAPK pathways to cause apoptosis and have anti-inflammatory properties (Kumar et al. 2022). In cancer cells, eugenol exhibits cytotoxic effects and inhibits the PI3K/Akt pathway, which lowers cell survival and proliferation (Hasan et al. 2023). Tulsi contains rosmarinic acid, which has anti-inflammatory and antioxidant qualities that may have an indirect effect on cellular signaling pathways (Kumar et al. 2022). It has been demonstrated that the tulsi flavonoid apigenin has anticancer properties and can alter several signaling pathways, such as PI3K/Akt and MAPK (Hasan et al. 2023).

Tulsi's bioactive compounds may affect other related signaling cascades in addition to directly modulating the PI3K/Akt and MAPK pathways (Bhalla and Iyengar 1999). For instance, tulsi extracts may have an indirect impact on the Sonic Hedgehog (SHh) signaling pathway, which is essential for development and linked to cancer (Mannan et al. 2023). Furthermore, dietary agents may modulate the Notch signaling pathway, which is frequently aberrantly activated in cancers, providing chemopreventive and anticancer effects (Kiesel and Stan 2022). To fully comprehend the range of tulsi's effects, it may also be important to consider how various signaling pathways interact, such as the Hippo pathway and Ca^{2+} signaling (Sayedyahossein et al. 2023).

12.5 Neuroprotective Effects of Tulsi: In Vitro and In Vivo Evidence

Studies conducted both in vitro and in vivo have shown that tulsi exhibits neuroprotective effects through a variety of mechanisms (Yadav et al. 2023). Respected in Ayurveda, this herb has demonstrated promise in treating metabolic, chemical, physical, and psychological stressors (Jamshidi and Cohen 2017). Tulsi's anti-inflammatory and antioxidant qualities are thought to be responsible for its neuroprotective qualities, which lessen the effects of stroke and other neurodegenerative diseases (Haiyi et al. 2024; Shukla et al. 2021).

12.5.1 In Vitro Studies

Studies conducted in vitro have demonstrated that tulsi can shield neurons from harm caused by a variety of stressors (Singh et al. 2013). For example, in primary neuronal cultures exposed to hydrogen peroxide, hypoxia, or oxygen-glucose deprivation, bone marrow mononuclear cells have demonstrated neuroprotective potential (Singh et al. 2013). Likewise, *Coeloglossum viride* var. *bracteatum* extract shielded cortical neurons from oxidative stress and amyloid toxicity (Garcia et al. 2015). Additionally, pretreatment with acetyl-L-carnitine and L-carnitine has demonstrated neuroprotective effects on ischemic injury in vitro (Zhou et al. 2016).

According to these findings, tulsi may have neuroprotective benefits by lowering oxidative stress and halting cell death (Garcia et al. 2015; Juul 2004).

12.5.2 In Vivo Studies

The neuroprotective properties of tulsi are further supported by in vivo research. *Ocimum sanctum* has been shown to modify the lipid composition of the brain cortex and plasma, thereby reducing the consequences of photothrombotic stroke in mice (Haiyi et al. 2024). *Ocimum sanctum* may act as a potential immunomodulator, shielding the brain from harm, according to research done on animal models of ischemic stroke (Sari et al. 2020). Using a transient middle cerebral artery occlusion model, pretreatment with L-carnitine and acetyl-L-carnitine has also demonstrated neuroprotective effects on ischemic injury in vivo (Zhou et al. 2016). Following traumatic brain injury, the flavonoid compound fisetin has shown neuroprotective effects in vivo by preventing ferroptosis and oxidative stress (Dicou et al. 2001). According to these results, tulsi may shield the brain by lowering oxidative stress, regulating inflammatory responses, and averting neuronal damage (Haiyi et al. 2024; Sari et al. 2020).

Mechanisms of Action: There are probably a number of mechanisms underlying tulsi's neuroprotective benefits. Its antioxidant qualities aid in scavenging free radicals and lowering oxidative stress, which is a primary cause of neurodegenerative illnesses (Haiyi et al. 2024; Canudas et al. 2003; Zhang et al. 2011; Nelson et al. 2012). Tulsi's anti-inflammatory properties may aid in lowering brain inflammation, which may also be a factor in neuronal damage (Cinar et al. 2025). Tulsi may also reduce neurodegeneration and alter the cholinergic system (Li et al. 2024; Gu et al. 2024).

Stroke: By changing the lipid profile and lowering inflammation, *Ocimum sanctum* has demonstrated promise in lessening the effects of stroke-like brain injuries (Haiyi et al. 2024; Sari et al. 2020).

Neurodegenerative Diseases: In experimental models of neurodegeneration, seed extracts, including those with antioxidant qualities, have demonstrated neuroprotective effects (Manochkumar et al. 2021). Certain seeds contain carotenoids, which have also shown neuroprotective potential in in vitro and in vivo studies (Beal 2011).

12.6 Potential Therapeutic Applications of Tulsi in Neurodegenerative Diseases

12.6.1 Possibilities of Using Tulsi as a Complementary or Adjunctive Therapy

Ocimum sanctum has been proven to exhibit neuroprotective properties through different mechanisms, as evidenced by many experimental studies as discussed earlier.

Along with its antioxidant, anti-inflammatory, and anti-apoptotic activity, it also possesses other neurodegeneration-specific properties like inhibiting protein aggregation, modulating neurotransmission, preventing cellular dysfunction, silencing selective mutations, etc., which would be utilized for the treatment of different neurodegenerative diseases. Although preclinical studies had reported the effectiveness of tulsi against these diseases as monotherapy, clinical evidence is still lacking. However, some of the studies have demonstrated its efficacy as an adjunctive therapy, as it slows down neurodegeneration and improves cognition and other motor functions in patients suffering from neurodegenerative diseases. Moreover, it also delays the onset of these age-related diseases and preserves the integrity of neurons in different regions of the brain, like the hypothalamus, substantia nigra, etc. (Rodrigues et al. 2022).

Possessing multiple neuroprotective properties fits tulsi into the model of a multitargeted approach for the treatment of neurodegenerative diseases, which are considered complex and multifactorial due to their vast pathophysiology. Moreover, it is also considered safe and has a better tolerability profile as compared to conventional treatments of these disorders, which possess multiple intolerable side effects. Being herbal in nature, it is easy to afford and access this drug. Moreover, adding tulsi as an adjunct treatment will also reduce the dose of the conventional treatments, which in turn will reduce their dose-related side effects and improve the quality of life of the patients.

12.6.2 Challenges and Limitations of Using Tulsi in Clinical Settings

Mostly, the preclinical evidence has proven tulsi as a neuroprotective agent. Limited literature is available on its efficacy in clinical settings. Well-planned randomized controlled clinical trials are needed to elicit their efficacy in real-world scenarios.

Another concern is the standardization of its dose and dosage forms for its clinical efficacy. Being an herbal drug, the potency and composition of its active constituents vary with season, species, and extraction procedures. Thus, it should be standardized as per the concentration of its active constituents (Verma 2016). Further, preclinical and clinical safety and toxicity studies are warranted.

Also, the controlled clinical trial should focus on herb-drug interactions, which play an important role in determining the efficacy of tulsi as an adjunctive treatment in neurodegenerative diseases.

Generally, tulsi is considered safe and efficacious when consumed in traditional preparations. However, it has very mild and rare adverse effects. However, at high doses, it affects fertility, and moreover, its supplementation should be considered cautiously when prescribed to lactating and pregnant women, individuals with autoimmune conditions, and during complex medication regimens. Its drug interactions with anti-coagulants and hypoglycemic agents should be considered while taking the history of the patients (Siva et al. 2016).

12.7 Conclusion

Tulsi is a traditional herbal medicine in the Ayurvedic system of medicine in India. It is available in different formulations and is even undergoing research on newer drug delivery systems of medicine like nanoparticles, liposomes, etc., which are under investigation for their efficacy and safety. It has a wide spectrum of uses in many diseases of different organ systems, like cardiovascular diseases, skin diseases, blood disorders, neurological diseases, endocrine disorders, etc. It has been proven to be neuroprotective in neurodegenerative diseases both experimentally and clinically, as discussed in the chapter. It possesses anti-inflammatory, antioxidant, immunomodulatory, antimicrobial, adaptogenic, and cytoprotective properties to be used as an adjunctive treatment in neurodegenerative diseases like AD, PD, and HD.

12.8 Future Directions for Research on Tulsi's Neuroprotective Effects

- Clinical trials with a large number of patients for evaluating the dose, duration, efficacy, and safety of tulsi are required.
- Understanding of the molecular mechanisms of its phytochemical constituents is needed to be used in the treatment of neurodegenerative diseases.
- Elucidation of its interactions with conventional drugs used in the treatment of different diseases (both pharmacokinetic and pharmacodynamic interactions).
- Formulating standard and validated tulsi-based products for integration into conventional healthcare facilities.

Thus, tulsi, having broad-spectrum pharmacological properties, affirms its reputation as an "herb for all reasons." It augments immunity, manages metabolic and inflammatory disorders, and prevents various infections and stress-related diseases. Therefore, its integrative synergy with proven traditional applications and emerging valuable adjunctive treatment options in modern holistic medicine requires further elaborative research on its molecular mechanisms for its fruitful clinical adoption.

References

Almatroodi SA, Alsahli MA, Almatroudi A, Rahmani AH (2020) Ocimum sanctum: role in diseases management through modulating various biological activity. Pharm J 12(5):1198–1205. https://doi.org/10.5530/pj.2020.12.168

Alonso AD, Cohen LS, Corbo C, Morozova V, ElIdrissi A, Phillips G, Kleiman FE (2018) Hyperphosphorylation of tau associates with changes in its function beyond microtubule stability. Front Cell Neurosci 12:338

Arya R, Faruquee HM, Shakya H, Rahman SA, Begum MM, Biswas SK, Apu MAI, Islam MA, Sheikh MMI, Kim JJ (2024) Harnessing the antibacterial, anti-diabetic and anti-carcinogenic properties of Ocimum sanctum Linn (Tulsi). Plants (Basel) 13(24):3516

Baliga MS, Jimmy R, Thilakchand KR, Sunitha V, Raghavendra BN et al (2013) Ocimum Sanctum L (Holy Basil or Tulsi) and its phytochemicals in the prevention and treatment of cancer. Nutr Cancer 65(sup1):26–35. https://doi.org/10.1080/01635581.2013.785010

Bawankule DU, Kumar A, Agarwal K, Maurya AK, Shanker K, Bushra U et al (2015) Pharmacological and phytochemical evaluation of Ocimum sanctum root extracts for its anti-inflammatory, analgesic and antipyretic activities. Pharmacogn Mag 11(42):217. https://doi.org/10.4103/0973-1296.157743

Beal MF (2011) Neuroprotective effects of creatine. Amino Acids 40(5):1305–1313. https://doi.org/10.1007/s00726-011-0851-0

Bhalla US, Iyengar R (1999) Emergent properties of networks of biological signaling pathways. Science 283(5400):381–387. https://doi.org/10.1126/science.283.5400.381

Canudas AM, Pubill D, Sureda FX, Verdaguer E, Camps P et al (2003) Neuroprotective effects of (±)-huprine Y on in vitro and in vivo models of excitoxicity damage. Exp Neurol 180(2):123–130. https://doi.org/10.1016/s0014-4886(02)00029-8

Chaudhari LY, Chaudhari SP, Chavan GM (2022) A brief review on tulsi: a holy plant with high medicinal values and therapeutic uses. Int J Res Ayurveda Pharm 13(3):118–125. https://doi.org/10.7897/2277-4343.130372

Cinar I, Yayla M, Bayram P, Binnetoglu D (2025) Comparison of the neuroprotective effects of gossypin on cisplatin-induced neurotoxicity in vitro and in vivo. Mol Biol Rep 52(1). https://doi.org/10.1007/s11033-025-10501-7

Dahms-Verster S, Nel A, van Vuren JHJ, Greenfield R (2020) Biochemical responses revealed in an amphibian species after exposure to a forgotten contaminant: an integrated biomarker assessment. Environ Toxicol Pharmacol 73:103272. https://doi.org/10.1016/j.etap.2019.103272

De Luca A, Maiello MR, D'Alessio A, Pergameno M, Normanno N (2012) The RAS/RAF/MEK/ERK and the PI3K/AKT signalling pathways: role in cancer pathogenesis and implications for therapeutic approaches. Expert Opin Ther Targets 16(sup2):S17–S27. https://doi.org/10.1517/14728222.2011.639361

Dicou E, Attoub S, Gressens P (2001) Neuroprotective effects of leptin in vivo and in vitro. Neuroreport 12(18):3947–3951. https://doi.org/10.1097/00001756-200112210-00019

Fanlo-Ucar H, Picón-Pagès P, Herrera-Fernández V, ILL-Raga G, Muñoz FJ (2024) The dual role of amyloid Beta-peptide in oxidative stress and inflammation: unveiling their connections in Alzheimer's disease etiopathology. Antioxidants 13:1208

Fantasma F, Samukha V, Saviano G, Chini MG, Iorizzi M, Caprari C (2024) Nutraceutical aspects of selected wild edible plants of the Italian Central Apennines. Nutraceuticals 4(2):190–231. https://doi.org/10.3390/nutraceuticals4020013

Gadhave DG, Sugandhi VV, Kokare CR (2024) Potential biomaterials and experimental animal models for inventing new drug delivery approaches in the neurodegenerative disorder: multiple sclerosis. Brain Res 1822:148674

Garcia MA, Alonso J, Melgar MJ (2015) Radiocaesium activity concentrations in macrofungi from Galicia (NW Spain): influence of environmental and genetic factors. Ecotoxicol Environ Saf 115:152–158. https://doi.org/10.1016/j.ecoenv.2015.02.005

Garg P, Sardana S (2016) Pharmacological and therapeutic effects of Ocimum sanctum. https://doi.org/10.0410/cata/44b7971a26a929d12b0fddd089e80a94

Ghosh S, Chatterjee D, Das S, Bhattacharjee P (2013) Supercritical carbon dioxide extraction of eugenol-rich fraction from Ocimum sanctum Linn and a comparative evaluation with other extraction techniques: process optimization and phytochemical characterization. Ind Crop Prod 47:78–85. https://doi.org/10.1016/j.indcrop.2013.02.030

Gu J, Yan C, Yin S, Wu H, Liu C et al (2024) Erythrocyte membrane-coated nanocarriers modified by TGN for Alzheimer's disease. J Control Release 366:448–459. https://doi.org/10.1016/j.jconrel.2023.12.030

Guo J, Li G, Yang L (2020) Role of H2S in pain: growing evidences of mystification. Eur J Pharmacol 883:173322. https://doi.org/10.1016/j.ejphar.2020.173322

Haiyi Y, Hong Y, Gong M, Cai S, Yuan Z et al (2024) Fisetin exerts neuroprotective effects in vivo and in vitro by inhibiting ferroptosis and oxidative stress after traumatic brain injury. Front Pharmacol 15:1480345. https://doi.org/10.3389/fphar.2024.1480345

Hasan MR, Alotaibi BS, Althafar ZM, Mujamammi AH, Jameela J (2023) An update on the therapeutic anticancer potential of Ocimum sanctum L.: "Elixir of Life.". Molecules 28(3):1193. https://doi.org/10.3390/molecules28031193

Hu W-H, Li F, Mahavadi S, Murthy KS (2009) Upregulation of RGS4 expression by IL-1β in colonic smooth muscle is enhanced by ERK1/2 and p38 MAPK and inhibited by the PI3K/Akt/GSK3β pathway. Am J Phys Cell Phys 296(6):C1310–C1320. https://doi.org/10.1152/ajpcell.00573.2008

Jamshidi N, Cohen MM (2017) The clinical efficacy and safety of tulsi in humans: a systematic review of the literature. Evid Based Complement Alternat Med 2017(1). https://doi.org/10.1155/2017/9217567

Juul S (2004) Recombinant erythropoietin as a neuroprotective treatment: in vitro and in vivo models. Clin Perinatol 31(1):129–142. https://doi.org/10.1016/j.clp.2004.03.004

Kammula SV, Tripathi S, Wang N, Dawson VL, Dawson TM, Mao X (2023) Unraveling the tau puzzle: a review of mechanistic targets and therapeutic interventions to prevent tau pathology in Alzheimer's disease. Ageing Neurodegener Dis 3:22

Kiesel VA, Stan SD (2022) Modulation of notch signaling pathway by bioactive dietary agents. Int J Mol Sci 23(7):3532. https://doi.org/10.3390/ijms23073532

Krsek A, Baticic L (2024) Nanotechnology-driven therapeutic innovations in neurodegenerative disorders: a focus on Alzheimer's and Parkinson's disease. Fut Pharmacol 4(2):352–379

Kumar P, Patel D (2022) Ocimum Sanctum: an all-round treatment for cancer? Altern Ther Health Med. https://doi.org/10.0410/cata/35efe9486ec585697c5bc998e453910e

Kumar R, Saha P, Lokare P, Datta K, Selvakumar P, Chourasia A (2022) Systemic review of Ocimum sanctum (Tulsi): morphological characteristics, phytoconstituents and therapeutic applications. Int J Res Appl Sci Biotechnol 9(2):221–226. https://doi.org/10.31033/ijrasb.9.2.15

Lee J, Meijer E, Langa KM, Ganguli M, Varghese M, Banerjee J, Khobragade P, Angrisani M, Kurup R, Chakrabarti SS, Gambhir IS, Koul PA, Goswami D, Talukdar A, Mohanty RR, Yadati RS, Padmaja M, Sankhe L, Rajguru C, Gupta M, Kumar G, Dhar M, Chatterjee P, Singhal S, Bansal R, Bajpai S, Desai G, Rao AR, Sivakumar PT, Muliyala KP, Bhatnagar S, Chattopadhyay A, Govil D, Pedgaonkar S, Sekher TV, Bloom DE, Crimmins EM, Dey AB (2023) Prevalence of dementia in India: national and state estimates from a nationwide study. Alzheimers Dement 19(7):2898–2912

Li J, Huang Z, Jin Y, Liang L, Li Y (2024) Neuroprotective effect of Tauroursodeoxycholic acid (TUDCA) on in vitro and in vivo models of retinal disorders: a systematic review. Curr Neuropharmacol 22(8):1374–1390. https://doi.org/10.2174/1570159x21666230907152207

Lukianenko N, Kens O, Nurgaliyeva Z, Toguzbayeva D, Sakhipov M (2021) Finding a molecular genetic marker for the incidence of recurrent episodes of acute obstructive bronchitis in children. J Med Life 14(5):695–699. https://doi.org/10.25122/jml-2021-0052

Macedo C, Costa PC, Rodrigues F (2024) Bioactive compounds from Actinidia arguta fruit as a new strategy to fight glioblastoma. Food Res Int 175:113770. https://doi.org/10.1016/j.foodres.2023.113770

Mandal AK, Poudel M, Neupane NP, Amita V (2022) Phytochemistry, pharmacology, and applications of Ocimum sanctum (Tulsi). In: edible plants in health and diseases. Springer Singapore, pp 135–174. https://doi.org/10.1007/978-981-16-4959-2_4

Mannan A, Dhimann S, Gargn N, Singh TG (2023) Pharmacological modulation of sonic hedgehog signaling pathways in angiogenesis: a mechanistic perspective. Dev Biol 504:58–74. https://doi.org/10.1016/j.ydbio.2023.09.009

Manochkumar J, Doss CGP, El-Seedi HR, Efferth T, Siva R (2021) The neuroprotective potential of carotenoids in vitro and in vivo. Phytomedicine 91:153676. https://doi.org/10.1016/j.phymed.2021.153676

Marchiş M, Iorga M (2025) Exploring non-pharmacologic adjunctive therapies for patients with neurodegenerative diseases. Medicina 61(7):1224

Monga S, Dhanwal P, Kumar R, Kumar A, Chhokar V (2017) Pharmacological and physicochemical properties of Tulsi (Ocimum gratissimum L.): an updated review. Pharma Innovation 6(4):181–186

Mukherjee PK, Wahile A (2006) Integrated approaches towards drug development from Ayurveda and other Indian system of medicines. J Ethnopharmacol 103(1):25–35

Nelson ED, Ramberg JE, Best T, Sinnott RA (2012) Neurologic effects of exogenous saccharides: a review of controlled human, animal, and in vitro studies. Nutr Neurosci 15(4):149–162. https://doi.org/10.1179/1476830512y.0000000004

Neto EMFL, Almeida ALS, Peroni N, Castro CC, Albuquerque UP (2013) Phenology of Spondias tuberosaArruda (Anacardiaceae) under different landscape management regimes and a proposal for a rapid phenological diagnosis using local knowledge. J Ethnobiol Ethnomed 9(1):10. https://doi.org/10.1186/1746-4269-9-10

Nita M, Grzybowski A (2016) The role of the reactive oxygen species and oxidative stress in the patho mechanism of the age-related ocular diseases and other pathologies of the anterior and posterior eye segments in adults. Oxid Med Cell Longev 2016:3164734. https://doi.org/10.1155/2016/3164734

Oliveira VDC, Rodrigues SDO, Souto SMT, da Silva GA, Vilegas W, Ferri BG et al (2024) Chemical profile and evaluation of the pharmacological activity of the dry extract and fraction of ethyl acetate obtained from the leaves of Mimosa caesalpiniifolia. J Ethnopharmacol 323:117716. https://doi.org/10.1016/j.jep.2024.117716

Pappalardo F, Russo G, Candido S, Pennisi M, Cavalieri S et al (2016) Computational modeling of PI3K/AKT and MAPK signaling pathways in melanoma cancer. PLoS One 11(3):e0152104. https://doi.org/10.1371/journal.pone.0152104

Pathak C, Borah G (2022) Cu_2O/CuO@mont K 10 promoted one-pot synthesis of 1,2,3-triazoles through azide-alkyne cycloaddition reaction. Chem Pap 76(8):4749–4761. https://doi.org/10.1007/s11696-022-02203-2

Patwardhan B, Mashelkar RA (2009) Traditional medicine-inspired approaches to drug discovery: can Ayurveda show the way forward? Drug Discov Today 14(15–16):804–811

Pooja P, Kumar A (2023) A systemic review of Tulsi (Ocimum tenuiflorum or Ocimum sanctum): phytoconstituents, ethnobotanical and pharmacological profile. Res J Pharmacogn Phytochem:179–188. https://doi.org/10.52711/0975-4385.2023.00028

Prasongdee P, Posridee K, Oonsivilai A, Oonsivilai R (2024) A culinary and medicinal gem: exploring the phytochemical and functional properties of Thai Basil. Foods 13(4):632. https://doi.org/10.3390/foods13040632

Preeti P, Ramamurthy A, Chaudhary K (2021) Tulsi (Ocimum sanctum Linn.): an ayurvedic and contemporary review. Int Ayurvedic Med J 9(12):3134–3138. https://doi.org/10.46607/iamj3409122021

Ravindra KB, Murugesh Babu K (2016) Study of antimicrobial properties of fabrics treated with Ocimum sanctum L (tulsi) extract as a natural active agent. J Nat Fibers 13(5):619–627. https://doi.org/10.1080/15440478.2015.1093577

Richard EJ, illuri R, Bethapudi B, Senthilkumar A, Anirban B, Velusami CC et al (2016) Anti-stress activity of Ocimum sanctum: possible effects on hypothalamic–pituitary–adrenal axis. Phytother Res 30(5):805–814. https://doi.org/10.1002/ptr.5584

Rodrigues V, Rao MS, Rao GS, Mohandas Rao KG (2022) Neuroprotective potential of Ocimum sanctum (Linn) leaf extract in preventing and attenuating stress induced substantia nigral neuronal damage in rats. J Ayurveda Integr Med 13(4):100651. https://doi.org/10.1016/j.jaim.2022.100651

Sahoo DD, Tabassum Y, Sharma D (2022) Multiple health benefits of tulsi plants. J Med Plants Stud 10(5):95–102. https://doi.org/10.22271/plants.2022.v10.i5b.1471

Santwani S (2023) Int J Res Appl Sci Eng Technol 11(3):859–861. https://doi.org/10.22214/ijraset.2023.49494

Sarfraz H, Ahmad IZ (2023) A systematic review on the pharmacological potential of Linum usitatissimum L.: a significant nutraceutical plant. J Herbal Med 42:100755. https://doi.org/10.1016/j.hermed.2023.100755

Sari FA, Sandhika W, Yuliawati TH (2020) Tulsi (Ocimum sanctum) leaf ethanol extract reduces inflammatory cell infiltration in aspirin-induced gastritis rats. J Kedokt Brawijaya 31(1):49–52. https://doi.org/10.21776/ub.jkb.2020.031.01.10

Savova MS, Mihaylova LV, Tews D, Wabitsch M, Georgiev MI (2023) Targeting PI3K/AKT signaling pathway in obesity. Biomed Pharmacother 159:114244. https://doi.org/10.1016/j.biopha.2023.114244

Sayedyahossein S, Thines L, Sacks DB (2023) Ca2+ signaling and the hippo pathway: intersections in cellular regulation. Cell Signal 110:110846. https://doi.org/10.1016/j.cellsig.2023.110846

Sen K, Goyal M, Mukopadayay S (2022) Review on phytochemical and pharmacological, medicinal properties of holy basil (Ocimum sanctum L.). Int J Health Sci:7276–7286. https://doi.org/10.53730/ijhs.v6ns4.10165

Shaohui G, Jingqi W, Chunli S, Li L, Yijin J, Jingyuan F et al (2025) Dose-dependent immunotoxic mechanisms of celastrol via modulation of the PI3K-Akt signaling pathway. Front Pharmacol 16. https://doi.org/10.3389/fphar.2025.1567193

Shukla MR, Kibler A, Turi CE, Erland LAE, Sullivan JA et al (2021) Selection and micropropagation of an elite melatonin rich tulsi (Ocimum sanctum L.) germplasm line. Agronomy 11(2):207. https://doi.org/10.3390/agronomy11020207

Singh D, Chaudhuri PK (2018) A review on phytochemical and pharmacological properties of Holy basil (Ocimum sanctum L.). Ind Crop Prod 118:367–382. https://doi.org/10.1016/j.indcrop.2018.03.048

Singh N, Megha A, Dore S (2013) Neuroprotective properties and mechanisms of resveratrol in in vitro and in vivo experimental cerebral stroke models. ACS Chem Neurosci 4(8):1151–1162. https://doi.org/10.1021/cn400094w

Singh A, Singh A, Singh J, Singh MP (2022) Phytochemical analysis & antimicrobial activity of the leaves of Ocimum sanctum. J Pharmacogn Phytochem 11(3):174–176

Sinoriya SK, K Singh (2024) Review on therapeutic uses of Tulsi with its phytochemical constituents in different kind of extracts. https://doi.org/10.0410/cata/d63b31e205bcf55d535824b6d54249ed

Siva M, Shanmugam KR, Shanmugam B, Venkata G, Ravi S, Sathyavelu K et al (2016) Ocimum sanctum: a review on the pharmacological properties. Int J Basic Clin Pharmacol:558–565. https://doi.org/10.18203/2319-2003.ijbcp20161491

Siwecka N, Saramowicz K, Galita G, Rozpe Ãdek-Kaminska W, Majsterek I (2023) Inhibition of protein aggregation and endoplasmic reticulum stress as a targeted therapy for alpha synucleinopathy. Pharmaceutics 15:2051

Srinivasan E, Ram V, Rajasekaran R (2022) A review on Huntington protein: insight into protein aggregation and therapeutic interventions. Curr Drug Metab 23:260–282

Srivastav Y, Thakur G, Tiwari S (2024) Study of medicinal plants: Ocimum sanctum, Tinospora cordifolia, bioresources to characterize bioactive compounds using soxhlet apparatus. Asian J Biotechnol Bioresour Technol 10(3):23–44. https://doi.org/10.9734/ajb2t/2024/v10i3209

Toader C, Tataru CP, Munteanu O, Serban M, Covache-Busuioc RA, Ciurea AV, Enyedi M (2024) Decoding neurodegeneration: a review of molecular mechanisms and therapeutic advances in Alzheimer's, Parkinson's, and ALS. Int J Mol Sci 25:12613

Van Hoyweghen L, Karolien DB, Guy H, Dieter D, Arne H (2013) In vitro inhibition of the transcription factor NF-κB and cyclooxygenase by bamboo extracts. Phytother Res 28(2):224–230. https://doi.org/10.1002/ptr.4978

Van Schependom J, D'haeseleer M (2023) Advances in neurodegenerative diseases. J Clin Med 12(5):1709

Venuprasad MP, Kandikattu HK, Khanum F (2013) Neuroprotective effects of hydroalcoholic extract of Ocimum sanctum against H_2O_2 induced neuronal cell damage in SH-SY5Y cells via its antioxidative defence mechanism. Neurochem Res 38(10):2190–2200. https://doi.org/10.1007/s11064-013-1128-7

Verma S (2016) Chemical constituents and pharmacological action of Ocimum sanctum (Indian holy basil-Tulsi). J Phytopharmacol 5(5):205–207

Verma AK, Raj J, Sharma V, Singh TB, Srivastava S, Srivastava R (2017) Epidemiology and associated risk factors of Parkinson's disease among the north Indian population. Clin Epidemiol Global Health 5(1):8–13

Viola KL, Klein WL (2015) Amyloid β oligomers in Alzheimer's disease pathogenesis, treatment, and diagnosis. Acta Neuropathol 129:183–206

Wang T, Li Q, Zhou Z, Liu J, Tao Y, Zhang Y et al (2024) Review of traditional uses, phytochemistry and pharmacology of Tibetan Medicine tangchong. J Ethnopharmacol 319:117129. https://doi.org/10.1016/j.jep.2023.117129

Yadav I, Sharma N, Velayudhan R, Zeeshan F, Singh MJ (2023) Ocimum sanctum alters the lipid landscape of the brain cortex and plasma to ameliorate the effect of photothrombotic stroke in a mouse model. Life 13(9):1877. https://doi.org/10.3390/life13091877

Zhang Z, Lu T-S, Yuan H-Y (2011) Neuroprotective effects of ultra-low-molecular-weight heparin in vitro and vivo models of ischemic injury. Fundam Clin Pharmacol 25(3):300–303. https://doi.org/10.1111/j.1472-8206.2010.00845.x

Zhang Z, Richmond A, Yan C (2022) Immunomodulatory properties of PI3K/AKT/mTOR and MAPK/MEK/ERK inhibition augment response to immune checkpoint blockade in melanoma and triple-negative breast cancer. Int J Mol Sci 23(13):7353. https://doi.org/10.3390/ijms23137353

Zhang Y, Chen H, Li R, Sterling K, Song W (2023) Amyloid β-based therapy for Alzheimer's disease: challenges, successes and future. Sig Transduct Target Ther 8:248

Zhou T, Zu G, Zhang X, Wang X, Li S et al (2016) Neuroprotective effects of ginsenoside Rg1 through the Wnt/β-catenin signaling pathway in both in vivo and in vitro models of Parkinson's disease. Neuropharmacology 101:480–489. https://doi.org/10.1016/j.neuropharm.2015.10.024

Neuro-Resilience with *Emblica officinalis* (Amla): Targeting Oxidative Stress and Neuroinflammation

13

Anuradha Sharma, Anuj Babbar, and Kokkonda Jackson Sugunakara Chary

Abstract

Neuro-resilience refers to the brain's capacity to adapt to and recover from stress, trauma, and other challenges. It plays an important role in delaying the onset of neurodegenerative diseases, which damage "neurons"—the fundamental units of the brain and spinal cord. According to WHO data from 2021, over 3 billion people are affected by neurological disorders worldwide. Conditions such as neurodegenerative diseases, epilepsy, and dementia are major health concerns, often incurable and progressively worsening over time. Although several drugs are available for the treatment of these disorders, they primarily slow neuronal damage or act on the symptoms of the disease. Phytochemicals, however, not only perform these functions but also promote the regeneration of neurons. Because they are derived from plants, they tend to have fewer side effects, making them a safer alternative to conventional therapies. This chapter explores the neuroprotective potential of phytochemicals found in *Emblica officinalis* (Amla), including flavonoids, hydrolyzable tannins, simple phenolics, and tannin-related compounds. It highlights their mechanisms of action—such as antioxidant, anti-inflammatory properties, as well as enhancement of mitochondrial function and cellular signaling. The chapter also discusses future aspects, focusing on improved advanced delivery methods to enhance bioavailability through synergistic formulations, integrative approaches, and phytonanotechnology. With the rising global demand for safer and cost-effective neurotherapeutic option, *E. officinalis* emerges as a potent candidate.

A. Sharma (✉) · A. Babbar · K. J. S. Chary
Department of Molecular Biology and Genetic Engineering, School of Bioengineering and Biosciences, Lovely Professional University, Phagwara, Punjab, India
e-mail: anuradha.28927@lpu.co.in

H. Dilnashin, S. P. Singh (eds.), *Phototherapeutic Approaches to Neurodegeneration*,
https://doi.org/10.1007/978-981-95-4751-7_13

Keywords

Neurodegeneration · *Emblica officinalis* · Phytochemicals · Antioxidant · Alternative therapy

13.1 Introduction

With the increasing prevalence of stress and various diseases worldwide, both physical and mental health are deteriorating significantly. The concept of neuro-resilience, which is characterized by adaptation, recovery, and strengthening the body against stress, is emerging as a vital mechanism to deal with such health conditions. Neuro-resilience harnesses the brain's plasticity, the ability of neuronal cells in mammals to rewire and reorganize in response to environmental stressors, thereby restoring normal brain function. This neuronal rewiring also helps in maintaining memory retention, cognitive functioning, and recovery from injuries due to the associated property of plasticity. Advanced research in neurosciences has shown that various brain parts and neurotransmitter systems are responsible for neuro-resilience. For example, the prefrontal cortex, which helps in decision-making and impulse control, plays a crucial role in regulating our emotional responses and modulating stress. Similarly, neurotransmitters like dopamine and serotonin contribute to resilience by influencing mood and emotional well-being (DeNike 2024).

However, with aging or persistent exposure to stress, the levels of oxidative stress increases significantly, resulting in neuroinflammation and diminished neuroplasticity. These two processes—oxidative stress and neuroinflammation—are deeply interlinked and are playing a pivotal role in the development of neurodegenerative diseases and their progression. Excessive production of reactive oxygen species (ROS) and reactive nitrogen species (RNS), coupled with an overwhelmed antioxidant defence system, disrupts intracellular pathways, resulting in cellular damage and neuronal death. The imbalanced redox homeostasis affects brain functioning by activating chronic activation of astrocytes and glial cells along with damaged neurons, thus developing chronic neuroinflammation. This sustained inflammation establishes a vicious cycle of neuronal damage and death by exacerbation of oxidative damage via proinflammatory cytokine production and vice versa (Ayuwardani et al. 2024). Under such conditions, neuroresilient mechanisms—such as antioxidant defence, neuroplasticity, and cellular repair—are compromised, leading to the development of brain function impairments, including neurodegenerative diseases.

13.2 Neurodegenerative Diseases and Their Treatment

Neurological disorders are now the leading cause of global morbidity and disability, currently affecting nearly three billion individuals worldwide. Disability, illness, and premature deaths caused by neurological conditions have increased by 18% since 1990, and 80% of these deaths and health issues are being reported from

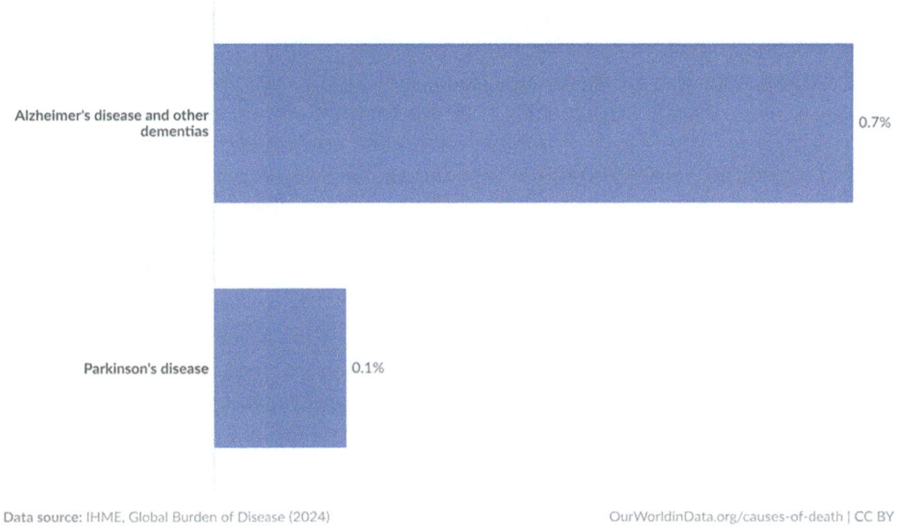

Fig. 13.1 Prevalence of Alzheimer's and Parkinson's disease among various neurological disorders

developing countries (Kaur et al. 2013). Top 10 neurological conditions that cause death and disability include stroke, migraine, dementia, epilepsy, autism, diabetic neuropathy, etc., with the most common neurodegenerative diseases being Alzheimer's, Parkinson's disease (Fig. 13.1), ALS, Huntington's disease, and spinal muscular atrophy (SMA) (WHO newsroom, Over 1 in 3 people affected by neurological conditions, the leading cause of illness and disability worldwide 2024; Lamptey et al. 2022).

Current therapeutic approaches for neurodegenerative diseases (NDs) target disease pathogenesis or only the improvement of the symptoms of the associated disease. Monoclonal antibodies like *aducanumab*, which target amyloid beta plaques and remove them, are used in AD. Cholinesterase inhibitors, a class of medication used as first-line medication for AD, prevent the degradation of acetylcholine and diminish the cognitive decline. Other classes for AD medication include glutamate regulators, which are uncompetitive NMDA receptor antagonists and also protect from inhibitory effects due to overstimulation. The challenge for managing NDs is that the drugs are symptom-targeting drugs and cannot eliminate the root cause of neurodegenerative diseases. The biggest challenge here is that these are synthetic drugs, and the blood-brain barrier (BBB) prevents substances from entering the brain from the bloodstream. For a compound to cross the BBB, physicochemical properties, including size, molecular weight, solubility, lipid solubility, and charge, are major factors. Molecules like ethanol, CO_2, and barbiturates can freely cross the BBB through passive diffusion. Another way of transportation through BBB is via receptor-mediated transport, which also aids the transport of hydrophilic molecules such as peptides and proteins (Lamptey et al. 2022). Further, current approaches for

PD management also provide symptomatic relief, including dopamine-replenishing drugs, which can help in restoring motor functions resulting from the loss of dopamine. Along with loss of motor functions, non-motor symptoms like depression, psychosis, sleep disorders, dementia, and olfactory deficits are also observed. Drugs like rivastigmine, donepezil, clozapine, quetiapine, melatonin, and clonazepam are used to treat non-motor symptoms. Whereas these 2 diseases have drugs targeting their symptoms and neurodegeneration, ALS has no disease-targeting therapies available. The only drugs available for ALS are glutamate-receptor antagonists and free radical scavenging intravenous drugs, which delay the progression of the disease.

This palliative approach to disease management offers modest benefits in neurological disease management; however, it is associated with limited efficacy in reversing the disease progression and is typically constrained by adverse effects. In such conditions, the supportive therapies having the potential to enhance/reprogram the neuro-resilience, like mindfulness and meditation, somatic therapy, social support, self-care, cognitive restructuring, and herbal interventions, are warranted. Herbal interventions are well reported to be effective against oxidative stress, neuroinflammation, and mitochondrial dysfunction, thus offering a multi-targeted approach and enhancing neuroplasticity (DeNike 2024; Hampton 2018).

13.3 *Emblica officinalis* and Its Phytoconstituents

E. officinalis, commonly known as Amla, is a member of the Euphorbiaceae family, the traditional Ayurvedic medicinal plant of India. A popular name for the Indian gooseberry is *Phyllanthus emblica*. It is deep-rooted in Ayurveda, Unani, and other traditional systems; the fruit is valued as a "Rasayana," or rejuvenator. Traditional uses encompass treatment of common cold, fever, cough, asthma, diabetes, digestive ailments, and skin diseases, and as a tonic for liver and hair health (Saini et al. 2022; Variya et al. 2016). Modern research has also revealed antioxidant, anti-inflammatory, antimicrobial, hepatoprotective, neuroprotective, and anticancer properties of *E. officinalis* (Mirunalini and Krishnaveni 2010).

E. officinalis possesses a rich array of phytochemicals, including vitamin C, tannins, phenolic acids, alkaloids, flavonoids, epigallocatechin, etc., which are responsible for the pharmacological properties of this plant.

13.3.1 Vitamin C

E. officinalis is one of the best natural sources of vitamin C. It is an effective antioxidant that supports healthy skin and hair, strengthens the immune system, and fights against free radicals in the body. The synthesis of collagen and the healing of wounds are both aided by vitamin C (Scartezzini et al. 2006). It is involved in modulating different metabolic pathways, also boosting the adaptive immune response. Studies suggest that high vitamin C intake from *E. officinalis* contributes to improving blood glucose metabolism in type 2 diabetes mellitus, improves iron absorption in cases of anemia, and strengthens immunity through increased white blood cells. Therapeutic reports show vitamin C from *E. officinalis* protects against infections, accelerates healing of ulcers, and helps mitigate oxidative stress, which underlies chronic conditions such as cardiovascular disease and diabetes (Gul et al. 2022).

Additionally, vitamin C also improves brain health and cognitive function by modulating different neurotransmitter pathways like glutamatergic, cholinergic, and catecholinergic systems. These systems play an important role in neuronal differentiation, myelin sheath formation, and neuronal maturation (Travica et al. 2017).

13.3.2 Polyphenols

Gallic acid is a principal polyphenol antioxidant found in *E. officinalis*, which imparts strong antioxidant activity, scavenging free radicals that drive inflammation and tissue damage. In vitro and in vivo studies show that gallic acid tends to protect organs such as the liver, heart, and brain from injuries caused by toxins and ischemia. It has shown efficacy in improving metabolic disturbances like dyslipidemia and hepatic steatosis induced by high-fat diets and reduces oxidative stress in experimental models. Importantly, gallic acid also showed potential anticancer properties by inducing apoptosis in cancer cells, suggesting a potential adjuvant role alongside chemotherapeutics (Prananda et al. 2023; Sawant et al. 2010).

Another polyphenol, ellagic acid, is studied for its multifaceted benefits, like inhibiting oxidative stress, inflammation, and fibrosis, and showing notable renoprotective effects against chemical-induced nephrotoxicity. Clinical trials highlight ellagic acid's ability to improve skin hyperpigmentation, support cognitive function, ameliorate UV-induced skin damage, and provide cardiovascular protection by decreasing blood pressure and arterial thickness. In vivo, ellagic acid from *E. officinalis* demonstrates neuroprotective potential and reduces markers of inflammation through modulation of key cytokines and enzymes (Sharif et al. 2023; Dasaroju et al. 2014).

Both of these polyphenols have shown significant neuroprotective properties against neurodegenerative conditions. Ellagic acid has a neuroprotective effect due to its strong antioxidant status, along with its mitochondrial protective effect and iron-chelating properties. Gallic acid exhibited neuroprotective effects, resulting in memory and LTP enhancement by ameliorating the cytokine increase in a rat model of traumatic brain injury (Mirshekar et al. 2018). Gallic acid is further reviewed to exhibit anti-inflammatory properties via modulating pathways like NF-κB and MAPK, which are key pathways of inflammation (Bai et al. 2021).

13.3.3 Tannins

Corilagin, geraniin, chebulagic acid, and elaeocarpusin are some of the complex tannins and ellagitannins found in *E. officinalis* that are the most active and have strong antioxidant activity as well as anti-inflammatory characteristics. Tannins have antibacterial properties and can help shield cells from oxidative damage. Tannins are recognized for their antioxidant capabilities, which can prevent vitamin C from being degraded and support a healthy digestive system and the body's natural cleansing process (Chambial et al. 2013). The activity of tannic acid (TA) against the influenza virus compared to gallic acid is 12 times more potent. The antiviral activity of galloyl is dependent on the total number of residues. The suppression of neuraminidase activity and influenza A virus (IAV) receptor binding is linked to TA (high molecular weight tannin) activity. A TA-enriched extract has been shown to prevent the infection of the human papillomavirus (HPV) type (Kaczmarek et al. 2020). Tannins also exhibit antibacterial activities, as suggested by a study from Dabbaghi et al. (2019), where TA was incorporated in hydrogels to study the action of TA against antibacterial activity against *Escherichia coli* and *Staphylococcus aureus* (Dabbaghi et al. 2019).

Tannin-enriched fractions of *E. officinalis* have been found to ameliorate the high salt and cholesterol-rich diet-induced cognitive impairment. These fractions exhibited strong antioxidative potential and regulated the Nrf2-ARE pathway (Husain et al. 2018).

13.3.4 Flavonoids

E. officinalis has higher concentrations of bioactive flavonoids such as naringin, quercetin, kaempferol, and rutin than other fruits like lemon and cucumber peels. These substances enhance the overall health advantages of *E. officinalis* because they contain anti-inflammatory and antioxidant characteristics (Rahman et al. 2020). These compounds have strong antioxidant activities, which contribute to *E. officinalis*'s role in protecting against oxidative stress and rejuvenating tissues. Both quercetin and kaempferol act as natural free radical scavengers, helping shield cells from damage caused by inflammation and environmental toxins. They are also effective at regulating immune responses, reducing LDL oxidation, and mitigating risk factors for chronic diseases such as cardiovascular and metabolic syndromes (Ahmed et al. 2021).

Quercetin suppresses the production of pro-inflammatory cytokines like TNF-α, IL-6, and NF-κB, molecules that drive many inflammatory and autoimmune conditions. It enhances overall immunity, modulates allergic responses, and provides protection against various infectious diseases (Sun et al. 2021). Evidence suggests quercetin can shield liver cells from toxins, as shown in studies investigating its ability to mitigate drug-induced hepatotoxicity and support liver function. Preclinical research shows the highlights of quercetin's ability to induce apoptosis in cancer cells and act as a chemopreventive agent by disrupting cellular proliferation and angiogenesis. Quercetin demonstrates broad-spectrum activity against bacteria (including resistant strains) and fungi, making it useful as a supportive agent in infection control (Xu et al. 2024). In the scenario of neuroinflammation, quercetin activates the energy sensor molecule, AMPK, and manages the homeostasis. It further interacts with NLRP3 inflammasome and NF-κB and is helpful in the management of neuroinflammation-related toxicity in various neuroinflammatory and neurodegenerative conditions (Chiang et al. 2023).

13.4 Evidence of *E. officinalis* in Brain Health Improvement

13.4.1 *E. officinalis* for Neurodegenerative Diseases

Parkinson's disease is characterized by a gradual and selective loss of dopaminergic neurons in the substantia nigra pars compacta (SNPC) region, having symptoms like muscular rigidity, tremors, bradykinesia, postural instability, and trouble in swallowing. PD is a multifactorial disease that may include excitotoxicity, protein aggregation (Lewy bodies), and inhibition of mitochondrial complex-1, all collectively leading to the formation of ROS, disturbing dopamine metabolism, and reducing levels of endogenous antioxidant compounds, resulting in neurodegeneration.

Few studies have explored the neuroprotective effect in Parkinson's disease models. In a study, symptoms of haloperidol-induced catalepsy and Parkinson's disease were significantly reduced upon pretreatment of animals (Wistar albino rats) with the methanolic *E. officinalis* fruit extract. The pretreated animals with *E. officinalis* extract exhibited a significant reduction in malondialdehyde (MDA) levels and significant increase in superoxide dismutase (SOD), glutathione (GSH), and catalase (CAT) levels as compared to the non-treated group (Singh et al. 2020). Another model used Drosophila flies, which were given rotenone to induce Parkinson's and were treated with both low (0.25%) and high (0.50%) doses of CO_2 extracts of *E. officinalis* for 7 days. Both groups showed a significant increase in climbing ability of flies compared to the disease group, with the higher dose group showing more significant results. Biochemical assays revealed a significant decrease in levels of MDA and an increase in levels of dopamine (DA) in the brain of the treated group when compared to the diseased control group (Patil and Date 2025). A polyherbal formulation containing *E. officinalis* exhibited neuroprotective effects against MPTP intoxication in Swiss albino rats intoxicated with MPTP and enhanced dopaminergic neuronal population in the treated group (Srivastava et al. 2020). In a study, H_2O_2-induced DNA damage was ameliorated when cells were treated with aqueous or methanolic extracts of *E. officinalis* in a dose-dependent manner after exposure to H_2O_2 for 24 hours (Ramakrishna et al. 2014).

Alzheimer's disease is an age-influenced and irreversible neurodegenerative disorder (ND), which can be identified by short-term memory loss and, in advanced stages, by confusion, aggression, mood changes, long-term memory loss, and social withdrawal. Pathophysiology of AD shows that there is an accumulation of amyloid-beta (Aβ) and tau proteins within the brain. Tannoids, a bioactive class of phytochemicals found in *E. officinalis* fruits, have reportedly attenuated aluminum-induced neurotoxic effects on Wistar rats. Pretreatment with the tannoid principle of *E. officinalis* before treatment with $AlCl_3$ attenuated $AlCl_3$-induced memory deficits, ameliorated AlC13-induced aluminum overload, nullified AlC13-mediated acetylcholinesterase (AChE) hyperactivity, and attenuated $AlCl_3$-induced apoptosis (Justin Thenmozhi et al. 2016).

A hallmark feature of AD is low levels of neurotransmitter acetylcholine, which is broken down or inhibited by acetylcholinesterase (AChE) and butyrylcholinesterase (BuChE) enzymes. In vitro analysis of crude methanol extracts of dry fruit of *E. officinalis* showed inhibitory activity against acetylcholinesterase (AChE) and butyrylcholinesterase (BuChE) by Ellman's method with IC_{50} = 53.88 μg/mL and 65.12 μg/mL, respectively (Biswas et al. 2015). A study on *Dracocephalum moldavica* L. has reported that flavonoid extract has bioactive compounds like apigenin and luteolin, which are also found in *E. officinalis*, and showed amelioration of cognitive impairments and neurodegeneration and improved the antioxidant defense system in APP/PS1 mice. A decrease was seen in Aβ deposition, a decrease in insoluble Aβ levels, and inhibition of the β-amyloid processing pathway involving down-regulation of β-secretase and β-C-terminal fragments. The in vitro model of the same study used copper-induced APP Swedish mutation overexpressing SH-SY5Y cells, which were protected with treatment of TFDM (Chen et al. 2020; Liu et al. 2018). A test on amnesia induced by scopolamine and diazepam in a mouse model was used to check memory-improving, cholesterol-lowering, and anticholinesterase activity by administering *E. officinalis* at doses of 50, 100, and 200 mg/kg to Swiss mice for 15 days. *E. officinalis* extract showed significant inhibition of cholinesterase activity when compared to the amnesia-induced group. Memory enhancement was also significantly improved, which was evaluated by behavioral studies (Vasudevan and Parle 2007).

13.4.2 *Emblica officinalis* for Epilepsy

Epilepsy, the fourth most common neurological disorder, is characterized by seizures, loss of consciousness, and cognitive and social impairments in patients. Epilepsy, though mechanistically distinct from chronic neurodegenerative disorders, shares oxidative stress and excitotoxicity pathways, making antioxidant phytochemicals relevant. Epileptic seizures can originate due to an imbalance of glutamate-GABA neurotransmission. This imbalance facilitates depolarization of the neuronal membrane by lower-intensity stimuli, which leads to rapid propagation and action potentials. Other reasons can be neuronal hyperexcitability and oxidative damage resulting from excessive free radical production. A study involving male Wistar rats when treated with 500 and 700 mg/kg of *E. officinalis* orally for 28 days daily showed delayed seizures and, in 3 cases (of 6), even prevented the onset of seizures. The seizures were induced by lithium-pilocarpine, and the study reported that neither of the treatments alters any biochemical or hematological parameters

(Rivadeneyra-Domínguez et al. 2024). A computational study investigated genes related to epilepsy by use of GeneCard, OMIM, DrugBank, TTD, and DisGeNET. Then phytochemicals from *E. officinalis* were identified, and molecular docking was performed using target proteins and ligands (bioactive compounds from *E. officinalis*). Docking results revealed that key compounds from *E. officinalis*, such as 2-ethylhexyl, phyllanthin, luteolin, and kaempferol, target critical proteins such as SRC, AKT1, APP, MAPK3, and MAPK1 molecules involved in epilepsy (Xiao et al. 2025). Another study, including male Wistar rats, investigated treatment with 300, 500, and 700 mg/kg doses of hydroalcoholic extract of *E. officinalis* for 7 days. The seizure was induced using PTZ. The 500 and 700 mg/kg dosees of hydroalcoholic *E. officinalis* completely abolished the generalized tonic seizure state (Golechha et al. 2010).

13.4.3 *E. officinalis* for Traumatic Brain Injury

Traumatic brain injury (TBI) refers to an injury caused by an external force to the brain, which can be caused by a forceful bump, blow, or jolt to the head (National Institute of Neurological Disorders and Stroke 2025). TBI has been associated with a cascade of oxidative stress, neurotoxicity, mitochondrial dysfunction, and neuronal loss, particularly following the secondary phase of injury, which can result in cognitive decline and neurodegeneration (Bhargavi et al. 2025). Recent research has focused on designing and employing natural plant extracts and herbal products against TBI-triggered cognitive impairments and degenerative processes in experimental models. *E. officinalis* has various phytochemicals—flavonoids, tannins, polyphenols, vitamins—that have been reported to have therapeutic activity against TBI. Reports suggest that these phytochemicals can inhibit the progression caused by secondary injury and have shown neuroprotective effects by activating the neuroregenerative signaling pathways. These compounds have the potential to alter brain functions, enhance antioxidant potential, and reduce the inflammatory response, making them a potential candidate for TBI management. Luteolin, a flavonoid present in *E. officinalis*, has reduced neuronal degeneration, alleviated brain edema, preserve BBB disruption, and mitigate oxidative stress in a weight drop TBI rat model (Xu et al. 2014). In TBI mouse models, apigenin, another flavonoid, reportedly prevented microglial activation, lowered levels of TNF-α and IL-1β, and attenuated the NF-kB pathway. By modulation of Bax/Bcl-2 expression and caspase-3 activation, neuronal apoptosis was also inhibited, overall preserving hippocampal synaptic plasticity, upregulation of BDNF, and antioxidative activity (Hasan et al. 2024). Multiple polyphenols, like TA, have ameliorated behavioral defects, oxidative damage, mitochondrial dysfunction, and inflammation in the controlled cortical impact (CCI) rat model (Salman et al. 2020). Ellagic acid, another polyphenol found in *E. officinalis*, has reportedly reduced neuroinflammation, restored BBB permeability, and prevented cognitive deficits in a weight drop TBI rat model (Farbood et al. 2015; Mashhadizadeh et al. 2017). Resveratrol, a polyphenol, can be found in a wide variety of plants and can improve the neurological function in TBI by reducing miR-21 expression, hence avoiding neuronal apoptosis. Improvement in post-TBI cognitive function was observed due to an increase in synaptophysin and activation of the SIRT1/PGC-1 pathway. Reports suggest that resveratrol can suppress neutrophil infiltration, microglial activation, and cytokine production, therefore protecting against neuroinflammation (Hasan et al. 2024). Neuroprotective effects of resveratrol were observed in TBI, which were due to improved antioxidant potential, scavenging of reactive species, and targeting p38/Nrf2 signaling in the brain (Shi et al. 2018). Epigallocatechin (EGCG), which also belongs to the class of polyphenols, when administered, showed a reduction in brain infarct volume and edema, along with inhibition of microglial activation, neutrophil infiltration, reduction in levels of TNF-α, IL-1β, and IL-6, and activation of the AMPK pathway. EGCG has also been reported to attenuate neuronal apoptosis in the hippocampus by upregulating Bcl-2 and reducing Bax and cleaved caspase-3 expression, showing therapeutic potential in reducing secondary brain damage post-TBI (Hasan et al. 2024). Gallic Acid (GA), which has anti-tyrosine potential, can reportedly protect the brain against increased antioxidant enzymes and reduce inflammation in the cerebral hypoperfusion model. In TBI rats, administration of GA inhibited pro-inflammatory cytokines, reduced lipid peroxidation, and enhanced memory (Mirshekar et al. 2018).

13.4.4 Lifestyle-Associated Cognitive Impairment

Modern diets consist of a high number of processed foods that contain high quantities of salt and cholesterol, causing adverse effects on our health. Such diets, when consumed for a longer time, can result in various disorders, including cognitive deficits (Husain et al. 2018). A diet containing higher amounts of cholesterol is a basic risk factor for inducing Alzheimer's disease, as a high-fat diet stimulates deposition of amyloid beta plaques in the cerebral cortex (Saha et al. 2013). Neurological disorders can also be induced by various adulterations in the food. For example, lead is a non-biodegradable, ubiquitously present food contaminant that induces oxidative stress and neurotoxicity, resulting in learning loss, memory, and cognitive loss (Paul et al. 2025). Studies showed that a high salt and cholesterol diet (HSCD), which induced oxidative stress in rats, was attenuated by tannin-enriched fractions of *E. officinalis* (EOT). After the treatment, significant alterations in levels of MDA, GSH, thiobarbituric acid reactive substances (TBARS), glutathione peroxidase (GPx), mitochondrial adenosine triphosphate (ATP), and mitochondrial membrane potential were observed, which were possibly due to significantly increased expression of Nrf2 in the hippocampus and cortex (Husain et al. 2018). Another study reported that escape latency time (ELT) was improved when compared to the non-treated group, with the groups treated with *E. officinalis* in a high-fat diet-induced memory loss model of rats. Rats treated with *E. officinalis* showed a significant decrease in cholesterol levels, which ultimately stops the neuronal damage, and neurotransmission is increased (Saha et al. 2013). Lead toxicity, which induces severe degenerative changes in the brain by enhancing oxidative stress, was significantly ameliorated when these groups were treated with *E. officinalis*. A significant improvement in antioxidant enzymes like CAT, SOD, and GPx was observed, and TBARS levels were decreased in *E. officinalis*-treated groups (Paul et al. 2025).

13.5 Possible Molecular Mechanism Targeted by *E. officinalis*

The neuroprotective potential of *E. officinalis* is primarily attributed to its potent antioxidant and anti-neuroinflammatory properties of this plant. Additionally, its ability to modulate neurotransmitter synthesis and promote neurogenesis also contributes significantly in maintaining brain health and preventing neurodegeneration.

13.5.1 Antioxidant Pathway

As discussed, *E. officinalis* is rich in hydrolysable tannins like emblicanin, which was discovered first to be responsible for its antioxidative properties. Further investigations revealed that tannoids—including emblicanin A, emblicanin B, punigluconin, pedunculagin, and gallic acid—exhibit structural similarities to ascorbic acid and are responsible for its antioxidant properties (Variya et al. 2016). When cells are under stress, they produce free radicals like ROS and RNS, which can damage cells by lipid peroxidation, protein denaturation, and DNA damage (Fig. 13.2). Various in vitro studies and biochemical assays—including ABTS, DPPH, phosphomolybdenum, lipid peroxidation, and deoxyribose degradation tests—have revealed the free radical scavenging activity of *E. officinalis* fruit extracts (Middha et al. 2015).

Excess levels of polyunsaturated fatty acid (PUFA) can damage the brain by inducing oxidative stress (Farbood et al. 2015). A study on aged rats confirmed the antioxidant activity of ethyl acetate extract of *E. officinalis*, as oral administration of this extract at 300 mg/kg of concentration for 100 days in these rats significantly increased the hepatic peroxisome proliferator-activated receptor alpha (PPAR-α) protein levels, which is involved in lipid and cholesterol metabolism. The same treatment also reduced thiobarbituric acid reactive substances (TBARS) levels in both serum and hepatic mitochondria, effectively preventing oxidative stress (Yokozawa et al. 2007; Ramakrishna et al. 2014). Thus, the different phytoconstituents of *E. officinalis* reinforce the strong antioxidant system as a defense against ROS/RNS production.

13.5.2 Anti-inflammatory Pathway

Neuroinflammation serves as a root cause of neurodegenerative diseases, and several exogenous and endogenous factors (like trauma, stroke, chronic infections, abnormal protein aggregation like Aβ plaques, tau fibrils, or synthesis of misfolded

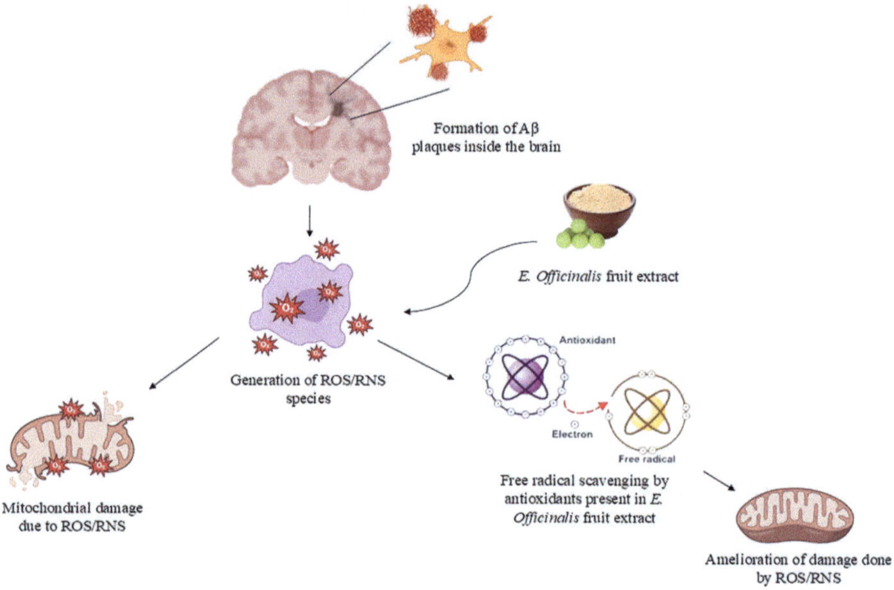

Fig. 13.2 Targeting of reactive oxygen species (ROS)/reactive nitrogen species (RNS) induced by oxidative stress

proteins like α-synuclein) can lead to microglial activation. Microgliosis further triggers several signal transduction pathways, including PI3K/AKT, MAPK, and mTOR, leading to activation of NF-κB in activated B-cells and chemokine and cytokine production. However, *E. officinalis*-based in vitro studies suggest the amelioration of cytokine (IL-6, TNF-α) production in the BV2 microglial cell line and enhancement of neurite outgrowth in N2A cells via MAP2 and TuJ1 expression (Phochantachinda et al. 2021). Further, the in vivo studies suggest *E. officinalis's* ability to inhibit systemic and brain specific inflammation, restore the antioxidant (GST, GR) system and enhancement of synaptic plasticity by modulating the serotonergic receptors (HT1D and 5-HT2A) (Gouda et al. 2024). Additionally, Phyto-polyphenols of this plant are found to activate the Nrf2 signaling pathway, which is critical in inhibiting inflammation. Protein expression and molecular docking studies revealed the compatibility of fruit polyphenols of this plant with Nrf2 as a ligand and stimulation of the Nrf2 pathway in the hippocampus (Li et al. 2023). Thus, *E. officinalis* quenches the excessive ROS/RNS production and stabilizes the redox balance and inhibits microglial activation via Nrf2 pathway (Fig. 13.3).

13.6 The Need for Advanced Delivery for Phytochemicals

Although *E. officinalis* possesses wonderful antioxidative and anti-inflammatory properties, supplementing its neuroprotective potential, the clinical translation of these beneficial effects is hindered by considerable challenges, including

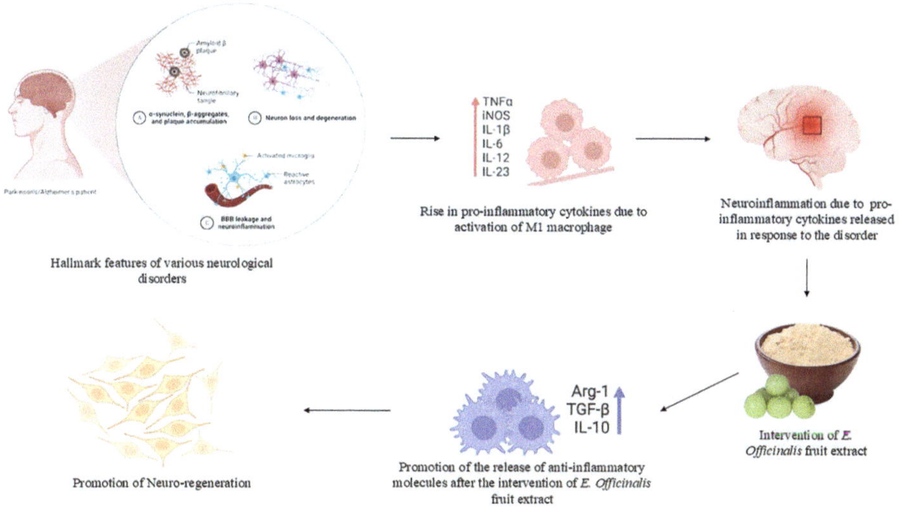

Fig. 13.3 Targeting of inflammatory pathway by *E. officinalis* by lowering levels of pro-inflammatory factors like TNF-α, iNOS, IL-1β, IL-6, IL-12, IL-23, and promotion of anti-inflammatory molecules like Arg-1, TGF-β, and IL-10

bioavailability, poor absorption, limited stability, and restricted BBB permeability (Prananda et al. 2023). Furthermore, a variety of factors, such as the food matrix's composition, processing methods, the digestive environment, and interactions with other dietary substances, affect how bioavailable these compounds are—that is, how much they are liberated from the plant matrix and made available for absorption (Bode and Dong 2015). These challenges highlight the necessity for advanced delivery systems designed to enhance both release and absorption of phytochemicals while minimizing adverse effects (Imam et al. 2021). Traditional methods that simply increase dosage to overcome poor bioavailability frequently raise the risk of toxicity, emphasizing the importance of innovative synergistic formulations, integrative approaches, and nanotechnology-based nanoparticles. These carriers enable targeted and controlled release by protecting phytochemicals from degradation, improving their solubility and stability, and thereby enhancing therapeutic outcomes (Fig. 13.4) (Zhang et al. 2018).

The convergence of nanotechnology, biotechnology, and intelligent AI systems is revolutionizing drug delivery systems, marking the beginning of dynamic new era. Developing more accurate, patient-centred, and flexible delivery systems that can get past persistent issues like inadequate targeting, bioavailability, and drug resistance is key to the field's future (Fernandes 2025). Aiming delivery methods like synergistic formulations, integrative approaches, and nanotechnology, which offer biocompatible advanced scientific techniques. Possessing the ability to biodegrade and enter tissue and cells without causing particular toxicity, immunogenicity, or accumulation in organs other than the tumor (Rana and Bhatnagar 2021). In order to achieve these objectives, multidisciplinary research teams and solid industry and

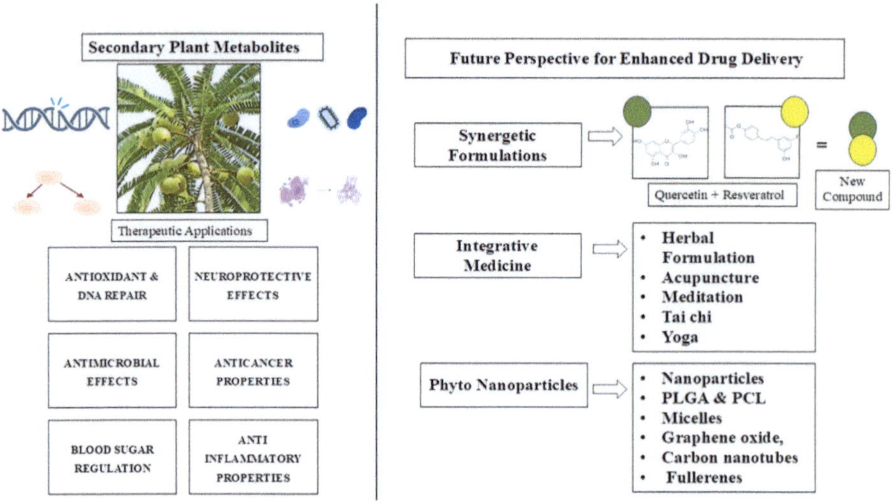

Fig. 13.4 Various biotechnological techniques employed in the advancement of delivery methods of phytochemicals

clinical partnerships must work together. They must strive for both innovation and safe, scalable, and useful clinical applications (Rana and Bhatnagar 2021).

13.6.1 Synergetic Formulations

Synergistic formulations have gained momentum in the medicine and pharmacology fields due to their ability to improve the bioavailability and safety profile of therapeutics. The approach of combining different agents or medications, also termed "cocktail formulations," enables higher potency with reduced toxicity, acting on particular conditions like cancer or multidrug-resistant infections. This is because many illnesses necessitate concoctions for treatment (Greco et al. 1996). However, a particular medication dose may be required, which may enable the therapeutic effect to be maximized while the adverse effect, or side effects, is minimized.

A study on oral carcinogenesis revealed that combination therapy of *E. officinalis* extract with cisplatin exerts inhibitory effects through induction of apoptosis and suppression of molecules involved in cell proliferation and angiogenesis, suggesting enhanced anticancer efficacy compared to cisplatin alone (Mirunalini et al. 2016). *E. officinalis* combined with ampicillin tested against *S. aureus*, *E. coli*, and *Klebsiella pneumoniae*, with minimum inhibitory concentration (MIC) values between 3.13 and 6.25 μg/mL, resulted in an increased zone of inhibition (ZOI) (Tiwana et al. 2025). A study by Chauhan et al. (2023) formulated a lyophilized slurry (LS) of ethanolic extracts of *Lycopersicon esculentum*, *Beta vulgaris L.*, *Phyllanthus emblica L.*, and *Daucus carota L.* against the MKN-45 cell line gastric adenocarcinoma. Cells showed that the LS exhibited significantly higher content of phytochemicals and antioxidant activity. Also showing reduced cell viability cytotoxically affected by the LS extract, which at 320 μg/mL reduced cell viability to about 23.33% and reached cytotoxic levels that were comparable to those of doxorubicin at a higher concentration reported in the literature, is responsible for this enhanced anticancer effect (Chauhan et al. 2023).

Additionally, a recent study by Jaisamut et al. (2021) revealed that quercetin flavonoids, when combined with resveratrol, stabilized the latter compound, preventing its degradation and thereby enhancing its anti-inflammatory and antioxidant effects. Their formulations involve processes like cell cycle arrest, DNA repair, and immune system activity, resulting in stronger and more effective anticancer effects. The future of enhanced delivery systems is expected to bring significant change in the field of medicine and human health through the integration of synergistic formulations.

13.6.2 Phyto Nanoparticles

Central to these advancements are phyto-nanoparticles, which represent a new frontier in delivery technologies. Synthesized using various green synthesis techniques that employ plant extracts, gaining attention for their biocompatibility, low toxicity, eco-friendly production, and enhanced targeting capability. Phyto-nanoparticles can be functionalized with specific ligands to direct drugs precisely to diseased cells, such as cancer or infected cells, thus reducing collateral damage to healthy tissues. Their ability to selectively accumulate in target sites, biodegrade in a controlled way, and provide a versatile application of carrying both traditional and novel therapeutics of drug delivery. These nanoparticles have demonstrated promising potential in cancer therapy, antimicrobial treatment, diagnostics, and chronic disease management applications (Teleanu et al. 2019).

The treatment of NDs is significantly challenged by the presence of the BBB, which restricts drug penetration into the central nervous system (CNS). Current therapies often suffer from poor efficacy, limited brain delivery, and systemic toxicity, underscoring the pressing need for novel treatment strategies (Kim et al. 2017). Among various approaches, nanotechnology has emerged as a highly promising and safe platform for targeted drug and gene delivery to the CNS. Nanoparticles, typically ranging from 1 to 1000 nm in size, are capable of interacting at the molecular level with biological systems, making them ideal carriers for crossing the BBB. Diverse materials have been employed for nanoparticle synthesis, including natural polymers (e.g., proteins and polysaccharides), synthetic polymers such as poly(lactic-co-glycolic acid) (PLGA) and polycaprolactone (PCL), and inorganic substances including gold, silver, and cerium (Kim et al. 2017). These nanocarriers provide several advantages, including high drug loading capacity, enhanced drug permeability across the BBB, improved stability, and reduced systemic toxicity, all contributing to their suitability for managing and treating neurodegenerative diseases (Table 13.1).

Among inorganic nanocarriers, metal-based nanoparticles have attracted significant attention because of their ease in crossing the BBB and modifiable properties, which facilitate efficient brain targeting. Functionalization with brain-specific ligands—such as antibodies, proteins, and small molecules—further enhances targeted delivery, theranostics, and imaging applications (Lamptey et al. 2022). The unique physicochemical and optical properties of gold nanoparticles (AuNPs), coupled with their biocompatibility, make them particularly suitable as carriers for small molecules and biomacromolecules within the CNS (Martins et al. 2017; Masoudi et al. 2020).

E. officinalis silver nanoparticles (AgNPs) have been investigated for brain delivery due to their accumulation ability in critical brain regions such as the hippocampus. *E. officinalis*-based AgNPs exhibited anti-inflammatory and antioxidant effects in microglia by upregulating the antioxidative enzymes (Akhter et al. 2024). Cerium oxide nanoparticles (CNP), known for their potential antioxidant activity for their

Table 13.1 Types of nanoparticles and their therapeutic applications

S. no.	Nanoparticle type	Source	Particle size (nm)	Therapeutic role	References
1	Silver nanoparticles (AgNPs)	Fruit extract	10–20, 50, and 188	Antimicrobial against Gram-positive and Gram-negative bacteria; antioxidant, antihypercholesterolemic via HMG-CoA reductase inhibition, showed anticancer properties against the human MCF-7 breast cancer cell line and Hep2 cell line (laryngeal carcinoma cells)	Gandhi et al. (2023), Sivalingam and Pandian (2024), Rosarin et al. (2013)
2	Palladium nanoparticles (PdNPs)	Seed extract	28 ± 2	Showed antibacterial, cytotoxicity to HeLa cells; hemolytic profile evaluation	Dinesh et al. (2017)
3	Gold nanoparticles (AuNPs)	Fruit extract	5–60	Showed anticancer activity against gastric carcinoma AGS cell line via apoptosis	Wang et al. (2021)
4	Magnesium oxide (MgO) nanoparticles	Fruit extract	100–200	Showed antimicrobial (e.g., *Pseudomonas aeruginosa*, *Acinetobacter baumannii*, *Klebsiella pneumoniae*)	Thatyana et al. (2023)
5	Iron nanoparticles (Fe NPs)	Fruit extract	4–15	Showed anticancer activity (lung cancer A549 cell line)	Thatyana et al. (2023)
6	Nitrogen-doped carbon dots (NCDs)	Hydrothermal fruit extract	4	Showed catalytic (textile effluent reduction); potential antioxidant/catalytic applications	Gandhi et al. (2023)
7	Selenium nanoparticles (SeNPs)	Fruit extract	15–40	Showed antimicrobial; antioxidant; antifungal; free radical scavenging	Gandhi et al. (2023)

ability to switch oxidation states (Ce^{3+} and Ce^{4+}), have shown considerable promising effects in neuroprotection. A study by Javid et al. (2022) demonstrated the therapeutic potential of *E. officinalis*-based cerium oxide nanoparticles (CNP) in YM1 oesophageal squamous cell carcinoma (ESCC) and cancer stem cell-like cells (CSC-LC).

E. officinalis-based CNP therapy increased gene expression and boosts antioxidant enzyme activities—including SOD, CAT, thiol content, and total antioxidant capacity [TAC]—while simultaneously decreasing ROS and MDA levels (Javid et al. 2022). CeO2 NPs have also been observed to scavenge peroxynitrite radicals in ischemic stroke models and to improve motor functions in mouse models of multiple sclerosis and amyotrophic lateral sclerosis (Naz et al. 2017).

Further, *E. officinalis*-based magnesium oxide (MgO), selenium, and iron nanoparticles have demonstrated significant therapeutic effects. All the therapeutic applications of *E. officinalis*-based nanoparticles extend beyond their neuroprotective effects, encompassing antimicrobial, antioxidant, anti-inflammatory, and anticancer properties, as summarized in Table 13.1. These nanoparticles synthesized from *E. officinalis* represent a multifaceted strategy for various diseases, inflammation, and anticancer treatment.

13.6.3 Artificial Intelligence (AI) in Drug Delivery

AI is revolutionizing the field of drug delivery and holds immense potential to advance phytotherapeutics, including *E. officinalis* and its phytocompound. AI can significantly contribute to formulation design, targeted delivery, and enhanced therapeutic efficacy (Jena et al. 2024). AI-driven complex algorithms analyze different chemical and biological datasets to design and optimize drug delivery systems like nanofibers, nanodots, or nanoparticles, which can prove beneficial in targeted delivery and enhanced bioavailability of *E. officinalis*-based phytotherapeutics. Furthermore, machine learning platforms can predict synergistic formulations by identifying optimal combinations of E. officinalis phytochemicals with conventional drugs to enhance overall treatment outcomes. By increasing the speed of phytochemical screening and analyzing the large datasets, AI has accelerated the speed of phytochemical research and may pave the way for effective and tailored *E. officinalis*-based formulations for brain health promotion.

13.7 Conclusion and Future Perspective

E. officinalis (Amla) emerges as the potent natural agent in the pursuit of neuro-resilience with a strong foundation of antioxidant and anti-inflammatory properties. Extensive literature suggests the effectiveness of *E. officinalis*—in various forms, such as crude extracts, tannoid fractions, powders, pastes, ointments, or syrups—against various neurological disorders. Its bioactive phytochemicals, specifically flavonoids and polyphenols, are well-known for neutralizing the ROS and maintaining the redox balance, which play a crucial role in preserving cognitive function, increasing neuronal plasticity, and preventing neurodegeneration. Advanced drug design and delivery approaches could help *E. officinalis* achieve its neurodrug status by dealing with the issues related to BBB permeability, bioavailability, and precision. As oxidative imbalance and neuroinflammation are central to the majority of neurological disorders, multifaceted bioactivity with strong antioxidant and anti-inflammatory properties makes this plant a promising candidate for adjunct/alternative therapy for the promotion of neuro-resilience.

In the current era of AI and rapid biotechnological progress, the limitations of conventional phytotherapy emphasize the need for the development and integration of advanced drug design and delivery systems to enhance *E. officinalis*'s targeted action and therapeutic efficacy. Meticulous efforts in terms of clinical trials and translational approach-based research are required for optimal drug formulations, dosage frequency, and persistent safety profiles. Further, synergistic and integrative approach combinations of *E. officinalis* with other neurotherapeutics and understanding their molecular mechanism could unravel their potential to manage neurological disorders and to reinstate neuro-resilience.

References

Akhter MS, Rahman MA, Ripon RK, Mubarak M, Akter M, Mahbub S et al (2024) A systematic review on green synthesis of silver nanoparticles using plants extract and their bio-medical applications. Heliyon 10(11). https://doi.org/10.1016/j.heliyon.2024.e29766

Ahmed OM, AbouZid SF, Ahmed NA, Zaky MY, Liu H (2021) An up-to-date review on citrus flavonoids: Chemistry and benefits in health and diseases. Curr Pharm Des 27(4):513–530. https://doi.org/10.2174/1381612826666201127122313

Ayuwardani N, Sofro ZM, Fakhrudin N, Ikawati Z (2024) Antioxidant and neuroprotective activitics of Phyllanthus emblica L.: A review. Trop J Nat Prod Res 8(10). https://doi.org/10.26538/tjnpr/v8i10.2

Bai J, Zhang Y, Tang C, Hou Y, Ai X, Chen X, Zhang Y, Wang X, Meng X (2021) Gallic acid: Pharmacological activities and molecular mechanisms involved in inflammation-related diseases. Biomed Pharmacother 133:110985. https://doi.org/10.1016/j.biopha.2020.110985

Barani M, Sangiovanni E, Angarano M, Rajizadeh MA, Mehrabani M, Piazza S, Gangadharappa HV, Pardakhty A, Mehrbani M, Dell'Agli M, Nematollahi MH (2021) Phytosomes as innovative delivery systems for phytochemicals: a comprehensive review of literature. Int J Nanomedicine 16:6983–7022. https://doi.org/10.2147/IJN.S318416

Bhargavi KM, Gowthami N, Chetan GK, Srinivas Bharath MM (2025) Neuroprotective effects of nutraceuticals and natural products in traumatic brain injury. Neurochem Int 182:105904. https://doi.org/10.1016/j.neuint.2024.105904

Biswas K, Islam MA, Sharmin T, Biswas PK (2015) In-vitro cholinesterase inhibitory activity of dry fruit extract of Phyllanthus emblica relevant to the treatment of Alzheimer's disease. J Phytopharmacol 4(1):5–8. https://doi.org/10.31254/phyto.2015.4102

Bode AM, Dong Z (2015) Toxic phytochemicals and their potential risks for human cancer. Cancer Prev Res Phila 8(1):1–8. https://doi.org/10.1158/1940-6207.CAPR-14-0160

Chauhan M, Garg V, Zia G, Dutt R, Alghamdi BS, Zawawi A, Ashraf GM, Farhana A (2023) Effect of extraction methods on the antioxidant potential and cytotoxicity of the combined ethanolic extracts of *Daucus carota L., Beta vulgaris L., Phyllanthus emblica L. and Lycopersicon esculentum* against gastric adenocarcinoma cells. Molecules (Basel, Switzerland) 28(18):6589. https://doi.org/10.3390/molecules28186589

Chen SY, Gao Y, Sun JY, Meng XL, Yang D, Fan LH, Xiang L, Wang P (2020) Traditional Chinese medicine: role in reducing β-amyloid, apoptosis, autophagy, neuroinflammation, oxidative stress, and mitochondrial dysfunction of Alzheimer's disease. Front Pharmacol 11:497. https://doi.org/10.3389/fphar.2020.00497

Chiang MC, Tsai TY, Wang CJ (2023) The potential benefits of quercetin for brain health: A review of anti-inflammatory and neuroprotective mechanisms. Int J Mol Sci 24(7):6328. https://doi.org/10.3390/ijms24076328

Chambial S, Dwivedi S, Shukla KK, John PJ, Sharma P (2013) Vitamin C in disease prevention and cure: An overview. Indian J Clin Biochem 28(4):314–328. https://doi.org/10.1007/s12291-013-0375-3

Dasaroju S, Gottumukkala K (2014) Current trends in the research of Emblica officinalis (amla): A pharmacological perspective. Int J Pharm Sci Rev Res 24:150–159. https://www.researchgate.net/publication/287524229_Current_trends_in_the_research_of_Emblica_officinalis_Amla_A_pharmacological_perspective

Dabbaghi A, Gholipour A, Kharaziha M, Rumi SS (2019) Tannic acid-based antibacterial hydrogels: Characterization and evaluation against Escherichia coli and Staphylococcus aureus. J Appl Polym Sci 136(38):47943. https://doi.org/10.1002/app.47943

DeNike C (2024) What is Neuro Resilience and How Can It Transform Your Life? An article. URL What is Neuro Resilience and How Can It Transform Your Life? | by Claire DeNike | Finding Self Journey | Medium (Accessed on 10.08.2025)

Dinesh M, Roopan SM, Selvaraj CI, Arunachalam P (2017) Phyllanthus emblica seed extract mediated synthesis of PdNPs against antibacterial, heamolytic and cytotoxic studies. J Photochem Photobiol B 167:64–71. https://doi.org/10.1016/j.jphotobiol.2016.12.012

Farbood Y, Sarkaki A, Dianat M, Khodadadi A, Haddad MK, Mashhadizadeh S (2015) Ellagic acid prevents cognitive and hippocampal long-term potentiation deficits and brain inflammation in rat with traumatic brain injury. Life Sci 124:120–127. https://doi.org/10.1016/j.lfs.2015.01.013

Fernandes R (2025) The convergence of nanotechnology and biotechnology in modern medicine. Nanomaterials (Basel, Switzerland) 15(3):182. https://doi.org/10.3390/nano15030182

Gul M, Liu ZW, Iahtisham-Ul-Haq, Rabail R, Faheem F, Walayat N, Nawaz A, Shabbir MA, Munekata PES, Lorenzo JM, Aadil RM (2022) Functional and nutraceutical significance of amla (Phyllanthus emblica L.): A review. Antioxidants 11(5):816. https://doi.org/10.3390/antiox11050816

Gandhi Y, Grewal J, Jain V, Rawat H, Mishra SK, Kumar V et al (2023) Emblica officinalis: a promising herb confining versatile applications. S Afr J Bot 159:519–531. https://doi.org/10.1016/j.sajb.2023.06.041

Global Burden of Disease (2021) Nervous System Disorders Collaborators (2024) Global, regional, and national burden of disorders affecting the nervous system, 1990-2021: a systematic analysis for the Global Burden of Disease Study 2021. Lancet Neurol 23(4):344–381. https://doi.org/10.1016/S1474-4422(24)00038-3

Golechha M, Bhatia J, Singh Arya D (2010) Hydroalcoholic extract of Emblica officinalis Gaertn. affords protection against PTZ-induced seizures, oxidative stress and cognitive impairment in rats. Indian J Exp Biol 48(5):474

Gouda B, Sinha SN, Sangaraju R, Huynh T, Patangay S, Venkata Mullapudi S, Mungamuri SK, Patil PB, Periketi MC (2024) Extraction, phytochemical profile, and neuroprotective activity of *Phyllanthus emblica* fruit extract against sodium valproate-induced postnatal autism in BALB/c mice. Heliyon 10(15):e34992. https://doi.org/10.1016/j.heliyon.2024.e34992

Greco WR, Faessel H, Levasseur L (1996) The search for cytotoxic synergy between anticancer agents: a case of Dorothy and the ruby slippers? J Natl Cancer Inst 88(11):699–700. https://doi.org/10.1093/jnci/88.11.699

Hampton D (2018) The Neurosciences of Building a Resilient Brain: an article. URL The Neuroscience of Building a Resilient Brain - The Best Brain Possible (Accessed on 10.08.2025)

Hasan MR, Islam MN, Islam MR (2016) Phytochemistry, pharmacological activities and traditional uses of Emblica officinalis: a review. Int Curr Pharm J 5(2):14–21. http://www.icpjonline.com/documents/Vol5Issue2/02.pdf

Hasan GM, Anwar S, Shamsi A, Sohal SS, Hassan MI (2024) The neuroprotective potential of phytochemicals in traumatic brain injury: mechanistic insights and pharmacological implications. Front Pharmacol 14:1330098. https://doi.org/10.3389/fphar.2023.1330098

Hossain MM, Hamza A, Polash SA, Tushar MH, Takikawa M, Piash AB et al (2024) Green synthesis of silver nanoparticles using Phyllanthus emblica extract: investigation of antibacterial activity and biocompatibility in vivo. RSC Pharmaceutics 1(2):245–258. https://doi.org/10.1039/D3PM00077J

Husain I, Akhtar M, Madaan T, Vohora D, Abdin MZ, Islamuddin M, Najmi AK (2018) Tannins enriched fraction of *Emblica officinalis* fruits alleviates high-salt and cholesterol diet-induced cognitive impairment in rats via Nrf2-ARE pathway. Front Pharmacol 9:23. https://doi.org/10.3389/fphar.2018.00023

Imam SS, Alshehri S, Ghoneim MM, Zafar A, Alsaidan OA, Alruwaili NK, Gilani SJ, Rizwanullah M (2021) Recent advancement in chitosan-based nanoparticles for improved oral bioavailability and bioactivity of phytochemicals: challenges and perspectives. Polymers 13(22):4036. https://doi.org/10.3390/polym13224036

Jaisamut P, Wanna S, Limsuwan S, Chusri S, Wiwattanawongsa K, Wiwattanapatapee R (2021) Enhanced oral bioavailability and improved biological activities of a quercetin/resveratrol combination using a liquid self-microemulsifying drug delivery system. Planta Med 87(4):336–346. https://doi.org/10.1055/a-1270-7606

Javid H, Hashemy SI, Heidari MF, Esparham A, Gorgani-Firuzjaee S (2022) The anticancer role of cerium oxide nanoparticles by inducing antioxidant activity in esophageal cancer and cancer stem-like ESCC spheres. Biomed Res Int 2022:3268197. https://doi.org/10.1155/2022/3268197

Justin Thenmozhi A, Dhivyabharathi M, William Raja TR, Manivasagam T, Essa MM (2016) Tannoid principles of Emblica officinalis renovate cognitive deficits and attenuate amyloid pathologies against aluminum chloride induced rat model of Alzheimer's disease. Nutr Neurosci 19(6):269–278. https://doi.org/10.1179/1476830515Y.0000000016

Jena GK, Patra CN, Jammula S, Rana R, Chand S (2024) Artificial intelligence and machine learning implemented drug delivery systems: A paradigm shift in the pharmaceutical industry. J Bio-X Res 7:0016. https://doi.org/10.34133/jbioxresearch.0016

Kaczmarek B (2020) Tannic acid with antiviral and antibacterial activity as a promising component of biomaterials—a minireview. Materials 13(14):3224. https://doi.org/10.3390/ma13143224

Kaur J, Kaur D, Singh H, Khan MU (2013) Emblica officinalis: a meritocratic drug for treating various disorders. Indo Am J Pharm Res 3(6)

Kim T, Lee N, Arifin DR, Shats I, Janowski M, Walczak P, Hyeon T, Bulte JWM (2017) In vivo micro-CT imaging of human mesenchymal stem cells labeled with gold-poly-L-lysine nanocomplexes. Adv Funct Mater 27(3):1604213. https://doi.org/10.1002/adfm.201604213

Lamptey RNL, Chaulagain B, Trivedi R, Gothwal A, Layek B, Singh J (2022) A review of the common neurodegenerative disorders: current therapeutic approaches and the potential role of nanotherapeutics. Int J Mol Sci 23(3):1851. https://doi.org/10.3390/ijms23031851

Li C, Long P, He M, Han F, Jiang W, Li Y et al (2023) Phyllanthus emblica Linn. fruit polyphenols improve acute paradoxical sleep deprivation-induced cognitive impairment and anxiety via Nrf2 pathway. J Funct Foods 110:105884. https://doi.org/10.1016/j.jff.2023.105884

Liu QS, Jiang HL, Wang Y, Wang LL, Zhang JX, He CH, Shao S, Zhang TT, Xing JG, Liu R (2018) Total flavonoid extract from Dracoephalum moldavica L. attenuates β-amyloid-induced toxicity through anti-amyloidogenesic and neurotrophic pathways. Life Sci 193:214–225. https://doi.org/10.1016/j.lfs.2017.10.041

Martins PA, Alsaiari S, Julfakyan K, Nie Z, Khashab NM (2017) Self-assembled lipoprotein based gold nanoparticles for detection and photothermal disaggregation of β-amyloid aggregates. Chem Commun (Camb) 53(13):2102–2105. https://doi.org/10.1039/c6cc09085k

Mashhadizadeh S, Farbood Y, Dianat M, Khodadadi A, Sarkaki A (2017) Therapeutic effects of ellagic acid on memory, hippocampus electrophysiology deficits, and elevated TNF-α level in brain due to experimental traumatic brain injury. Iran J Basic Med Sci 20(4):399–407. https://doi.org/10.22038/IJBMS.2017.8581

Masoudi Asil, S., , Ahlawat, J., , Guillama Barroso, G., , & Narayan, M., (2020). Nanomaterial based drug delivery systems for the treatment of neurodegenerative diseases. Biomater Sci, 8(15), 4109–4128. https://doi.org/10.1039/d0bm00809e

Middha SK, Goyal AK, Lokesh P, Yardi V, Mojamdar L, Keni DS, Babu D, Usha T (2015) Toxicological evaluation of Emblica officinalis fruit extract and its anti-inflammatory and free radical scavenging properties. Pharmacogn Mag 11(Suppl 3):S427–S433. https://doi.org/10.4103/0973-1296.168982

Mirshekar MA, Sarkaki A, Farbood Y, Gharib Naseri MK, Badavi M, Mansouri MT, Haghparast A (2018) Neuroprotective effects of gallic acid in a rat model of traumatic brain injury: behavioral, electrophysiological, and molecular studies. Iran J Basic Med Sci 21(10):1056–1063. https://doi.org/10.22038/IJBMS.2018.29639.7165

Mirunalini S, Krishnaveni M (2010) Therapeutic potential of Phyllanthus emblica (amla): The Ayurvedic wonder. J Basic Clin Physiol Pharmacol 21(1):93–105. https://doi.org/10.1515/JBCPP.2010.21.1.93

Mirunalini S, Krishnaveni M (2016) Therapeutic efficacy of Phyllanthus emblica L. fruit extract and cisplatin combination against 7,12-dimethylbenz(a)anthracene-induced hamster buccal pouch carcinogenesis. FoodSci Indian J Res Food Sci Nutr 3(1). https://doi.org/10.15613/fijrfn/2016/v3i1/108896

National Institute of Neurological Disorders and Stroke (2025) Traumatic Brain Injury [www document]. URL Traumatic Brain Injury (TBI) | National Institute of Neurological Disorders and Stroke (Accessed on 10.08.2025)

Naz S, Beach J, Heckert B, Tummala T, Pashchenko O, Banerjee T, Santra S (2017) Cerium oxide nanoparticles: a 'radical' approach to neurodegenerative disease treatment. Nanomedicine (Lond) 12(5):545–553. https://doi.org/10.2217/nnm-2016-0399

Our world in Data (2021) Share of the population with neurodegenerative diseases, World, 2021 [www document]. URL https://ourworldindata.org/grapher/alzheimers-parkinsons-prevalence?time=latest (accessed on 10.8.2025)

Prananda AT, Dalimunthe A, Harahap U, Simanjuntak Y, Peronika E, Karosekali NE, Hasibuan PAZ, Syahputra RA, Situmorang PC, Nurkolis F (2023) Phyllanthus emblica: A comprehensive review of its phytochemical composition and pharmacological properties. Front Pharmacol 14:1288618. https://doi.org/10.3389/fphar.2023.1288618

Patil KS, Date VC (2025) Evaluation of neuroprotective effects of Allium sativum and Phyllanthus emblica in rotenone-induced Parkinson's Disease in Drosophila melanogaster. https://doi.org/10.69605/ijlbpr_14.3.2025.38

Paul A, Sujatha K, Srilatha C, Kumar NV, Shikha D (2025) Oxidative stress and histomorphological changes in brain and effects of supplementation of Linumusitatissimum (flaxseed) and Emblica officinalis (amla) against lead toxicity in Wistar rats. A Anand Kumar 49(1):49

Phochantachinda S, Chatchaisak D, Temviriyanukul P, Chansawang A, Pitchakarn P, Chantong B (2021) Ethanolic fruit extract of *Emblica officinalis* suppresses neuroinflammation in microglia and promotes neurite outgrowth in Neuro2a cells. Evid Based Complement Alternat Med 2021:6405987. https://doi.org/10.1155/2021/6405987

Ramakrishna V, Gupta KP, Setty HO, Kondapi KA (2014) Neuroprotective effect of Emblica officinalis extract against H2O2 induced DNA damage and repair in neuroblastoma cells. J Homeop Ayurv Med S:1002. https://doi.org/10.4172/2167-1206.S1-002

Rana A, Bhatnagar S (2021) Advancements in folate receptor targeting for anti-cancer therapy: a small molecule-drug conjugate approach. Bioorg Chem 112:104946. https://doi.org/10.1016/j.bioorg.2021.104946

Rivadeneyra-Domínguez E, Zamora-Bello I, Rodríguez-Landa JF, Ortega-García AA, Rosales-Sánchez Ó (2024) *Emblica officinalis* Gaertn as a potential alternative therapy for the treatment of epilepsy: an animal study. Dose-response : a publication of International Hormesis Society 22(3):15593258241282018. https://doi.org/10.1177/15593258241282018

Rosarin FS, Arulmozhi V, Nagarajan S, Mirunalini S (2013) Antiproliferative effect of silver nanoparticles synthesized using amla on Hep2 cell line. Asian Pac J Trop Med 6(1):1–10. https://doi.org/10.1016/S1995-7645(12)60193-X

Rahman MM, Ferdous KU, Roy S, Nitul IA, Mamun F, Hossain MH, Subhan N, Alam MA, Haque MA (2020) Polyphenolic compounds of amla prevent oxidative stress and fibrosis in the kidney and heart of 2K1C rats. Food Sci Nutr 8(7):3578–3589. https://doi.org/10.1002/fsn3.1640

Saini R, Sharma N, Oladeji OS, Sourirajan A, Dev K, Zengin G, El-Shazly M, Kumar V (2022) Traditional uses, bioactive composition, pharmacology, and toxicology of Phyllanthus emblica fruits: A comprehensive review. J Ethnopharmacol 282:114570. https://doi.org/10.1016/j.jep.2021.114570

Sawant L, Pandita N, Prabhakar B (2010) Determination of gallic acid in Phyllanthus emblica Linn. dried fruit powder by HPTLC. J Pharm Bioallied Sci 2(2):105–108. https://doi.org/10.4103/0975-7406.67012

Sharif MA, Khan AM, Salekeen R, Rahman MH, Mahmud S, Bibi S, Biswas P, Nazmul Hasan M, Islam KMD, Rahman SMM, Islam ME, Alshammari A, Alharbi M, Hayee A (2023) Phyllanthus emblica (amla) methanolic extract regulates multiple checkpoints in 15-lipoxygenase-mediated inflammopathies: Computational simulation and in vitro evidence. Saudi Pharm J 31(8):101681. https://doi.org/10.1016/j.jsps.2023.06.014

Sun Y, Li C, Li Z, Shangguan A, Jiang J, Zeng W, Zhang S, He Q (2021) Quercetin as an antiviral agent inhibits the Pseudorabies virus in vitro and in vivo. Virus Res 305:198556. https://doi.org/10.1016/j.virusres.2021.198556

Scartezzini P, Antognoni F, Raggi MA, Poli F, Sabbioni C (2006) Vitamin C content and antioxidant activity of the fruit and of the Ayurvedic preparation of Emblica officinalis Gaertn. J Ethnopharmacol 104(1–2):113–118. https://doi.org/10.1016/j.jep.2005.08.065

Saha S, Dimri A, Chaudhary A, Durgapal S, Joshi N (2013) Effect of Embellica Officinalis on scopolamine and high fat diet induced memory loss. Int J Pharm Tech Res 5:1271–1280

Salman M, Tabassum H, Parvez S (2020) Tannic acid provides neuroprotective effects against traumatic brain injury through the PGC-1α/Nrf2/HO-1 pathway. Mol Neurobiol 57(6):2870–2885. https://doi.org/10.1007/s12035-020-01924-3

Shi Z, Qiu W, Xiao G, Cheng J, Zhang N (2018) Resveratrol attenuates cognitive deficits of traumatic brain injury by activating p38 signaling in the brain. Med Sci Monit 24:1097–1103. https://doi.org/10.12659/msm.909042

Singh SMW, Agarwal B (2020) Anti-Parkinson's activity of Emblica officinalis and Terminalia bellirica. https://www.researchgate.net/profile/Himani-Badoni-2/publication/347947836_Himani_Paper-Antiparkinson's/links/6450e262809a535021423205/Himani-Paper-Antiparkinsons.pdf

Sivalingam AM, Pandian A (2024) Characterization of silver nanoparticles (AgNPs) synthesized using polyphenolic compounds from Phyllanthus emblica L. and their impact on cytotoxicity in human cell lines. Carbohydrate Polymer Technologies and Applications 8:100535. https://doi.org/10.1016/j.carpta.2024.100535

Srivastava AK, Naseer A, Gupta A (2020) Evaluation of neurobehavioral and biochemical parameters of poly-herbal formulation on MPTP-intoxicated mice for the treatment of PD. Curr Bioact Comp 16(9):1290–1298. https://doi.org/10.2174/1573407215666191022114956

Travica N, Ried K, Sali A, Scholey A, Hudson I, Pipingas A (2017) Vitamin C status and cognitive function: A systematic review. Nutrients 9(9):960. https://doi.org/10.3390/nu9090960

Teleanu DM, Negut I, Grumezescu V, Grumezescu AM, Teleanu RI (2019) Nanomaterials for drug delivery to the central nervous system. Nanomaterials (Basel, Switzerland) 9(3):371. https://doi.org/10.3390/nano9030371

Thatyana M, Dube NP, Kemboi D, Manicum ALE, Mokgalaka-Fleischmann NS, Tembu JV (2023) Advances in phytonanotechnology: a plant-mediated green synthesis of metal nanoparticles using phyllanthus plant extracts and their antimicrobial and anticancer applications. Nanomaterials 13(19):2616. https://doi.org/10.3390/nano13192616

Thomas P, Patil R (2024) Exploring therapeutic potential of Indian ayurvedic plants for Parkinson's disease treatment. In: Medicinal and aromatic plants: current research status, value-addition to their waste, and agro-industrial potential, vol I. Springer Nature Switzerland, Cham, pp 125–141. https://doi.org/10.1007/978-3-031-60117-0_10

Tiwana G, Cock IE, Cheesman MJ (2025) *Phyllanthus emblica*: phytochemistry, antimicrobial potential with antibiotic enhancement, and toxicity insights. Microorganisms 13(3):611. https://doi.org/10.3390/microorganisms13030611

Variya BC, Bakrania AK, Patel SS (2016) Emblica officinalis (Amla): a review for its phytochemistry, ethnomedicinal uses and medicinal potentials with respect to molecular mechanisms. Pharmacol Res 111:180–200. https://doi.org/10.1016/j.phrs.2016.06.013

Vasudevan M, Parle M (2007) Memory enhancing activity of Anwala churna (Emblica officinalis Gaertn.): an Ayurvedic preparation. Physiol Behav 91(1):46–54. https://doi.org/10.1016/j.physbeh.2007.01.016

Wang R, Xu X, Puja AM, Perumalsamy H, Balusamy SR, Kim H, Kim YJ (2021) Gold nanoparticles prepared with *Phyllanthus emblica* fruit extract and *Bifidobacterium animalis* subsp. *lactis* can induce apoptosis via mitochondrial impairment with inhibition of autophagy in the human gastric carcinoma cell line AGS. Nanomaterials (Basel, Switzerland) *11*(5):1260. https://doi.org/10.3390/nano11051260

WHO (2024) Over 1 in 3 people affected by neurological conditions, the leading cause of illness and disability worldwide [www document]. URL https://www.who.int/news/item/14-03-2024-over-1-in-3-people-affected-by-neurological-conditions%2D%2Dthe-leading-cause-of-illness-and-disability-worldwide (accessed on 10.8.2025)

Xiao L, Chen W, Guo W, Li H, Chen R, Chen Q (2025) Exploring the mechanism of action of Phyllanthus emblica in the treatment of epilepsy based on network pharmacology and molecular docking. Medicine (Baltimore) 104(7):e41414. https://doi.org/10.1097/MD.0000000000041414

Xu J, Wang H, Ding K, Zhang L, Wang C, Li T, Wei W, Lu X (2014) Luteolin provides neuroprotection in models of traumatic brain injury via the Nrf2-ARE pathway. Free Radic Biol Med 71:186–195. https://doi.org/10.1016/j.freeradbiomed.2014.03.009

Xu K., Qu L, Li H, Ren X, Yan N, & Fu X (2024) Functional properties of dietary quercetin in cardiovascular health and disease. Food Frontiers, 5(5), 1951–1967

Yokozawa T, Kim HY, Kim HJ, Okubo T, Chu DC, Juneja LR (2007) Amla (Emblica officinalis Gaertn.) prevents dyslipidaemia and oxidative stress in the ageing process. Br J Nutr 97(6):1187–1195. https://doi.org/10.1017/S0007114507691971

Zhang RX, Li J, Zhang T, Amini MA, He C, Lu B, Ahmed T, Lip H, Rauth AM, Wu XY (2018) Importance of integrating nanotechnology with pharmacology and physiology for innovative drug delivery and therapy - an illustration with firsthand examples. Acta Pharmacol Sin 39(5):825–844. https://doi.org/10.1038/aps.2018.33

Phototherapeutics in Neurodegenerative Disorders: Present Insights and Emerging Horizons

14

Ravinesh Mishra, Priya Sharma, Bhartendu Sharma, and Archana Kumari

Abstract

Neurodegenerative disease management benefits from phototherapeutics, which function through light-based therapeutic approaches as a new and promising treatment strategy. This paper examines contemporary phototherapeutic research alongside development predictions for Alzheimer's disease and Parkinson's disease along with amyotrophic lateral sclerosis care. Recent advancements in photobiomodulation and optogenetics have demonstrated potential in mitigating neuronal damage, reducing inflammation, and enhancing cellular energy production. Present insights reveal the efficacy of near-infrared light in stimulating mitochondrial function and promoting neuroplasticity, while optogenetic techniques offer unprecedented precision in modulating specific neural circuits. Emerging horizons in phototherapeutics include the development of novel light-sensitive compounds, improved light delivery systems, and combination therapies integrating photobiomodulation with traditional pharmacological approaches. Additionally, ongoing research explores the potential of photodynamic therapy in targeting protein aggregates characteristic of neurodegenerative disorders. Further illumination of light-tissue interactions creates opportunities for phototherapeutics to transform neurodegenerative disorder treatments, thus providing better patient results and improved life quality. The paper reviews existing research data while emphasizing modern developments in addition to exploring potential future directions within this quickly expanding scientific sector.

R. Mishra · P. Sharma (✉) · B. Sharma · A. Kumari
School of Pharmacy and Emerging Sciences, Baddi University of Emerging Sciences and Technology, Baddi, Solan, Himachal Pradesh, India

H. Dilnashin, S. P. Singh (eds.), *Phototherapeutic Approaches to Neurodegeneration*,
https://doi.org/10.1007/978-981-95-4751-7_14

Keywords

Phototherapeutics · Neurodegenerative disorders · Photobiomodulation · Optogenetics · Near-infrared light · Neuroplasticity

14.1 Introduction

Neurodegenerative diseases (NDs) present as the continuous degradation of brain cellular structure, resulting in simultaneous deterioration of neuronal operating capacity alongside diminished performance and conduct of patients. These disorders are multifactorial in origin, involving genetic predisposition, environmental triggers, mitochondrial dysfunction, protein misfolding, oxidative stress, and neuroinflammation. Despite extensive research, therapeutic development has been hampered by the complex and heterogeneous nature of these diseases and the impermeability of the blood-brain barrier (BBB).

Phototherapeutics have emerged as a versatile platform offering spatial and temporal precision, minimal invasiveness, and potential disease-modifying effects. Phototherapeutic strategies encompass a wide range of techniques, including photobiomodulation, photoactivated drug delivery systems, optogenetics, and photothermal therapy. These modalities have shown the potential to influence neuronal function, modulate immune responses, reduce pathological protein accumulation, and restore disrupted neural circuits. As neurodegenerative diseases continue to rise globally, understanding the scope and limitations of phototherapeutics becomes essential in guiding future research and clinical translation.

14.2 Mechanistic Basis of Phototherapeutics

Phototherapeutics exploit the interaction of light with biological tissues to induce therapeutic effects. The primary modalities include:

14.2.1 PhotoBiomodulation (PBM)

Low-intensity red or near-infrared light at wavelengths from 600 to 1100 nm is applied through PBM for cellular process stimulation. The primary target of PBM functions as cytochrome c oxidase (CCO) located within the mitochondrial respiratory chain. Upon absorption of photons, CCO activity increases, leading to enhanced ATP production, modulation of reactive oxygen species (ROS), and activation of transcription factors such as nuclear factor-kappa B (NF-κB) (Hamblin 2017). This cascade promotes cell survival, reduces oxidative stress, and attenuates inflammation.

14.2.2 Photoactivated Drug Delivery

This modality involves the use of light-sensitive carriers or prodrugs that release active therapeutic agents upon exposure to specific wavelengths. The advantages include spatial and temporal control over drug release, improved targeting across the BBB, and reduced systemic toxicity (Lucky et al. 2015). Materials such as gold nanoparticles, upconversion nanoparticles, and photosensitive liposomes are employed for precise control.

14.2.3 Optogenetics

Scientists use optogenetics to control particular neuron activities through lights that result from genetic engineering technology. The genetic modification of neurons allows them to incorporate light-sensitive ion channels (e.g., channelrhodopsins, halorhodopsins), enabling their activation or inhibition by light of specific wavelengths. This approach allows dissection of neural circuits involved in neurodegeneration and offers therapeutic potential by restoring normal firing patterns (Deisseroth 2015).

14.2.4 Photothermal Therapy (PTT)

PTT utilizes photothermal agents that convert absorbed light into heat, resulting in localized hyperthermia. This can be used to ablate toxic protein aggregates, damaged neurons, or pathological structures such as amyloid plaques and Lewy bodies. The precision and minimally invasive nature of PTT make it an attractive option for targeted neurotherapeutic interventions (Huang et al. 2007).

- *Mechanisms of Phototherapeutics in Neurodegeneration*

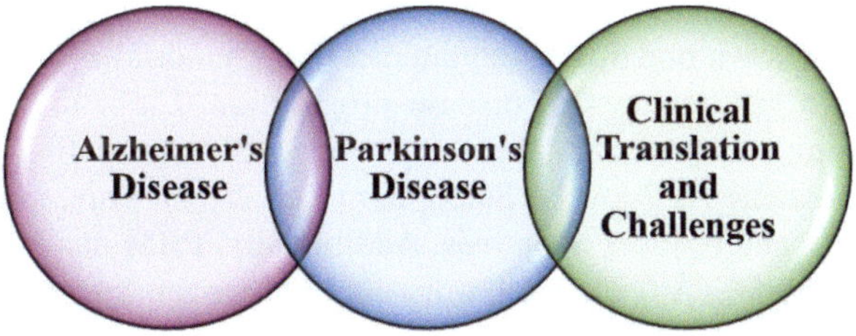

14.3 Photobiomodulation in Neurodegeneration

Photobiomodulation (PBM) has emerged as a particularly promising modality in the management of neurodegenerative disorders. Research studies both in animals and early phases of human trials show how this compound controls various pathways linked to diseases.

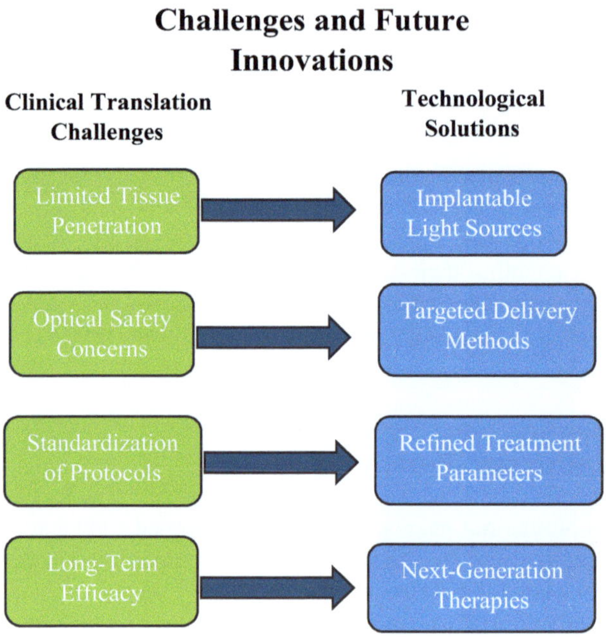

14.3.1 Alzheimer's Disease (AD)

In AD, PBM has shown potential to counteract mitochondrial dysfunction and oxidative stress, two key features of disease progression. Light in the red to near-infrared spectrum can penetrate the scalp and skull, stimulating mitochondrial activity and increasing ATP levels (Salehpour et al. 2021). This, in turn, enhances neuronal viability and reduces apoptosis. Additionally, PBM has been observed to decrease beta-amyloid (Aβ) plaque burden and reduce tau hyperphosphorylation, both hallmark features of AD pathology (Johnstone et al. 2016).

Animal studies using transcranial PBM have reported improvements in learning and memory functions, correlated with reduced Aβ levels in the hippocampus (Grillo et al. 2013). Some small-scale human trials have also shown cognitive improvements in AD patients following PBM, although larger randomized controlled trials are required to validate these findings (Saltmarche et al. 2017).

14.3.2 Parkinson's Disease (PD)

In PD, PBM has demonstrated neuroprotective effects on dopaminergic neurons in the substantia nigra, potentially through mechanisms involving anti-inflammatory cytokine modulation and inhibition of mitochondrial dysfunction (Mitrofanis 2017). Animal models have shown that PBM can enhance motor function and reduce alpha-synuclein accumulation, which is implicated in the pathogenesis of PD (Cheng et al. 2021).

Transcranial and intranasal PBM protocols have been trialed with encouraging outcomes in terms of improving mobility and quality of life in PD patients (Hamilton et al. 2019). However, optimization of parameters such as wavelength, intensity, and treatment duration remains an area of active investigation.

14.3.3 Clinical Translation and Challenges

Despite promising preclinical results, the translation of PBM to widespread clinical use faces several challenges. These include limited penetration depth of light, variability in patient response, and the need for standardization of dosimetry. Nevertheless, advances in light delivery technologies and wearable devices may overcome some of these barriers, enabling more consistent and effective applications.

14.4 Photoactivated Drug Delivery in Neurodegenerative Disorders

Photoactivated drug delivery systems offer an innovative strategy to overcome the limitations of conventional therapies in neurodegenerative diseases, particularly in enhancing drug penetration across the blood-brain barrier (BBB) and achieving site-specific action.

14.4.1 Principles and Materials

These systems typically involve light-responsive nanocarriers or prodrugs that are inert until exposed to a specific wavelength of light. Upon activation, they release the therapeutic compound at the target site, allowing for spatial and temporal precision. Commonly used materials include gold nanoparticles, which exhibit strong surface plasmon resonance; upconversion nanoparticles, which convert low-energy infrared light to higher-energy visible light; and photosensitive liposomes (Rwei et al. 2015; Idris et al. 2012).

14.4.2 Application in Alzheimer's Disease

In Alzheimer's disease, photoactivated systems have been employed to deliver anti-amyloid agents, antioxidants, or neuroprotective drugs. For instance, curcumin-loaded nanoparticles activated by near-infrared light have demonstrated enhanced brain delivery and reduced amyloid plaque deposition in animal models (Wang et al. 2020a). Another approach involves using photo-responsive liposomes encapsulating donepezil or memantine to achieve controlled drug release in specific brain regions.

14.4.3 Application in Parkinson's Disease

For Parkinson's disease, light-triggered systems have been designed to release dopamine or dopamine agonists in response to light exposure, thereby minimizing systemic side effects. Studies using ROS-sensitive nanoparticles have shown improved survival of dopaminergic neurons and motor function recovery in rodent models (Li et al. 2020).

14.4.4 Challenges and Future Directions

Despite the promise, several challenges remain. These include ensuring biocompatibility and long-term safety of the delivery materials, optimizing light penetration for activation in deep brain regions, and integrating smart targeting ligands for improved specificity. Research is ongoing to develop multifunctional systems that combine diagnostic and therapeutic capabilities (theranostics) to further personalize treatment approaches (Chen et al. 2017).

14.5 Optogenetics in Neurodegeneration

Optogenetics has transformed neuroscience by enabling millisecond-scale control of neuronal activity with light. This precision tool is now being explored for therapeutic intervention in neurodegenerative diseases, particularly for circuit modulation and restoration of impaired neuronal networks.

14.5.1 Mechanisms and Tools

Optogenetics uses microbial opsins such as channelrhodopsin-2 (ChR2), halorhodopsin (NpHR), and archaerhodopsin (Arch) to control neuronal excitability. By delivering opsin genes via viral vectors (e.g., AAV, lentivirus), neurons can be rendered responsive to specific wavelengths of light. Blue light activates ChR2 for depolarization, while yellow or green light activates NpHR or Arch to inhibit neuronal firing (Fenno et al. 2011; Zhang et al. 2006).

14.5.2 Application in Parkinson's Disease

In Parkinson's disease, optogenetic approaches have been used to modulate basal ganglia circuits. Deisseroth and colleagues demonstrated that stimulating ChR2-expressing neurons in the motor cortex or subthalamic nucleus improves motor function in rodent PD models (Gradinaru et al. 2009). This mimics the effects of deep brain stimulation (DBS) but with improved spatial resolution and fewer side effects.

Research investigators have directed their studies toward dopaminergic neurons located in the substantia nigra pars compacta (SNc). Light-induced activation of these neurons has shown to restore dopamine levels and alleviate motor deficits in experimental PD models (Kravitz et al. 2010).

14.5.3 Application in Alzheimer's Disease

In Alzheimer's disease, optogenetics has been employed to dissect hippocampal and cortical circuit dysfunctions. Studies have shown that modulating inhibitory interneurons can restore excitatory-inhibitory balance, which is often disrupted in AD (Verret et al. 2012).

Optogenetic stimulation has also been used to enhance memory retrieval and synaptic plasticity in transgenic AD models, suggesting a therapeutic role in cognitive rehabilitation (Roy et al. 2016).

14.5.4 Safety, Delivery, and Translational Barriers

Major challenges to clinical application include the invasive nature of opsin delivery and the need for implantable light sources. However, recent advances in wireless and minimally invasive optoelectronic systems offer promise for future translation. Immunogenicity and long-term expression of opsins also require careful assessment (Pisanello et al. 2017; Towne et al. 2013).

Despite these challenges, optogenetics holds great potential for precision neuromodulation and circuit-level intervention in neurodegenerative diseases.

14.6 Photothermal Therapy in Neurodegeneration

Photothermal therapy (PTT) uses light-sensitive agents to produce localized heat, which achieves specific thermal damage of pathological targets. In neurodegenerative diseases, PTT has shown promise in disrupting protein aggregates, reducing neuroinflammation, and modulating neural microenvironments.

14.6.1 Mechanism and Agents

PTT operates by converting near-infrared (NIR) light into heat through photothermal agents, including gold nanorods, carbon nanotubes, graphene oxide, and conjugated polymer nanoparticles (Liu et al. 2019). These agents can be engineered for targeted delivery to pathological sites, where controlled NIR irradiation induces localized hyperthermia (42–45 °C), resulting in structural disruption of toxic protein aggregates and modulated cellular responses (Huang et al. 2007).

14.6.2 Applications in Alzheimer's Disease

Amyloid-beta (Aβ) plaques are a hallmark of Alzheimer's disease (AD). PTT has been utilized to selectively ablate these aggregates using targeted gold nanorods functionalized with anti-Aβ antibodies. Upon NIR exposure, these nanorods generate sufficient thermal energy to disaggregate plaques without affecting surrounding tissues (Yang et al. 2012).

Experimental models have demonstrated improved memory function and reduced amyloid burden post-PTT treatment. Furthermore, the generated heat also activates heat shock proteins, which promote protein refolding and clearance pathways, enhancing therapeutic efficacy (Wang et al. 2020b).

14.6.3 Applications in Parkinson's Disease

In Parkinson's disease (PD), PTT has been used to target alpha-synuclein (α-syn) aggregates. Gold nanoshells and carbon-based nanomaterials designed to bind α-syn have successfully disrupted aggregates in vitro and in vivo, reducing neuronal toxicity and improving motor performance (Liao et al. 2020).

PTT also modulates glial cell activation, reducing neuroinflammation, which is a key contributor to PD progression. These anti-inflammatory effects further support its potential as a multifaceted therapeutic tool (Yang et al. 2018).

14.6.4 Delivery Strategies and Safety Considerations

Effective delivery of photothermal agents to the brain poses a challenge due to the BBB. Strategies such as receptor-mediated transcytosis, ultrasound-induced BBB disruption, and intranasal administration have shown promise in enhancing brain penetration (Dong 2018).

Concerns regarding overheating, off-target damage, and long-term biocompatibility remain. However, the development of stimuli-responsive and biodegradable nanomaterials, along with refined light delivery systems, is improving the safety profile of PTT for neurological applications (Jain and El-Sayed 2010).

14.7 Challenges and Future Perspectives

While phototherapeutics offers exciting opportunities for treating neurodegenerative disorders, several challenges must be addressed to facilitate clinical translation and long-term efficacy (Fig. 14.1).

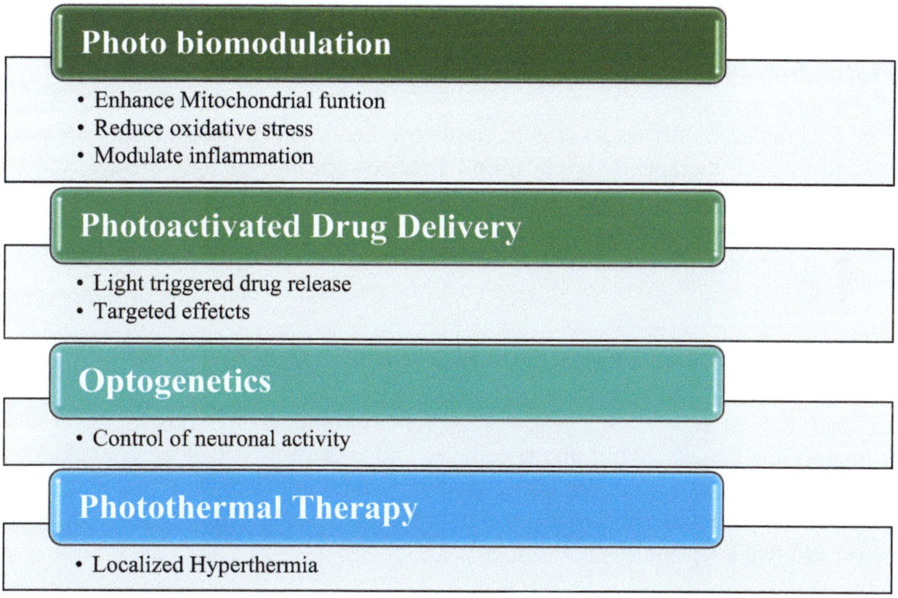

Fig. 14.1 Challenges and future innovations. A conceptual roadmap of clinical translation challenges and tech solutions

14.7.1 Light Penetration and Targeting

One of the primary limitations of phototherapeutics is the restricted penetration of light through biological tissues, particularly the skull and deeper brain regions. Near-infrared (NIR) light, serves as a common imaging tool that penetrates tissues up to a few centimeters while maintaining disadvantages during deep structure investigations. Advances in fiber-optic systems, transcranial phototherapy helmets, and minimally invasive light delivery devices are being explored to overcome this barrier (Hamblin 2016).

14.7.2 Blood-Brain Barrier (BBB) and Delivery Systems

The high barrier function of the BBB stands as an essential challenge in supplying phototherapeutic medications to the central nervous system. Nanoparticle-based delivery systems, ligand-mediated transport, and noninvasive techniques such as focused ultrasound have been employed to improve CNS access (Saraiva et al. 2016). However, issues related to long-term toxicity, immunogenicity, and stability of nanocarriers remain critical areas for optimization.

14.7.3 Safety and Off-Target Effects

Concerns regarding off-target effects, tissue overheating, and phototoxicity are paramount in phototherapeutic applications. Precise control over light dosage, wavelength, exposure duration, and localization is essential to ensure safety. Long-term studies are needed to assess chronic toxicity, immune responses, and cumulative effects of repeated treatments (Wang et al. 2020c).

14.7.4 Genetic and Ethical Considerations in Optogenetics

The use of viral vectors and gene-editing tools in optogenetics raises ethical and regulatory challenges. While the specificity and control provided by optogenetics are unparalleled, concerns regarding genetic manipulation in humans, long-term effects, and potential off-target gene expression must be thoroughly addressed before clinical adoption (Deisseroth 2015).

14.7.5 Translation from Bench to Bedside

Most current data on phototherapeutics stem from in vitro and animal studies. Human trials are limited, and there is a gap in translating preclinical findings into clinical outcomes. Standardization of protocols, improved light-delivery tools, real-time monitoring systems, and interdisciplinary collaboration between neuroscientists, clinicians, and engineers are vital for successful translation (Ruan et al. 2007).

14.7.6 Future Directions

Future research should focus on:

- Designing biodegradable and targeted nanomaterials for safe delivery.
- Combining phototherapy with immunomodulatory and gene-editing approaches.
- Developing wireless, implantable devices for long-term, closed-loop phototherapeutic interventions.
- Integrating real-time imaging and biosensing for feedback-controlled therapy.

The intersection of photonics, nanotechnology, and neurobiology holds immense promise for developing precision therapeutics for neurodegenerative diseases. Continued investments in multidisciplinary research, ethical regulation, and clinical validation will be critical in realizing the full potential of phototherapeutics.

14.8 Conclusion

Phototherapeutics represent a transformative and rapidly evolving frontier in the treatment of neurodegenerative disorders. Through diverse modalities such as photobiomodulation, photoactivated drug delivery, optogenetics, and photothermal therapy, light-based technologies offer unprecedented precision, noninvasiveness, and multifaceted therapeutic potential. These approaches have demonstrated the capacity to modulate mitochondrial function, target and eliminate pathological protein aggregates, restore neural circuit dynamics, and enhance drug delivery across the blood-brain barrier.

Despite these advances, significant hurdles remain in translating these promising findings into clinical practice. Challenges related to light delivery, brain targeting, biosafety, and ethical considerations necessitate robust interdisciplinary research and strategic innovation. Importantly, the synergy of phototherapeutics with other therapeutic domains—such as gene editing, immunotherapy, and artificial intelligence—could pave the way for personalized and adaptive treatment strategies.

As the burden of neurodegenerative diseases continues to escalate globally, phototherapeutics offer hope for interventions that not only alleviate symptoms but also modify disease progression. The coming years will be critical in defining their clinical relevance, with ongoing trials, technological refinements, and growing translational evidence set to shape their role in modern neuroscience and medicine.

References

Chen H, Wang GD, Chuang YJ et al (2017) Nanoscintillator-mediated X-ray inducible photodynamic therapy for deep-seated tumors: current status and future perspectives. Biomaterials 146:87–103

Cheng Y, MacHado AE, Gould LJ (2021) Photobiomodulation for treatment of neurodegenerative disorders: a systematic review. Neurobiol Dis 148:105193

Deisseroth K (2015) Optogenetics: 10 years of microbial opsins in neuroscience. Nat Neurosci 18(9):1213–1225

Dong X (2018) Current strategies for brain drug delivery. Theranostics 8(6):1481–1493

Fenno L, Yizhar O, Deisseroth K (2011) The development and application of optogenetics. Annu Rev Neurosci 34:389–412

Gradinaru V, Mogri M, Thompson KR, Henderson JM, Deisseroth K (2009) Optical deconstruction of Parkinsonian neural circuitry. Science 324(5925):354–359

Grillo SL, Duggett NA, Ennaceur A, Chazot PL (2013) Non-invasive infra-red therapy (1072 nm) reduces β-amyloid protein levels in the brain and improves memory in a rat model of Alzheimer's disease. Eur J Neurosci 37(3):324–335

Hamblin MR (2016) Shining light on the head: photobiomodulation for brain disorders. BBA Clin 6:113–124

Hamblin MR (2017) Mechanisms and applications of the anti-inflammatory effects of photobiomodulation. AIMS Biophys 4(3):337–361

Hamilton C, Hamilton D, Nicklason F, Mitrofanis J (2019) Transcranial photobiomodulation therapy for Parkinson's disease: a pilot study of safety and efficacy. J Photochem Photobiol B 203:111636

Huang X, Jain PK, El-Sayed IH, El-Sayed MA (2007) Gold nanoparticles: interesting optical properties and recent applications in cancer diagnostics and therapy. Nanomedicine 2(5):681–693

Idris NM, Gnanasammandhan MK, Zhang J et al (2012) In vivo photodynamic therapy using upconversion nanoparticles as remote-controlled nanotransducers. Nat Med 18(10):1580–1585

Jain PK, El-Sayed MA (2010) Plasmonic coupling in noble metal nanostructures. Chem Phys Lett 487(4–6):153–164

Johnstone DM, Moro C, Stone J et al (2016) Turning on lights to stop neurodegeneration: the potential of near infrared light therapy in Alzheimer's and Parkinson's disease. Front Neurosci 9:500

Kravitz AV, Freeze BS, Parker PR et al (2010) Regulation of parkinsonian motor behaviours by optogenetic control of basal ganglia circuitry. Nature 466(7306):622–626

Li Y, Liu G, Ma J et al (2020) ROS-responsive nanoparticles designed for dual-drug delivery: improving the therapeutic efficacy of Parkinson's disease. J Control Release 321:667–683

Liao J, Jia Y, Wu Y et al (2020) Targeting alpha-synuclein aggregation via near-infrared light-induced photothermal effect of graphene oxide in Parkinson's disease models. Nanoscale 12(17):9133–9145

Liu Y, Bhattarai P, Dai Z, Chen X (2019) Photothermal therapy and photoacoustic imaging via nanotheranostics in fighting cancer and brain diseases. Chem Soc Rev 48(7):2053–2108

Lucky SS, Soo KC, Zhang Y (2015) Nanoparticles in photodynamic therapy. Chem Rev 115(4):1990–2042

Mitrofanis J (2017) Why and how does light therapy offer neuroprotection in Parkinson's disease? Neural Regen Res 12(4):574–575

Pisanello F, Sileo L, Oldenburg IA et al (2017) Multipoint-emitting optical fibers for spatially addressable in vivo optogenetics. Neuron 93(4):1049–1064

Roy DS, Arons A, Mitchell TI et al (2016) Memory retrieval by activating engram cells in mouse models of early Alzheimer's disease. Nature 531(7595):508–512

Ruan G, Agrawal A, Marcus AI, Nie S (2007) Imaging and tracking of tat peptide-conjugated quantum dots in living cancer cells. Anal Chem 79(2):720–728

Rwei AY, Lee JJ, Zhan C et al (2015) Repeatable and adjustable on-demand sciatic nerve blockade with phototriggerable liposomes. Proc Natl Acad Sci U S A 112(50):15719–15724

Salehpour F, Majdi A, Pazhuhi M, Ghasemi F, Khademi M, Hamblin MR (2021) Transcranial photobiomodulation therapy for Alzheimer's disease: a review of mechanisms, devices, and clinical applications. Aging Dis 12(8):2046–2065
Saltmarche AE, Naeser MA, Ho KF, Hamblin MR, Lim L (2017) Significant improvement in cognition in mild to moderately severe dementia cases treated with transcranial and intranasal photobiomodulation: case series report. Photomed Laser Surg 35(8):432–435
Saraiva C, Praça C, Ferreira R et al (2016) Nanoparticle-mediated brain drug delivery: overcoming blood-brain barrier to treat neurodegenerative diseases. J Control Release 235:34–47
Towne C, Montgomery KL, Iyer SM, Deisseroth K, Delp SL (2013) Optogenetic control of targeted peripheral axons in freely moving animals. PLoS One 8(8):e72691
Verret L, Mann EO, Hang GB et al (2012) Inhibitory interneuron deficit links altered network activity and cognitive dysfunction in Alzheimer model. Cell 149(3):708–721
Wang Y, Cui H, Ma N et al (2020a) Near-infrared light-responsive biodegradable nanoparticles for enhanced drug delivery in Alzheimer's disease. ACS Appl Mater Interfaces 12(41):45902–45914
Wang J, Tian L, Yu X et al (2020b) Photothermal clearance of Alzheimer's disease amyloid-beta aggregates using polyoxometalate-functionalized nanoparticles. ACS Appl Mater Interfaces 12(5):5907–5914
Wang H, Yang W, Zhang C et al (2020c) The safety and toxicity of nanoparticles in neurodegenerative diseases. Chem Biol Interact 331:109273
Yang K, Xu H, Cheng L, Sun C, Wang J, Liu Z (2012) In vitro and in vivo near-infrared photothermal therapy of brain tumor using gold nanorods. Adv Mater 24(14):1868–1872
Yang Y, Wang Z, Wang Y et al (2018) Photothermal modulation of microglial activation using carbon-based nanomaterials in a model of Parkinson's disease. ACS Nano 12(2):1920–1934
Zhang F, Wang LP, Boyden ES, Deisseroth K (2006) Channelrhodopsin-2 and optical control of excitable cells. Nat Methods 3(10):785–792

Mechanistic Pathways of Bioactive Phytocompounds in Neurodegenerative Disorders

15

Jhanvi Soni and Neeraj Kumar

Abstract

The majority of neurodegenerative disorders, including Alzheimer's, Parkinson's, and Huntington's diseases, lead to gradual deterioration of neuronal structure and function, resulting in cognitive decline, motor dysfunction, and behavioral impairments. Despite substantial advances in understanding the molecular basis of neurodegeneration, no definitive cure or disease-modifying drug has yet been discovered. Most of these disorders share pathophysiological similarities such as oxidative stress, inflammation, protein aggregation, mitochondrial dysfunction, dysregulated proteostasis, and finally neurodegeneration. In this context, bioactive phytocompounds derived from medicinal plants have gained increasing attention due to their multitarget actions and diverse biological properties, notably antioxidant, anti-inflammatory, and neuroprotective effects. Various classes of phytochemicals have demonstrated neuroprotective mechanisms. Polyphenols, including flavonoids and phenolic acids, exhibit strong antioxidant activity by scavenging reactive oxygen species and modulating signaling pathways. Alkaloids like berberine and stilbenes such as resveratrol reduce inflammation and oxidative stress, while resveratrol also activates SIRT1, promoting mitochondrial biogenesis. Terpenoids and saponins support neurogenesis, regulate apoptosis, and inhibit neuronal damage.

The mechanistic pathways by which these compounds act include oxidative stress modulation, enhancing antioxidant enzymes like superoxide dismutase and catalase; anti-inflammatory effects, through inhibition of cytokines (e.g., IL-1β, TNF-α); microglial activation and apoptosis regulation, via mitochondrial

J. Soni (✉)
Parul Institute of Pharmacy, Parul University, Vadodara, India

N. Kumar
Parul College of Pharmacy and Research, Parul University, Ahmedabad, India

H. Dilnashin, S. P. Singh (eds.), *Phototherapeutic Approaches to Neurodegeneration*,
https://doi.org/10.1007/978-981-95-4751-7_15

stabilization and caspase inhibition; prevention of protein aggregation, including amyloid-beta clearance and tau protein stabilization; promotion of neurogenesis and synaptic plasticity, through upregulation of brain-derived neurotrophic factor (BDNF); and enhancement of synaptogenesis. Notable examples include curcumin, which inhibits amyloid-beta aggregation; resveratrol, which improves mitochondrial function; and quercetin, which reduces oxidative stress and inflammation.

Evidence from preclinical in vitro and animal studies consistently supports these neuroprotective actions, while limited human trials suggest cognitive and biochemical improvements. Overall, this chapter aims to highlight the possible mechanisms of phytocompounds in mitigating the complex pathophysiology of neurodegenerative disorders.

Keywords

Cytokine inhibition · BDNF · SIRT1 · Multi-target therapy · Neurogenesis · Synaptic plasticity

15.1 Overview of Neurodegenerative Disorders

Most neurodegenerative disorders (NDDs), such as Alzheimer's disease (AD), Parkinson's disease (PD), Huntington's disease (HD), and amyotrophic lateral sclerosis (ALS), involve progressive and irreversible neuron loss in specific brain regions. Normal cognitive and motor function, as well as a productive quality of life, are profoundly affected by these disorders. Moreover, they are also largely untreatable (Singh et al. 2025). As a leading cause of disability and the second leading cause of death in the global population, neurological disorders (ND) cannot easily be neglected. Although current pharmaceuticals can reduce symptoms, they are only effective as long as they are used. Further, they can have detrimental adverse side effects while lacking sufficient neuroprotective capacity, unless paired with a neuroprotective compound or adjunctive treatment, in which case the overall clinical outcome may be better. Phytochemicals have received recent recognition as future therapeutic candidates due to their ability to modulate various molecular pathways and their ability to induce neuroprotective mechanisms through a host of actions such as antioxidant, anti-inflammatory, anti-amyloid, mitochondrial, and neurotropic properties (Wang et al. 2016; Sharifi-Rad et al. 2022). This chapter will discuss recent advances in the molecular mechanisms by which a variety of phytocompounds have been shown to modulate neurodegeneration.

15.2 Pathophysiological Hallmarks of Neurodegeneration

Although the clinical presentations of NDDs differ, it is well established that NDDs share different pathogenic hallmarks from both a cellular and molecular perspective. The identified shared features in NDDs can be classified into multiple clusters.

15.2.1 Protein Aggregates/Disrupted Proteostasis

The major hallmark of most NDDs is the accumulation of proteins with altered physicochemical properties, also termed misfolded proteins or "aggregates" (Kaushik and Cuervo 2015). The stability of aggregates can vary with that of a mature aggregate present; however, it is the interaction of the intermediate oligomeric protein "assemblies" that are associated with neurotoxicity. This hallmark is referred to as "disrupted proteostasis," and occurs because the physiological processes clearing pathophysiological proteins decline with age, leading to an increased pathological protein accumulation. Commonly implicated associated protein aggregates in NDDs are as follows:

1. Beta-amyloid (Aβ) accumulation is due to the cleaving of amyloid precursor protein (APP) via amyloidogenic pathways producing Aβ peptides that can polymerize into amyloid fibrils and plaques, where the Aβ pathology usually occurs prior to tau pathologies; however, pure taupathies exist that are likely due to an independent pathological pathway (Selkoe and Hardy 2016).
2. α-Synuclein aggregation is punctuated by aberrant cytoplasmic deposits, specifically Lewy bodies, which is a hallmark of PD neuropathology (Lotharius and Brundin 2002).
3. Huntington's disease is caused by the aggregating neurotoxic form of the protein huntingtin that occurs as a result of an expansion of CAG repeats in the HTT gene, producing an abnormally long polyglutamine domain in the huntingtin protein (Jimenez-Sanchez et al. 2017).
4. Ubiquitinated TDP-43 inclusions are present in motor neurons in ALS (Dugger and Dickson 2017).

Protein degradation systems are fundamental to clear these aggregates. Aggregated proteins undergo degradation via the ubiquitin–proteasome system by targeting ubiquitinated proteins, or alternatively, the autophagy–lysosome pathway degrades aggregated proteins through clearance of aggregated proteins and damaged organelles (Dugger and Dickson 2017). The longitudinal studies to date have demonstrated that both systems can be impaired in neurodegenerative diseases (NDDs), while the aberrant aggregated proteins, such as Aβ and tau, can also directly inhibit proteasome function. Some research has proposed that pathological proteins spread intercellularly between neurons designed like "seeds" for the further misfolding of native proteins (Tyson et al. 2016).

15.2.2 Neuronal and Synaptic Dysfunction

NDDs are mostly considered a consequence of neuronal network disruptions or synaptic failure. Neuronal network dysfunction may occur through excitotoxicity via elevated glutamate levels, which promote a calcium ion influx, initiate mitochondrial dysfunction, and lower Adenosine triphosphate (ATP) levels, ultimately resulting in the death of neurons. The collective evidence demonstrates that protein aggregation can mechanically strain across synaptic junctions. Physiological stress is determined through Aβ oligomers that disrupt normal synaptic signalling eventuating synapse loss or cognitive decline. Physiological brain events engaged in signaling include increased release of glutamate or ligand binding directly to NMDA receptors which promote excitotoxicity (Lall et al. 2021; Soukup et al. 2018; Verma et al. 2022).

15.2.3 Cytoskeletal Abnormalities/Deformations

Cytoskeletal structures are susceptible to defects in the structure and polymerization of building block monomers leading to NDDs. These structures include actin-based microfilaments, intermediate filaments, and tubulin-based microtubules which are specialized for maintaining neuronal architecture, intracellular cargo transport or trafficking, and energy homeostasis. Mutational change in genes that are linked with cytoskeletal abnormalities such as neuronal intermediate filament and microtubule-binding protein tau are known to contribute to NDDs such as Charcot–Marie–Tooth disease and tauopathies respectively (Dugger and Dickson 2017; Stone et al. 2021).

15.2.4 Altered Energy Homeostasis/Metabolic Dysfunction

Neurons exhibit high-energy demand, and defects in cellular energy metabolism pathways have been documented in NDDs. Mitochondrial dysfunction is a significant and initiated process of loss ATP and ionic imbalance, so potential cytoskeletal derangements which are compounded by uncoupled calcium homeostasis. This results in oxidative stress due to a burst of reactive oxygen species (ROS) and reactive nitrogen species (RNS), which can produce the macromolecule damage that contributes to neuronal dysfunction and cell death (Cunnane et al. 2020; Nunnari and Suomalainen 2012).

Metabolic dysfunction, which can manifest as obesity, metabolic syndrome and/or insulin resistance, are parallels directly associated with the aging process and with NDDs such as AD, PD, ALS, and HD. However, the shifts in abnormal metabolic situations caused by insulin resistance can impair insulin signaling and enhance Aβ deposition and tau phosphorylation, which are also highly correlated with severe increases in Aβ (Gomez et al. 2018; Roberts et al. 2014).

15.2.5 DNA and RNA Defects/Oxidative and Endoplasmic Reticulum (ER) Stress

DNA damage and alterations in RNA metabolism can lead to diseases that affect neuronal cells. Continuous changes in DNA lead to mutagenesis, chromosomal abnormalities, and changes in RNA transcription, resulting in a dysfunctional state of the cell and ultimately to death. Oxidative stress [the imbalance between the production and removal of ROS and RNS] causes oxidative damage to biomolecules (lipids, proteins, and nucleic acids). Neurons are especially vulnerable to oxidative stress due to their high metabolic rate and their limited ability to produce antioxidants. DNA damage from oxidative stress can promote aging and cause inflammation (Haeusler et al. 2016). Endoplasmic reticulum (ER) stress pathways also lead to neuronal apoptosis. The ER is an important part of neuronal function and is central to protein production and folding, and the "unfolded protein response" (UPR) is activated by misfolded proteins. Chronic UPR activation leads to ND pathology. The presence of misfolded protein aggregates can compromise ER function (Garthe and Maughan 2018).

15.2.6 Inflammation/Neuroimmune System Alterations

Neuroinflammation, an inflammatory process that involves neuronal cells, is a feature of virtually all ND diseases. Neuroinflammation involves the activation of microglial cells (microgliosis) and astrocytes (astrogliosis). Microglia are the innate immune cells of the brain that detect and protect the central nervous system from harmful agents. Chronic microglia activation is often stimulated by protein aggregations, low-fidelity protein misfolds, and damaged synaptic connections; this not only accelerates disease progression but also leads to maladaptive inflammation. Reactive astrocytes and microglia interact with other NDD hallmarks, influencing synaptic function, energy homeostasis, protein aggregation, and neuronal degeneration. Overactive microglia can eliminate synaptic junctions, leading to neuronal network dysfunction (Anderson et al. 1987; Heneka et al. 2018; Mandolesi et al. 2015).

The mechanisms described above lead to neuronal cell death, a significant characteristic of neurodegenerative diseases. Neurons are highly susceptible to cell death because they are postmitotic cells (accumulate damage with age), require large amounts of energy, produce large quantities of ROS, have greatly elongated axons and dendrites, and rely on glial cells for support. Oxidative stress and ER stress pathways trigger neuronal cell death through the induction of neuronal apoptosis (Bloomingdale et al. 2022).

In conclusion, while the NDDs often manifest unique clinical signs and symptoms, specific NDDs target different neuronal populations and share certain hallmarks: protein misfolding and aggregation, impaired waste clearance (proteostasis), energy deficiencies, alterations to our genetic/epigenetic makeup, oxidative stress, inflammation, and neuronal cell death.

15.3 Limitations of Current Therapies

The primary restriction of present neurodegenerative treatment approaches exists because they provide symptomatic relief but do not modify disease progression (Muqit and Alessi 2009). Thus, the current therapies fall short due to their inability to stop disease progression, while traditional symptomatic treatments become less effective with time and cause problematic side effects (Di Paolo et al. 2019; Rasool et al. 2014).

Parkinson's disease treatments that prove most beneficial for motor symptom management remain insufficient because they do not address nonmotor symptoms or prevent disease progression (Koo 2021). The blood–brain barrier functions as the primary barrier to developing effective treatments for NDDs because it prevents most potential therapeutics from entering brain tissue (Shafiee et al. 2020). The physiological barrier imposes major obstacles because it blocks the passage of 98% of small-molecule drugs together with 100% of large-molecule drugs into the central nervous system (De Gioia et al. 2020). The efflux transporter P-glycoprotein, together with multidrug resistant protein transporters, makes drug penetration more difficult by actively removing substances that reach the barrier (Rajput et al. 2022; Patel et al. 2017; Matsumoto et al. 2018).

The current neurodegenerative treatments have side effects that severely impact patient's quality of life while making treatment adherence challenging. Traditional symptomatic treatments lose their potency with time because they trigger additional side effects that become increasingly severe. Numerous pharmacological interventions generate unwanted side effects that reduce patient tolerance during extended treatment periods (Shashikant et al. 2024; Connolly and Lang 2014).

The main obstacle in treating NDDs exists because scientists have not fully understood their intricate underlying mechanisms. The ineffective performance of neurotherapies in clinical trials remains a problem for researchers because they have not identified the underlying reasons despite their incomplete comprehension of neurogenesis mechanisms and influencing factors (Mouhieddine et al. 2014). The complex mechanisms of neuronal loss, together with conflicting physiological factors that cause these diseases, make it difficult to understand pathogenic processes, thus limiting the development of effective treatments (Chang et al. 2020). The pathogenic mechanisms remain unclear, while therapeutic interventions remain limited because of incomplete targeting of complex pathological processes (Yang et al. 2024).

The neurodegenerative disorder drug development sector faces unprecedented failures despite spending massive amounts of research money. The Alzheimer's disease drug development field has invested $42.5 billion in clinical trials during the last 30 years, yet achieved minimal promising outcomes (Amartumur et al. 2024; Cummings et al. 2021). Key genetic discoveries made over the past 15 years have not produced any therapeutic benefits for patients despite extensive basic and clinical research (Muqit and Alessi 2009). The financial hurdles in neurodegenerative disease treatment pose a significant obstacle for advancement since substantial research expenditures have produced minimal therapeutic outcomes.

15.4 Role of Phytocompounds in Neuroprotection

The protection of neurons from different types of cellular injuries and neurodegenerative diseases is known as neuroprotection (Elufioye et al. 2017). Multiple disease mechanisms can be targeted by phytocompounds, which are natural plant-derived chemical substances, making them potential neuroprotective agents (Khan et al. 2020). Different types of biologically active secondary metabolites, including polyphenols, alkaloids, and terpenoids, have been recorded in numerous plant genera (Elufioye et al. 2017).

Multiple interconnected molecular pathways enable phytocompounds to defend the brain through the treatment of neurodegenerative disease mechanisms. Antioxidant effects represent the main action of phytocompounds because these compounds remove reactive oxygen species and activate the Nrf2-ARE system, which defends cells against oxidative stress by managing antioxidant gene expression (Hannan et al. 2020). Phytocompounds defend cellular life by stimulating the TrkB/PI3K/Akt pathway that enables Nrf2 nuclear migration and creates protective effects (Hannan et al. 2020).

Phytocompounds protect against inflammation by blocking traditional inflammatory signaling molecules while they affect NF-κB, TLR4, and NLRP3 inflammasome pathways (Nath et al. 2020; Khayatan et al. 2024). The compounds enhance tight junctional complexes between blood–brain barrier cells, which results in better barrier performance following cerebral injury (Nath et al. 2020). The protective mechanisms of mitochondria function as a vital pathway because polyphenols enhance electron transport chain activity while controlling redox states, biogenesis, and dynamics and autophagy processes to maintain proper mitochondrial number and quality (Naoi, Wu et al. 2019; Gibellini et al. 2015). The protective mechanisms of phytocompounds include the control of apoptotic pathways through the regulation of Bcl-2 family proteins (Naoi, Wu et al. 2019; Gibellini et al. 2015).

Phytocompounds protect the brain by ensuring protein homeostasis through their ability to control cell-clearing systems, which include autophagy and the ubiquitin–proteasome system (Limanaqi et al. 2019a, 2020; Momtaz et al. 2020). These compounds enhance neurotrophin expression while activating CREB and other intracellular pathways to improve synaptic function and neuronal integrity and learning and memory abilities (Fiore et al. 2025; Carito et al. 2016). The therapeutic potential becomes more effective through the regulation of vital signaling pathways such as Akt, Nrf2, STAT, and MAPK that support neuroprotection and immune response (Jalouli et al. 2025).

The therapeutic potential of phytocompounds extends across various neurodegenerative diseases because they target multiple disease mechanisms simultaneously. Scientists study plant-derived bioactive compounds because they have established medicinal value and their ability to affect protein expression along with various protein activities related to oxidative stress, neuroinflammation, apoptosis, and protein aggregation (Shoaib et al. 2023).

15.5 Mechanistic Pathways of Action of Phytochemicals

15.5.1 Oxidative Stress Modulation

Oxidative stress is a major pathological mechanism in neurodegenerative diseases and is characterized by an imbalance between the formation and elimination of reactive nitrogen species (RNS) and ROS (Rahman et al. 2020; Wang et al. 2016). This imbalance can cause cellular damage, ranging from lipid peroxidation to damage of nucleic acids, which leads to neuronal dysfunction, damage, and death.

Phytochemicals can counteract oxidative stress through systematized mechanisms, as outlined in the following subsections.

15.5.1.1 Scavenging Free Radicals and Quenching Singlet Oxygen

Dietary organic nonenzymatic antioxidants, including phytochemicals like polyphenols (e.g., quercetin, resveratrol), carotenoids (e.g., beta-carotene, lutein), and sulfur-containing plant antioxidants counter the impact of oxidative damage by directly neutralizing free radicals [e.g., superoxide anion (O^{2-}), hydrogen peroxide (H_2O_2), hydroxyl radicals (HO)]. The phytochemical curcumin has strong activity against pro-oxidants of superoxide and hydrogen peroxide radicals (Nebrisi 2021).

15.5.1.2 Chelating Metal Ions

Phytochemicals can chelate metal ions involved in radical formation by beta-amyloid (Aβ) aggregates thereby inhibiting oxidative damage. Here, curcumin has iron-chelating effects as well (Can and Sanlier 2025).

15.5.1.3 Strengthening Cellular Antioxidant Defense Systems

Phytochemicals stimulate the action of endogenous antioxidant enzymes, such as superoxide dismutase (SOD), catalase, glutathione peroxidase (GPx), and glutathione reductase. They usually do this by activating the Nrf2/ARE (nuclear factor erythroid-2-related factor 2/antioxidant response element) signaling pathway, thereby upregulating genes that encode these cytoprotective enzymes. Resveratrol, for instance, promotes SOD and GSH protein levels via Nrf2 nuclear translocation. Sulfuretin, likewise, protects against Aβ-mediated neurotoxicity through the Nrf2/HO-1 signaling pathway (Hossain et al. 2025).

Polyphenols have been shown to exert antioxidant effects and also can modulate gene expression (and epigenetic changes). Curcumin, for example, scavenges free radicals (ROS and RNS) and regulates antioxidant enzyme activity (Daverey and Agrawal 2020). Resveratrol reduces ROS production and enhances antioxidant enzymes SOD and HO-1 functions by activating the PI3K/Akt/Nrf2 pathway, while also reducing oxidative stress in rotenone-induced cells (Ali et al. 2018). In addition, epigallocatechin-3-gallate (EGCG), a polyphenol from green tea, protects against oxidative stress while enhancing tau degradation (Chesser et al. 2016). Capsaicin also prevents an increase in MDA levels in brain tissue, which is a measure of oxidative stress (Chung et al. 2017). Lycopene, a carotenoid, mediates against oxidative stress, attenuating lipopolysaccharide-induced amyloidogenesis and cognitive impairments (Wang et al. 2018a). Ferulic acid has been shown to reduce neurodegeneration through the Nrf2/ARE signaling pathway (Singh et al. 2022). Quercetin-mitigated mitochondrial dysfunction by reducing ROS levels and promoting AMPK expression (Wang et al. 2014). Luteolin also has antioxidant properties and protects neuronal cells from oxidative stress (He et al. 2023). Ginsenosides exhibit anti-oxidative properties and increase antioxidant enzyme activity (Zhou et al. 2015). Astaxanthin has been shown to reduce oxidative stress and increase antioxidant activity (Lee et al. 2011). *Urtica dioica* (stinging nettle) extracts confer neuroprotection against chlorpyrifos-induced abnormalities and facilitate redox homeostasis (Mhalhel et al. 2024).

15.5.2 Anti-inflammatory Effects

Neuroinflammation is a critical pathological component of neurodegenerative diseases, including Parkinson's and Alzheimer's. Neuroinflammation is mainly driven by neuroinflammatory microglial cells, which are the macrophages of the central nervous system. Chronic inflammation, in relation to neuroinflammation, is typified by the excess production of inflammatory cytokines, including interferon (IFN), tumor necrosis factor-α (TNF-α), IL-1, IL-6, IL-8, IL-13, transforming growth factor-β (TGF-β), and C-reactive protein (CRP), which contribute to chronic disease and accelerated aging (Wang et al. 2018b). Phytochemicals are important in reducing neuroinflammation through various mechanisms. They inhibit the expression and release of pro-inflammatory cytokines such as TNF-α, IL-1β, and IL-6, and inflammatory enzymes like nitric oxide synthase (NOS) and cyclooxygenase-2 (COX-2) (Wang et al. 2016).

A major component of their action is the ability to modulate the inflammatory signaling pathways. Phytochemicals are shown to inhibit many of the pathways involved in inflammatory processes such as the NF-κB pathway, a transcription factor that initiates the production of cytokines; curcumin has been shown to mitigate the TLR4/NF-κB axis (Daverey and Agrawal 2020; de Lima et al. 2025; He et al. 2020). Phytochemicals also regulate the MAPK and PI3K/Akt pathways which are key pathways involved in microglial activation and cytokine release; inhibiting the PI3K/Akt pathway reduced the production of pro-inflammatory cytokines (Rahman et al. 2020; Yoon 2025). In addition, phytochemicals affect microglial polarization by favoring the transition from the M1 pro-inflammatory phenotype to the M2 anti-inflammatory phenotype, enhancing neuroprotection (Wang et al. 2016). There are several bioactive compounds that illustrate these effects. Curcumin has strong anti-inflammatory properties and can inhibit amyloidosis. Resveratrol also modulates neuroinflammation through SIRT1 and NF-κB signaling and inhibition of IL-6, IL-1β, and TNF-α induction (Can and Sanlier 2025; Ghafouri-Fard et al. 2022).

Examples of phytochemicals that modulate gene expression and provide neuroprotection include polyphenols and flavonoids as flavones, like apigenin (Hossain et al. 2025), and omega-3 polyunsaturated fatty acids Eicosapentaenoic Acid (EPA) and Docosahexaenoic Acid (DHA)) (Shinto et al. 2009). Additional examples include ginsenosides, propolis constituents, Caffeic Acid Phenethyl Ester (CAPE) and chrysin, chlorogenic acid, gingerol, and flavonoids from licorice (Chung et al.

2017; Pázmándi et al. 2024; Wang et al. 2016, 2018b; Yoon 2025; Zhou et al. 2015), including recently detected compounds eleuhenryiside I, which targets TLR4, Src, MAPK, and NF-κB (Tang et al. 2025), and argyinolide S, a sesquiterpenoid that inhibits JAK1. Broadly, phytochemicals enhance gastrointestinal function and stimulate the gut–brain axis via the vagal reflex indirectly, ultimately promoting brain health and halting neuroinflammatory disruptions (Wang et al. 2018b).

15.5.3 Mitochondrial Protection

Mitochondrial dysfunction is a core feature and leading contributor to the pathophysiology of neurodegenerative diseases, defined by decreased energy production, increased ROS generation, oxidative stress, and neuronal apoptosis. Pathological proteins, including amyloid-β (Aβ) in the case of Alzheimer's disease (AD) and α-synuclein in the case of Parkinson's disease (PD), impair mitochondrial structure and function by inhibiting components of the electron transport chain and increasing ROS production (Bloomingdale et al. 2022; Wang et al. 2016).

Phytochemicals afford protective effects against mitochondrial impairment via many mechanisms. For example, phytochemicals increase mitochondrial biogenesis and modulate fission–fusion dynamics, which are vital to maintaining mitochondrial health, with anthocyanins modulating mitochondrial dynamics and resveratrol and EGCG promoting mitochondrial biogenesis via Sirt1/AMPK and PGC-1α activation (de Oliveira et al. 2016; Gorlach et al. 2015; Parrado-Fernández et al. 2016). Phytochemicals prevent apoptotic cell death induced by loss of mitochondrial membrane potential by keeping mitochondrial membrane potential and by exerting an antioxidant action to lessen oxidative stress, followed by resumption of ATP production (Rahman et al. 2020; Nebrisi 2021). Moreover, phytochemicals have the ability to regulate autophagy and proteolytic clearance mechanisms, since these compounds also modulate the ubiquitin–proteosome system and autophagy, promoting the degradation of pathological proteins. For instance, resveratrol has been reported to upregulate proteasome activity in AD, and phytochemicals, including resveratrol, work to inhibit mTOR signaling, providing a way to upregulate autophagy and induce clearance of Aβ and tau aggregates in AD (Pineda-Ramírez et al. 2020). Furthermore, phytochemicals exert a protective effect on mitochondrial structure and function by regulating pro-apoptotic pathways, thus preventing neuronal cell death. Various phytochemicals have exhibited these aforementioned protective effects, inclusive of: curcumin, which exhibited protective effects against Aβ-induced mitochondrial dysfunction and improved the activity of mitochondrial enzymes in models of PD obtained from rotenone (Nebrisi 2021); resveratrol, which inhibited oxidative stress through activation of mitophagy while also promoting biogenesis via AMPK and Sirt1 (de Oliveira et al. 2016; Pineda-Ramírez et al. 2020); quercetin, which reversed depolarization-induced loss of mitochondrial membrane potential and restored ATP production in isolated neuronal cells (Wang et al. 2014); astaxanthin, which reduced dopaminergic degeneration seen in botanically-induced PD (Lee et al. 2011); ginkgolides and limonene, both of which improved mitochondrial function and reduced deficits in mitochondrial function caused by complex I inhibition (de Lima et al. 2025). Together, this represents the contribution of phytochemicals toward mitochondrial health.

15.6 Inhibition of Protein Aggregation

The accumulation of insoluble protein aggregates is the fundamental pathological mechanism that links Alzheimer's to Parkinson's and Huntington's diseases (Hajieva 2017). Oxidative compounds together with sugar residues cause changes to protein secondary structures, which leads to misfolding and aggregation that becomes dangerous when autophagy and proteasome cellular clearing systems function improperly (Limanaqi et al. 2020).

The pathological cascade consists of multiple interconnected events, which include oxidative and inflammatory processes alongside mitochondrial damage, together with protein aggregation and dysfunction of both autophagy and proteasome systems (Limanaqi et al. 2019b, 2020; Pinelli et al. 2020). The combination of chronic oxidative stress and inflammation with protein aggregation produces a self-perpetuating cycle that disrupts cell-clearing systems and leads to more damaged mitochondria accumulation and toxic protein spreading between cells (Limanaqi et al. 2020; Thibaudeau et al. 2018; Lai et al. 2018).

Among the most investigated anti-aggregation compounds, EGCG from green tea shows strong capabilities to bind different misfolded proteins, including Aβ, tau, α-synuclein, transthyretin, and huntingtin (Gonçalves et al. 2021; Sonawane et al. 2020; Ehrnhoefer et al. 2006). The IC50 value of EGCG as an anti-tau aggregation agent is 64.2 μM, which enables both inhibition of tau formation and breaking down of existing fibrils (Sonawane et al. 2020). The compound curcumin displays notable anti-aggregation properties when interacting with multiple neurodegenerative proteins (Gonçalves et al. 2021; Fukutomi et al. 2021). The compound induces natural heat shock proteins that lead to chaperone-mediated autophagy, which degrades mutant protein complexes in patients with Huntington's disease (Labanca et al. 2020; Maiti et al. 2014). Curcumin binds to oligomeric forms of α-synuclein rather than monomers to accelerate nontoxic pathways of aggregation while decreasing surface hydrophobicity of toxic species (Singh et al. 2013; Thapa et al. 2016).

Resveratrol works to decrease Aβ aggregation while enhancing autophagic clearance mechanisms and protects tau from hyperphosphorylation through its effects on GSK-3β, CaMKII, and PP2A activities (Nahar et al. 2025; Arbo et al. 2020; He et al. 2017).

The compound baicalein found in *Scutellaria baicalensis* effectively blocks the aggregation processes of both α-synuclein and tau proteins (Meng et al. 2010; Sonawane et al. 2021). The compound functions by capturing oligomers and then dissociating them to break down preformed mature fibrils into nontoxic oligomers that protect neurons from damage (Grabska-Kobyłecka et al. 2023; Sonawane et al. 2021).

The phosphorylation of tau proteins at Ser396, Ser199, Thr231, and Thr205 sites is significantly reduced by quercetin through CDK5 inhibition and blocking the

Ca^{2+}-calpain-p25-CDK5 signaling pathway (Grabska-Kobyłecka et al. 2023; Shen et al. 2017).

The anti-aggregation processes of phytocompounds function through various molecular approaches to disrupt protein aggregation directly. The binding capacity of flavonoids to target proteins is particularly effective because they stabilize native unfolded conformations (Meng et al. 2010). The strong flavonoid inhibitors both stop new aggregate formation and break down preexisting fibrils, thereby altering the fundamental thermodynamic equilibrium between protein species (Meng et al. 2010).

The anti-aggregation efficacy of flavonoids depends on their structural elements since compounds with vicinal dihydroxyphenyl groups exhibit enhanced activity against α-synuclein fibrillation (Martins et al. 2023). Flavonoids containing three adjacent hydroxyl groups, like baicalein and EGCG, show better aggregation inhibition through their ability to stabilize monomers and soluble oligomers (Martins et al. 2023). The covalent binding of oxidized flavonoid species (quinones) to target proteins plays a fundamental role in maintaining prolonged fibrillation inhibition (Martins et al. 2023).

Phytocompounds fight protein aggregation by enhancing cellular protein degradation pathways in a coordinated manner. These compounds modify major protein degradation systems through multiple levels of influence, which results in neuron maintenance of proper misfolded protein formation and degradation balance (Hajieva 2017). The autophagy–lysosomal system acts as the main site where phytochemicals initiate their intervention. The protection of neurons through natural compound-mediated autophagy rescue functions as an essential mechanism to fight protein aggregation and toxic spread and prion-like transmission of α-synuclein (Limanaqi et al. 2019a). Quercetin activates both macro-autophagy and proteasomal pathways to enhance protein degradation, which prevents aggregate formation while protecting cellular function (Regitz et al. 2014).

Phytocompounds use Heat shock protein (HSP) activation as their main method to stop protein aggregation. The molecular chaperones function to determine which pathway will eliminate aggregates through lysosomes or proteasomes or phagosomes based on size and sequence and misfolding extent (Maiti et al. 2014). Curcumin serves as a natural HSP inducer that restores HSP levels in animal models with neurodegenerative transgenic conditions (Labanca et al. 2020; Maiti et al. 2014). The degradation of huntingtin protein aggregates through chaperone-mediated autophagy occurs when HSP40 and HSP70 bind to mutant protein precursors and activate specialized degradation mechanisms (Labanca et al. 2020; Maiti et al. 2014).

15.7 Regulation of Apoptosis

The neuroprotective effects of phytocompounds are achieved by influencing multiple vital apoptotic pathways that establish the fate between neuronal survival and death. These compounds demonstrate remarkable specificity in targeting key molecular switches that control programmed cell death.

The MAPK pathways function as primary targets that phytocompounds use to create their therapeutic effects. Multiple phytocompounds specifically inhibit the c-Jun N-terminal kinase (JNK) pathway, which acts as a stress-response mechanism that induces apoptosis (Kujawska and Jodynis-Liebert 2018). JNK hyperphosphorylation inhibition by curcumin stops Bax translocation to mitochondria and cytochrome c release, which causes apoptosis through mitochondrial mechanisms (Pan et al. 2012). Several plant extracts suppress p38 MAPK phosphorylation to block apoptotic signaling pathways in neural cells (Li et al. 2011; Raza et al. 2024).

The enhancement of phytocompounds results in the simultaneous activation of survival-promoting signaling cascades. Flavonoids activate the PI3K/AKT pathway together with ERK and NF-κB and endogenous antioxidant systems to support neuronal survival (Calderaro et al. 2022). The survival genes AKT, NF-κB, ERK1/2, and p38 MAPK of germinated brown rice function to support cell survival rather than cell death (Azmi et al. 2015). p53 acts as a key tumor suppressor protein that regulates apoptotic control in NDDs, leading to neuronal death. Neuroblastoma cells receive protection from EGb761 through the regulation of p53 transactivation that controls pro-apoptotic genes of the Bcl-2 family members as well as caspases and Fas (Crispi and Filosa 2020).

The Bcl-2 family of proteins serves as the primary regulatory system for how polyphenols steer cells toward apoptosis through mitochondrial mechanisms. These compounds increase the levels of anti-apoptotic Bcl-2 together with decreasing the amounts of pro-apoptotic Bax and Bad proteins (Kujawska and Jodynis-Liebert 2018; Naoi et al. 2019). Polyphenols establish direct mitochondrial connections through which they either stop or initiate apoptosis according to cellular conditions (Naoi, Wu et al. 2019). The restoration of protein kinase C (PKC) provides a new pathway that polyphenols use to block apoptotic signaling because these enzymes control both cell death and survival processes (Kujawska and Jodynis-Liebert 2018).

The molecular intervention with transcriptional regulators represents a fundamental strategy for phytocompound action. EGb761 protects neuroblastoma cells through its direct modulation of the tumor suppressor p53, which controls the expression of pro-apoptotic genes including Bcl-2 family members, caspases, and Fas (Crispi and Filosa 2020). Phytocompounds elevate forkhead box O (FoxO) 3a levels, which increases catalase activity and anti-apoptotic Bcl-2 protein expression to promote neuronal survival (Saeed et al. 2023).

Phytocompounds protect neurons through complex networks of multiple interconnected signaling cascades which work together to determine cell fate. The compounds work through complex cellular communication networks instead of targeting single pathways to establish strong neuroprotective effects against neurodegeneration.

The regulation of MAPK pathways operates as a fundamental component that protects against neurodegeneration. Germinated brown rice implements this multifaceted approach through AKT, NF-κB, ERK1/2, and p38 MAPK survival gene regulation to promote cell survival rather than death by modulating APP metabolism genes and apoptosis pathways through JNK, ERK, and p53 coordination (Azmi et al. 2015). Polyphenols regulate four essential signaling pathways including Akt, Nrf2, STAT, and MAPK to protect neurons while activating immune response and reducing inflammation through the regulation of pro-inflammatory genes (Jalouli et al. 2025).

Phytocompounds exhibit impressive abilities to modify mitochondrial apoptotic mechanisms directly, which leads to neuroprotection. Multiple death and survival signals converge within these organelles to establish them as essential targets for neurodegenerative disease treatment.

The direct regulation of mitochondrial apoptotic systems functions as the primary method by which polyphenols protect neurons. The compounds demonstrate dual redox properties through their ability to either enhance or suppress the mitochondrial apoptosis system in direct ways (Naoi, Wu et al. 2019). Polyphenols achieve neuroprotection by directly controlling mitochondrial functions, particularly electron transport chain activities, while they adjust redox states and block apoptosis pathways (Naoi, Wu et al. 2019).

The protection of mitochondria relies on Bcl-2 family protein control as a central mechanism. Bacosides found in *Bacopa monnieri* act through Bcl-2 family protein regulation to stop apoptosis by blocking cytochrome c release from mitochondria in Alzheimer's disease models (Nahar et al. 2025; Abdul Manap et al. 2019). Through this mechanism the cell prevents the sequence of events which results in the mitochondrial death of neurons.

The prevention of cytochrome c release serves as a primary protective mechanism which various phytocompounds have demonstrated. Through JNK hyperphosphorylation inhibition, curcumin prevents Bax from moving to mitochondria which blocks cytochrome c release that leads to mitochondria-mediated apoptosis (Pan et al. 2012). Through JNK pathway suppression, curcumin creates a protective effect which blocks nigrostriatal degeneration and mitochondrial dysfunction (Pan et al. 2012).

The complete protection of mitochondria needs simultaneous engagement of multiple therapeutic targets. Plant food-derived phytochemicals protect neurons by directly controlling mitochondrial apoptotic machinery, altering cellular signaling pathways and by increasing antiapoptotic Bcl-2 protein family members, and boosting pro-survival neurotrophic factors like brain- and glial cell line-derived neurotrophic factor (Naoi et al. 2019). Polyphenols both clean up reactive oxygen and nitrogen species and trigger redox-responsive transcription factors which regulate the expression of genes encoding antioxidants and anti-apoptotic Bcl-2 proteins as well as pro-survival neurotrophic factors (Naoi, Wu et al. 2019). Phytocompounds protect cells through their multiple mechanisms that target both the proximate causes of neuronal death and the core mitochondrial defects which occur in neurodegenerative diseases.

15.8 Neurogenesis and Synaptic Plasticity Enhancement

The activation of several interconnected signaling cascades plays a central role in the molecular mechanisms that drive phytocompound-induced neural stem cell proliferation, differentiation, and survival. The brain-derived neurotrophic factor

(BDNF) pathway functions as the central regulator of adult neurogenesis mainly in the hippocampus (Socci et al. 2017). The activation of BDNF expression by phytocompounds together with their regulation of downstream signaling pathways such as GSK-3β, MAPK/ERK, PI3K/Akt/mTOR, CREB, and Wnt/β-catenin occurs through flavonoids, terpenoids, alkaloids, and phenolic compounds (Singh et al. 2025).

Neurogenic signals that stem from bioactive phytocompounds converge at the PI3K/Akt signaling pathway which functions as a crucial crossroads. Flavonoids bind to signaling proteins such as ERK and PI3-kinase/Akt to modify their signaling activities thus creating neuroprotective effects that enhance hippocampal neurogenesis (Ayaz et al. 2019). Natural products activate neuroprotective pathways such as the PI3K/AKT and BDNF-TrkB-CREB signaling axes which protect neurons and enhance synaptic plasticity and prevent neurodegeneration (Lim et al. 2024).

The BDNF expression, along with its release, increases due to bioactive agents which include flavonoids, terpenes, saponins, and lignans while these agents activate signaling pathways like TrkB, ERK, Akt, and mTOR leading to increased spine maturation and synaptic numbers in neuronal cells (Yoon et al. 2023). The TrkB receptor pathway holds significant importance since Brassicaceae vegetable compound 3,3′-diindolylmethane (DIM) protects neurons from oxidative stress through BDNF and antioxidant enzyme activation while enhancing TrkB signaling (Singh et al. 2024; Lee et al. 2019).

Phytocompounds activate the essential kinase cascades that direct neural progenitor cell (NPC) proliferation and differentiation. Research shows that bioactive substances control Cdc42 expression which manages cell cytoskeletal changes and cell polarization necessary for NPC cell regeneration (Zong et al. 2021). Phytocompounds affect neurotrophic pathways by modifying fundamental proteins including tropomyosin receptor kinase (Trk) and glycogen synthase kinase 3 (GSK3βser9) along with phosphatidylinositol 3-kinase (PI3K) and extracellular signal-regulated protein kinase (Erk1/2) (Habiba et al. 2024).

Phytocompound-induced neurogenesis depends on epigenetic mechanisms as a fundamental factor. Polyphenols such as curcumin, quercetin, and resveratrol boost BDNF transcription through epigenetic modifications that include histone acetylation and DNA methylation changes which enhance cognitive function and protect neurons from stress damage (Jie et al. 2025; Prasanth et al. 2024). These compounds influence the expression of the *heme oxygenase-1* gene, activate the Nrf2 signaling pathway, and modify genes involved in neurogenesis and synaptic plasticity (Morante-Carriel et al. 2024).

The therapeutic potential of phytocompounds becomes more powerful because they can operate through multiple pathways at once. Ginsenosides protect neurons by activating the Nrf2/HO-1 antioxidant pathway and blocking NF-κB-mediated neuroinflammation together with promoting neurogenesis through PI3K/Akt signaling pathway activation (Zheng and Zhou 2025). Phytocompounds apply a multipathway therapeutic strategy to tackle NDDs because they simultaneously combat oxidative stress and neuroinflammation while restoring synaptic plasticity (Table 15.1).

Table 15.1 Comparative summary

Phytocompound type	Mechanism of action	Neurodegenerative disorders targeted	Molecular targets	Experimental model	Effects on neuroinflammation	References
Polyphenols, flavonoids, curcumin; relevant for neurodegenerative disorders like Alzheimer's and Parkinson's	Polyphenols activate SIRT1, Nrf2, antioxidant, and anti-inflammatory pathways, aiding neuroprotection in disorders like AD and PD	Alzheimer's, Parkinson's, and synucleinopathies are addressed	NF-κB, tau protein, Aβ aggregation, Nrf2, SIRT1, NLRP3 inflammasome, HO-1, JNK, p38-MAPK	Cell, organismal, animal, and human models	Polyphenols reduce neuroinflammation by modulating inflammasomes and cytokine levels, enhancing antioxidant pathways	Leri et al. (2020)
Resveratrol, a SIRT1 activator, modulates neuroinflammation in Alzheimer's disease	SIRT1 activation, decreased CSF MMP9, modulated neuroinflammation, induced adaptive immunity	Alzheimer's disease	SIRT1 activation, decreases CSF MMP9, modulates neuroinflammation	Clinical trial with human participants ($N = 119$)	Decreases CSF MMP9, increases MDC, IL-4, and FGF-2	Moussa et al. (2017)
Polyphenol; neuroinflammation protection in Alzheimer's disease	Inhibits TXNIP/TRX/NLRP3 pathway, suppresses inflammatory response in microglia	Alzheimer's disease	TXNIP/TRX/NLRP3 pathway, NF-κB phosphorylation, IL-1β, IL-6, TNF-α	BV-2 murine microglial cells, in vitro model	Inhibits Aβ-induced proliferation and activation of microglial cells	Feng and Zhang (2019)
Flavonoids; neuroprotective properties through antioxidant and anti-inflammatory activity	Antioxidant activity, modulation of neuroinflammatory pathways, interference with Aβ aggregation, tau pathology	Alzheimer's disease	Amyloid-β aggregation, tau pathology, microglial polarization, oxidative stress reduction	Preclinical and clinical models	Modulation of neuroinflammatory pathways, effects on microglial polarization	Kruszka et al. (2025)
Eugenol, α-humulene, β-caryophyllene, gallic acid, quercetin, luteolin; neuroprotective, antioxidant, anti-inflammatory	Scavenging free radicals, modulating calcium channels, reducing neuroinflammation, oxidative stress, NF-κB pathway inhibition	Alzheimer's disease	NF-κB pathway inhibition, modulation of calcium channels, antioxidant and anti-inflammatory pathways	N/A	Inhibition of NF-κB pathway, reduction of neuroinflammation	Sargsyan et al. (2025)

References

Abdul Manap AS, Vijayabalan S, Madhavan P, Chia YY, Arya A, Wong EH, Rizwan F, Bindal U, Koshy S (2019) Bacopa monnieri, a neuroprotective lead in Alzheimer disease: a review on its properties, mechanisms of action, and preclinical and clinical studies. Drug Target Insights 13:117739281986641

Ali T, Kim T, Rehman SU, Khan MS, Amin FU, Khan M, Ikram M, Kim MO (2018) Natural dietary supplementation of anthocyanins via PI3K/Akt/Nrf2/HO-1 pathways mitigate oxidative stress, neurodegeneration, and memory impairment in a mouse model of Alzheimer's disease. Mol Neurobiol 55(7):6076–6093. https://doi.org/10.1007/s12035-017-0798-6

Amartumur S, Nguyen H, Huynh T, Kim TS, Woo R-S, Oh E, Kim KK, Lee LP, Heo C (2024) Neuropathogenesis-on-chips for neurodegenerative diseases. Nat Commun 15:2219

Anderson J, Blair JA, Armstrong R (1987) The effect of age on tetrahydrobiopterin metabolism in the human brain. J Neurol Neurosurg Psychiatry 50(2):231. https://doi.org/10.1136/jnnp.50.2.231

Arbo BD, André-Miral C, Nasre-Nasser RG, Schimith LE, Santos MGD, Costa-Silva DG, Muccillo-Baisch AL, Hort MA (2020) Resveratrol derivatives as potential treatments for Alzheimer's and Parkinson's disease. Front Aging Neurosci 12

Ayaz MM, Sadiq A, Junaid M, Ullah F, Ovais M, Ullah I, Ahmed J, Shahid M (2019) Flavonoids as prospective neuroprotectants and their therapeutic propensity in aging associated neurological disorders. Front Aging Neurosci 11

Azmi NH, Ismail M, Ismail N, Imam MU, Alitheen NBM, Abdullah MA (2015) Germinated brown rice alters aβ(1-42) aggregation and modulates Alzheimer's disease-related genes in differentiated human SH-SY5Y cells. Evid Based Complement Alternat Med 2015:153684

Bloomingdale P, Karelina T, Ramakrishnan V, Bakshi S, Véronneau-Veilleux F, Moye M, Sekiguchi K, Meno-Tetang G, Mohan A, Maithreye R, Thomas VA, Gibbons F, Cabal A, Bouteiller JM, Geerts H (2022) Hallmarks of neurodegenerative disease: A systems pharmacology perspective. CPT Pharmacometrics Syst Pharmacol 11(11):1399–1429. https://doi.org/10.1002/psp4.12852

Calderaro A, Patanè GT, Tellone E, Barreca D, Ficarra S, Misiti F, Laganà G (2022) The neuroprotective potentiality of flavonoids on Alzheimer's disease. Int J Mol Sci 23

Can B, Sanlier N (2025) Alzheimer, Parkinson, dementia, and phytochemicals: insight review. Crit Rev Food Sci Nutr 65(9):1706–1728. https://doi.org/10.1080/10408398.2023.2299340

Carito V, Ceccanti M, Tarani L, Ferraguti G, Chaldakov GN, Fiore M (2016) Neurotrophins' modulation by olive polyphenols. Curr Med Chem 23(28):3189–3197

Chang CY, Ting H-C, Liu C-A, Su HL, Chiou TW, Lin S-Z, Harn H-J, Ho TJ (2020) Induced pluripotent stem cell (iPSC)-based neurodegenerative disease models for phenotype recapitulation and drug screening. Molecules 25

Chesser AS, Ganeshan V, Yang J, Johnson GVW (2016) Epigallocatechin-3-gallate enhances clearance of phosphorylated tau in primary neurons. Nutr Neurosci 19(1):21–31. https://doi.org/10.1179/1476830515Y.0000000038

Chung YC, Baek JY, Kim SR, Ko HW, Bok E, Shin W-H, Won S-Y, Jin BK (2017) Capsaicin prevents degeneration of dopamine neurons by inhibiting glial activation and oxidative stress in the MPTP model of Parkinson's disease. Exp Mol Med 49(3):e298–e298. https://doi.org/10.1038/emm.2016.159

Connolly BS, Lang AE (2014) Pharmacological treatment of Parkinson disease: a review. JAMA 311(16):1670–1683

Crispi S, Filosa S (2020) Novel perspectives for neurodegeneration prevention: effects of bioactive polyphenols. Neural Regen Res 16:1411–1412

Cummings JL, Goldman DP, Simmons-Stern NR, Ponton E (2021) The costs of developing treatments for Alzheimer's disease: a retrospective exploration. Alzheimers Dement 18:469–477

Cunnane SC, Trushina E, Morland C, Prigione A, Casadesus G, Andrews ZB, Beal MF, Bergersen LH, Brinton RD, de la Monte S, Eckert A, Harvey J, Jeggo R, Jhamandas JH, Kann O, la Cour CM, Martin WF, Mithieux G, Moreira PI et al (2020) Brain energy rescue: an emerging therapeutic concept for neurodegenerative disorders of ageing. Nat Rev Drug Discov 19(9):609–633. https://doi.org/10.1038/s41573-020-0072-x

Daverey A, Agrawal SK (2020) Curcumin protects against white matter injury through NF-κB and Nrf2 cross talk. J Neurotrauma 37(10):1255–1265. https://doi.org/10.1089/neu.2019.6749

De Gioia R, Biella F, Citterio G, Rizzo F, Abati E, Nizzardo M, Bresolin N, Comi GP, Corti S (2020) Neural stem cell transplantation for neurodegenerative diseases. Int J Mol Sci 21

de Lima EP, Laurindo LF, Catharin VCS, Direito R, Tanaka M, Jasmin Santos German I, Lamas CB, Guiguer EL, Araújo AC, Fiorini AMR, Barbalho SM (2025) Polyphenols, alkaloids, and terpenoids against neurodegeneration: evaluating the neuroprotective effects of phytocompounds through a comprehensive review of the current evidence. Metabolites 15(2):124. https://doi.org/10.3390/metabo15020124

de Oliveira MR, Nabavi SF, Manayi A, Daglia M, Hajheydari Z, Nabavi SM (2016) Resveratrol and the mitochondria: from triggering the intrinsic apoptotic pathway to inducing mitochondrial biogenesis, a mechanistic view. Biochim Biophys Acta Gen Subj 1860(4):727–745. https://doi.org/10.1016/j.bbagen.2016.01.017

Di Paolo M, Papi L, Gori F, Turillazzi E (2019) Natural products in neurodegenerative diseases: a great promise but an ethical challenge. Int J Mol Sci 20

Dugger BN, Dickson DW (2017) Pathology of neurodegenerative diseases. Cold Spring Harb Perspect Biol 9(7):a028035. https://doi.org/10.1101/cshperspect.a028035

Ehrnhoefer DE, Duennwald ML, Markovic P, Wacker JL, Engemann S, Roark M, Legleiter J, Legleiter J, Marsh JL, Thompson LM, Lindquist S, Muchowski PJ, Muchowski PJ, Wanker EE (2006) Green tea (−)-epigallocatechin-gallate modulates early events in huntingtin misfolding and reduces toxicity in Huntington's disease models. Hum Mol Genet 15(18):2743–2751

Elufioye TO, Berida TI, Habtemariam S (2017) Plants-derived neuroprotective agents: cutting the cycle of cell death through multiple mechanisms. Evid Based Complement Alternat Med 2017:3574012

Feng L, Zhang L (2019) Resveratrol suppresses aβ-induced microglial activation through the TXNIP/TRX/NLRP3 signaling pathway. DNA Cell Biol 38:874

Fiore M, Terracina S, Ferraguti G (2025) Brain neurotrophins and plant polyphenols: a powerful connection. Molecules 30:2657

Fukutomi R, Ohishi T, Koyama Y, Pervin M, Nakamura Y, Isemura M (2021) Beneficial effects of epigallocatechin-3-O-gallate, chlorogenic acid, resveratrol, and curcumin on neurodegenerative diseases. Molecules 26:415

Garthe I, Maughan RJ (2018) Athletes and supplements: prevalence and perspectives. Int J Sport Nutr Exerc Metab 28(2):126–138. https://doi.org/10.1123/ijsnem.2017-0429

Ghafouri-Fard S, Bahroudi Z, Shoorei H, Hussen BM, Talebi SF, Baig SG, Taheri M, Ayatollahi SA (2022) Disease-associated regulation of gene expression by resveratrol: special focus on the PI3K/AKT signaling pathway. Cancer Cell Int 22(1):298. https://doi.org/10.1186/s12935-022-02719-3

Gibellini L, Bianchini E, De Biasi S, Nasi M, Cossarizza A, Pinti M (2015) Natural compounds modulating mitochondrial functions. Evid Based Complement Alternat Med 2015:527209

Gomez G, Beason-Held LL, Bilgel M, An Y, Wong DF, Studenski S, Ferrucci L, Resnick SM (2018) Metabolic syndrome and amyloid accumulation in the aging brain. J Alzheimer's Dis 65(2):629–639. https://doi.org/10.3233/JAD-180297

Gonçalves PB, Sodero ACR, Cordeiro Y (2021) Green tea epigallocatechin-3-gallate (EGCG) targeting protein misfolding in drug discovery for neurodegenerative diseases. Biomolecules 11

Gorlach S, Fichna J, Lewandowska U (2015) Polyphenols as mitochondria-targeted anticancer drugs. Cancer Lett 366(2):141–149. https://doi.org/10.1016/j.canlet.2015.07.004

Grabska-Kobyłecka I, Szpakowski P, Król A, Ksiazek-Winiarek D, Kobyłecki A, Głąbiński A, Nowak D (2023) Polyphenols and their impact on the prevention of neurodegenerative diseases and development. Nutrients 15

Habiba SU, Choi HJ, Munni YA, Yang I-J, Haque MN, Moon IS (2024) Neurotrophic effects of Foeniculum vulgare ethanol extracts on hippocampal neurons: role of anethole in neurite outgrowth and synaptic development. Int J Mol Sci 25

Haeusler AR, Donnelly CJ, Rothstein JD (2016) The expanding biology of the C9orf72 nucleotide repeat expansion in neurodegenerative disease. Nat Rev Neurosci 17(6):383–395. https://doi.org/10.1038/nrn.2016.38

Hajieva P (2017) The effect of polyphenols on protein degradation pathways: implications for neuroprotection. Molecules 22:159

Hannan MA, Dash R, Sohag AAM, Haque MN, Moon IS (2020) Neuroprotection against oxidative stress: phytochemicals targeting TrkB signaling and the Nrf2-ARE antioxidant system. Front Mol Neurosci 13

He X, Li Z, Rizak JD, Wu S, Wang Z, He R, Su M, Qin D, Wang J, Hu X (2017) Resveratrol attenuates formaldehyde induced hyperphosphorylation of tau protein and cytotoxicity in N2a cells. Front Neurosci 10

He Z, Ning N, Zhou Q, Khoshnam SE, Farzaneh M (2020) Mitochondria as a therapeutic target for ischemic stroke. Free Radic Biol Med 146:45–58. https://doi.org/10.1016/j.freeradbiomed.2019.11.005

He Z, Li X, Wang Z, Cao Y, Han S, Li N, Cai J, Cheng S, Liu Q (2023) Protective effects of luteolin against amyloid beta-induced oxidative stress and mitochondrial impairments through peroxisome proliferator-activated receptor γ-dependent mechanism in Alzheimer's disease. Redox Biol 66(July):102848. https://doi.org/10.1016/j.redox.2023.102848

Heneka MT, McManus RM, Latz E (2018) Inflammasome signalling in brain function and neurodegenerative disease. Nat Rev Neurosci 19(10):610–621. https://doi.org/10.1038/s41583-018-0055-7

Hossain MS, Wazed MA, Asha S, Amin MR, Shimul IM (2025) Dietary phytochemicals in health and disease: mechanisms, clinical evidence, and applications—a comprehensive review. Food Sci Nutr 13(3):1706–1728. https://doi.org/10.1002/fsn3.70101

Jalouli M, Rahman MA, Biswas P, Rahman H, Harrath AH, Lee I-S, Kang S, Choi J, Park MN, Kim B (2025) Targeting natural antioxidant polyphenols to protect neuroinflammation and neurodegenerative diseases: a comprehensive review. Front Pharmacol 16

Jie S, Fu A, Wang C, Rajabi S (2025) A comprehensive review on the impact of polyphenol supplementation and exercise on depression and brain function parameters. Behavioral Brain Funct 21:10

Jimenez-Sanchez M, Licitra F, Underwood BR, Rubinsztein DC (2017) Huntington's disease: mechanisms of pathogenesis and therapeutic strategies. Cold Spring Harb Perspect Med 7(7):a024240. https://doi.org/10.1101/cshperspect.a024240

Kaushik S, Cuervo AM (2015) Proteostasis and aging. Nat Med 21(12):1406–1415. https://doi.org/10.1038/nm.4001

Khan A, Jahan S, Imtiyaz Z, Alshahrani S, Antar Makeen H, Mohammed Alshehri B, Kumar A, Arafah A, Rehman MU (2020) Neuroprotection: targeting multiple pathways by naturally occurring phytochemicals. Biomedicine 8

Khayatan D, Razavi SM, Arab ZN, Khanahmadi M, Samanian A, Momtaz S, Sukhorukov VN, Jamialahmadi T, Abdolghaffari A, Barreto GE, Sahebkar A (2024) Protective effects of plant-derived compounds against traumatic brain injury. Mol Neurobiol 61:7732–7750

Koo EH (2021) Preclinical and clinical understanding: major gaps in our understanding and capabilities

Kruszka J, Martyński J, Szewczyk-Golec K, Woźniak A, Nuszkiewicz J (2025) The role of selected flavonoids in modulating neuroinflammation in Alzheimer's disease: mechanisms and therapeutic potential. Brain Sci 15

Kujawska M, Jodynis-Liebert J (2018) Polyphenols in Parkinson's disease: a systematic review of in vivo studies. Nutrients 10

Labanca F, Ullah H, Khan H, Milella L, Xiao J, Dajić-Stevanović Z, Jeandet P (2020) Therapeutic and mechanistic effects of curcumin in Huntington's disease. Curr Neuropharmacol 19:1007–1018

Lai M, Yao H, Shah SZA, Wu W, Wang D, Zhao Y, Wang L, Zhou X, Zhao D, Yang L (2018) The NLRP3-caspase 1 inflammasome negatively regulates autophagy via TLR4-TRIF in prion peptide-infected microglia. Front Aging Neurosci 10

Lall D, Lorenzini I, Mota TA, Bell S, Mahan TE, Ulrich JD, Davtyan H, Rexach JE, Muhammad AKMG, Shelest O, Landeros J, Vazquez M, Kim J, Ghaffari L, O'Rourke JG, Geschwind DH, Blurton-Jones M, Holtzman DM, Sattler R, Baloh RH (2021) C9orf72 deficiency promotes microglial-mediated synaptic loss in aging and amyloid accumulation. Neuron 109(14):2275–2291.e8. https://doi.org/10.1016/j.neuron.2021.05.020

Lee D-H, Kim C-S, Lee YJ (2011) Astaxanthin protects against MPTP/MPP+-induced mitochondrial dysfunction and ROS production in vivo and in vitro. Food Chem Toxicol 49(1):271–280. https://doi.org/10.1016/j.fct.2010.10.029

Lee BD, Yoo J-M, Baek SY, Li FY, Sok DE, Kim MR (2019) 3,3′-diindolylmethane promotes BDNF and antioxidant enzyme formation via TrkB/Akt pathway activation for neuroprotection against oxidative stress-induced apoptosis in hippocampal neuronal cells. Antioxidants 9

Leri M, Scuto M, Ontario ML, Calabrese V, Calabrese EJ, Bucciantini M, Stefani M (2020) Healthy effects of plant polyphenols: molecular mechanisms. Int J Mol Sci 21

Li B-y, Yuan Y-h, Hu J-f, Zhao Q-G, Zhang D m, Chen N-h (2011) Protective effect of Bu-7, a flavonoid extracted from Clausena lansium, against rotenone injury in PC12 cells. Acta Pharmacol Sin 32:1321

Lim DW, Lee J-E, Lee C, Kim YT (2024) Natural products and their neuroprotective effects in degenerative brain diseases: a comprehensive review. Int J Mol Sci 25

Limanaqi F, Biagioni F, Busceti CL, Ryskalin L, Polzella M, Frati A, Fornai F (2019a) Phytochemicals bridging autophagy induction and alpha-synuclein degradation in parkinsonism. Int J Mol Sci 20

Limanaqi F, Biagioni F, Gaglione A, Busceti CL, Fornai F (2019b) A sentinel in the crosstalk between the nervous and immune system: the (immuno)-proteasome. Front Immunol 10

Limanaqi F, Biagioni F, Mastroiacovo F, Polzella M, Lazzeri G, Fornai F (2020) Merging the multi-target effects of phytochemicals in neurodegeneration: from oxidative stress to protein aggregation and inflammation. Antioxidants 9

Lotharius J, Brundin P (2002) Pathogenesis of parkinson's disease: dopamine, vesicles and α-synuclein. Nat Rev Neurosci 3(12):932–942. https://doi.org/10.1038/nrn983

Maiti P, Manna J, Veleri S, Frautschy S (2014) Molecular chaperone dysfunction in neurodegenerative diseases and effects of curcumin. BioMed Res Int 2014

Mandolesi G, Gentile A, Musella A, Fresegna D, De Vito F, Bullitta S, Sepman H, Marfia GA, Centonze D (2015) Synaptopathy connects inflammation and neurodegeneration in multiple sclerosis. Nat Rev Neurol 11(12):711–724. https://doi.org/10.1038/nrneurol.2015.222

Martins GF, Nascimento CC d, Galamba N (2023) Mechanistic insights into polyphenols' aggregation inhibition of α-synuclein and related peptides. ACS Chem Neurosci 14:1905–1920

Matsumoto J, Stewart TH, Banks WA, Zhang J (2018) The transport mechanism of extracellular vesicles at the blood-brain barrier. Curr Pharm Des 23(40):6206–6214

Meng X, Munishkina L, Fink AL, Uversky VN (2010) Effects of various flavonoids on the α-synuclein fibrillation process. Parkinson's Disease 2010

Mhalhel K, Kadmi Y, Ben Chira A, Levanti M, Pansera L, Cometa M, Sicari M, Germanà A, Aragona M, Montalbano G (2024) Urtica dioica extract abrogates chlorpyrifos-induced toxicity in zebrafish larvae. Int J Mol Sci 25(12):6631. https://doi.org/10.3390/ijms25126631

Momtaz S, Memariani Z, El-Senduny FF, Sanadgol N, Golab F, Katebi M, Abdolghaffari AH, Farzaei MH, Abdollahi M (2020) Targeting ubiquitin-proteasome pathway by natural products: novel therapeutic strategy for treatment of neurodegenerative diseases. Front Physiol 11

Morante-Carriel J, Nájera H, Samper-Herrero A, Živković S, Martínez-Esteso MJ, Martínez-Márquez A, Sellés-Marchart S, Obrebska A, Bru-Martínez R (2024) Therapeutic potential of prenylated flavonoids of the fabaceae family in medicinal chemistry: an updated review. Int J Mol Sci 25

Mouhieddine TH, Kobeissy FH, Itani MM, Nokkari A, Wang KKW (2014) Stem cells in neuroinjury and neurodegenerative disorders: challenges and future neurotherapeutic prospects. Neural Regen Res 9:901–906

Moussa C, Hebron ML, Huang X, Ahn J, Rissman RA, Aisen PS, Turner RS (2017) Resveratrol regulates neuro-inflammation and induces adaptive immunity in Alzheimer's disease. J Neuroinflammation 14:1

Muqit MMK, Alessi DR (2009) A meeting of minds: overcoming roadblocks in the development of therapies for neurodegenerative disorders. EMBO Mol Med 1:139–141

Nahar L, Charoensup R, Kalieva K, Habibi E, Guo M, Wang D, Kvasnica M, Onder A, Sarker SD, Tsim K, Erekat NS, Heinrich M (2025) Natural products in neurodegenerative diseases: recent advances and future outlook. Front Pharmacol 16

Naoi M, Shamoto-Nagai M, Maruyama W (2019) Neuroprotection of multifunctional phytochemicals as novel therapeutic strategy for neurodegenerative disorders: antiapoptotic and antiamyloidogenic activities by modulation of cellular signal pathways. Future Neurol 14

Naoi M, Wu Y, Shamoto-Nagai M, & Maruyama W (2019) Mitochondria in Neuroprotection by Phytochemicals: Bioactive Polyphenols Modulate Mitochondrial Apoptosis System, Function and Structure. Int J Mol Sci 20(10):2451. https://doi.org/10.3390/ijms20102451

Nath D, Khanam N, Ghosh A (2020) Role of phytocompounds en route blood-brain barrier in cerebral ischemia. Eur J Med Plants

Nebrisi EE (2021) Neuroprotective activities of curcumin in Parkinson's disease: A review of the literature. Int J Mol Sci 22(20):11248. https://doi.org/10.3390/ijms222011248

Nunnari J, Suomalainen A (2012) Mitochondria: in sickness and in health. Cell 148(6):1145–1159. https://doi.org/10.1016/j.cell.2012.02.035

Pan J, Li H, Ma J, Tan Y-y, Xiao Q, Ding J, Chen S-D (2012) Curcumin inhibition of JNKs prevents dopaminergic neuronal loss in a mouse model of Parkinson's disease through suppressing mitochondria dysfunction. Transl Neurodegener 1

Parrado-Fernández C, Sandebring-Matton A, Rodriguez-Rodriguez P, Aarsland D, Cedazo-Mínguez A (2016) Anthocyanins protect from complex I inhibition and APPswe mutation through modulation of the mitochondrial fission/fusion pathways. Biochim Biophys Acta 1862(11):2110–2118. https://doi.org/10.1016/j.bbadis.2016.08.002

Patel AA, Patel RJ, Patel S (2017) Nanomedicine for intranasal delivery to improve brain uptake. Curr Drug Deliv 15(4):461–469

Pázmándi K, Ágics B, Szöllősi AG, Bácsi A, Fekete T (2024) Ginger-derived bioactive compounds attenuate the Toll-like receptor mediated responses of human dendritic cells. Eur J Pharmacol 967(February):176399. https://doi.org/10.1016/j.ejphar.2024.176399

Pineda-Ramírez N, Alquisiras-Burgos I, Ortiz-Plata A, Ruiz-Tachiquín M-E, Espinoza-Rojo M, Aguilera P (2020) Resveratrol activates neuronal autophagy through AMPK in the ischemic brain. Mol Neurobiol 57(2):1055–1069. https://doi.org/10.1007/s12035-019-01803-6

Pinelli R, Biagioni F, Limanaqi F, Bertelli M, Scaffidi E, Polzella M, Busceti CL, Fornai F (2020) A re-appraisal of pathogenic mechanisms bridging wet and dry age-related macular degeneration leads to reconsider a role for phytochemicals. Int J Mol Sci 21

Prasanth MI, Sivamaruthi BS, Cheong CSY, Verma K, Tencomnao T, Brimson JM, Prasansuklab A (2024) Role of epigenetic modulation in neurodegenerative diseases: implications of phytochemical interventions. Antioxidants 13

Rahman MA, Rahman MH, Biswas P, Hossain MS, Islam R, Hannan MA, Uddin MJ, Rhim H (2020) Potential therapeutic role of phytochemicals to mitigate mitochondrial dysfunctions in Alzheimer's disease. Antioxidants 10(1):23. https://doi.org/10.3390/antiox10010023

Rajput A, Pingale PL, Dhapte-Pawar VS (2022) Nasal delivery of neurotherapeutics via nanocarriers: facets, aspects, and prospects. Front Pharmacol 13

Rasool M, Malik A, Qureshi MS, Manan A, Pushparaj PN, Asif M, Qazi MH, Qazi AM, Kamal MA, Gan SH, Sheikh IA (2014) Recent updates in the treatment of neurodegenerative disorders using natural compounds. Evid Based Complement Alternat Med 2014

Raza C, Mohsin S, Faheem M, Hanif U, Alkhathlan HZ, Shaik MR, Riaz HA, Anjum R, Jurrat H, Khan M (2024) In vivo study of Moringa oleifera seed extracts as potential sources of neuroprotection against rotenone-induced neurotoxicity. Plants 13

Regitz C, Dußling LM, Wenzel U (2014) Amyloid-beta (aβ_{1-42})-induced paralysis in Caenorhabditis elegans is inhibited by the polyphenol quercetin through activation of protein degradation pathways. Mol Nutr Food Res 58(10):1931–1940

Roberts RO, Knopman DS, Przybelski SA, Mielke MM, Kantarci K, Preboske GM, Senjem ML, Pankratz VS, Geda YE, Boeve BF, Ivnik RJ, Rocca WA, Petersen RC, Jack CR (2014) Association of type 2 diabetes with brain atrophy and cognitive impairment. Neurology 82(13):1132–1141. https://doi.org/10.1212/WNL.0000000000000269

Saeed F, Imran M, Hussain G, Suleria HAR, Imran A (2023) Editorial: polyphenol-based dietary intervention against neurodegenerative disorders. Front Nutr 10

Sargsyan T, Simonyan HM, Stepanyan L, Tsaturyan A, Vicidomini C, Pastore R, Guerra G, Roviello GN (2025) Neuroprotective properties of clove (Syzygium aromaticum): state of the art and future pharmaceutical applications for Alzheimer's disease. Biomolecules 15

Selkoe DJ, Hardy J (2016) The amyloid hypothesis of Alzheimer's disease at 25 years. EMBO Mol Med 8(6):595–608. https://doi.org/10.15252/emmm.201606210

Shafiee A, Ahmadi H, Taheri B, Hosseinzadeh S, Fatahi Y, Soleimani M, Atyabi F, Dinarvand R (2020) Appropriate scaffold selection for CNS tissue engineering. Avicenna J Med Biotechnol 12:203–220

Sharifi-Rad J, Rapposelli S, Sestito S, Herrera-Bravo J, Arancibia-Diaz A, Salazar LA, Yeskaliyeva B, Beyatli A, Leyva-Gómez G, González-Contreras C, Gürer ES, Martorell M, Calina D (2022) Multi-target mechanisms of phytochemicals in Alzheimer's disease: effects on oxidative stress, neuroinflammation and protein aggregation. J Personal Med 12(9):1515. https://doi.org/10.3390/jpm12091515

Shashikant K, Sanjana M, Gayatri K, Rakhi G (2024) Navigating neurodegenerative disorders: A comprehensive review of current and emerging therapies for neurodegenerative disorders. J Neurosci Neurol Disord 8:033

Shen X-Y, Luo T, Li S, Ting O, He F, Xu J, Wang H-Q (2017) Quercetin inhibits okadaic acid-induced tau protein hyperphosphorylation through the Ca2+-calpain-p25-CDK5 pathway in HT22 cells. Int J Mol Med 41(2):1138–1146

Shinto L, Marracci G, Baldauf-Wagner S, Strehlow A, Yadav V, Stuber L, Bourdette D (2009) Omega-3 fatty acid supplementation decreases matrix metalloproteinase-9 production in relapsing-remitting multiple sclerosis. Prostaglandins Leukot Essent Fat Acids 80(2–3):131–136. https://doi.org/10.1016/j.plefa.2008.12.001

Shoaib S, Ansari MA, Fatease AA, Safhi AY, Hani U, Jahan R, Alomary MN, Ansari MN, Ahmed NR, Wahab S, Ahmad W, Yusuf N, Islam N (2023) Plant-derived bioactive compounds in the management of neurodegenerative disorders: challenges, future directions and molecular mechanisms involved in neuroprotection. Pharmaceutics 15

Singh PK, Kotia V, Ghosh D, Mohite GM, Kumar A, Maji SK (2013) Curcumin modulates α-synuclein aggregation and toxicity. ACS Chem Neurosci 4(3):393–407

Singh S, Arthur R, Upadhayay S, Kumar P (2022) Ferulic acid ameliorates neurodegeneration via the Nrf2/ARE signalling pathway: a review. Pharmacol Res Modern Chin Med 5(August):100190. https://doi.org/10.1016/j.prmcm.2022.100190

Singh AA, Yadav D, Khan F, Song M (2024) Indole-3-carbinol and its derivatives as neuroprotective modulators. Brain Sci 14

Singh AA, Katiyar S, Song M (2025) Phytochemicals targeting BDNF signaling for treating neurological disorders. Brain Sci 15

Sivandzade F, Cucullo L (2018) In-vitro blood–brain barrier modeling: A review of modern and fast-advancing technologies. J Cereb Blood Flow Metab 38:1667–1681

Socci V, Tempesta D, Desideri G, de Gennaro L, Ferrara M (2017) Enhancing human cognition with cocoa flavonoids. Front Nutr 4

Sonawane SK, Chidambaram H, Boral D, Gorantla NV, Balmik AA, Dangi A, Ramasamy S, Marelli UK, Chinnathambi S (2020) EGCG impedes human tau aggregation and interacts with tau. Sci Rep 10

Sonawane SK, Uversky VN, Chinnathambi S (2021) Baicalein inhibits heparin-induced tau aggregation by initializing non-toxic tau oligomer formation. Cell Commun Signal 19

Soukup S, Vanhauwaert R, Verstreken P (2018) Parkinson's disease: convergence on synaptic homeostasis. EMBO J 37(18):1–16. https://doi.org/10.15252/embj.201898960

Stone EJ, Kolb SJ, Brown A (2021) A review and analysis of the clinical literature on Charcot–Marie–Tooth disease caused by mutations in neurofilament protein L. Cytoskeleton 78(3):97–110. https://doi.org/10.1002/cm.21676

Tang S, Suo Z, Liu D, Wei K, Xu Y, Huang H, Liu X, Li X (2025) Phytochemical constituents from root barks of Eleutherococcus henryi Oliv. and their anti-neuroinflammatory effect. J Ethnopharmacol 337:118877. https://doi.org/10.1016/j.jep.2024.118877

Thapa A, Jett SD, Chi EY (2016) Curcumin attenuates amyloid-β aggregate toxicity and modulates amyloid-β aggregation pathway. ACS Chem Neurosci 7(1):56–68

Thibaudeau TA, Anderson R, Smith DM (2018) A common mechanism of proteasome impairment by neurodegenerative disease-associated oligomers. Nat Commun 9

Tyson T, Steiner JA, Brundin P (2016) Sorting out release, uptake and processing of alpha-synuclein during prion-like spread of pathology. J Neurochem 139(S1):275–289. https://doi.org/10.1111/jnc.13449

Verma M, Lizama BN, Chu CT (2022) Excitotoxicity, calcium and mitochondria: a triad in synaptic neurodegeneration. Transl Neurodegen 11(1):3. https://doi.org/10.1186/s40035-021-00278-7

Wang D-M, Li S-Q, Wu W-L, Zhu X-Y, Wang Y, Yuan H-Y (2014) Effects of long-term treatment with quercetin on cognition and mitochondrial function in a mouse model of Alzheimer's disease. Neurochem Res 39(8):1533–1543. https://doi.org/10.1007/s11064-014-1343-x

Wang J, Song Y, Gao M, Bai X, Chen Z (2016) Neuroprotective effect of several phytochemicals and its potential application in the prevention of neurodegenerative diseases. Geriatrics 1(4):29. https://doi.org/10.3390/geriatrics1040029

Wang J, Li L, Wang Z, Cui Y, Tan X, Yuan T, Liu Q, Liu Z, Liu X (2018a) Supplementation of lycopene attenuates lipopolysaccharide-induced amyloidogenesis and cognitive impairments via mediating neuroinflammation and oxidative stress. J Nutr Biochem 56:16–25. https://doi.org/10.1016/j.jnutbio.2018.01.009

Wang J, Song Y, Chen Z, Leng SX (2018b) Connection between systemic inflammation and neuroinflammation underlies neuroprotective mechanism of several phytochemicals in neurodegenerative diseases. Oxidative Med Cell Longev 2018(1):1706–1728. https://doi.org/10.1155/2018/1972714

Yang X, Gao X, Jiang X, Yue K, Luo P (2024) Targeting capabilities of engineered extracellular vesicles for the treatment of neurological diseases. Neural Regen Res 20:3076–3094

Yoon C-S (2025) Natural products in the treatment of neuroinflammation at microglia: recent trend and features. Cells 14(8):571. https://doi.org/10.3390/cells14080571

Yoon S, Iqbal H, Kim SM, Jin M (2023) Phytochemicals that act on synaptic plasticity as potential prophylaxis against stress-induced depressive disorder. Biomol Ther 31:148–160

Zheng S-y, Zhou X-q (2025) A perspective on the mechanisms of herbal medicine for cognitive impairment. Front Neurol 16

Zhou T, Zu G, Wang X, Zhang X, Li S, Liang Z, Zhao J (2015) Immunomodulatory and neuroprotective effects of ginsenoside Rg1 in the MPTP(1-methyl-4-phenyl-1,2,3,6-tetrahydropyridine)-induced mouse model of Parkinson's disease. Int Immunopharmacol 29(2):334–343. https://doi.org/10.1016/j.intimp.2015.10.032

Zong W, Gouda MM, Cai E, Wang R-L, Xu W, Wu Y, Munekata PES, Lorenzo J (2021) The antioxidant phytochemical Schisandrin A promotes neural cell proliferation and differentiation after ischemic brain injury. Molecules 26

MIX
Papier aus verantwortungsvollen Quellen
Paper from responsible sources
FSC® C105338

If you have any concerns about our products,
you can contact us on
ProductSafety@springernature.com

In case Publisher is established outside the EU,
the EU authorized representative is:
Springer Nature Customer Service Center GmbH
Europaplatz 3, 69115 Heidelberg, Germany

Printed by Libri Plureos GmbH
in Hamburg, Germany